Electricity and Electronics Today

W. J. Haynie, III
Assistant Professor
Industrial Arts/Technology Education
North Carolina State University

Reviewers

Mike Warbritton
Eastern Hills High School
Fort Worth, Texas

William Strong
Kearning High School
San Diego, California

Joe S. Miller
Greensboro City Schools
Greensboro, North Carolina

Changing Times Education Service
EMC Publishing
Saint Paul, Minnesota

Design and photography:
Slater Studio, Minneapolis, MN

Library of Congress Catalog Number: 87-6749

ISBN 0-8219-0179-6

© 1988 by EMC Corporation
All rights reserved. Published 1987

No part of this publication can be adapted, reproduced, stored in a retrieval system or transmitted in any form or by any means, electronic, mechanical, photocopying, recording, or otherwise without permission from the publisher.

Published by EMC Publishing
300 York Avenue
St. Paul, Minnesota 55101

Printed in the United States of America
0 9 8 7 6 5 4 3 2

TABLE OF CONTENTS

SECTION 1 — Getting Started in Electricity-Electronics

Chapter 1 Careers in Electronics: A Photo Essay 2
         2 Tools and Safety 6
         3 What is Electric Current? 13
         4 Electric Circuits 20
         5 Sources of Electricity 26

SECTION 2 — Principles of Electricity

Chapter 6 Voltage 34
         7 Current 40
         8 Resistance 45
         9 Resistor Color Code 51
        10 Relationship of E, I, and R 59
        11 Computing Electrical Values 65
        12 Using Calculators to Solve Problems 69

SECTION 3 — Electricity and Magnets at Work

Chapter 13 Electrical Power 76
        14 Series Circuits 82
        15 Parallel Circuits 89
        16 Series Parallel Circuits 96
        17 Safety Devices 102
        18 Magnetism 107
        19 Electromagnetism 113

SECTION 4 Meters and Measurements

Chapter 20	Meters	121
21	Simple Current Meters	126
22	Simple Voltage Meters	134
23	Simple Resistance Meters	139
24	Multimeters	144
25	Electronic Meters	150

SECTION 5 Direct and Alternating Currents

Chapter 26	Batteries	157
27	Generators and Alternators	163
28	AC Currents	171
29	Oscilloscopes	177
30	Transformers	182
31	Inductance	189
32	RL Time Constant	195

SECTION 6 Reactances and Resonance in AC Circuits

Chapter 33	Inductive Reactance	203
34	Capacitance	208
35	RC Time Constant	216
36	Capacitive Reactance	223
37	Resonance	229
38	Series Resonant Circuits	236
39	Parallel Resonant Circuits	243

SECTION 7 Control Devices

| Chapter 40 | Diodes | 251 |
| 41 | Power Supplies | 259 |

42	Vacuum Tubes	267
43	Transistors	274
44	Light Sensitive Elements	281
45	Other Semiconductor Devices	287

SECTION 8 Systems and Applications

Chapter 46	Digital Electronics	294
47	Integrated Circuits	304
48	Computer Hardware and Software	309
49	Amplifiers	315
50	Putting Amplifier Stages Together	321
51	Oscillators	328

SECTION 9 Power and Communication

Chapter 52	Power for Heating	335
53	Power for Lighting	341
54	Power for Motors	347
55	Communication by Telephone	352
56	Communication by Radio	357
57	Communication by Television	366
58	Other Forms of Electronic Communication	371

SECTION 10 Control and Other Applications

Chapter 59	Industrial Control	376
60	Data Processing	383
61	Robotics	387
62	Electronics in the Automobile	393
63	House Wiring	401
64	Appliance Repair	408

LIST OF PROJECTS   TABLE OF CONTENTS

#	Name	Page
1	Fire and Ice Voltage Tester	19
2	Darkness Detector	39
3	Electronic Battleship Game	57
4	Motor Speed Control/Light Dimmer	73
5	TV Bugger	95
6	Bike Burglar Alarm (or Scream Box)	111
7	"Magnutic" Game	119
8	Easy Use Multitester	133
9	Multimeter from Surplus Parts	154
10	Automatic Rocket Launcher	169
11	Three-Channel Adjustable Color Organ	187
12	Pulsed Alarm Circuit	201
13	Bargraph Lie Detector	221
14	AM Transmitting Siren	242
15	Variable Regulated Power Supply	265
16	Light Meter with Bargraph Display	285
17	The Simplest Logic Probe	308
18	Light Activated Control Circuit	327
19	Flashing Numerals	346
20	Fiber Optic Transmission System	364
21	Easy, Accurate Transistor Tester	381
22	AC Timing Light	399

PROJECTS

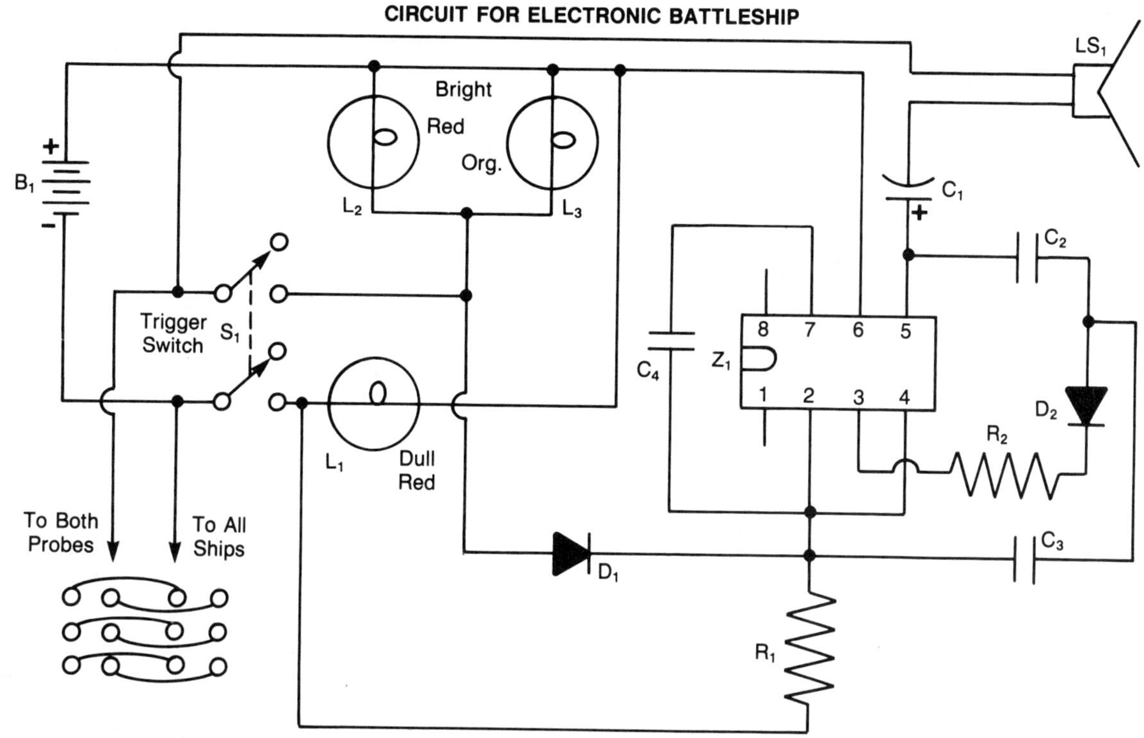

CIRCUIT FOR ELECTRONIC BATTLESHIP

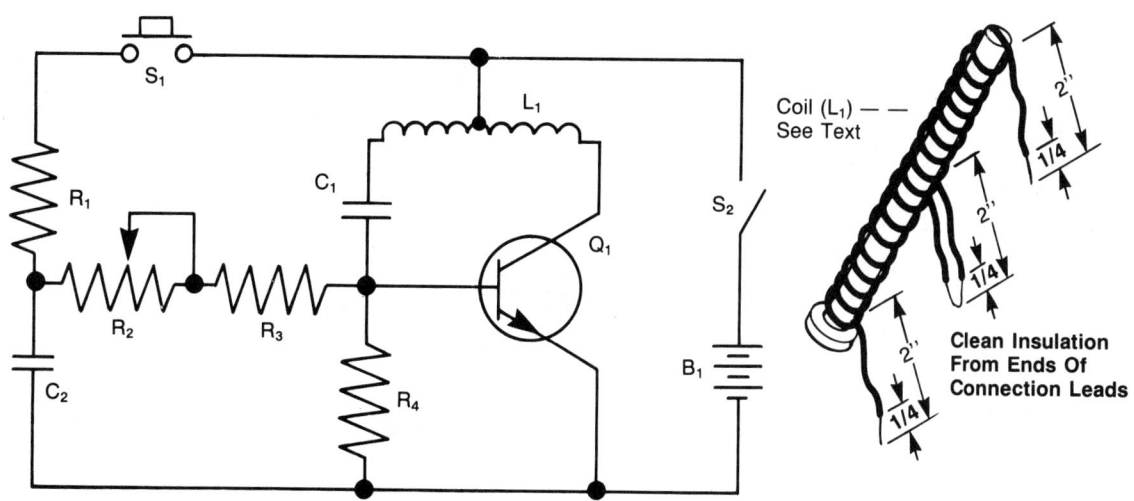

Electronic Battleship Game, see pg. 57

AM Transmitting Siren, see pg. 242

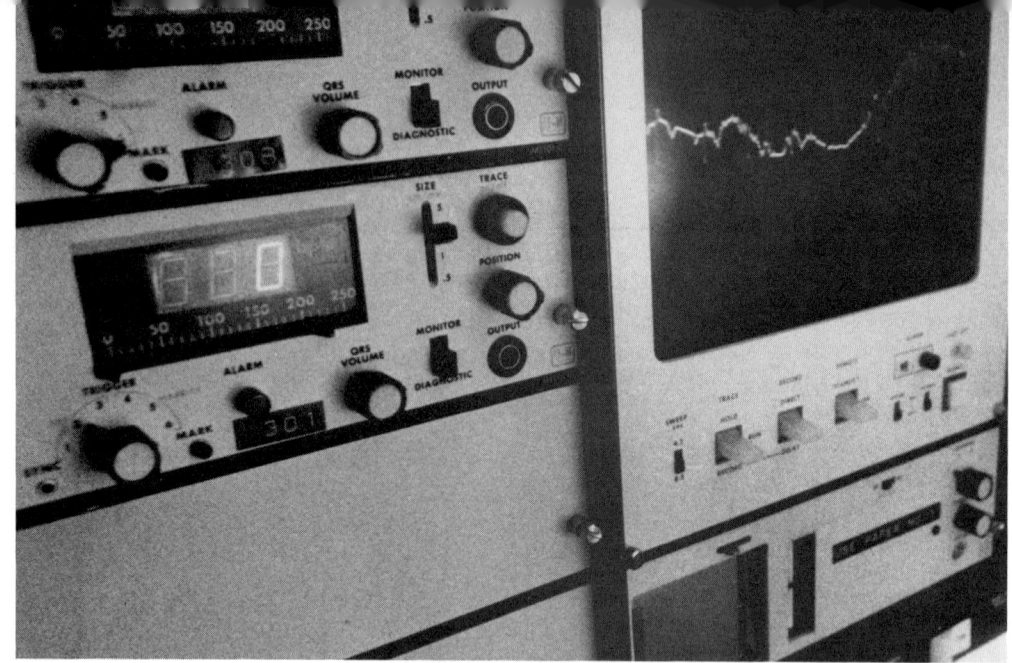

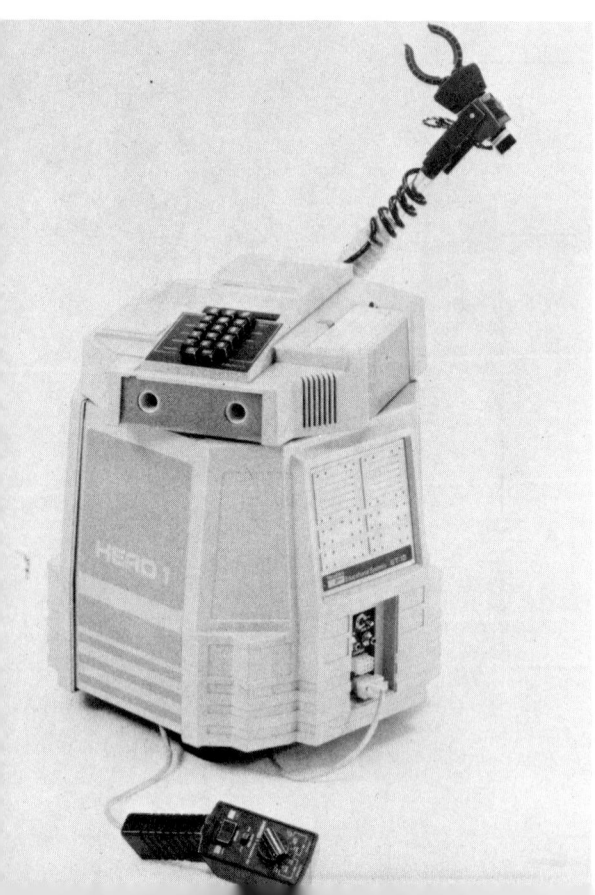

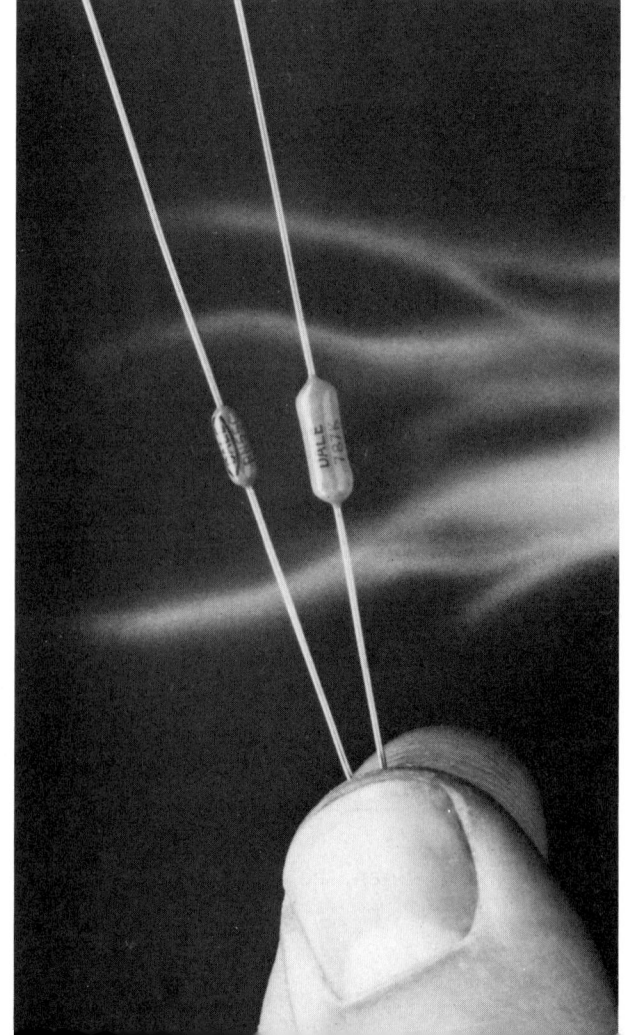

Getting Started in Electricity-Electronics

Section 1

Chapter 1
Careers in Electronics: A Photo Essay

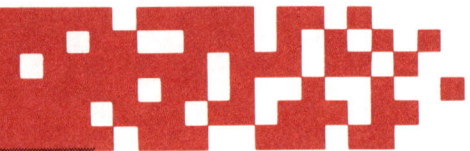

Imagine that you are from the Stone Age and have just been discovered frozen in a glacier. Further, imagine that a new, quick-melting process can thaw you out and revive you.

The first thing you see is a laboratory full of electronic equipment. Remember, you have never encountered anything more technologically advanced than a rock used to kill an animal — and that rock didn't even have a handle! If your first view of the laboratory doesn't scare you to death, you will be confronted with television, automobiles, video games, electronic music, electric lighting, portable radios, jet airplanes, telephones, beeping electronic watches, electric can openers, washing machines, soft drink machines, microwave ovens, air conditioning, electrically powered subway trains, power saws, and even the flashlight the doctor shines in your eyes.

We do not have to go back that far in time to find someone who is totally ignorant of our modern technology. Only 110 years ago, these devices could not have been imagined.

Electronics is an exciting field of study. Many of the convenience items you enjoy today would not be possible without the progress made in electronics. This chapter will introduce you to electronics by examining many career opportunities. It will help open your eyes (and your mind) to just how broad this field of technology has become — and where it may lead in the future. Many career fields seem to be shrinking, but opportunities in electronics are likely to increase in the years to come. You made a wise selection when you chose to study electronics.

Repair and Service Careers

Many service and repair careers involve electronics. If you like to fix things, you can have your choice of jobs which take you outdoors or keep you inside. There are service jobs in which the technician must stay as clean as a surgeon performing an

Fig. 1-1 Great advances in electronics have made possible today's medical technology, including this high-powered microscope.

operation. Others permit the repairperson to get really grubby. Some service workers carry their tools in a briefcase. Others need a truck. Service work in electronics can require very little education or a great deal of it — depending on the equipment to be serviced.

Engineering Careers

Many engineering positions deal with electronics. In fact, a number of specialties within the electrical engineering field didn't even exist just 50 years ago. There are countless careers available in engineering, drafting, and related fields.

Active Jobs in Electrical Work

If you like to work outdoors, a career as a lineworker may be for you. Telephone and electrical lineworkers and installation workers lead active lives and earn high wages compared with other construction workers. Have you ever envied the view of the world electricians have from the top of a telephone pole?

Computer Careers

Computer-related fields have grown faster than any other career area of our technological age. Think of this: Computers must be designed, built, shipped, advertised, sold, installed, programmed, operated, repaired, tested, and evaluated. Qualified people are needed to perform all those jobs. In addition, people write computer operation manuals, teach others how to use computers, even search for new ways to use them. It is becoming harder to find a job that has nothing to do with computers!

Fig. 1-2 Do you enjoy working with your hands? If so, a career in electronics may be in your future. Note how this technician must wear plastic gloves in order to handle this detailed work.

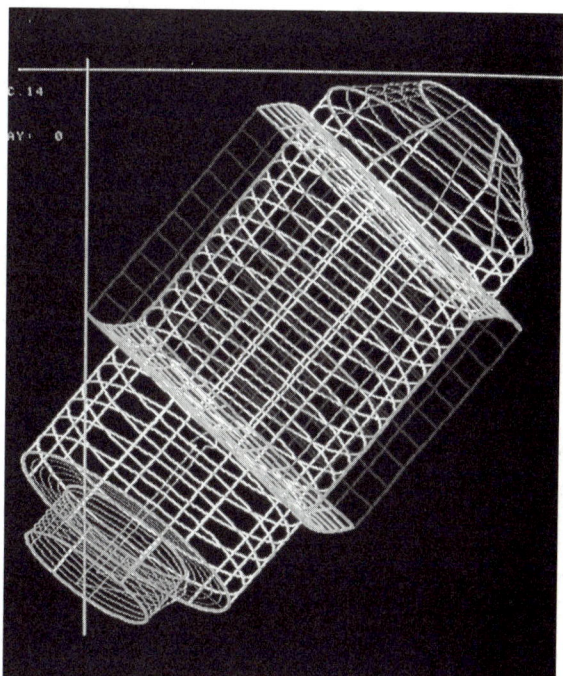

Fig. 1-3 Many of today's products are first designed and drafted on a computer.

Chapter 1   *Careers in Electronics: A Photo Essay*

Communication Careers

Television is an important part of our lifestyle. The improvements made in television in the last 20 years have made them work better, cost less, and last longer than ever before. However, a career in television is not the only electronic communication area available.

Your newspaper uses electricity and electronic communication from start to finish. The news reporter writes the story using a computer. The story is typeset on a computerized phototypesetter. The plates used for the printing are produced with electronic equipment. Finally, the presses are powered and controlled by electricity.

Many careers in entertainment would not be possibile without today's electronic mass communication system. How would a rock music star become famous if there were no stereo systems, radios, televisions, computer booking systems for concerts, electric lights for stages, microphones, amplifiers, electric guitars, and synthesizers? Even rock stars depend on the people who know electronics. When the amplifier dies in the middle of a concert, the technician who can get things working again becomes the "star."

Scientific and Medical Careers

Doctors, scientists, and researchers depend upon electronic equipment to do their work. Sometimes, important research has to wait until equipment can be developed which has the necessary capabilities. Electronics is at the center of this developing technology. Operations which were far too risky a few years ago are now commonplace because accurate electronic monitoring equipment has been developed. And, the space research program would not be conceivable without highly advanced electronic systems. Many career opportunities within the space exploration program depend on electronics.

Summary

Fascinating career opportunities are open to people with a knowledge of electronics — and those opportunities promise to be plentiful for many years to come. This chapter has included a look at electronics in our daily lives and careers available in related fields. A series of short features on selected careers is also included in this book. By learning about electronics, you may be preparing for an exciting career with a real future.

Fig. 1-4 Without electricity and electronics, you wouldn't be able to watch television at night. Even the music you enjoy on your radio is possible because of electricity and electronics.

Fig. 1-5 Imagine how many electricity and electronics careers are available in this city. But don't think that careers in this field are limited only to the cities. Suburban and rural areas also offer a vast array of job opportunities. How many can you name?

Review Questions

1. What are some electronic devices in your home which were not mentioned here? Which ones do you think were invented before 1950?

2. Look in the classified advertising section of the newspaper and circle all of the jobs which involve electronics. How many of these seem interesting to you?

3. Try to list five career choices (which are still valid jobs today) that do not involve electronics or the use of any electrical-electronic devices. Think carefully about all of the worker's duties in each job before you list it. Does he or she really not use any electrical devices? This may be a tough task in our technological age.

Chapter 2
Tools and Safety

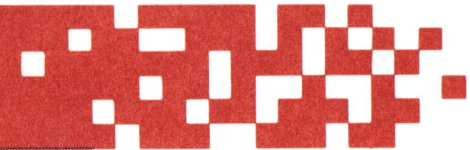

Before beginning study of the exciting field of electricity-electronics, you need to become familiar with the tools and equipment you will use. Take time to get to know your way around the laboratory or shop and become acquainted with new friends who share your interest in electronics.

Safety in the Shop

A measure of good old commonsense is important when working in any type of shop. Because there are many people around who could be injured by careless actions, a school shop must be run with a strict code of conduct.

The following general shop safety rules are intended to make the laboratory a safe place to work:

1. **Always wear your safety glasses in the laboratory.** — Eye injuries can be caused by a number of hazards in the electronics shop. There are chemicals for etching printing circuit boards and dangers of getting hot solder splashed in your eyes. There are always risks involved when many people work with sharp tools in crowded areas.

   In many states, it is now a law that safety glasses must be worn by everyone in a school shop or laboratory any time the room is being used for work. This means you must wear your glasses even if you are drawing a project plan

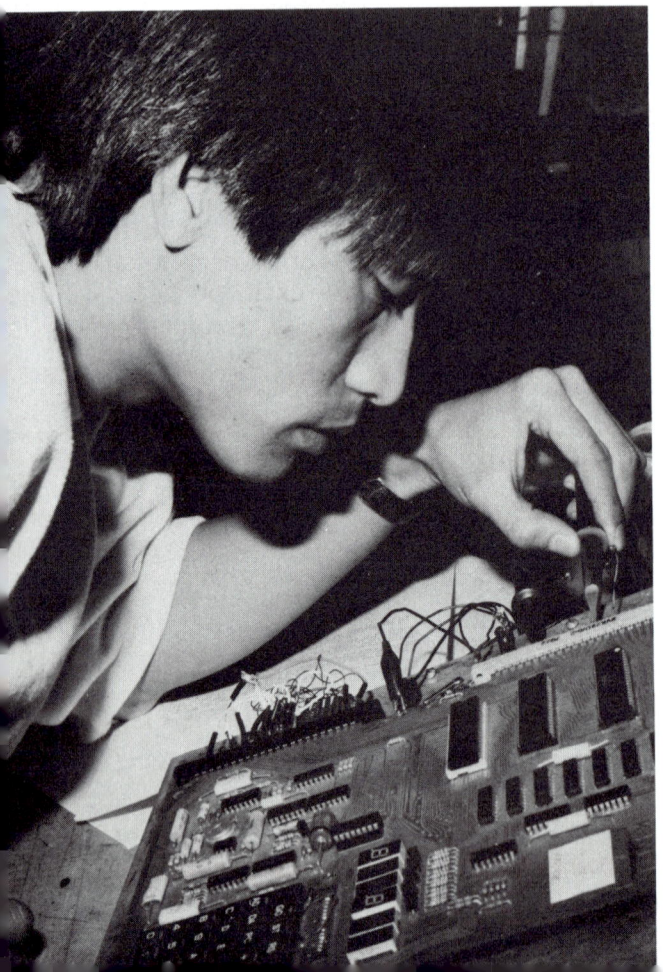

Fig. 2-1 Safety in the electronics lab requires following a general set of safety rules. Can you find anything wrong in this photo?

while another student is working on a lab experiment at another bench. This may seem to be overkill, but many shop accidents injure people who are "innocent" bystanders. If the other student wired a circuit incorrectly and exploded a piece of equipment, fragments could easily travel far enough to injure you. An ounce of prevention is worth more than a pound of cure — and wearing your glasses any time you are in the shop is certainly worth the inconvenience they may cause. The glasses may seem uncomfortable at first, but you will get used to them in time.

You may want to buy your own safety glasses which can be adjusted for maximum comfort, but get them approved by your teacher.

2. **Never run in the shop.** — Your electronics laboratory has many tables or benches with fragile equipment and some heavy tools. If you slip or bump into another student while running in the shop, you may cause a shock, a cut from a sharp tool, or some other injury.
3. **No "horseplay" allowed.** — It is a lot of fun when done in its proper place. That place is *not* the electronics shop.
4. **Never operate equipment until you have been checked out on it.** — Don't take a chance by using a machine or instrument until you have been taught how to use it properly. Some teachers require that students pass safety tests on certain pieces of equipment before they are allowed to use those machines.
5. **Clean up spills.** — In any shop, liquids occasionally will be spilled and scraps naturally will accumulate on the floor because of the work being done. It is important, however, to clean up these hazards immediately.
6. **Ask permission before using equipment.** — There are certain machines in your shop that are more hazardous. Ask your teacher for permission before you use them.
7. **Never carry more than you can handle safely.** — You may be tempted to load up with as many tools as possible to save an extra trip. This can result in injury or damage to the tools. Don't be impatient. You also can hurt your back by carrying too much.
8. **Do not talk to anyone while operating a machine.** — This is a double-sided rule. People cannot pay full attention to the machine they are operating and talk to a friend at the same time. Don't interrupt someone when he or she is operating a machine.
9. **Never use a dull or damaged tool.** — Report any damaged or dull tools to the teacher immediately. You might apply extra pressure to make a dull tool cut, and this might make you slip and jab the tool into yourself or another student. Tools with broken handles, loose parts, and other defects can cause serious injury. Don't take the chance.
10. **Don't work when you are impaired.** — If you have to use medicine which causes drowsiness, do not work in the shop with hazardous equipment. Likewise, if you have an injury that affects the use of your hands, consult your teacher about special safety arrangements. Most important, do not work in a shop of any type while under the influence of alcohol or any other drug. You are a danger to yourself and everyone else in the room if you do.
11. **Remove loose clothing and jewelry.** — Long, loose sleeves and hair must be bound, covered, or tied back when working with machines. You could be dragged into a rotating machine by them. Remove all jewelry. Rings and watch bands may become red-hot because of nearby AC currents or magnetic fields. This can cause you to be severely burned.

12. **Use commonsense.** — Most of these safety rules are direct applications of good commonsense. Although there are some hidden hazards in a shop, you will be fine if you just use good commonsense. If you are not sure about something, ask your teacher. Don't take chances in any situation. Don't rush, and keep a serious attitude about safety.

Special Safety Rules for Electronics

In addition, here are special rules for working with electronics. Your teacher may add more to this list. Likewise, other safety precautions are presented as the need arises in later chapters.

1. **Always check circuits carefully according to the wiring diagram.** — Before turning on the power, double-check circuit wiring to be sure you have not made mistakes.
2. **Avoid working on live circuits.** — Although some tests must be made with power in the circuit, get in the habit of disconnecting the power from circuits before changing any components or touching any wires.
3. **Be sure that safety devices and fuses are used properly.** — Chapter 17 covers this topic. It is very foolish to try to get around using safety devices.
4. **Read instruction manuals and repair manuals carefully.** — Many hidden hazards lurk in equipment for the tinkerer who does not read instructions.
5. **Be careful when handling chemicals and hot soldering irons.** — Wear protective clothing and safety glasses. Take extra care to avoid quick motions.
6. **Some components store a charge.** — Capacitors and television picture tubes can store high voltage electric charges even after the circuit's power is off. Be very careful when working with them.
7. **Beware of wet areas.** — Never work on electrical circuits near water. Water provides a direct path for the current to go through you to the ground.
8. **Follow the "one-hand" rule.** — When you must work on a live circuit, use one hand to do the work and keep the other one in your pocket. This way, there is less danger of current finding a path through your head and chest region (where it does the most damage).
9. **Inspect electrical equipment for defects.** — Do not use equipment with frayed wires, loose connections, and damaged cases. Repair such problems before using the equipment. And use the right tool for each job.
10. **Assume all circuits are live and dangerous.** — Before reaching into an electrical circuit, make the proper tests to ensure that it is safe to do so. Repairing a meter may cost a few dollars, but you cannot be replaced at any cost.

Tools for Electronics

Figures 2-2 to 2-7 show common tools and equipment you will learn to use. Familiarize yourself with the equipment available in your electronics shop and its locations. Your teacher will demonstrate the proper use of each piece of equipment when it is time for you to use it.

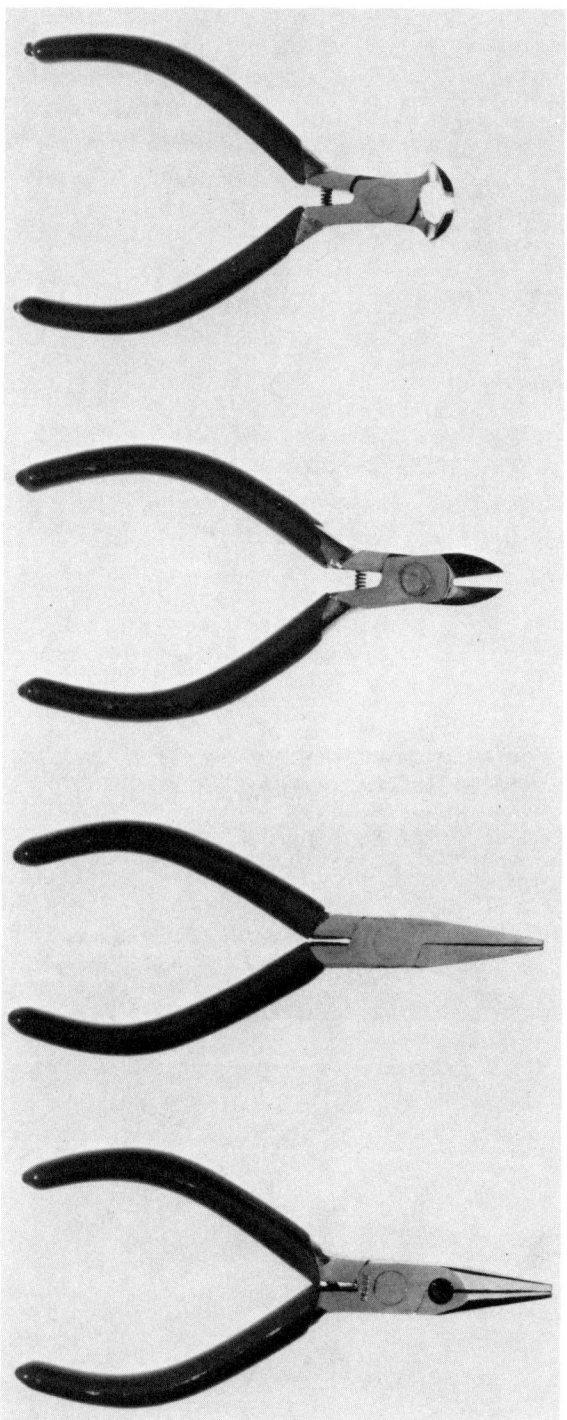
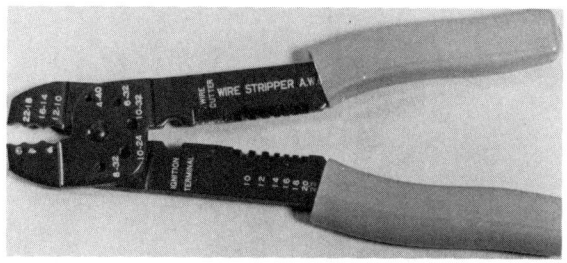
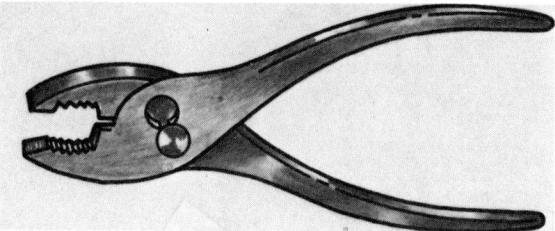
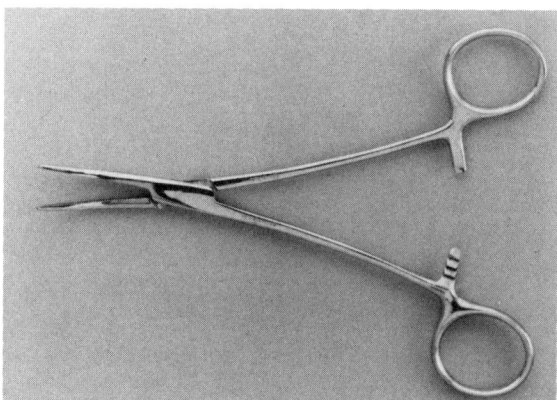
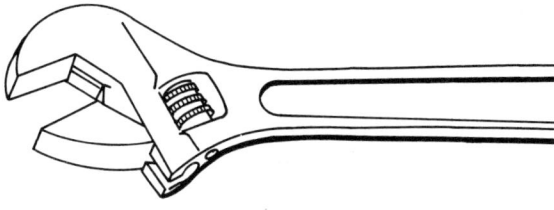

Fig. 2-2 Gripping and cutting tools: end nippers, diagonal cutters, long-nose pliers, long-nose cutters (left column, top to bottom), crimping tool, adjustable pliers, forceps, and adjustable wrench (right column, top to bottom).

Chapter 2 *Tools and Safety*

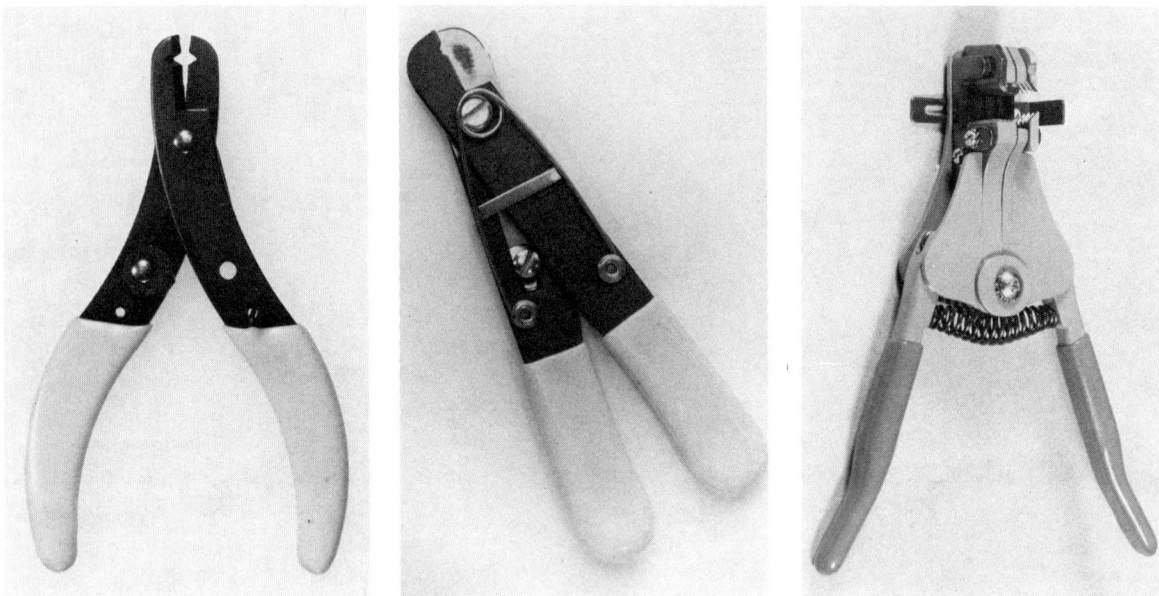

Fig. 2-3 Three different types of wire strippers.

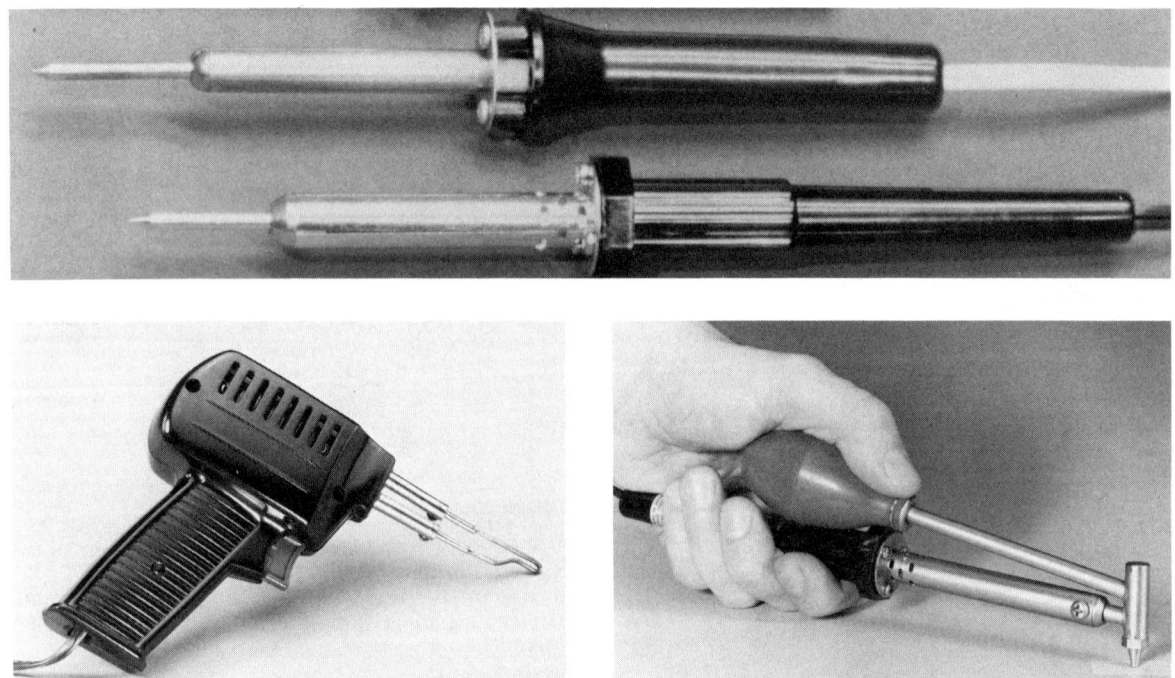

Fig. 2-4 Heat-process tools: soldering iron (top), soldering gun (bottom left), and desoldering tool.

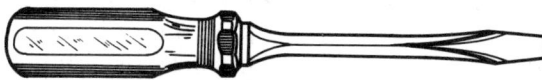 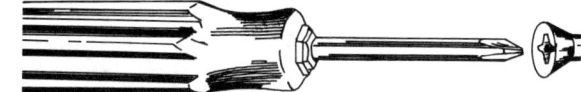

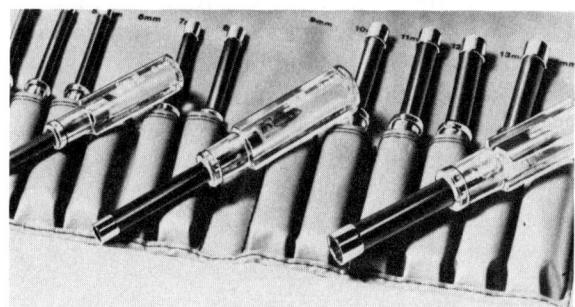 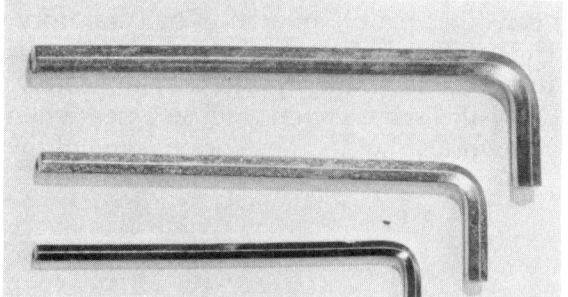

Fig. 2-5 Other hand tools: straight screwdriver (top left), Phillips screwdriver (top right), nut driver (bottom left), and allen wrench.

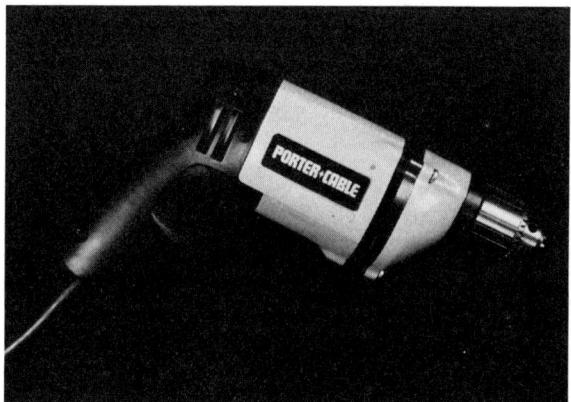

Fig. 2-6 An electric drill.

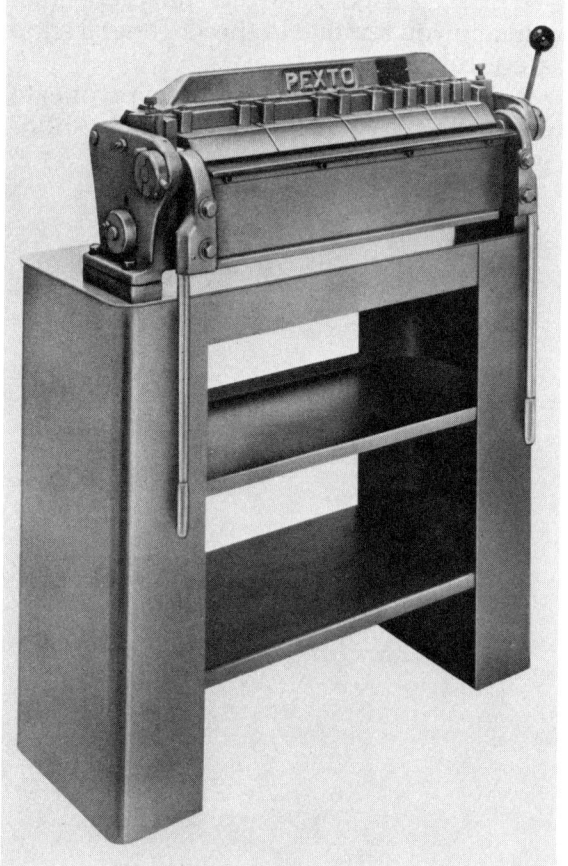

Fig. 2-7 The box and pan break is used to bend sheet metal.

Chapter 2 *Tools and Safety*

Summary

Safety is very important when working near other people in a shop or laboratory. The safety rules in this chapter are for your and your classmates' protection and safety. Obey the same commonsense rules when you work at home, too. Locate some of the common tools you will learn to use in your school electronics shop. You will begin your study of this "mysterious phenomenon" called electric current and how it works in the next chapter. Remember that most people really do view electronics as a mystery — you are about to gain knowledge which can open many doors for you. We hope you enjoy learning about this technological frontier.

Review Questions

1. Make a list of keyword statements that will help you remember the 12 general shop safety rules.
2. Why is it important for everyone to follow the safety rules in your school shop? Why should you follow them when you work at home?
3. What new tools and equipment did you see that you are anxious to use?

Chapter 3

What Is Electric Current?

Much of the technology we take for granted would be impossible without electricity. To understand electricity, you must first understand what creates it.

Matter and Atoms

All of the world is made up of matter. **Matter** can be defined as any substance that occupies space. Matter does not have to be a solid. Liquid and gas are matter, too.

The simplest forms of matter are **elements**. There are over 100 elements in nature. Some of the more familiar elements are hydrogen, oxygen, iron, copper, gold, and silver but not all of matter consists of simple elements. Elements can join together to form compounds. Water is a compound. Salt is another.

Elements can be broken down into smaller parts called **molecules**. Molecules retain the characteristics of the element. The molecules consist of atoms. Elements are made up of atoms which are all alike.

The simplest element is hydrogen. The atoms that make up the element hydrogen are all the same. All of the molecules in a sample of hydrogen will have only hydrogen atoms. Figure 3-1 shows a molecule of hydrogen. Notice that, unlike the water molecule in Fig. 3-2, it has only hydrogen atoms.

When different types of atoms are joined together chemically, they form **compounds**.

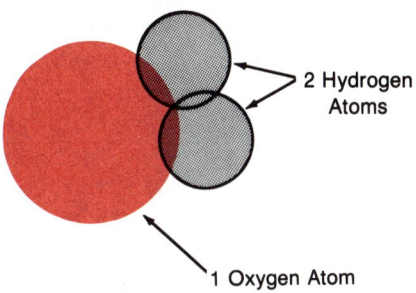

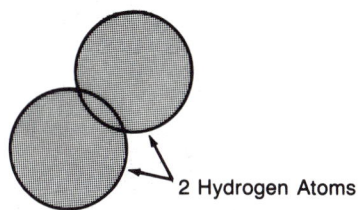

Fig. 3-1 Two hydrogen atoms may combine chemically to form a molecule. The molecule will still be hydrogen, and it will have no atoms of other elements.

Fig. 3-2 When two atoms of hydrogen are chemically combined with one oxygen atom, the result is one molecule of the compound H_2O — water.

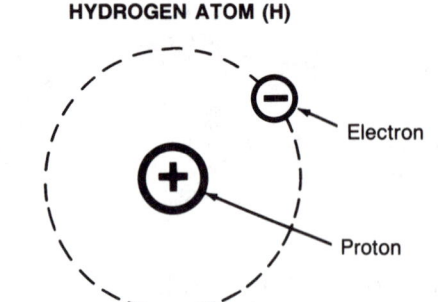

Fig. 3-3 The atom of the element hydrogen is the simplest of all atoms.

Water is a compound of two atoms of hydrogen (H_2) and one atom of oxygen (O). See Figure 3-2. The proper way to express this chemical combination is H_2O.

Parts of the Atom

The **atom** is the smallest particle of an element that still retains all the characteristics of that element. Atoms are made up of still smaller particles, however. When a molecule of hydrogen is broken down, the result is two or more atoms of hydrogen. When an atom of hydrogen is broken down, the result is *not* smaller particles of hydrogen. However, you will get a collection of atomic particles. All there is in an atom of hydrogen is two charged particles (Fig. 3-3). One of the particles, the **proton** carries a positive (+) charge. The other one, the **electron**, is charged negatively (−). Protons are found in the center of atoms. Electrons orbit around the center of the atom. All electrons are the same and all protons are the same. The differences in atoms occur in the number of protons and electrons they contain.

Figure 3-4 shows an atom of carbon (C). This atom is more complex than the simple hydrogen atom. One obvious difference is that there are some new particles. The little gray particles in the center are **neutrons**. The neutron is neutral — it carries no electrical charge. It mainly makes the atom heavier and has little effect on the electrical behavior of the atom. The cluster of protons and neutrons at the center of the carbon atom is called the **nucleus**.

The second difference that you probably noticed about the carbon atom is that the electrons are not all in the same ring around the nucleus. The rings represent the different paths or orbits of the electrons. The term **shells** refers to these paths of electron orbit. Atoms which are more complex than hydrogen and carbon have even more shells with electrons in them. They also have more protons and neutrons in their nuclei.

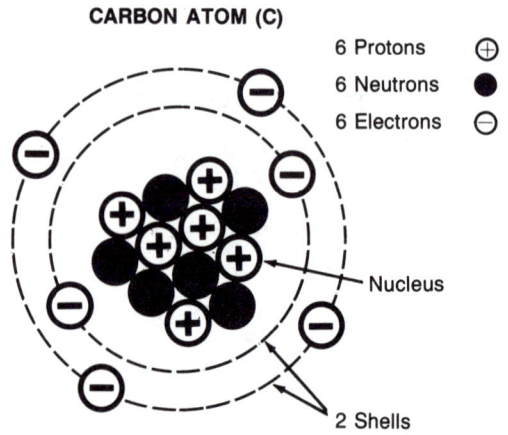

Fig. 3-4 The carbon atom has two shells and a nucleus made up of protons and neutrons.

Charges, Attraction, and Repulsion

Have you ever heard the saying, "Opposites attract"? That holds true with atoms. The negatively charged electrons are attracted to the positively charged protons. The electrons are repelled by other electrons. Like charges repel each other and unlike charges attract each other.

The electrons are very light compared to the protons, and they move very fast in an orbit around the nucleus. This combination of forces and particles keeps the atom stable and balanced under normal circumstances.

A piece of copper wire (made up of many copper atoms) is normally not charged positively or negatively because there are as many protons as there are electrons. The total charge is zero; there is a balance. In order to use these positive and negative charges in electrical devices, the balance must be upset.

Free Electrons

The shells of electrons are not haphazardly arranged. Each shell has the capability to hold a certain number of electrons. The shell that is farthest from the nucleus is the most important in electronics. The number of electrons in this outermost shell determines how useful the material will be in electronics.

Figure 3-5 shows a copper atom (Cu) and an atom of argon (Ar). The outermost shell of the argon atom has eight electrons. The copper atom only has one electron in the outermost shell. When there are many electrons in the last shell, it is very difficult for any electrons to become dislodged from the atom. However, when there are only one or two electrons in the farthest shell, it is relatively easy to knock an electron from its orbit. Would it be easier to dislodge an electron from the copper atom or argon atom?

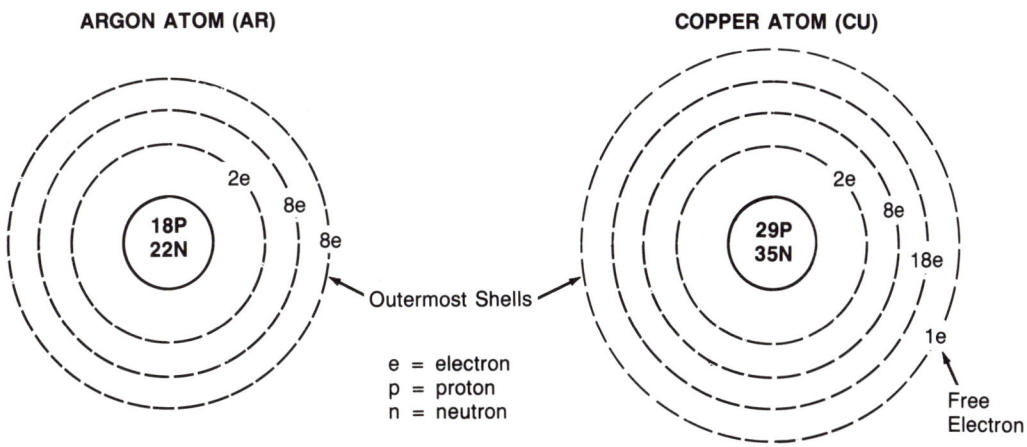

Fig. 3-5 Only one of the copper atom's 29 electrons is in the outermost shell. This is a free electron. The argon atom has no free electrons.

Chapter 3 *What Is Electric Current?*

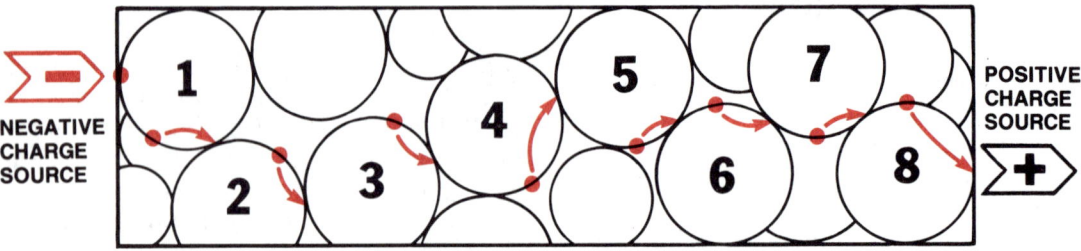

Fig. 3-6 The extra electron from the negative charge source bumps an electron from atom 1 to atom 2, and this chain reaction continues very quickly throughout the wire to the positive charge source.

If electrons are crowded in the outermost shells of two neighboring atoms, it is easy for them to "see" where they belong and to keep on track. The electrons are said to be bound, and the atoms are stable. On the other hand, the single electrons in the outermost shells of copper atoms do not have their paths so clearly defined. A little extra charge can confuse them into thinking they belong in orbit around their neighbor instead of their "home" nucleus. These electrons which can be easily knocked or drawn from their orbits are called **free electrons.**

Electron Flow

Where do electrons go when they are pushed or pulled from their normal orbits? Usually, they will join the nearest atom. If they are drawn away from their normal orbits by a positive charge, they will migrate toward that charge. If, however, they are knocked out of orbit by a negative charge, they will travel away from the source of the negative charge. Remember, like charges repel and unlike charges attract.

Look at Fig. 3-6 to see what would happen if a piece of copper wire were negatively charged on one end and positively charged on the other end. The free electron of the first atom will be repelled from the negative charging source and will jump to the second atom. The second atom will then become very unstable and will try to balance itself by giving up one of the two electrons it now has in its outer shell. In fact, those two electrons which share the space meant for only one will repel each other, and they will have less attraction to the nucleus than normal. These forces are all small, but they will be enough to push one of the two electrons away from the second atom on to the third. There the process will be repeated. A chain reaction will occur and energy will be transmitted. Finally, an electron from the last atom will be forced out to the positive charge at the other end.

In reality, an electron flow of one electron would not be useful. Increase the positive and negative charge sources, and many electrons will be forced to move at the same time. This is electric current! That's right. Electric current is free electrons being knocked from one atom to the next, flowing from negative to positive. This is the secret, almost magical force that does so much for us. In electronics, we call this phenomenon **electron flow**.

Conductors, Insulators, and Semiconductors

Conductors are materials that have few electrons in their outer shells. Conductors are able to transfer electrical energy easily by passing electron flow. Figure 3-7 shows that the best conductors for practical use are silver (which is rarely used because of its cost), copper, gold, and aluminum. All of these conductors (except aluminum) have only one electron in the outer shell. In other words, they all contain some free electrons.

Materials that will not allow electrons to flow easily are called **insulators**. In insulators, the outer shell has several electrons bound much tighter than in conductors. Electrons can be forced out of orbit even in these insulating materials, but it takes tremendously more force. Typical insulators include glass, plastic, rubber, paper, and air. Insulators are useful to stop the flow of electric current. They also may be used to store a charge without letting the electrons actually pass through. Realize, too, that most useful insulators are compounds rather than elements.

There are other materials that were once thought to be relatively unimportant in terms of electricity because they are not good conductors or insulators. These materials, such as silicon, germanium, and carbon, are known as **semiconductors**. We now realize that semiconductors are the most important of all materials in electronics.

Transistors, the major breakthrough of modern electronics, can be made only from semiconductor materials.

Common Materials Used in Electronics

Conductors	Notes
*Silver	Best conductor
*Copper	Copper and aluminum are used most
*Gold	Gold and silver are too expensive
*Aluminum	
*Iron	
Water	Earth and human bodies contain water

Insulators	Notes
Glass	
Plastic	The top three are best and most used
Rubber	
Paper	
Air	

Semiconductors	Notes
*Carbon	Used to make transistors
*Silicon	
*Germanium	

*These materials are elements

Fig. 3-7 These materials have special qualities that make them useful in electronics.

Summary

The atom is the smallest form of elements. Atoms contain protons, neutrons, and electrons. The negatively charged electrons are attracted to the positively charged protons, and the electrons orbit around the nucleus in paths called shells. Materials that have only a few electrons in the outer shell are called conductors because it is easy for those free electrons to become dislodged and travel from one atom to another. Insulating materials have few free electrons. Materials that are not good conductors or insulators are called semiconductors.

Important Terms

matter
elements
molecules
compounds
atoms

proton
electron
neutron
nucleus
shells

free electron
electron flow
conductors
insulators
semiconductors

Review Questions

1. Explain what elements and atoms are.
2. Explain the meaning and importance of free electrons.
3. Why do we say that electricity flows from negative to positive?
4. List some conductors, insulators, and semiconductors.
5. Why will a balloon cling to the wall after the balloon is rubbed with cloth?

Careers: Research Scientist

Scientists are at the forefront of technology. They strive to discover how our universe works and our place in it. Their challenging work requires long hours of study and careful examination of many different types of evidence. Scientists work for universities, government, and industry. The space exploration program employs many scientists and engineers.

Research scientists must have a college degree and advanced degrees as well. A strong background in science and mathematics at both the high school and college level is required. Salary varies with type of training and the institution or agency for which the scientist works.

You will read about outstanding people who made great discoveries in electricity-electronics; many were scientists or engineers. Electronics and science are close relatives, and nearly all scientists must have some knowledge of electronics.

Project: Fire and Ice Voltage Tester

Parts List:
- 1 NE-2 neon glow lamp
- 1 270 K ohm resistor, ½ watt
- 2 test leads, 16 ga. or larger, in contrasting colors
- 2 test probes, 10–14 ga. (stiff) T-W wires colored as above
- 2 1" pieces of heat shrink tubing
- 2 small pieces, electrical tape

polyester casting resin with catalyst
small paper cups for mixing resin
molds to pour resin (plastic ice trays will work well; should be polyethylene or silicon)

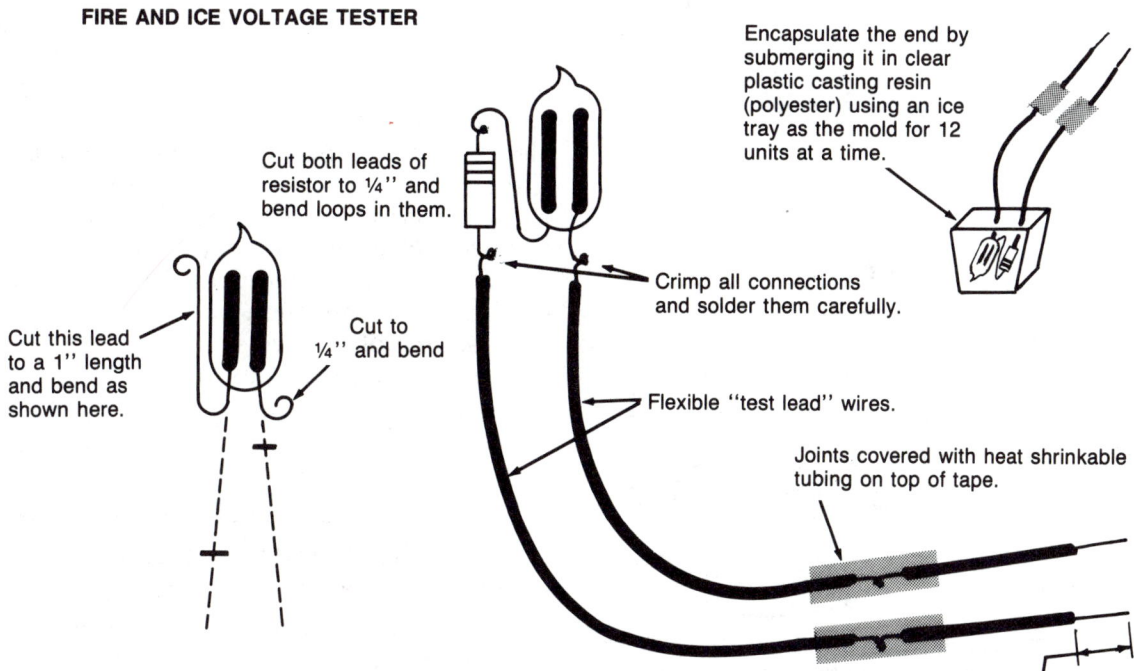

FIRE AND ICE VOLTAGE TESTER

Cut this lead to a 1" length and bend as shown here.

Cut both leads of resistor to ¼" and bend loops in them.

Cut to ¼" and bend

Encapsulate the end by submerging it in clear plastic casting resin (polyester) using an ice tray as the mold for 12 units at a time.

Crimp all connections and solder them carefully.

Flexible "test lead" wires.

Joints covered with heat shrinkable tubing on top of tape.

Test probes are 10-14 ga. T-W wires with ends skinned ½".

Chapter 4

Electric Circuits

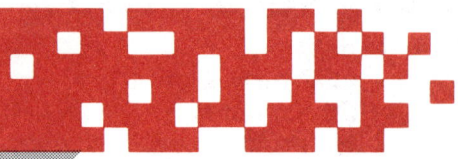

Source of Current

The source of current is the first part of the circuit. The six sources of electricity will be discussed in a later chapter.

For now, it is important to realize there must be a source from which the electrons can come. Figure 4-1 shows a battery connected by two wires to a light bulb. The battery is the source in this very simple circuit.

The source has two jobs. First, it must supply the circuit with the extra free electrons needed to set up an electron flow. Second, it must pump those electrons with enough force to push them around the whole circuit. Current will not flow if it cannot move completely around the circuit. The extra electrons are pumped out of the negative terminal of the source while the positive terminal acts as a vacuum cleaner sucking them back into the source from the other end of the circuit.

Circuits Control the Flow

Free electrons do not flow just anywhere. To make the energy useful, these electrons flow through **circuits**. You will learn about many different types of circuits in this course. Circuits are made of conductors and electronic parts usually called **components**. Circuits prevent wasting the energy it takes to get electrons flowing. They also make sure the electrons do not flow in places where they could harm or injure. Useful circuits have four main parts:
1. A source of current
2. A path for current to follow
3. Loads
4. Control devices

Path for Current

The conductors connect the components and the source. They are the path for the current. Anything with free electrons that is connected between negative and positive

SIMPLE CIRCUIT

Fig. 4-1 This circuit is complete. There is a path for electrons to follow from the negative battery terminal, through the lamp, and back to the positive terminal.

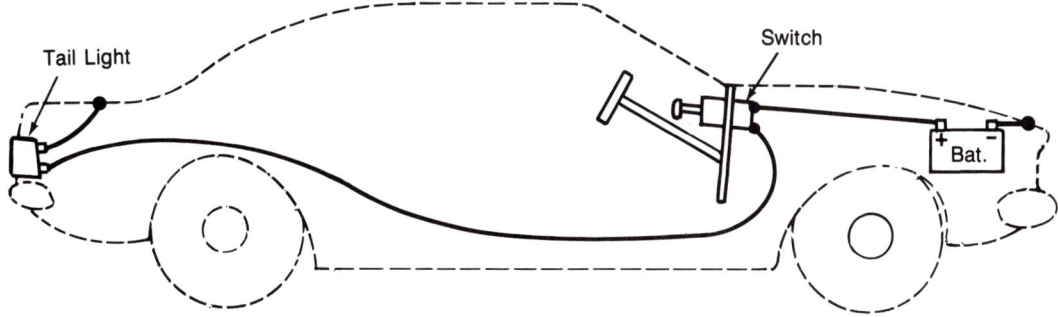

Fig. 4-2 The tail light circuit in a car uses the body of the car as a conductor. Electrons leave the negative battery terminal and flow through the metal car body to the light. From the light, they flow through the wire to the switch and back to the positive terminal.

charges can permit electron flow. If these conditions are met, you have a conductor in a circuit.

In useful circuits, conductors could be wires or other metal parts. For instance, a metal case that encloses other components is a very important conductor. The metal body, frame, and engine of a car act as conductors in this way. See Fig. 4-2. Earth and living bodies can be conductors.

Load

Circuits have at least one device which uses the energy. This is called the **load**. A load could be a light bulb, motor, radio, heater, or any number of other devices. Resistance is another word sometimes used for the load.

Look at the circuit in Fig. 4-1. The battery is the source, the wires are the conductors, and the light bulb is the load (or resistance). Later chapters discuss loads in more detail. Can you identify the loads in Figure 4-2?

Some circuits have several loads. The circuit in an automobile has many loads: lights, horn, gauges, starter motor, engine ignition, radio, and other accessories. Some of these devices, such as the radio, are complicated electronic circuits themselves and have smaller loads within them.

Control Devices

The fourth part of a practical circuit is the control device. Control devices allow you to tell the circuit what to do.

The circuit in Fig. 4-1 does not have any control device. To turn it off or on, you must disconnect or reconnect one of the wires. Most useful circuits, however, do have at least one control device.

A simple switch is a control device. The volume control on a stereo set is a control device, too. These examples show two ways you can control electronic circuits and currents. The switch can only be turned on or off. There is no in-between setting. The loudness control, though, has a wide range of settings from "off" to "low" to "medium" to "loud."

The switch is the simplest form of **digital control** (the type computers use). The

Chapter 4 *Electric Circuits*

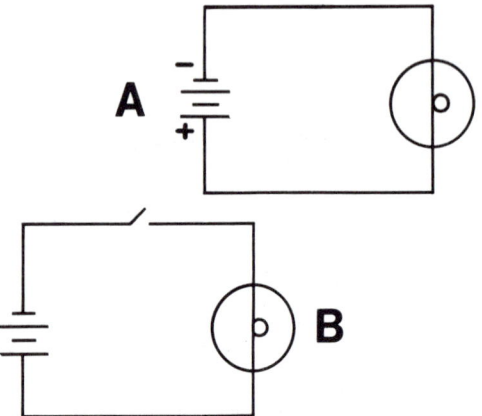

Fig. 4-3 A) The source, in this case a battery, is usually shown on the left. The load is a single light bulb. B) The control device (a switch) is placed between the source and the load. Switches are drawn in the open position.

Schematic Symbols and Drawings

The illustrations used so far in this book are pictorial drawings of the devices and circuits. It takes a long time to create these drawings and they take up a lot of space. For these reasons, a system of symbols and simplified drawings, called **schematics**, is usually used. Figure 4-3A is the schematic drawing of the circuit shown in Fig. 4-1. The same circuit is shown again in Fig. 4-3B with a control device added.

The lines in the schematics represent the conductors. Each load, control device, and source has its own symbol.

Since all loads are resistances, you can represent any load as a resistor if its proper symbol is not known. You can even simplify a complex device such as a television set by drawing it as a resistance.

Figure 4-4 shows the schematic for the working circuit in a room which has a television, fan, and two lamps. Schematic drawings will be used hereafter, so you should begin to learn the symbols as they are introduced.

volume control, however, is used where you wish to vary the amount. The volume control is an example of **analog control**.

Switches and volume controls are manual circuit controls. There are also special electronic components which control circuits automatically. These components can be either digital or analog.

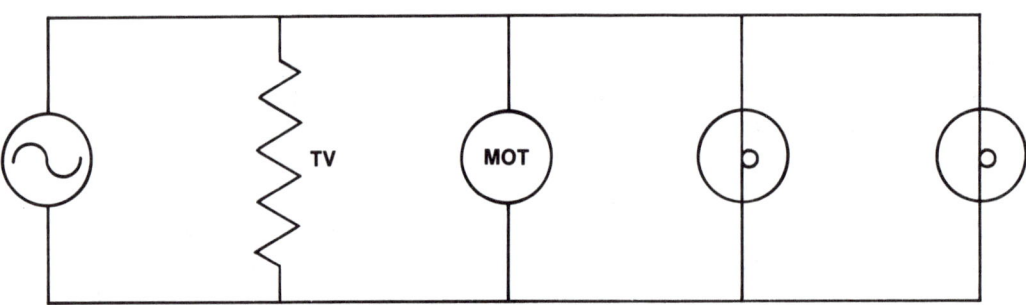

Fig. 4-4 The source in this circuit is the generator on the left. The first load is the television which is just shown as a simple resistance. Only the motor of the fan is depicted in a schematic.

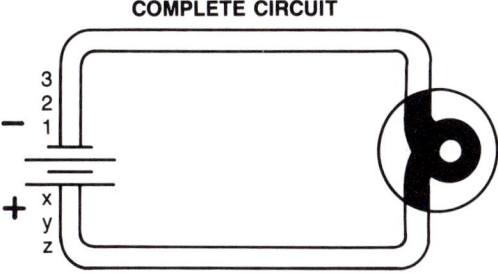 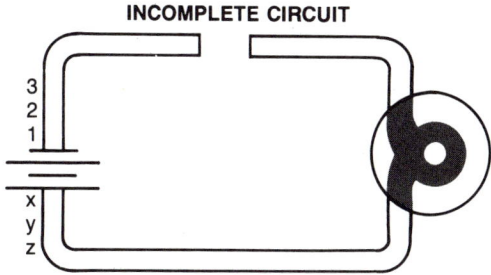

Fig. 4-5 The conductors and components have electrons line up like marbles in a tube. If electron number 1 moves, it will bump 2 which will bump 3 and so forth until electron X gets pushed into the + terminal. In the incomplete circuits, electron number 1 cannot move, and neither can any others, because of the dead ends at the open places in the circuits. The lamp will not burn (even in the right circuit) because none of the electrons can get moving.

Complete and Incomplete Circuits

Figure 4-5 shows a complete circuit. The circuit contains all the parts found in most circuits (source, path, load, and control device). But that is not what makes it a *complete* circuit. The words "complete" and "incomplete" have special meanings when applied to electrical circuits.

Whenever you have a path for electrons to flow out of the negative terminal of a source and get all the way back to the positive terminal, you have a **complete circuit**. It does not matter whether or not the circuit can do anything useful or whether the circuit was completed intentionally, if the electrons can make a full round trip, there is a complete circuit. If the electrons cannot go all the way from the negative terminal through the circuit and back to the positive terminal, then you do not have a complete circuit. In this situation you have an **incomplete circuit** and no current will flow. That's right, electrons will not flow in any part of the circuit if they can not make it all the way around the circuit. Figure 4-5 illustrates this important fact.

A broken or loose connection is an **open circuit**. An open circuit is an incomplete circuit, so the device will not operate. Many people confuse an open circuit with a **short circuit**, but they are opposites. The term "short" is often misused to explain why electrical devices do not work. The term "short" is really an abbreviation for short circuit. A short circuit is illustrated in Fig. 4-6. What has happened is that there is a shortcut for the electrons to travel from the negative pole to the positive pole without going through the load. This *is* a complete

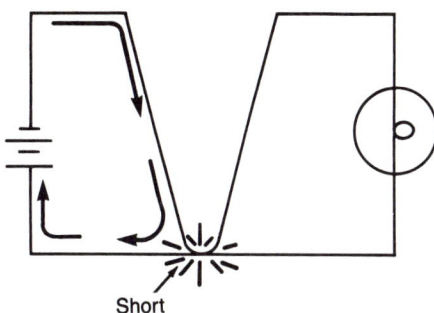

Fig. 4-6 In a short circuit, no electrons flow through the useful part of the circuit because they find an easier path back to the battery.

circuit because current will flow, but the short makes the circuit useless. The device will not work.

How do you tell the difference between an open circuit and a short circuit? In a short circuit the shorted wires may burn and start a fire, a fuse or circuit breaker will "blow," the source will be totally used up, or some other potentially dangerous problem will occur. Shorts are dangerous.

Fuses and safety devices, such as circuit breakers, open the circuit if a short occurs. This stops the high current flow so no damage is done. Chapter 17 discusses these important safety devices more fully.

Safety Considerations

Always keep in mind that you can become part of a complete circuit if you touch components or wires in such a way that the circuit is complete. Do not grab a wire with one hand and another wire with the other hand while the current is on.

Wires are not the only conductors, however. Water, metals, and the ground are often used as main conductors in a circuit. At the generating plant which produces the current you use in your home, one of the terminals is connected to the ground. If you touch something connected to the other generator terminal and at the same time touch anything that goes to the ground (water pipe, ground wire of another circuit, grounded piece of metal, water, etc.), your body would make a path for the electrons to flow through to get to the other pole. This could kill you!

You must be careful to avoid becoming part of a complete circuit. Serious injury or death can result from electrical shock. Be careful and use insulated tools. Wear appropriate insulated gloves when working with high voltages. Always turn the power off and disconnect circuits when possible. Never work with high-voltage circuits with the power on.

Summary

Electronic components are connected to form circuits. Circuits control the flow of electrons, and they have four main parts: source, conductor, load, and a control device. A circuit which does not allow electrons to make a full round trip is incomplete and is open. A short circuit is a complete circuit which is useless because the electrons take a shortcut. Shorts can cause damage to equipment and start fires. People who work with electricity must always be on guard to prevent completing circuits with parts of their bodies.

Important Terms

circuits
components
load
digital control
analog control
schematics
complete circuit
incomplete circuit
open circuit
short circuit

Review Questions

1. List the four parts of a useful circuit. Give an example of each.
2. What does the term complete circuit mean?
3. List the components in this chapter and draw their schematic symbols.
4. What is the difference between a short and an open?
5. Explain what is wrong with this statement, "As long as you only touch one wire, it is all right to work on hot circuits."
6. Is the speed control for a slot car an example of digital or analog control? Explain your answer.

Pep Talk: Science

If you wish to work in any area of electricity-electronics, a good knowledge of science would be beneficial. Sometimes students will study subjects in school which they enjoy and will let other subjects "drift" past them. If you have been doing that with your science classes and you wish to enter the field of electronics, you are cheating yourself!

Electronics is based on scientific principles — many of which are well beyond the scope of this book — and you have the opportunity to get a good headstart in your electronics training by simply doing your best in your science classes. This is true whether you become an electrical repairperson, electronics engineer, or any other electricity-electronics employee — or if you just enjoy electronics as a hobby.

Chapter 5
Sources of Electricity

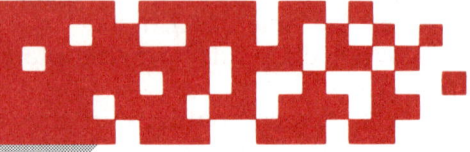

EMF from Magnetism

Magnetism is the most important source of voltage or **electromotive force** (EMF).

Magnetism is the principle on which all generators and alternators work. Huge generators produce the electrical current we use in our homes and factories. (A detailed explanation of how generators work will be covered in Chapter 27.)

As Fig. 5-1 illustrates, when a magnetic field moves past a wire that is part of a complete circuit, electrons move in the wire. The magnet does not have to touch the wire. When the magnetic field stops moving, the electrons stop moving too. This is true even if the magnet stops in contact with the wire!

A single magnet moving past a single wire does not make much current flow. However, if the wire is wound into a coil, as in Fig. 5-2, the effects of a few electrons moving in each loop of the coil are combined to make a useful amount of electron flow (current).

The playback heads in tape recorders work on this same principle. As the magnetized tape passes by the coil in the head, it sets up an ever-changing magnetic field around the coil. This moving magnetic field creates currents in the coil which electrically match the sound information on the tape.

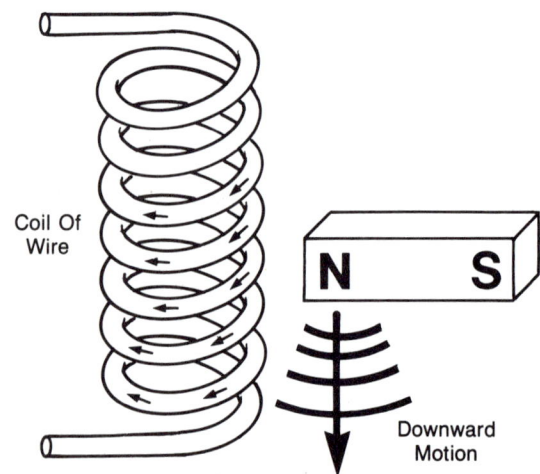

Fig. 5-1 As the magnet is moved, its field causes electrons in the wire to move also. The direction of current flow depends on which pole of the magnet is used and which way it moves. The magnet must be in motion and it need not touch the wire.

Fig. 5-2 Each loop of the coil produces a small EMF. These individual currents are added to produce a large current flow.

TAPE RECORDER PLAYBACK HEAD OPERATION

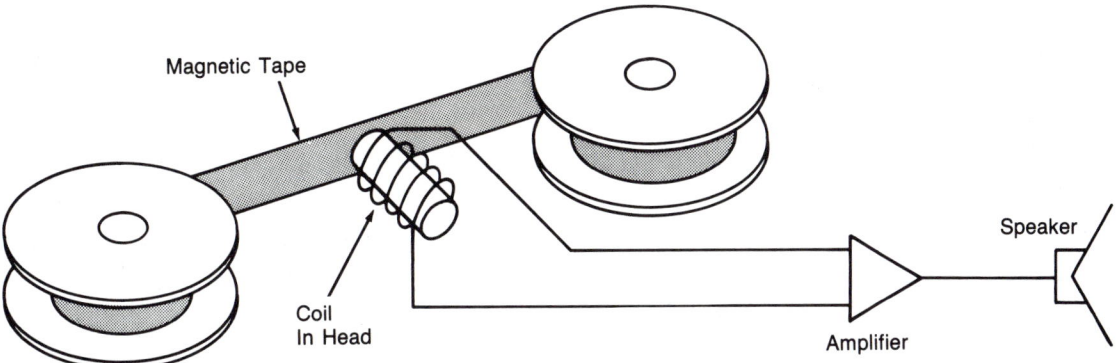

Fig. 5-3 The tapes used in tape recorders have sound information stored on them in areas of differing magnetic intensity. As the tape passes by the head, it generates an electron flow in the coil within the head that matches the sound information.

See Fig. 5-3. When these electrical impulses are amplified and sent to the speaker, they reproduce the original sound information.

You may have heard someone talk about getting electricity from coal, water, or nuclear energy. Actually, none of these sources can be converted directly into current. Coal or other fossil fuels and nuclear energy are used to make heat which produces steam to power a steam engine. The steam engine (usually a turbine) then turns a generator. Water pressure created when water is backed up behind large hydroelectric dams can also be used to turn turbines which power large generators (Figure 5-4). In all of these cases, magnetism is the source of EMF. The fuel, water, or nuclear energy is merely used to make the magnetic fields move.

EMF from Chemical Reactions

One of the most common sources of EMF is the battery. Actually, the word battery has been used rather loosely thus far. The little round things that go into a flashlight

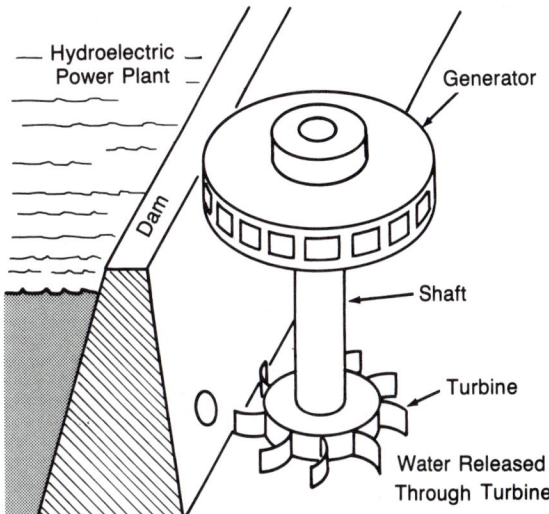

Fig. 5-4 The pressure of the water coming through the dam powers the generator. Magnetism is actually the "source of EMF" — the water pressure just turns the generator.

Chapter 5 *Sources of Electricity*

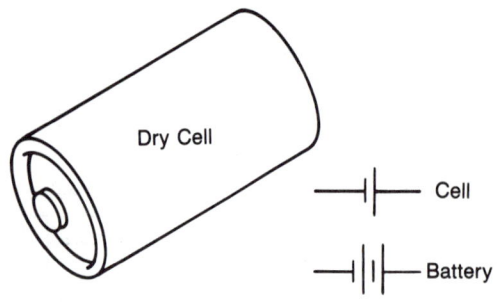

Fig. 5-5 Dry cells are usually used together to make a battery. The schematic symbols for a single cell and a battery are shown. The end of the symbol which has the shorter crossing line is the negative terminal.

and produce about 1.5 volts each are **cells** — not batteries! When two or more cells are put together to produce a larger voltage, that is a **battery**. Figure 5-5 shows a typical dry cell and its schematic symbol. Some cells have liquids in them. Others are called dry cells because they use a paste. Some types can be recharged over and over while others are used once and discarded.

The main thing to know about cells and batteries is that they produce electrical energy by chemical reaction. Refer to Fig. 5-6. Two different metals called **electrodes** are placed in a chemical solution called **electrolyte**. The metals and chemicals react with each other. This causes electrons to line up at one of the electrodes. This is the negative terminal. At the same time, the other electrode (the positive terminal) becomes deficient in electrons. The negative terminal pumps the extra electrons into the circuit.

The positive terminal tries to get electrons to make up for the ones the chemicals have driven away. The extra electrons at the negative terminal cannot take a shortcut and go backward through the cell to the positive terminal because the chemical reaction will not permit it. Thus, a battery forces electrons to move through the circuit.

EMF from Light

With other sources of energy, such as oil, becoming scarcer, solar energy has become more important. One way to use solar energy is with **photovoltaic cells** (fot-o-val-tay-ik). *Photo* means light and *voltaic* denotes that the cells produce voltage. Photocells are made of materials that give off electrons when light strikes them.

Figure 5-7 shows a calculator which is powered by photovoltaic cells. At the present time, these "solar cells" are limited to uses requiring fairly low levels of power output. They also have been used for power in space exploration.

There are also some other photo-sensitive devices used in electronics that do not produce currents. These devices are either

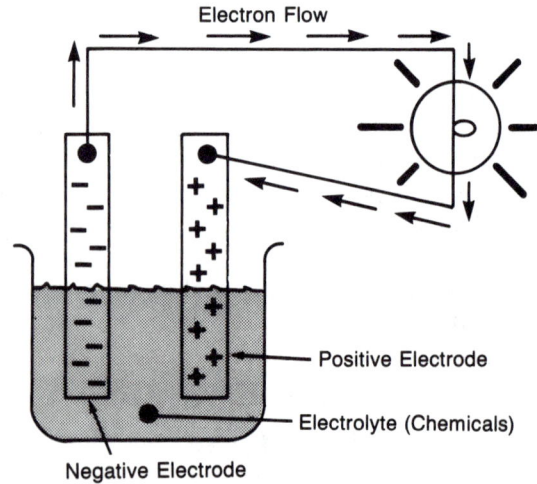

Fig. 5-6 The chemical reaction between the electrodes and electrolyte forces electrons to flow in the circuit.

resistances which change in value when struck by light or semiconductor devices which use light to control currents from other sources. Remember, however, only photovoltaic cells produce an EMF in reaction to light.

EMF from Heat

There are two very different methods of making electrons flow by means of heat. When some materials are heated, electrons can be made to boil off of their surfaces. These materials are very important parts of the vacuum tubes found in older model radios and televisions. In fact, television picture tubes still depend on this principle to supply the electrons which are fired at the screen to produce the picture.

The thermocouple is the other method of achieving an electron flow directly from heat. A **thermocouple** is a heat-sensing device used in very hot kilns and ovens in industry and homes. Thermocouples are made by joining two dissimilar metals. See Fig. 5-8. When the junction is heated, electrons are driven out of one of the metals and into the other one. That makes the first metal "hungry" for electrons (because it has given some away). At the same time, the other metal has extra electrons to get rid of, but they cannot go back through the junction! Therefore, if a circuit is connected to the wires leading from the junction, the extra electrons will flow around the circuit to get back to their home metal.

EMF from Pressure

Still another source of EMF is mechanical pressure or vibration of a crystal. This source of EMF is known as the **piezoelectric effect** (pea-a-zoe-e-lec-trick). When quartz and some other crystals are squeezed or vibrated, they produce a current. This effect

Fig. 5-7 Photovoltaic cells are used in sensing and power producing settings. This calculator will operate on the light available in an average room.

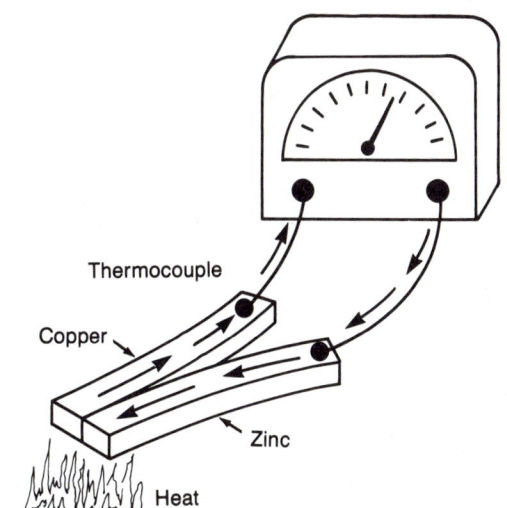

Fig. 5-8 When the junction of the two dissimilar metals is heated, electrons will flow through the meter and make its pointer move. This circuit, called a pyrometer, is used to measure the temperature of molten metals. The dissimilar metals used depend on the temperature range desired.

Chapter 5 *Sources of Electricity*

is used in numerous ways. The simplest example is in the crystal or ceramic cartridge of an inexpensive record player. See Fig. 5-9. When the needle rides in the groove of the moving record, it vibrates in reaction to small hills and valleys in the groove which carry the sound information. The vibrating needle shakes the crystal to which it is attached. The crystal, in turn, produces very small electrical impulses which are amplified (just as in the tape recorder discussed earlier) to produce sound in the speaker.

Another extremely important use of the piezoelectric effect is used in a quartz watch. As a crystal vibrates, it produces tiny electrical impulses which are in perfect time with its frequency of vibration. In other words, piezoelectric impulses can be used as perfect time bases. They are used in numerous solid state circuits inside all sorts of devices (from televisions to computers) in which accurate frequency or timing is essential. For instance, they are used to keep electronic (quartz) watches on time and to make sure that CB radio transmitters stay on the correct frequency.

EMF from Friction

This last major source of electrical pressure is friction. **Static electricity** is produced in this manner and so is lightning. The best demonstration of this source of electricity is when you walk across a carpeted floor and receive a painful shock as you touch a metal doorknob or another person (any conductor which is grounded).

The pain you feel is not at all the same as the sensation you would receive if you were shocked by an electrical appliance connected to the 120 volt line. What you feel in this case is not just a shock, but a small burn along with the shock. If the event takes place in a dark room, you can even see the bright blue spark as the current jumps from your finger to the grounded object.

Think about Chapter 4 for a moment. Since air is a good insulator, it has extremely high resistance. What would it take to make a current flow against this high resistance? That's right, very high voltage (pressure).

Static electricity is caused by friction. It is harmful in many industrial settings. Any time air or fluids travel or materials rub together, there is danger of static charges. Serious explosions have occurred in grain elevators due to sparks created as static charges overcame the resistance of the air.

Even slight static electrical charges can seriously damage semiconductor devices which are used in modern electronic circuits. Get in the habit of discharging yourself (touching a grounded metal surface) before you work on circuits to avoid this problem.

Static electricity can also be useful. When one object is charged positively and another is charged negatively, they attract each other just as magnets do. Perhaps you have heard of anodized painting. Anode really means positively charged pole. If a piece of metal is made to have a strong positive charge (anodized), then it will attract negative charges. Paint is sprayed from a negatively charged gun. As the paint is sprayed, it forms a mist of tiny negatively charged droplets which are attracted to the positively charged metal. The paint coats the metal evenly with little waste.

This same technique has been used to spray crops in experimental settings. The spray is not wasted on the ground because the water-filled stem of the plant is a better conductor to the water beneath the Earth's surface than is the crusty topsoil. In addition, the spray is attracted to the underside of the leaves, where bugs hide from traditionally applied sprays. Static charges are also used to clean pollutants from the air. Static electricity offers many interesting potentials and challenging problems for the future. See Fig. 5-10.

Other Possible Sources of EMF

There are other minor sources of electricity which are special examples of the sources already discussed. The electric eel and the electrical impulses found in the bodies of animals are sometimes labeled **bio-electricity.** Actually, both of these are essentially special cases of chemical sources of EMFs. The voltages and currents in humans and most other animals are very small values. Electric eels, however, have been known to produce violent shocks of up to 900 volts which could kill animals as large as a horse. The shocks come from special cells in the eel's tail.

When considered in the broadest possible sense, anything that can cause a static charge to build up, or can force even a few electrons to flow in a circuit could be a source of EMF. We have considered only those sources which have proven to be of real importance in light of present technology. Other sources can, for the most part, be identified as special cases of the six categories discussed in this chapter.

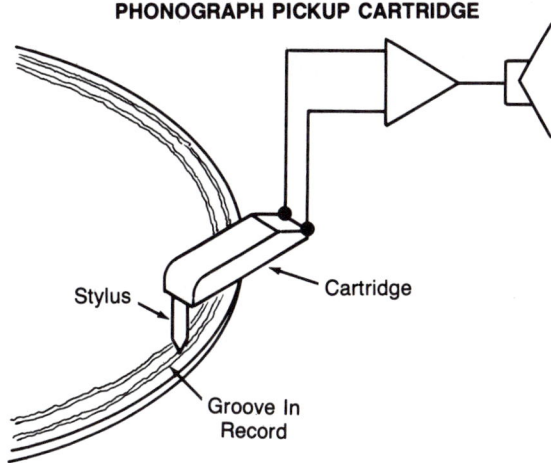

Fig. 5-9 As the stylus (needle) vibrates the crystal in the cartridge, electrical impulses are produced which match the sound information stored in the groove of the record.

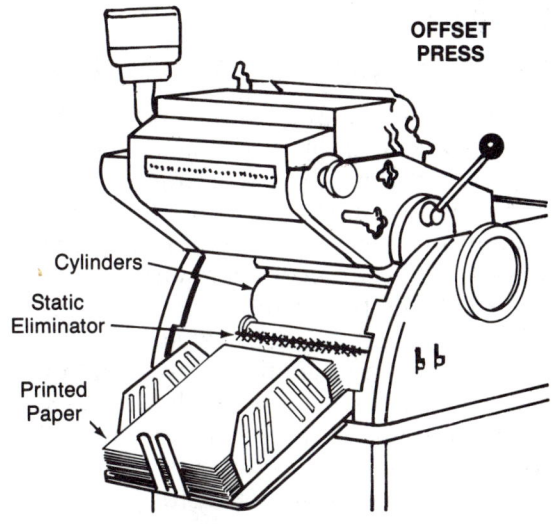

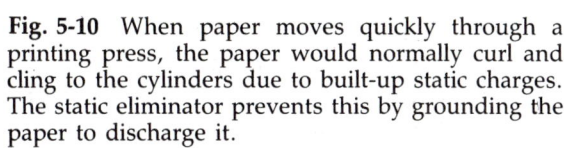

Fig. 5-10 When paper moves quickly through a printing press, the paper would normally curl and cling to the cylinders due to built-up static charges. The static eliminator prevents this by grounding the paper to discharge it.

Summary

There are six important sources of electrical pressure (EMF). The sources are chemical reaction, magnetism, light, heat, pressure, and friction. Each of these sources has proven to be useful in some way. Batteries (chemical) and generators (magnetic) are the two most important sources in terms of how much we use them. Heat, light, and pressure have been used far more often in sensing devices than in power producing applications. Static electricity (friction) has become more important in recent decades as static charges have been put to use with high-speed machinery. Bio-electricity is, generally, a special case of EMF produced by chemical reaction.

Important Terms

electromotive force
cell
battery
electrodes
electrolyte
photovoltaic cell
thermocouple
piezoelectric effect
static electricity
bio-electricity

Review Questions

1. List the six sources of EMF and give an example of each.
2. Explain how the head in a tape recorder and the cartridge in a phonograph are similar in function. How are they different?
3. When you are eating candy and a silver tooth filling contacts a piece of forgotten aluminum foil wrapper, you experience a painful sensation. What would account for this experience?
4. If there were no metal fibers in the hose (or separate grounding cables) on a gasoline pump, there would be serious danger of explosions when the nozzle is inserted into the gas tank fill tube of a car. Why is this so? (Hint: remember the car has been moving.)

Careers: TV Station Engineer

All television and radio stations must have station engineers who ensure the equipment is in good working order, and that it is operated in accord with Federal Communications Commission (FCC) regulations. The engineers regularly check to see that the transmitters are staying on their assigned frequencies so as not to interfere with the broadcasts of other stations. Station engineers also make sure other equipment in the station is in top condition so that the signals transmitted will be clean and free of distortion.

Station engineers must pass FCC examinations which test their knowledge of electronics, radio and television transmission principles and circuits, and FCC regulations. Many station engineers have college degrees, but this is not the most essential qualification. The high school subjects which are most important for preparing to be a station engineer are electronics, math, science, English, and government. Special courses in television and radio are offered in some schools. These are typically journalism courses which could be helpful.

Principles of Electricity

Section 2

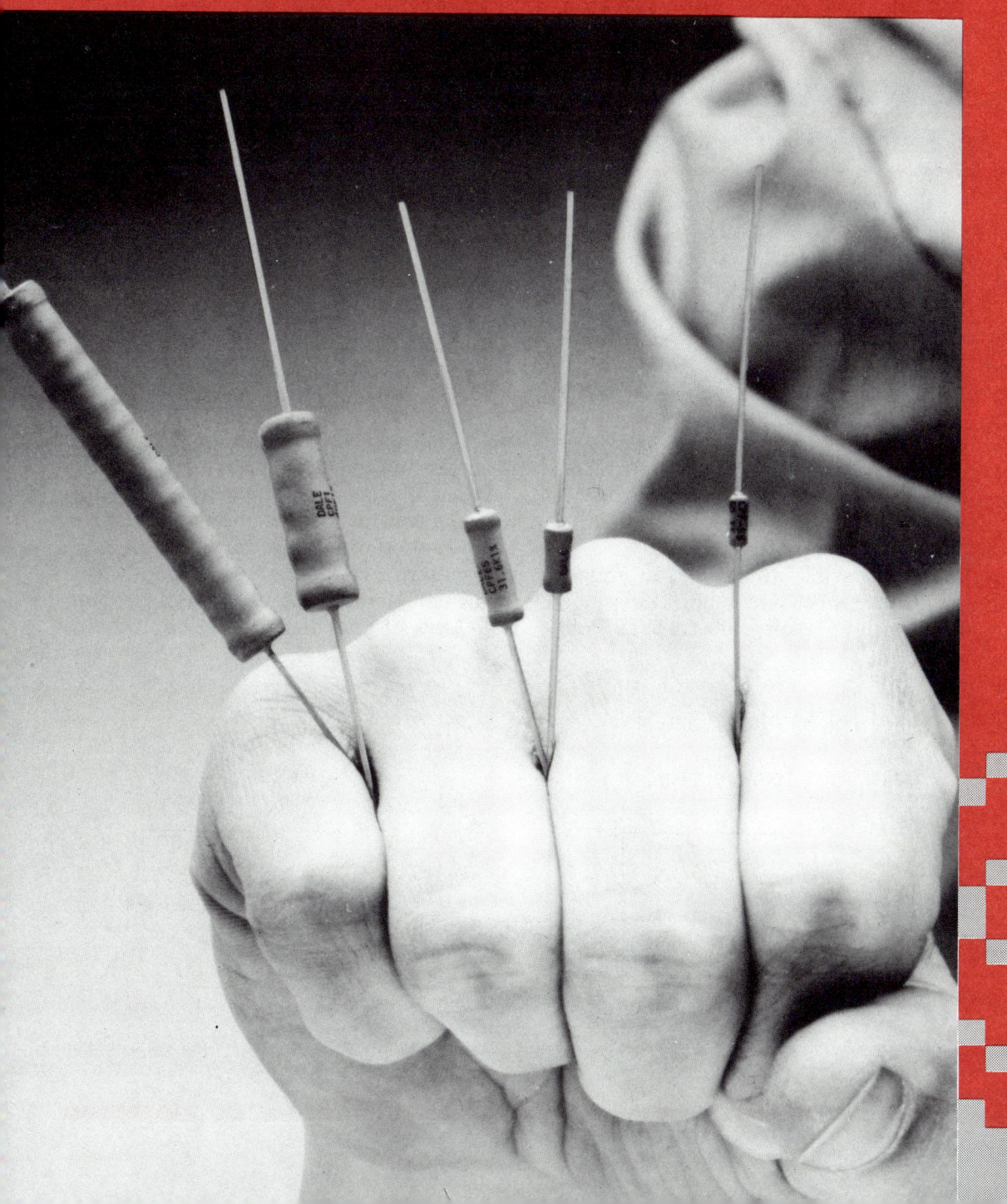

Chapter 6
Voltage

Electrical Pressure

In a useful circuit, the source creates an electrical pressure which pushes electrons through the circuit. This electrical pressure is known as **voltage**. The voltage must be great enough to push the electrons all the way around the circuit. Figure 6-1 shows that high voltage really means large amounts of pressure.

Some circuits and devices require more voltage to work properly than other devices. For instance, a large motor from a washing machine will not even pretend to turn when you connect it to a flashlight battery.

It is also possible to have too much voltage! For instance, what should happen if you connect a 6 volt bulb to a 12 volt battery? The pressure from the battery will be so great that it will force too many electrons through the bulb and burn it out quickly. In fact, if the voltage is far too great, very dangerous situations can occur. Always check to see how much voltage a circuit is supposed to operate on before you connect it.

Other Names For Voltage

Several names have been used for electrical pressure. The name used most often is voltage, chosen in honor of Alessandro Volta, inventor of the first electric cell. *Voltage*, though a common word, is not a very descriptive one. Another name which means the same thing as voltage is **potential**. One of the dictionary definitions of

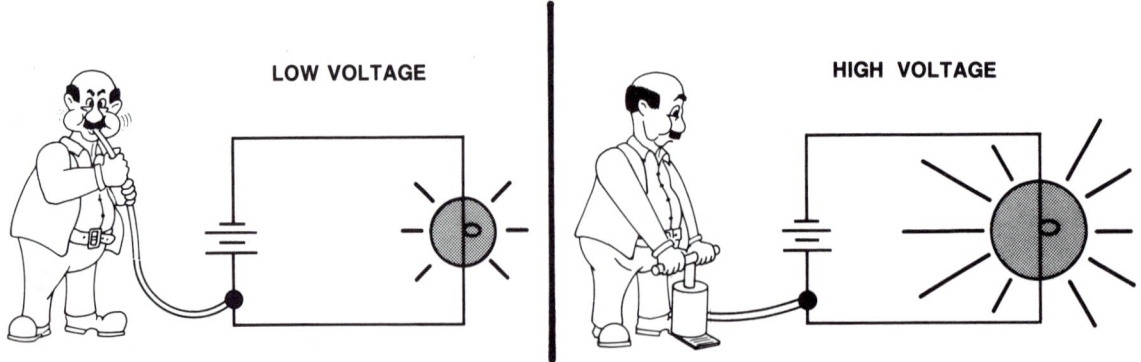

Fig. 6-1 Voltage is another way of saying "electrical pressure." High voltages can move more electrons than low voltages can.

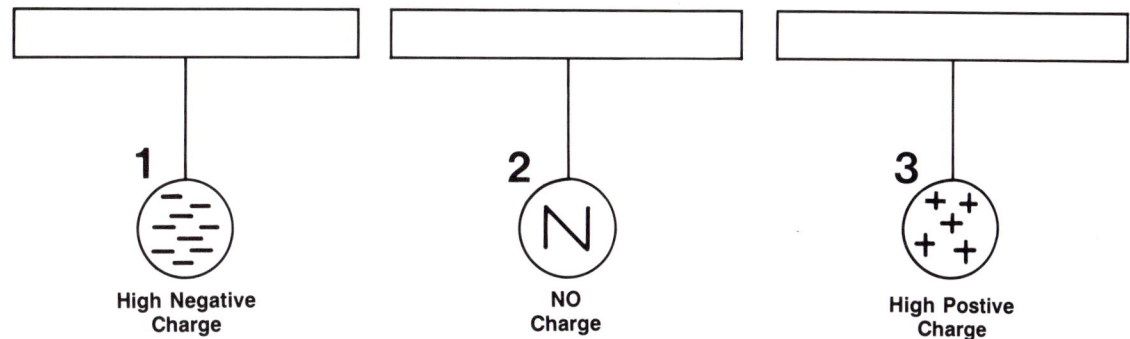

Fig. 6-2 There is a "difference" between any two of these spheres. Obviously, 1 is negative when compared with 3. But, if only 2 and 3 were here, there would still be a difference, and 2 would be negative as compared to 3. Likewise, 2 would be *positive* if compared only to 1. As long as there is a difference, voltage (EMF) exists.

the word "potential" is "having force or power." The term potential helps you see that this force can be stored up, for later use, as in a battery. A 12 volt automobile battery stores 12 volts of "potential" whether you are using it or not. Even when the automobile battery is sitting on the shelf in the auto parts store, it has 12 volts of potential! Electrons do not have to be in motion for there to be voltage (potential). Another term which basically means voltage is **difference**. If there were a large number of extra electrons stored in a terminal, you would say that it was negatively charged. Likewise, another terminal could be positively charged by removing free electrons from it. Figure 6-2 illustrates the relationship between two charged spheres and a third one which is uncharged (neutral). There is a "difference" in the charges (amounts of electrons) of the three spheres. The negatively charged sphere has more electrons than either of the other two. In fact, the neutral sphere has more electrons than the positively charged one does! You could say that there is a "difference" between any two of the spheres. If you touch any two of the spheres together, the extra electrons will flow from one to the other one so the charges are equalized. There would then be no "difference." As long as there is a difference in the charges, potential (or voltage) exists.

Still another term which can be used to mean voltage is **drop**. The meaning of this word will become clearer as you study circuits and resistance networks. Essentially, if some of the voltage produced by a source is used by a load, you would say that a "drop" or "voltage drop" has occurred. Figure 6-3 shows a voltage drop.

Remember the schematics you learned about in Chapter 4? Just as there are symbols to represent components and devices, there is also a letter symbol for voltage.

The most common letter used to designate voltage is E. You may ask, "Why E instead of V?" Not all books and references agree on this. Some books do use V for voltage, but E is much more common and it comes from another name for voltage — electromotive force (EMF). In further explanations the terms *V*, *voltage*, *pressure*, *E*, and *electromotive* force will be used interchangeably to help you get used to them all.

Measuring Voltage

We have spoken vaguely about sources and drops of "more" or "less" voltage, but we have not attempted to actually measure electrical pressure. In electricity and electronics, it is usually important to know exactly how much voltage exists in a circuit.

Just as temperature is measured in units, called degrees, electrical pressure can also be measured. Voltage is measured with a voltmeter and the units are called **volts**.

The higher the voltage, the more pressure there is. The lights, horn, and most accessories in an automobile usually require 12 volts to operate properly. Some flashlights may operate on as little as 1½ to 3 volts. The lights and appliances in your home need about 120 volts of pressure for their proper operation. Inside a color television set, approximately 26,000 volts are used to force the electrons to jump all the way from the back of the picture tube to the screen to make the picture.

Larger and Smaller Units

When you talk about very high or very low amounts of voltage, it is sometimes convenient to use other units related to the volt. The volt is the basic unit. Prefixes are added to volt to express larger or smaller quantities. For instance, it is common practice to use the term **kilovolts** when considering voltages larger than 1000 volts. The prefix *kilo* means 1000. When the potential gets even higher, it is expressed in megavolts. One **megavolt** equals 1,000,000 volts and it would be written as 1 MV. Notice that the letter M is used for *mega* and K for *kilo*. The abbreviations used in electronics come from the metric measurement system. The television picture tube mentioned earlier has a pressure of 26 KV.

Some circuits are so delicate they would be destroyed with potential differences as

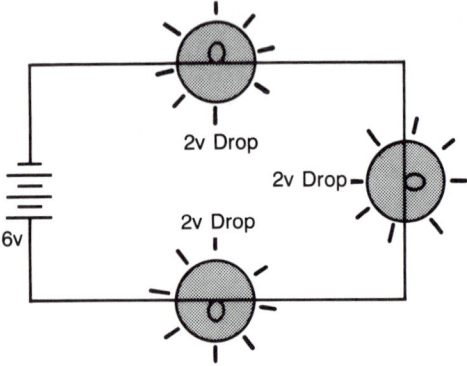

Fig. 6-3 Each lamp in this circuit "drops" (uses) 2 volts. Notice that, in this type of circuit, the drops add up to the voltage of the source.

low as 1 volt. When dealing with these circuits, it is easier to use smaller units than it is to write the measurements as fractions of a volt. The expression 1 mV or **millivolt** means one thousandth of 1 volt (1/1000 V). The small letter m stands for *milli* which means "thousandth of."

Be very careful to notice whether a value is written as 16 MV or 16 mV. Remember, the capital letter M means million while the small letter m stands for thousandth of. There would be a disaster if you mistakenly connected a 16 MV source to a circuit which only needed 16 mV! Sixteen millivolts would be a very small value (0.016 V), but 16 MV means 16,000,000 V.

In some circuits, still smaller units of measurement are used. The **microvolt** is one millionth of a volt (1/1,000,000 V). Abbreviate *micro* with the Greek letter μ which is known as mu. A table of the metric abbreviations which are used with voltage, their meanings, and abbreviations appears in Fig. 6-4.

Common Values of Voltage

Most circuits and devices in the home which operate on the power from the receptacles in the house use about 120 V.

Metrics

Terms:	Meaning:	Symbols:
Microvolt	1µV = 1/1,000,000 V	Greek letter Mu "µ"
Millivolt	1mV = 1/1,000 V	Small letter "m"
Kilovolt	1kV = 1,000 V	Small letter "k"
Megavolt	1MV = 1,000,000 V	Capitol letter "M"

Fig. 6-4 Abbreviations are used to label large and small values of EMF.

Sometimes these devices will have labels specifying from 110 to 125 V, but the nominal value is actually 120 V. When you are dealing with this high a potential, a variation of only a few volts is not much of a problem for most uses. Certain circuits in the home need much more electrical pressure to operate properly. This would include electric cooking ranges, some large air conditioners, and many electric clothes dryers. These devices are connected to special outlets which supply about 240 V. Voltages this high are very dangerous. Never work on these 120 and 240 V circuits without removing the power source first.

Small devices which are battery operated (such as calculators, flashlights, portable radios, and cameras) only need from 1 to 18 V. Many logic circuits, like the ones in computers, use 5 V differences. Batteries can be made which produce hundreds of volts, but it is not very practical to use them today. Generally, large devices which do tough jobs require higher voltages than small, light duty ones do — but this is not always so (Fig. 6-5).

Fig. 6-5 Though large devices usually require higher voltages, a real automobile uses only 12 V while the simple table lamp needs a 120 V source. This toy automobile uses about 9 V.

Summary

The electrical pressure which forces electrons to move through circuits is voltage. Voltage is sometimes called potential, difference, potential difference, drop, potential drop, electromotive force, or EMF. Each of these terms for electrical pressure refers to a way to understand voltage or its use. Electrical pressure is measured in units called volts. Units smaller than the volt are millivolts and microvolts. Large values of voltage are expressed as kilovolts and megavolts.

Important Terms

voltage
potential
difference
drop
volt
kilovolt
megavolt
millivolt
microvolt

Review Questions

1. Explain how the terms potential and difference point toward special meanings of voltage.
2. List the common metric prefixes for large and small values of voltage and tell the meaning of each.
3. List the proper operating voltages of 10 common devices around your home. How many of these can operate from two different voltage sources?

History: Count Alessandro Volta

In 1801, Napoleon made Alessandro Guiseppe Antonio Anastasio Volta a count for his invention of the first electric battery. As you probably can tell by the length of his name, Alessandro was born into a noble family in Italy.

Volta experimented a great deal with electrochemistry and held distinguished positions in the universities at Pavia and Padua. Other scientists of his time thought animal tissue was needed to make electricity. Volta led a group that was convinced electrical pressure could be produced by simple chemical reaction — without using animal tissue. He was right! Volta's discovery and successful demonstration of it are so important that the unit of measure of electrical pressure was named after this great figure. Remember him the next time you measure a volt.

Project: Darkness Detector

Parts List:
- 1 B_1 6-volt battery
- 1 B_2 silicon solar cell
- 1 S_1 SPST switch
- 1 R_1 resistor (about 10 ohms, determine if needed by experimentation)
- 1 Z_1 LM-386 integrated circuit (audio amp)
- 1 C_1 10 µf electrolytic capacitor, 10 V
- 1 C_2 220 pf. to .005 µf ceramic disk capacitor (determine by experimentation)
- 1 L_1 small loudspeaker, 8 ohms
- 1 socket for the integrated circuit
- 1 suitable enclosure

Circuit Description:

This circuit will produce an alarm when the solar cell is shadowed from light. If the solar cell were mounted across from a light source, the alarm would sound any time the beam of light was broken. For instance, you could mount the light source on one side of a doorway, and the solar cell on the other side, to sound the alarm whenever someone entered the door. Use your imagination to find other possible uses for the circuit. The values of the resistor and C_2 may need to be adjusted to fit your needs. The resistor adjusts the sensitivity of the circuit. The capacitor's value determines the tone of the alarm sound.

You could build a "magic sound box" by using a mercury switch for S_1 and mounting all parts except the solar cell inside a plastic or metal box. Then, if the box was turned to the proper position for the switch to be "on," the sound could be produced by shadowing the solar cell with your hand; but your friends would not realize that the position was also important, so when you give the box to them, they will not be able to produce the sound. The switch should be mounted at an unusual angle.

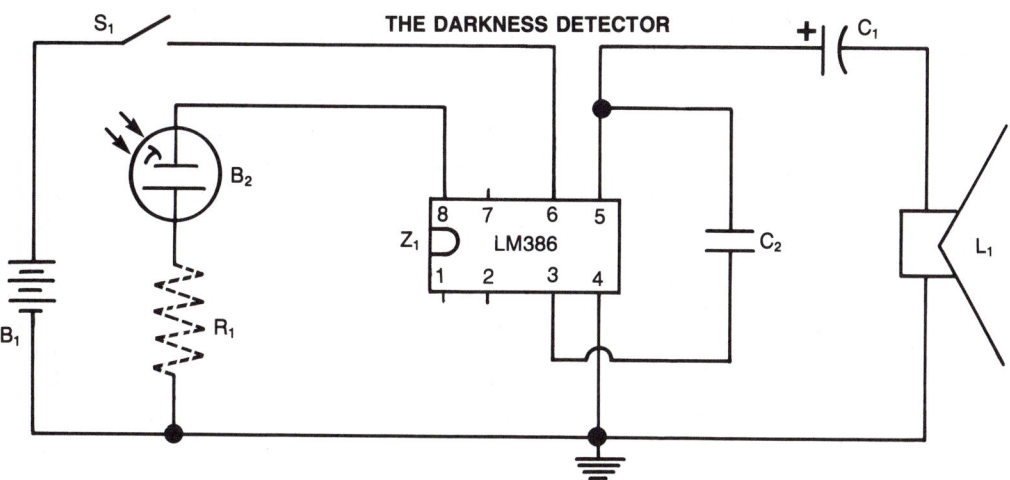

Fig. 6-6 This simple circuit uses one of the six sources of EMF you have studied by using a solar cell as a sensing device. You can probably think of several uses for such a circuit — just imagine, you could finally prove whether the light in the refrigerator actually does go off or not!

Chapter 7

Current

Rate of Electron Flow

Voltage is a very vital part of electricity. The second value that is important is the rate of flow. The flow rate is known as **current**. Electrons must be in motion for current to exist.

Current is related to voltage, but it is a different quality. The concept of current is basically a way of saying "number of electrons flowing per second." This concept is illustrated in Fig. 7-1. Even though the illustration shows the character counting electrons as he times them, this is stretching the truth a great deal. First, you cannot see the electrons. Second, they flow far too quickly to count them as they go by. Third, there are far too many of them. In fact, over 6,280,000,000,000,000,000 electrons can flow past the light every second!

So does that mean current cannot be measured? If voltage (electrical pressure) can be measured, shouldn't you be able to do the same with current? You can, and you will learn how. First, though, you need to understand how a charge is measured.

Static Charges

Sometimes, electrons are stored up and not moving. When electrons are stored, there is no "current," but there is a charge. The amount of charge can be discussed in terms of measurement units called **coulombs**. A coulomb is about 6.28 quintillion electrons (6,280,000,000,000,000,000). Since this is a very difficult number to write, scientific notation (a type of shorthand) is used to write it as 6.28×10^{18}.

When you read this notation aloud, you might say, "six point two-eight times ten to the eighteenth power." It means the original number is abbreviated by using the most informative parts, called significant digits. The decimal, which was originally behind the last zero, is moved to a new position behind the first significant digit. Since the decimal was moved 18 places to the left, the significant digits are multiplied by 10 to the positive 18th power.

You will not use the unit coulombs very often in this class, but you will need to learn how to write and read numbers in scientific notation. How would you write the number 7,000,000,000 in **scientific notation?** Think

Fig. 7-1 Current has to do with the number of electrons which flow through the circuit each second.

Scientific Notation

Number	Scientific Notation
20,000	2×10^4
25,000	2.5×10^4
250,000	2.5×10^5
250,000,000	2.5×10^8
2,500	2.5×10^3
205,000	2.05×10^5
.025	2.5×10^{-2}
.00025	2.5×10^{-4}
6,280,000,000,000,000,000	6.28×10^{18}

Fig. 7-2 Scientific notation is a shorthand way to write large or awkward numbers. The last entry is the number of electrons in a coulomb.

It would be easy to become confused at this point and say that current means the speed of electron flow. This would be a very bad mistake. It is generally accepted that, under normal conditions, electrical energy is transferred at 186,000 miles per second (mps). Figure 7-4 shows what this means. It does not mean the electron labeled E will travel at a speed of 186,000 mps from the negative pole to the positive pole of the battery. What it does mean is that when E bumps its neighbor and starts a chain reaction of neighbor bumping, the energy transfer will be almost instantaneous.

about it and then try it on a piece of scratch paper. First, write the digit 7. Then, count how many places there are after that first digit. Continue by writing "$\times 10^9$" just after the 7. This last part means to multiply the 7 times 1,000,000,000 (one with nine zeros behind it). Figure 7-2 gives other examples of large numbers in scientific notation.

Measurement of Current

Remember the main difference between a static charge and current is that the electrons must be moving for there to be current. If 1 coulomb of electrons were to pass by a certain point in 1 second, then you would say that 1 ampere of current was flowing in the circuit. The **ampere,** named after French physicist and mathematician André Marie Ampère, is the basic unit for measuring current. Abbreviate ampere by saying "amp" or simply using the capital letter A. To say that 1 amp of current is flowing means there are 6.28×10^{18} electrons passing a given point in the circuit every second. The difference between an amp and a coulomb is pointed out in Fig. 7-3.

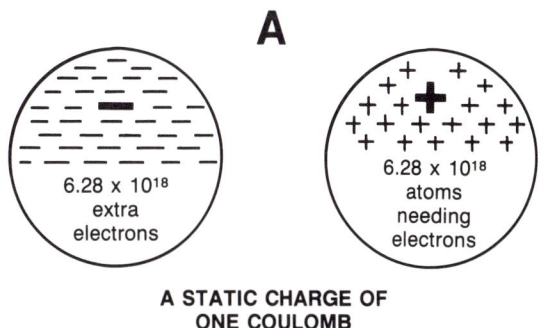

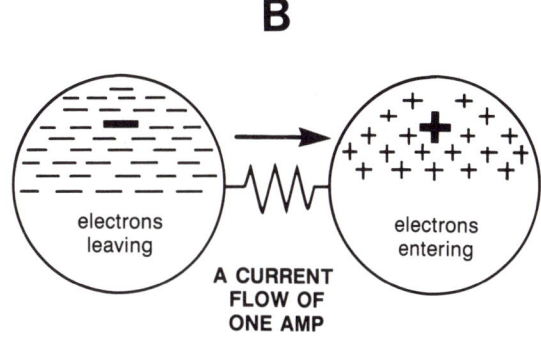

Fig. 7-3 A) 6.28×10^{18} stored electrons is a static charge of one coulomb. B) If those electrons flow through the resistor in one second, that would be one amp of current. If it took 2 seconds, the current would be .5 A

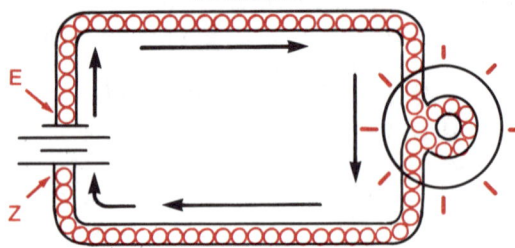

Fig. 7-4 If electron E moves, it will bump the next one which will bump its neighbor, and so on, until electron Z will be bumped back into the source. E will take a long time to travel (itself) all the way around the circuit, but Z will move almost the instant that E first does. If the wires in this circuit were 186,000 miles long, Z would move about one second after E did. The speed of energy transfer is 1.86×10^5 miles per second.

In other words, if you had a wire that was 186,000 miles long and you forced extra electrons into one end of it, it would only take one second for other electrons to be forced out the other end of the wire. Since the speed of electrical energy transfer is a constant value, current does not refer to that energy transfer. Rather, current refers to **rate**. In other words, as shown in Fig. 7-5, the number of electrons passing by a point in a circuit in one second determines current.

The symbol which is used to designate current is I. Intensity is the term from which the letter I was taken. A current of 400 electrons per second would be twice as intense as one of 200 electrons per second. That is, the flow rate would be twice as great with 400 electrons per second. The terms current, flow rate, and I will be used for this quality of intensity.

Smaller Units

One amp is a fairly large amount of current, so smaller units are often used for measuring current. As with voltage, use the standard metric prefixes. What should be the value of 1 mA (1 **milliamp**)? You are right if you said a milliamp is 1/1000 A. When even less current is required in a circuit, it may be expressed in μA values (**microamp**). A microamp is 1/1,000,000 A. Values large enough to require the prefixes K and M are seldom encountered, though not impossible. Therefore, amps, milliamps, and microamps are the units which are commonly used to measure and quantify current flow.

Typical Values of Current

Just as with voltage, some circuits require far more current for proper operation than others do. A common table lamp with a standard 100 watt bulb connected to the 120 V line supply will draw a little less than 1 amp of current (about 833 mA). A full-size solid state color television set requires only a little more — 1.2 A. In contrast, a transistorized, battery operated radio may

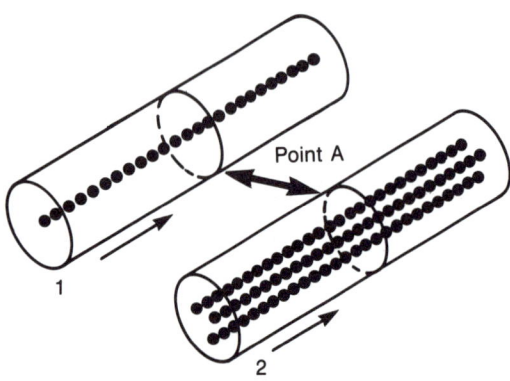

Fig. 7-5 Wire number 2 is carrying three times as much current as wire 1. Current (rate) depends on the number of electrons flowing, not the speed of energy transfer which is a constant value of about 186,000 m/s. The electrons do not really travel in straight lines as this figure suggests.

need as little as 15 mA (.015 A) while a large frostfree freezer or refrigerator could require up to 15 A. When a large V-8 automobile engine is being started, the battery may have to deliver as much as 400 A for a short time.

Voltage Without Current

When a source of EMF, a battery for instance, is not connected to a circuit, it has the potential to deliver its rated voltage, but no current is flowing. This means voltage can exist without electrons being in motion. But, if no electrons move, there is no current. Remember that current means electrons in motion. You can have voltage with no current.

You may then ask if you could also have current with no voltage. The answer is no. Any form of pressure which *could* make electrons flow is basically a source of EMF. When electrons flow, something has to be making them do so and that something (whatever it is) would be producing voltage.

Different Types of Current

Battery operated and portable equipment usually use **direct current** (DC). Direct current means the electrons always travel in one direction. They move from the negative terminal of the source to the positive terminal. Many devices in the home, however, can work with the electrons moving either way (forward and backward). When you cause electric currents to go back and forth, they are called AC or **alternating currents**. The alternating current which is available from the receptacles in your home changes directions (alternates) 120 times every second. Figure 7-6 illustrates an alternating current.

Special sources of EMF (like AC generators and alternators) are needed to produce these alternating currents. You will learn much more about AC and how it differs from DC in Chapter 28. For now, just get used to seeing these two types of currents in example circuits.

Summary

Current means the rate of electron flow. Current is measured in amps, milliamps, and microamps. One amp is 1 coulomb of electrons (6.28×10^{18}) passing by a given point in a circuit every second. Electrons do not travel at the speed of light, but energy transfer due to their motion does travel at about 186,000 miles per second. In AC circuits, the electrons flow first in one direction and then back in the other direction over and over again.

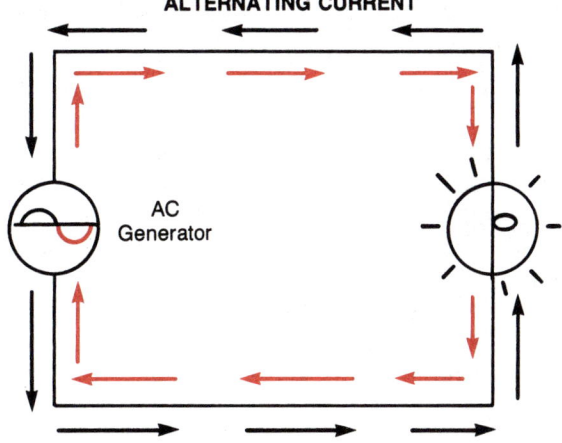

Fig. 7-6 The black arrows indicate the direction of electron flow in the first half of the generator's cycle and the other ones show the second half cycle. It does not matter to the lamp which direction the current flows — as long as enough electrons move through it either way, it will light.

Important Terms

current
coulombs
scientific notation
ampere
rate
milliamp
microamp
direct current
alternating current

Review Questions

1. Explain the difference between voltage and current. How can there be voltage, but no current?
2. What is wrong with this statement "Electricity flows at 186,000 miles per second"? Write a similar statement which is more nearly correct.
3. How much current would be flowing if 4 coulombs passed through a light bulb in 2 seconds?
4. What is AC and where is it found?

Careers: Motor Repair Person

People who repair electric motors and coils are found in several industries. Their jobs include disassembling, inspecting, repairing, and reassembling motors. They must be able to solder, make connections, use test equipment, use machine lathes, read schematics, and diagnose motor problems.

Motor repairers need a good knowledge of the fundamentals of electricity. They should have a high school education which includes mathematics, general sciences, and shop courses such as metals machining and electricity. Education from a community college or training in the military is valuable. Some on-the-job training is usually required.

Many motor repairers work mostly in a shop while others do a good bit of "on site" work. Pay varies with location, experience, and level of responsibility.

Chapter 8
Resistance

Restriction to Electron Flow: Resistance

One of the parts of a useful complete circuit is the load. In Chapter 3 you learned that all loads are, in some way, resistances. This is true because the load makes it harder for the electrons to flow through the circuit. Figure 8-1 shows how a load restricts water flow. As the load does its work — makes light, heat, motion, sound, etc. — electron flow is restricted in a way very similar to this. Large loads (high values of resistance) allow fewer electrons to flow than do small loads.

Resistance is found in other parts of a circuit, too. As electrical components and devices do their jobs, they oppose the flow of current. Everything, all materials, have some resistance. Some materials have far more resistance than others. Even copper and silver, two of the very best conductors, resist the flow of electrons through them.

In the case of conductors, however, the amount of resistance is so small you can practically ignore it. You could say the resistance of a good conductor is **negligible**. That is, it is so small it is not important to the operation of most circuits. Remember, though, resistance exists in all parts of every circuit, even in the wires. The

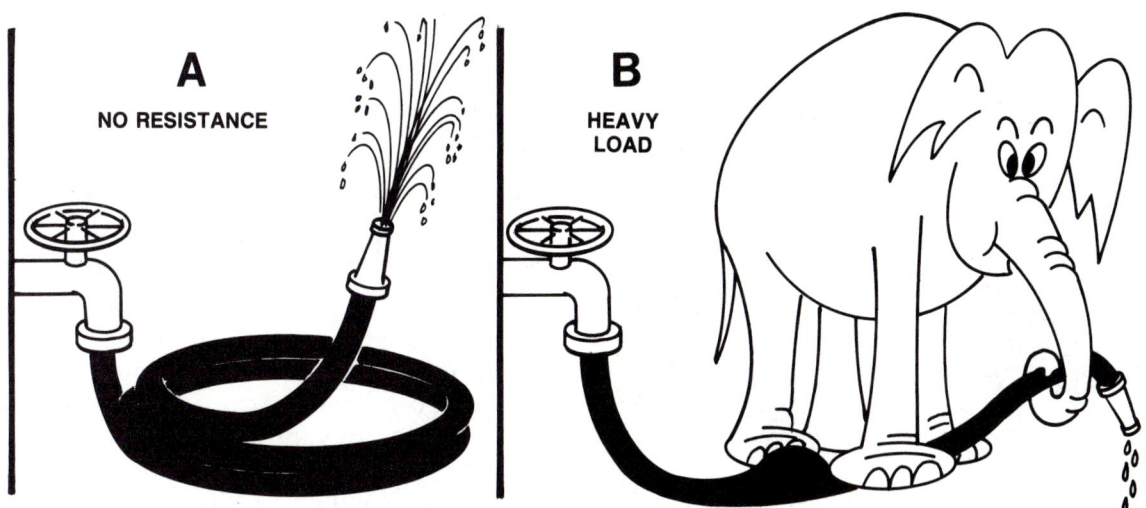

Fig. 8-1 A) Water flows freely from a water hose with no restriction. B) Here there is some resistance (restriction) and, even though the pressure pushing the water is the same, only a little water flows.

amount of resistance a conductor or device has depends on four things:

1. Material. Some materials have more resistance than others.
2. Length. The longer an item is, the more resistance it will have.
3. Diameter. The thicker the item is, the less resistance it will have.
4. Temperature. Resistance of most materials goes up when they are heated, but some special materials do the opposite.

Measuring Resistance

Just as there are units of measurement for voltage and current, there are also special units which are used to measure resistance. The basic unit of resistance is the **ohm**. Abbreviate ohm with the Greek letter omega (Ω). To write 47 ohms in shortened form, write 47 Ω. As with voltage, use K and M before the Ω to designate large values of resistance. A value of 68 MΩ would be read "68 megohms" or simply "68 meg." Values much smaller than 1 ohm will seldom be encountered in this course, so you will not use the prefixes *milli* and *micro* with resistance. There is also a symbol for resistance — R. As with voltage and current, resistance will be referred to in various ways. R and resistance will be used as well as descriptive phrases such as "opposition to current flow" and "restriction of flow."

Resistors

Some components are made just to cause resistance in electric circuits. Their job is to limit electron flow. You can use them to drop the voltage. Figure 8-2 shows some common resistors. Carbon is used to make most resistors because it is a semiconductor. It has too much resistance to be a very good conductor, but it has some free electrons which can be made to flow if the voltage is high enough. Other materials are usually mixed with the carbon to make resistors.

Types of Resistors

There are two types of resistors in electronic circuits: Fixed resistors and variable resistors. **Fixed resistors** are produced with one value of resistance, and they should not change in value during their useful lifetime. **Variable resistors**, on the other hand, are designed to allow you to change their value of resistance. The volume control in a radio is a variable resistor.

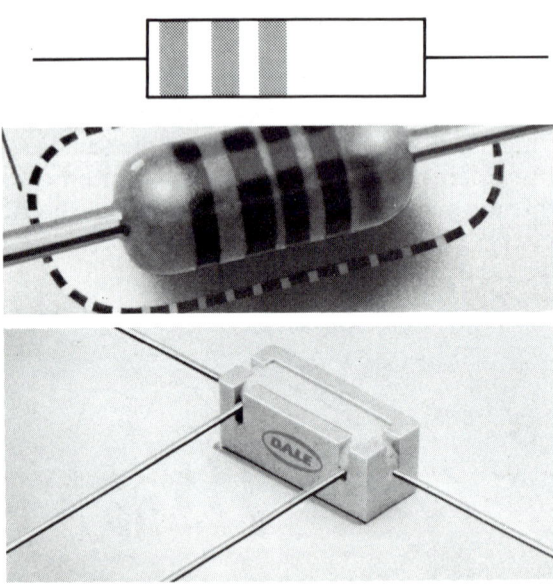

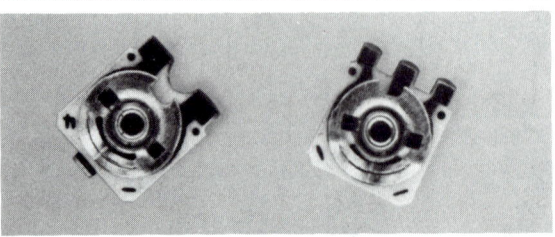

Fig. 8-2 Resistors are among the most common electric components. Four types are shown: carbon composition (top), metal film (second), wirewound (third), and variable.

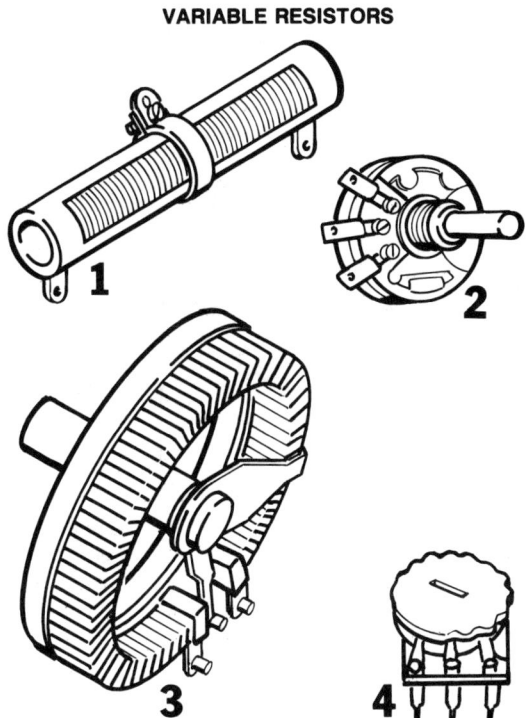

Fig. 8-3 Variable resistors can be changed in value. The types shown are: 1) adjustable, 2) carbon potentiometer, 3) wirewound rheostat, and 4) trimmer pot.

material onto a glass or ceramic core. The enlarged ends are due to the metal caps which attach the leads to the resistor. Film resistors are also coated with insulation and labeled just as composite ones are.

Still another type of resistor which is commonly found is the **wirewound resistor**. See again Fig. 8-2. These resistors are produced by winding special resistance wire around a ceramic core. The leads are added, and then the whole unit is given a thick ceramic coating for protection and insulation. Wirewound resistors are needed in high current circuits because they can withstand the heat generated when high rates of electrons flow.

Variable Resistors

Variable resistors are used as control devices. They may be wirewound or carbon types. Variable resistors are used to control (change) voltage and current levels in certain circuits. Nearly all of the controls on the front of a typical television, except the on-off switch and the channel selector, are variable resistors. Figure 8-3 shows variable

Fixed Resistors

There are three main types of carbon fixed resistors: composition, film, and wirewound. These three types differ in the way they are constructed. **Carbon composition resistors** were shown in Fig. 8-2. They are made by mixing powdered carbon and insulating materials together and then molding them into the desired shape. A plastic case often encloses the resistive element. Wires, called **axial leads**, are attached to the ends. The value is printed on the body in color coded bands.

Thin film resistors look similar to composition resistors, but they generally have a slightly enlarged area at each end. They are made by coating a thin film of resistive

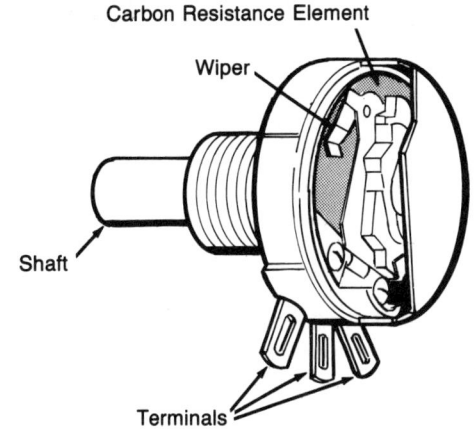

Fig. 8-4 When the shaft of the potentiometer is turned, the wiper moves around the resistance element to change the resistance of the device.

Chapter 8 *Resistance*

resistors. Some variable resistors are easily changed, while others (adjustable resistors and trimmers) need to be set by skilled technicians and not tampered with by the consumer. The internal parts of a variable resistor are shown in Fig. 8-4.

There are two ways to connect variable resistors into circuits. The difference between these modes of use is shown by the schematic symbols of each in Fig. 8-5. When all three leads are used (both ends of the resistor and the movable **wiper**), the variable resistor is called a **potentiometer**. This allows two paths for the electrons. When only two of the three leads are used (one end and the wiper), the device is called a **rheostat**. The rheostat has only one path for electrons.

Examine Fig. 8-6 and see if you can tell which resistors are rheostats and which ones are **pots** (potentiometers) before you look at the caption. A device may be used as either a pot or a rheostat if it has all three terminals. It is just connected differently. However, rheostats typically are used in higher current applications, so you must be

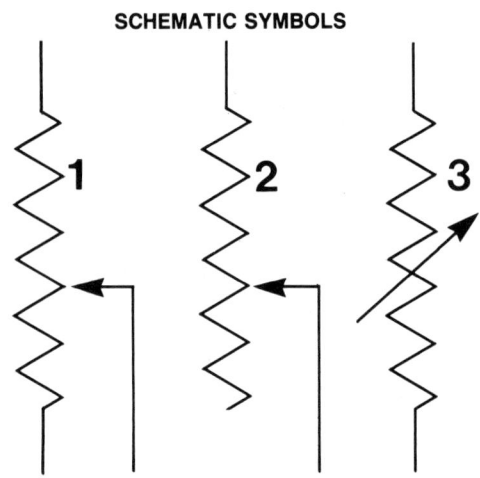

Fig. 8-5 The schematic symbols of variable resistors are shown: 1) Potentiometer — three leads, 2) Rheostat — leads only connected on one end and the wiper, 3) Alternate symbol used occasionally for rheostats.

certain the device can handle the current in the circuit. Rheostats are used to control current and potentiometers are used to control voltage.

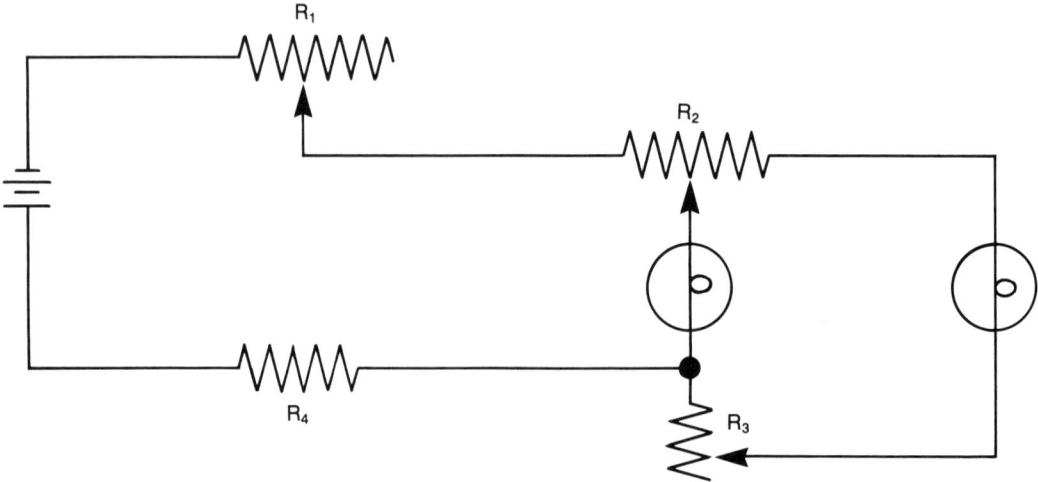

Fig. 8-6 This is not a practical circuit, it is used to illustrate how pots and rheostats are connected. The parts are: R_1 — Rheostat, R_2 — Potentiometer, R_3 — Rheostat (even though this one has three circuit wires connected to it, only one end of the resistive element and the wiper are used), R_4 — Fixed resistor.

Uses of Resistors

Resistors, both fixed and variable types, can be used in circuits to drop voltage and to limit current flow. If you wanted to operate a small lamp which required 3 V and you only had a 6 V battery, it would be possible to use a carefully selected resistor to drop the voltage to the desired level. See Fig. 8-7. A rheostat could be used to limit the current and provide speed control for some types of motors. The volume control in a stereo is usually a pot which varies the voltage of part of the circuit.

Problems with Resistance

Resistance can be harmful. Remember that all materials have some resistance. Even though copper wire conductors have very little resistance — usually negligible amounts — their resistance can become important in some situations. If the wire used in a circuit is too small in diameter to carry the current, it will have considerably more resistance than it should. This could result in improper circuit operation and, in extreme cases, even damage or fire.

The connections are still another matter to consider. Poorly made electrical connections create unwanted resistance (and some other unwanted effects as well). They can prevent devices and circuits from working. Any mechanic knows the first thing to check when a car will not even turn over (the starter motor will not crank the engine) is the battery connections. Corrosion can build up between the terminal and the cable clamp and cause so much extra resistance that the car does not get enough voltage or current to do anything.

In fact, poor electrical connections are the cause of many operational problems in student-built electronics projects. If one of your electricity projects does not work, one of the first things to look for will be bad solder connections.

Summary

Every material and device has some resistance. Sometimes we wish to add resistance to circuits to drop voltage or limit current. Unwanted resistance — resulting from wires that are too long or too thin, poor choice of materials, temperature changes, or poor connections — can cause circuit failure. Resistors can be fixed or variable. Carbon composition, thin film, and wirewound are common types of fixed resistors. Potentiometers and rheostats are variable resistances which are used primarily as control devices. We measure resistance, the opposition to electron current flow, in ohms.

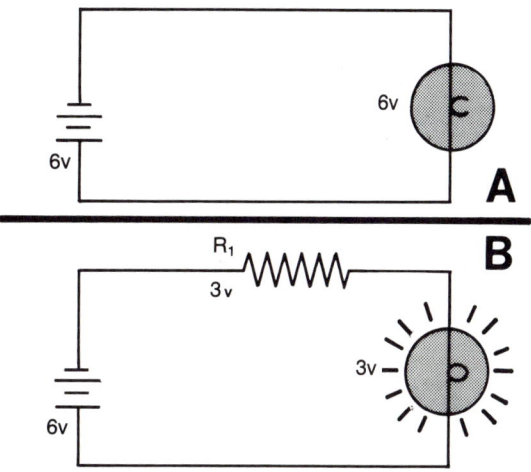

Fig. 8-7 The battery voltage is too high for the 3 V lamp, so the lamp burns out in diagram A. We could use a resistor, R_1, to drop some voltage as in circuit B, then the lamp would glow normally.

Important Terms

negligible
ohm
fixed resistor
variable resistor

carbon composition resistor
axial leads
thin film resistor
wirewound resistor

wiper
potentiometer
rheostat
pots

Review Questions

1. What is resistance? Give an example of a use for resistance.

2. List and describe three types of fixed resistors.

3. What is the difference between a pot and a rheostat? Draw their schematic symbols.

4. Briefly discuss unwanted resistance and some factors which can cause it.

Chapter 9
Resistor Color Code

Why Use a Color Code?

The resistor was developed when the field of electronics was in its infancy. Since resistors tend to be small, they are difficult to print on. In the early days, it was impractical, perhaps impossible, to print the actual values of resistors well enough so that they were legible and durable. Therefore, a **color code** was developed to label resistors and other components.

Advances in materials technology and the printing industry have made it possible to print the actual values on most devices today. But some components, especially resistors, are still marked with the traditional color code. One of the reasons for this is the high cost of retooling factories for printing processes. Also, the code simply and efficiently conveys such information in a small space. Finally, the code is an accepted standard worldwide. (Printed markings could differ from one language to another.)

The Colors

Ten colors are used to represent the digits 0 through 9. The colors are listed, with their numeric values, in Fig. 9-1. Some people think the code colors should be memorized; others believe you can pick it up as you use resistors in circuits. A simple saying may help you to memorize the order of the colors. Each word of this silly jingle begins with the same first letter as the color it represents from the code, except "they" which is not used. The poem is:

> Bright Boys Read Only Yesterday's Good Books, they Value Great Writing.

There are other versions as well. Whatever version you use, just remember that the first "B" is black, for 0, and then count to 9 (represented by white). After repeating this saying for a few days, you will know the colors without even thinking about them.

How to Read Coded Values

Resistors may be identified by three, four, or five color bands. Regardless of how many bands there are, however, the first

Values of Colors

Color	Value
Black	0
Brown	1
Red	2
Orange	3
Yellow	4
Green	5
Blue	6
Violet	7
Gray	8
White	9

Fig. 9-1 Each color of the code represents a numeric value.

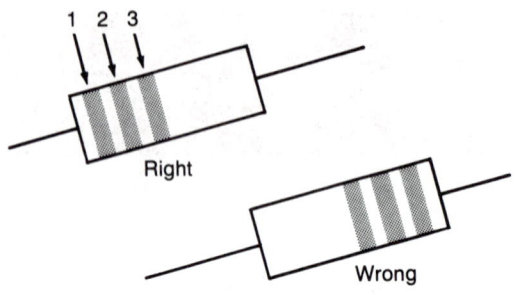

Fig. 9-2 The first step in reading the color code is to turn the resistor so that the end closest to a color band is toward the left. The band then at the left end is the *first* band.

three bands indicate the resistor's value in ohms. To read the coded value on a resistor, first turn the resistor so that the color band nearest the end of the resistor is on your left. See Fig. 9-2. The far left band represents the first digit of the resistor's value. The next band tells the second digit. The third band tells you how many zeros to add after the first two digits.

Here is an example: Pretend that you are trying to read the value of a resistor which has yellow, violet, and red bands. The first two bands (yellow and violet) tell you the two **significant digits**: 4 and 7. (If you are just learning the code, you should write these digits on a piece of paper.) Next, determine how many zeros to write behind the 47 by reading the third band. Since the third band is red (meaning 2), you would write two zeros just behind the 47 to give a value of 4700. Thus, a resistor coded yellow-violet-red has a value of 4700 ohms (4.7 kΩ).

Now try this one: what is the value of a resistor with brown-green-orange bands? The two significant digits are 1 (brown) and 5 (green). Then there are three zeros (orange), so the value is 15,000 Ω, or 15 kΩ. Other examples are given in Fig. 9-3 and 9-4. Figure 9-3 shows the coded colors, while Fig. 9-4 tells the values. Try to figure them out on your own.

Special Cases

The basic color code can be used only for resistors which have a value of 1 ohm or more (black-brown-black = 01 and no zeros behind the 1, or simply 1 Ω). There are, though, resistors with values lower than 1 ohm. To code them, use the first two bands just as with any other resistor (as significant digits) and then make the third band either silver or gold. Silver and gold third bands do not tell you to add zeros. Instead they tell you to move the decimal to the left.

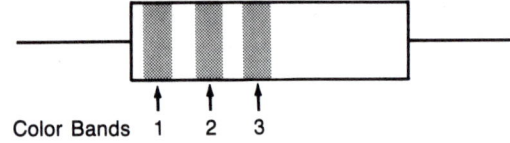

EXAMPLE RESISTORS

Color Bands 1 2 3

RESISTOR	COLORS OF BANDS		
	1	2	3
A	BROWN	RED	ORANGE
B	BLUE	BLACK	BROWN
C	RED	RED	BLACK
D	BLACK	WHITE	BLACK
E	ORANGE	ORANGE	GREEN

Fig. 9-3 Try to determine the values of these resistors. The answers are given in Fig. 9-4.

Values of Example Resistors

Resistor	Value	Comments
A	12,000 Ω = 12 kΩ	Leave answer in shortest form.
B	600 Ω	The black second band is the first "O" and then the brown third band tells us to add *one* more zero.
C	22 Ω	The black third band tells us to add *zero* zeros (add no zeros).
D	09 Ω = 9 Ω	A black band (zero) in the first position means there can only be one significant digit. The black third band tells us there are no zeros after the significant digit.
E	3,300,000 Ω = 3.3 MΩ	The green band gives you five zeros.

Fig. 9-4 These are the answers to the examples given in Fig. 9-3. Try to do some more for practice.

Even though you do not usually write a decimal after whole numbers, there really is one after the last digit to the right. For example, the number 52 actually means 52.0. If you had a resistor with the code green-red-silver, the two significant digits would be 5 and 2. The third band (silver) means to move the decimal two places to the left, so you would change 52.0 to 0.52, and our value would be 0.52 Ω. A gold third band means move only one place to the left. Another way to think of it would be to divide the significant digits by 100 if the third band is silver or by 10 if it is gold.

Tolerances

You already know that some resistors have more than three color bands and that the first three will always tell its value. The remaining bands, if any, tell about the resistor's manufacture. When resistors are produced, some batches turn out to be almost perfectly the correct value while others are a little 'off" in value. These resistors will still work very well in most circuits. Some circuits, however, must have almost perfect resistors in order to function correctly. The fourth color band tells you how much a resistor may be off in value. This fourth band is called the **tolerance** band. If there is no tolerance band, it is assumed that the resistor has a tolerance of plus or minus 20 percent (±20%).

Resistors which have a silver band in the fourth position have a tolerance of ±10%. A gold tolerance band indicates that the resistor actually has a value within 5% of the value given by the first three bands. A silver tolerance band indicates a higher quality resistor which is more nearly correct in value than one with no fourth band. Five percent resistors (gold band) are even better, but most circuits do not require this degree of precision.

Let's look again at one of the examples used before. The brown-green-orange resistor has an indicated value of 15 kΩ. If you measured its resistance with an ohm-

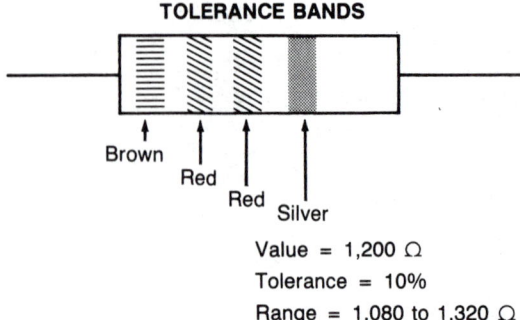

Fig. 9-5 The tolerance band tells us how much a resistor may vary in value and still be considered acceptable.

meter, it should be close to that value, but it probably would not be exactly 15 k. Since there is no fourth band, its tolerance is 20%. This means that it could be higher or lower in value by 20% and still be acceptable. To figure out how much it could actually vary, take 20% of 15,000. Remember that 20% = .20 and the word "of" means to multiply, so:

```
   15,000
 ×    .20
   ------
        0
   300 00
   ------
  3000.00  =  3,000 Ω
```

Thus, the resistor could be high or low in value by 3000 ohms. To find out its range of possible values, add 3 k to the indicated value and subtract it from that value also:

```
Subtract:  15,000     Add:  15,000
           - 3,000         + 3,000
           ------          ------
           12,000          18,000
```

You can now see that this resistor has a true value that lies between 12 k and 18 kΩ. If there had been a silver fourth band, the tolerance would have been 10%. In that case, the resistor could have been high or low in value by 1.5 k, so the range would have been 13.5 k to 16.5 kΩ. Figure out the tolerance and range of a resistor with the following bands: brown-red-red-silver. Try it on your own and then look at Fig. 9-5 to see if you were right.

Failure Rate

When resistors have five bands, or when the fourth band is a color other than silver or gold, then the last band tells the failure rate. The **failure rate** indicates how many of the resistors will fail per thousand hours of use. The manufacturer determines the failure rate by testing some of the resistors to see how they perform. The four colors used for failure rate bands are:

Brown — 1%
Red — 0.1%
Orange — 0.01%
Yellow — 0.001%

If a fifth brown band were added to the original example, making it brown-green-orange-silver-brown, then its value would still be 15 kilohms and its tolerance would still be 10%. The fifth band would not affect the reading of the other four bands. It simply tells that about 1% (1 out of every 100) of the resistors made in that particular batch will fail in a thousand hours of service. The table shown in Fig. 9-6 summarizes the whole color code. You may use it for quick reference as you learn.

Wattage Rating

You may have noticed that some resistors are physically larger than others. Perhaps you even mistakenly thought the big ones had the most resistance. Now you should realize this is not so. The color bands tell how much resistance a resistor has; there

Resistance Color Code

Colors	Bands 1 or 2	Band 3	Band 4 or 5
	Significant Digits	Add Zeros	
Black	0	0	--------
Brown	1	1	Failure Rate =1%
Red	2	2	0.1%
Orange	3	3	0.01%
Yellow	4	4	0.001%
Green	5	5	--------
Blue	6	6	--------
Violet	7	7	--------
Gray	8	8	--------
White	9	9	--------
No Band	-----	-----	Tolerances: 20%
Silver	-----	Move Decimal to: Left 2 Places	10%
Gold	-----	Left 1 Place	5%

Fig. 9-6 The resistance color code provides a lot of information in a condensed form.

could be a large resistor with the same color bands as a much smaller one. The size is important, though, for a very different reason. Even though size has nothing to do with the value of resistance, it does tell how much power, in watts, the resistor can handle. You do not need to be too concerned with power just yet, but it is an indication of how much voltage and current are *together* in a circuit. Larger resistors can carry larger currents than smaller ones. Thus, a large carbon resistor will have a higher **power rating** than a smaller one.

Carbon composition resistors which are about 1 cm long (just a little over ⅜ inch) are "half watt" resistors. Larger carbon resistors could carry more current and smaller ones could not carry as much current as these half watt resistors. Half watt carbon resistors are a very common size. Carbon resistors are generally available in ⅛, ¼, ½, 1, and 2 watt sizes. Thin film resistors are usually rated at ¼ watt. High current circuits require wirewound resistors which can handle five or more watts. The power rating is generally printed on wirewound resistors.

It is usually permissible to use a resistor

with a higher wattage rating than needed in a circuit. It will just take up a little more space and give an extra margin of safety. This is not safe, however, if the resistor is being used as a fuse in the circuit. If, on the other hand, a resistor with too *low* a power rating is used in a circuit, it will not withstand the current and will fail quickly.

Summary

The value of many resistors is printed on them by means of a color code. The code is easy to read. The first three bands tell the resistance value, and any other bands give information about the manufacture of the resistor. This additional information could include tolerances and/or failure rate. In addition to the resistance, the power rating in watts should be noted when using resistors in circuits. Now that you have studied about voltage, current and resistance separately, the next chapter will bring them all together in a meaningful way.

Important Terms

color code
significant digits
tolerance
failure rate
power rating

Review Questions

1. Why was the color code developed, and why is it still used?
2. List the bands in order from left to right, and tell what each band means.
3. What does the power rating deal with? Which type of resistors should be used in very high current circuits?
4. Obtain 10 assorted resistors from your teacher and identify their values and other information by reading the color code.

Careers: Electronics Assembly Line Worker

Factories which produce electronic components and devices employ many people. Some assemblyline workers operate or supervise machines which do specific jobs, while others actually do the jobs themselves. Some helpful skills include the ability to read schematics and blueprints, make and solder connections, read a meter, use basic hand tools, and identify components. You should learn most of these skills in the electronics/electricity class you are taking.

Many manufacturing companies provide on-the-job training for their employees. Good manual dexterity, willingness to work, and dependability are the most important qualifications for employment in this family of occupations. A high school education and work experiences are helpful as well. The opportunity for advancement and pay depend upon the talent and responsibility of the employee and the company for which he or she works.

Project: Electronic Battleship Game

Parts List:

1	B_1	6-volt lantern battery
1	S_1	DPDT switch
3	$L_{1,2,3}$	6-volt lamps with sockets
1	D_1	1 amp diode
1	D_2	silicon diode
1	LS_1	loudspeaker, 8 ohms
1	C_1	electrolytic capacitor, 10 µf, 10 V
2	$C_{2,3}$	.01 capacitors
1	C_4	.0015 capacitor
1	R_1	resistor, 15 ohms, 2 watts
1	R_2	resistor, 4.7 K ohms, ½ w
1	Z_1	LM 386 (IC), audio amp.
400		small wire nails
10		LEGO blocks (deep, single-row type)
		fine steel wool
		assorted wire
		plywood and contact paper or other materials for enclosure
100		washers, 10 painted red
1		small magnet

Circuit Description:

This electronic game is fun to make and interesting to play. Your teacher may have some suggestions to make it simpler or more realistic — ask for advice. Much of the work, such as building the enclosure, may even be done at home.

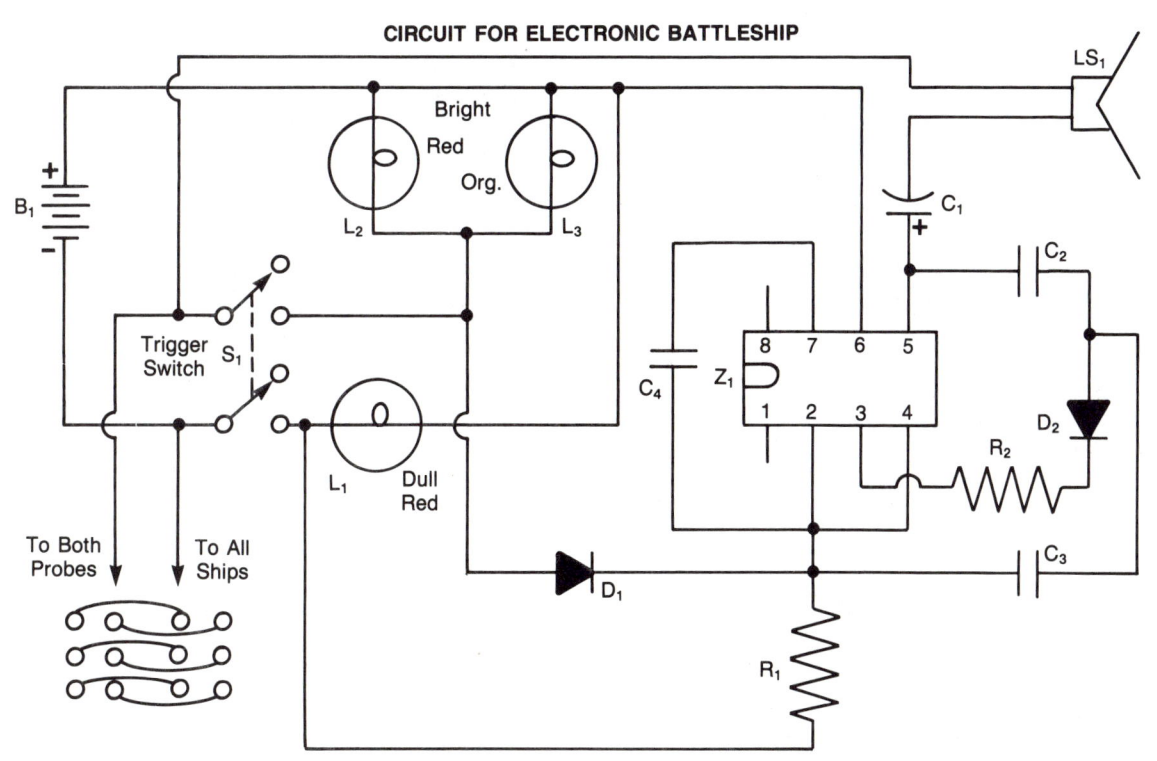

CIRCUIT FOR ELECTRONIC BATTLESHIP

Chapter 9   *Resistor Color Code*

Electronic Battleship Game

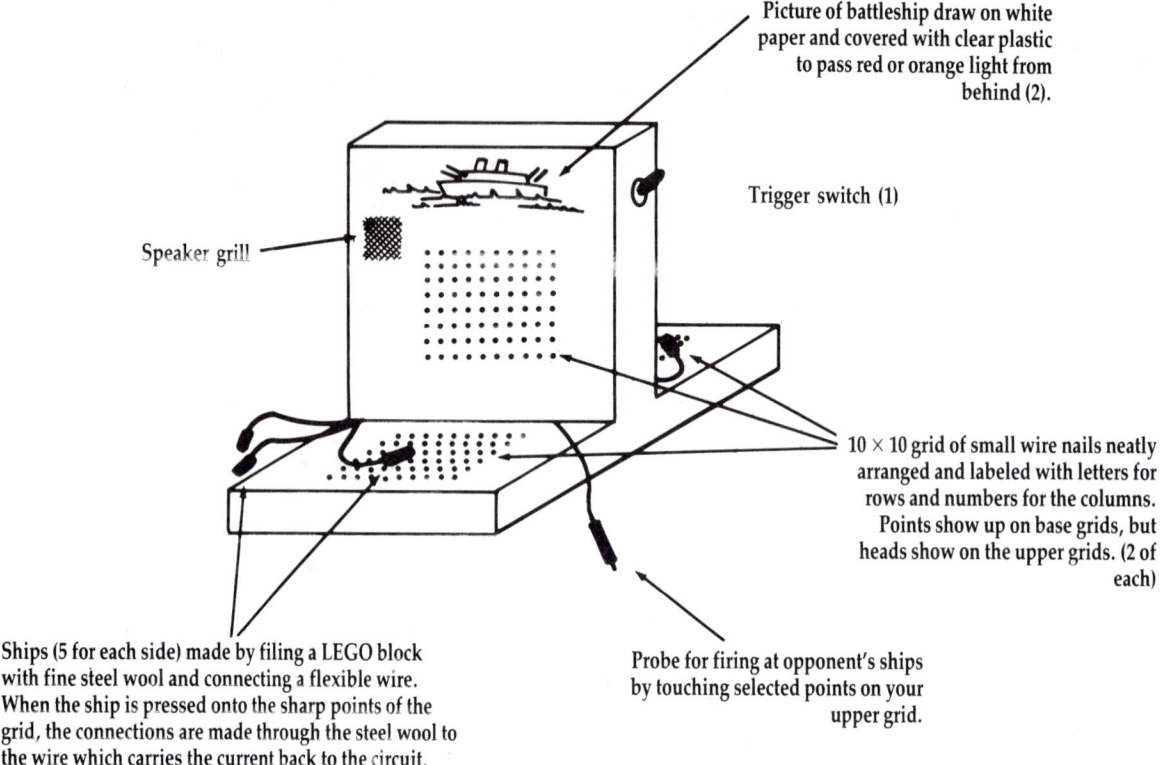

Picture of battleship draw on white paper and covered with clear plastic to pass red or orange light from behind (2).

Trigger switch (1)

Speaker grill

10 × 10 grid of small wire nails neatly arranged and labeled with letters for rows and numbers for the columns. Points show up on base grids, but heads show on the upper grids. (2 of each)

Ships (5 for each side) made by filing a LEGO block with fine steel wool and connecting a flexible wire. When the ship is pressed onto the sharp points of the grid, the connections are made through the steel wool to the wire which carries the current back to the circuit.

Probe for firing at opponent's ships by touching selected points on your upper grid.

Wires connect from each point on one side's base grid to the same point on the opponent's upper grid. So, if a shot is a miss, when the trigger is fired, there will be only a soft sound and a dim red light will light in the water near the ship. But, if a shot is a hit, then the explosion sound will be much louder and there will be a bright orange-red light in the middle of the ship. Each shot should be recorded by hanging a plain or red-painted washer on the nail for that shot on the upper grid. A small magnet will help in removing the washers.

The base and vertical divider can be made of inexpensive plywood or similar material and then covered with contact paper. A blank area must be left behind the picture of the ship so the light can pass. The ships should be of different lenghts — the PT boat should only cover 2 nails on the grid, the cruiser and submarine, 3; the battleship, 4; the aircraft carrier, 5. The LEGO blocks may be covered or decorated as you wish. The spacing of the nails on the base grid must fit the LEGO blocks exactly; that means 5/16" apart both vertically and horizontally. The lamps may be colored by covering them with scraps of cellophane from candy wrappers.

Chapter 10

Relationship of E, I, and R

The Relationship

How do the three important qualities — voltage, current, and resistance — relate to each other in circuits?

The electromotive force which pushes electrons through circuits is identified by the letter E. This quality is measured in units called volts. The more pressure there is, the more current will flow.

Electric current is measured in amperes. Actually, current is nothing but electrons moving through a circuit. The more electrons that are moving, the more current there is. One way to make more current flow in a circuit is to increase the voltage (the pressure that pushes the electrons). If you raise the number of volts, then more amps of current will flow.

Resistance, the opposition to electron flow, is measured in ohms. All devices used in electronics have some resistance, so every part of a circuit will try to limit current flow to some degree. The second way to increase current flow in a circuit is to lower the resistance (R).

Since current flow can be increased by raising voltage or by lowering resistance, it stands to reason that current could be lowered by doing the opposite of either of these. This is precisely the case. If you wanted to lower the current flow (I) in a circuit, you could either lower the voltage (E) or raise the resistance (R). Previous chapters have dealt with these qualities and the concepts of this relationship. Because this relationship is basic to electronics, you must understand it thoroughly. If you have questions about E, I, or R at this point, review chapters 6, 7, and 8.

Basically, if you want more electrons to flow, you must either push them harder or make their path easier for them to travel. In electronics, you would say to increase current flow (I), increase voltage (E) or decrease resistance (R). Likewise, if you want fewer electrons to flow, you must stop pushing so hard or put some restrictions in the path: to decrease I, either decrease E or increase R.

Quantifying Values

So far, E, I, and R have been increased or decreased in general terms without paying any attention to how much they were being increased or decreased. You have learned how to quantify each of these important qualities. You can use the units of measurement discussed previously to find out exactly how much a change in one value will affect another one. If you had a circuit with 1 ohm of resistance, and you applied 1 volt of pressure, 1 amp of current would flow through the circuit. If you raised the voltage to 2 volts, but left the resistance at 1 ohm as before, then the current would go up to 2 amps (Fig. 10-2). If another circuit had E = 6 V and R = 2 Ω, then 3 amps

Qualities of a Circuit

Quality	Symbol	Other Names and Descriptions
Voltage	E	Electrical Pressure Electromotive Force
Current	I	Rate of electron flow Intensity
Resistance	R	Opposition to electron flow Restriction

Fig. 10-1 The three most important qualities of a circuit are E, I, and R.

would flow (1 = 3 A). See Fig. 10-3. If the resistance in this circuit is increased to 12 ohms, then the 6 volt battery would only be able to force ½ A of current. All that you have to do to figure out what is going on in circuits when values are changed is be able to multiply and divide.

Ohm's Law

The relationship between E, I, and R was first discovered by Georg Simon Ohm in 1827. The relationship is called **Ohm's Law** to honor him. Ohm's Law simply states that the current is **directly proportional** to the voltage and **inversely proportional** to the resistance in a circuit. In other words, as voltage goes up or down, current will do the same thing. When resistance goes up or down, the current will do the opposite. More resistance causes less current.

You can use Ohm's Law to find unknown values in circuits and to predict what will happen to current flow if voltage or resistance is changed. All you have to do is multiply or divide! Using the symbols E, I, and R, Ohm's Law is written as $E = I \times R$.

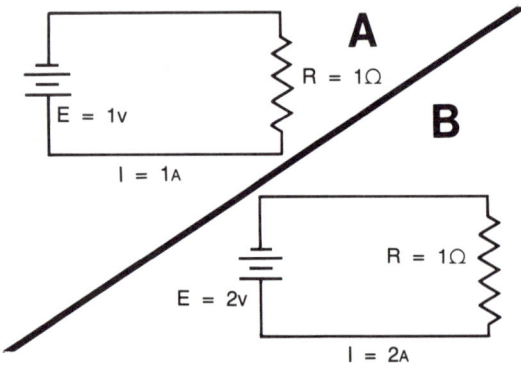

Fig. 10-2 When the voltage is increased (as in B), the current goes up proportionally.

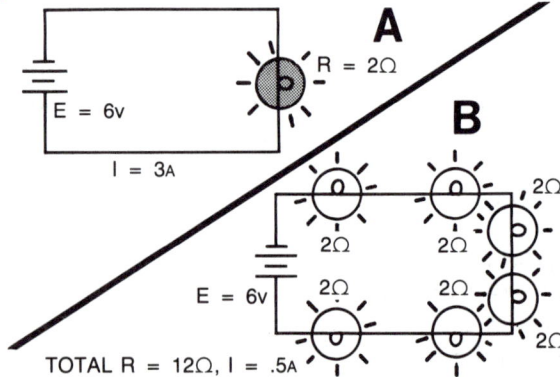

Fig. 10-3 If the resistance is increased, the current is decreased. This is an inversely proportional relationship.

Look at Fig. 10-4 for a moment. Use Ohm's Law to show how E, I, and R are related in this circuit. First, write Ohm's Law using the letter symbols. Next, just "plug in" the actual values of those qualities from this circuit. If 10 = 5 × 2 (and it has so far!), then you have demonstrated that Ohm's Law can be used to account for the relationship of these qualities in a circuit.

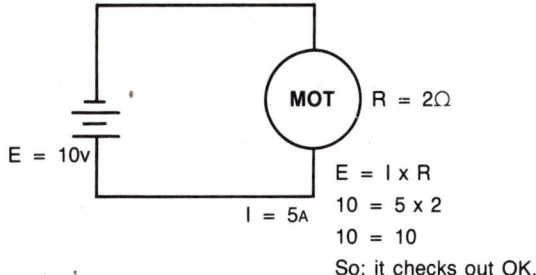

Fig. 10-4 A motor with a resistance of 2 ohms should draw 5 amps of current when connected to a 10-volt source. How much current should flow if we increase the EMF to 20 V?

Finding Unknown Values

Sometimes it is important to know a value in an electric circuit that is impossible or inconvenient to measure. Other times, as in designing circuits, you may have to find one of the needed values mathematically. Either of these two situations can be dealt with by using Ohm's Law. As long as the values of any two of the three qualities (E, I, and R) are known, the third one can be found by simple multiplication or division.

Suppose you wanted to make up a circuit for a motor that requires 4 amps of current and has 6 Ω of resistance when running. Figure out what battery to use in your circuit. Here are the steps to use in solving such a problem:
1. Write Ohm's Law   E = I × R
2. Substitute the actual values you know   E = 4 × 6
3. Solve the problem   E = 24
4. Label your answer   E = 24 V

You will need a 24 volt battery to make this motor run correctly.

This form of Ohm's Law is good to use if the voltage is the **unknown** quality. Sometimes, however, the voltage is known and you want to find either the resistance or the current. In such cases, you need to use a different **form of the equation**. There are two things you can do to choose the proper form. If mathematics is easy for you but memorizing is difficult, you should use method one. If you have a tough time with math but are great at memorizing, method two will work better for you. Here are the two methods:

Method One: To find either I or R when E and the other value are known, convert Ohm's Law to the form which leaves the needed unknown quality alone on the left side of the equal sign. For example, if you wish to find the current, start with E = I × R

then you convert to $I = \dfrac{E}{R}$

by dividing both sides by R and then exchanging sides for convenience.

Likewise, unknown values of R could be found by converting to the form $R = \dfrac{E}{I}$.

Method Two: The second method is simply to memorize the three forms of the **equation** for Ohm's Law. Then use the form which suits the problem you are trying to solve. The three forms are:

To find unknown voltage   E = I × R

To find unknown current $I = \dfrac{E}{R}$

To find unknown resistance $R = \dfrac{E}{I}$

A memory aid which could help you memorize the three forms of the equation appears in Fig. 10-5. To use it, simply cover the symbol for the unknown in the problem; the positions of the other two symbols will help you remember the proper form to use. After using the memory aid repeatedly, you will be able to remember the forms easily.

Practical Examples

What is the resistance of the lamp in a circuit if a 120 V generator causes 4 A to flow in the circuit? First, choose or convert to the proper form of the equation. The unknown here is resistance (R), so use the equation $R = \frac{E}{I}$. When you solve the equation, you find that R = 30 Ω.

Suppose you have an extension cord rated to safely carry 10 A. Would it be safe to use this cord to operate a drilling machine (motor) which has a resistance of 9 ohms on the 120 volt line? What is the unknown? The resistance is 9 Ω and the EMF is 120 V, so these two values are known. You want to know how much current this circuit will draw. Once the circuit current is found, compare that value to the safety rating of the extension cord (10 A) to see if it is possible to use it. So, the unknown in this problem is I, the actual current that the circuit would draw. Pro-

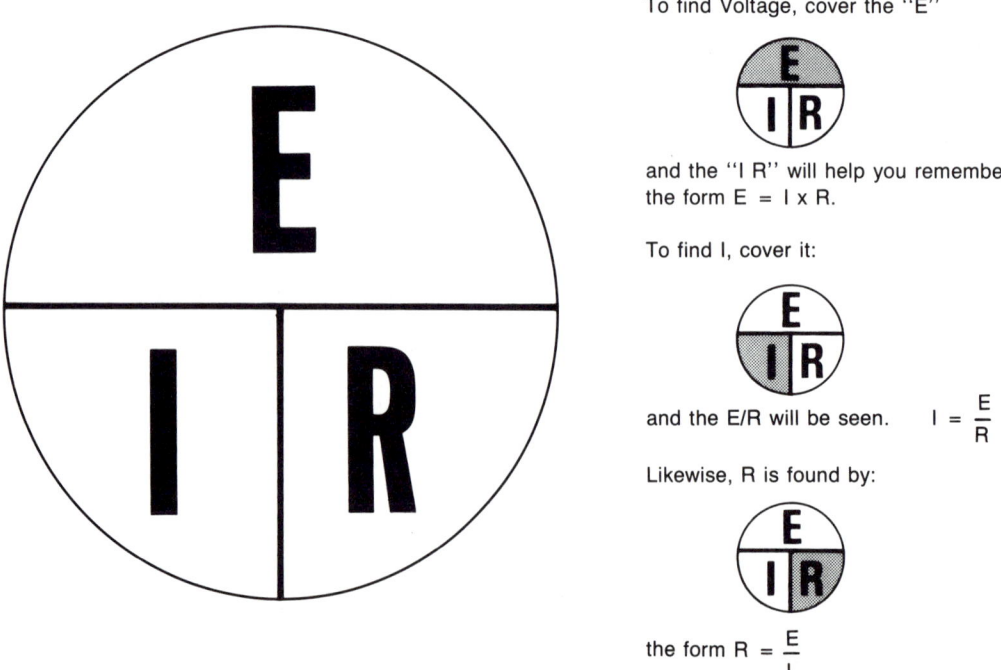

Fig. 10-5 This simple aid may help you to remember which form of Ohm's Law to use when solving for unknown values.

ceed as before: find the proper form of the equation, $I = \frac{E}{R}$, and then solve the problem by simple division. The answer is 13.3 A. Since this is more than 10 amps, do not use the extension cord. You must either move the machine closer to the outlet or get a heavier cord.

This is a good example of why it is important to be able to solve simple problems using Ohm's Law. There would be no possible way to measure the current in this example, because current (electron movement) can only be measured while it is flowing. If you tried to connect the circuit and then measure the current, damage could occur before you even got the measurements made. Being able to predict what will happen in a circuit (by using Ohm's Law) is a very important and practical ability for those who work with electricity and electronics.

Here is one more example: if the circuit shown in Fig. 10-6 causes the lamp to burn dimly, what should happen if you increase the voltage to 20 V? What is the unknown? You are changing the voltage, so that is known. The new voltage is 20 V. The bulb (the source of resistance in this case) is not going to change. The only thing left is the current flow. Before you solve the problem mathematically, always try to logically predict the type of change you expect to occur. In this case you are raising the voltage and keeping the resistance at a constant value. What would you expect to happen to the current? According to the relationship expressed by Ohm's Law, the current should increase when you increase the voltage. That is, when you push with a higher pressure, more electrons will move. Now solve the problem using the proper form of the equation: $I = \frac{E}{R}$. The increase of EMF from 10 V to 20 V does, indeed, cause the current flow to rise from .25 A to .50 A. The lamp should burn brighter when you

Fig. 10-6 What will happen if we put in a 20 V battery? **Hint:** If the current goes down, the lamp will dim; if the current goes up, it will brighten.

increase the voltage because that higher pressure will force more electrons (current) through it — and the current is what makes it burn in the first place.

Summary

The relationship between voltage, current, and resistance is expressed by Ohm's Law. Problems dealing with this relationship may be solved by simple multiplication and division. Basically, current depends directly on voltage and inversely on resistance. Ohm's Law is one of the most important principles in all of electronics. You can use it to predict or find values which are unknown and cannot be measured. The following chapter will explore other circuits and help you become skilled at finding unknown values and predicting the effects of changes in values.

Important Terms

Ohm's Law
directly proportional
inversely proportional
unknowns
form of the equation
equation

Review Questions

1. Explain the relationship between E, I, and R. What should happen to I if we increase E or R?

2. Why are there three forms of the Ohm's Law equation? What are they?

3. If you knew the resistance and voltage of a device, how could you find the current that would flow in the circuit without actually connecting it and measuring the current?

History: Georg Simon Ohm

Ohm's Law, which states the relationship between voltage, current, and resistance in electric circuits, was discovered by the man for whom it is named: Georg Simon Ohm. He was born in Bavaria, now a part of West Germany, in 1787. He studied physics and mathematics, and discovered his Law when he was a professor of mathematics at Jesuit's College at Cologne. His findings appear in a pamphlet titled *Die Galvanische Kette, Mathematisch Bearbeitet* (Mathematical Investigation of the Galvanic Circuit). When his work was not well received at Cologne he went on to Nuremberg and later to Munich. Today, we realize how important Ohm's Law is to the understanding of electrical circuits. In addition to the Law, the unit of measurement of resistance is named after this outstanding scientist. His memory lives each time you refer to the "ohms of resistance" in a circuit.

Chapter 11
Computing Electrical Values

Solving Problems

There are four steps to solving problems dealing with Ohm's Law or any other calculations in electronics, for that matter. Take the time to memorize the steps. You will be glad you did when the time comes to solve more difficult electronics problems. The steps are:
1. Draw the circuit.
2. Use the formula.
3. Label your answer.
4. Check the **logic** of answers.

Use these four steps, in order, as you solve this simple problem:

> What is the resistance of an automobile headlamp if it draws 4 A of current when connected to a 12 V battery?

Here are the steps to solve the problem:

1. Always draw the circuit. Even though this circuit is very simple, it is still important to draw the schematic diagram of the circuit for two reasons. First, it will help you to get into the *habit* of drawing circuits. Second, it will help you to organize your thinking and see which values go where in the formula. Figure 11-1 shows how a student might draw this circuit. Notice that all of the information in the problem is contained in the drawing — including the values which were **given** in the problem (4 amps and 12 volts). The schematic can be very simple. You do not need to draw switches, fuses, and other parts which are not a part of the problem. Thus, your

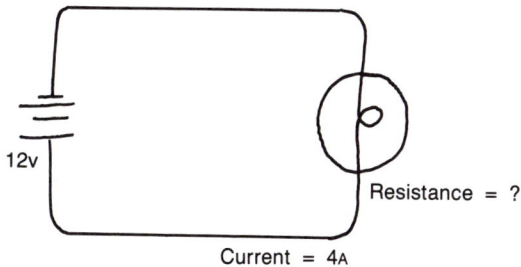

Fig. 11-1 The first step to solving a problem is to sketch the circuit. This helps you to organize the problem even if it is a simple one.

1. Write proper form of equation $R = \dfrac{E}{I}$

2. Substitute known Values $R = \dfrac{12}{4}$

3. Simplify (solve the arithmetic) $R = 3$

4. Label your answer $R = 3\,\Omega$

Fig. 11-2 Finally solving the problem is very simple. Follow this easy pattern to avoid getting confused. Step 4 should really be done on the same line as step 3.

drawing can include only the battery, lamp, and wires.

2. Always use the formula. Even when you become more skilled at working with electrical formulas, it is always a good idea to write the formula. This helps you to check your logic. It also helps to ensure that you **substitute** the correct **known** values into the formula and use the proper form of equation. See Fig. 11-2.

3. Always label your answer. As soon as the arithmetic part of the problem is done, you should immediately **label** the answer in its proper units. In this problem the answer came out to be 3, but 3 what? Simply look at the last line of the arithmetic. This last line will always say E = ____, or I = ____, or R = ____. The letter to the left tells you whether the answer you found is voltage, current, or resistance. So, behind the answer, you must write the symbol for the unit of measure of that quality — write V for voltage (E), A for current (I), or Ω for resistance (R).

4. Always check the logic of answers. This last step is also very important in two ways. First, it helps you learn how the mathematical formula represents the principles of Ohm's Law. Second, it self-checks against getting wrong answers.

Values in Nonbasic Units

So far, all of the examples have used the basic units of measurement for E, I, and R — volts, amps, and ohms. What if the resistance in a problem were stated in kilohms (kΩ)? Try this: what voltage does the battery supply if a resistor of 2 kΩ draws 3 A of current? Figure 11-3 shows both the correct and an incorrect way to approach this problem. If you plug 2 into the formula for the resistance value (ignoring the "k"), then the answer would be 6 volts. That is not correct! To solve this problem correctly, you must convert the resistance value to its basic unit. The 2 k really means 2000 ohms! Using 2000 for R, you get the correct answer of 6,000 V.

Here is another example: what is the resistance in a circuit that draws 750 mA from a 10 V source? Check Fig. 11-4 to see if you solved the problem correctly.

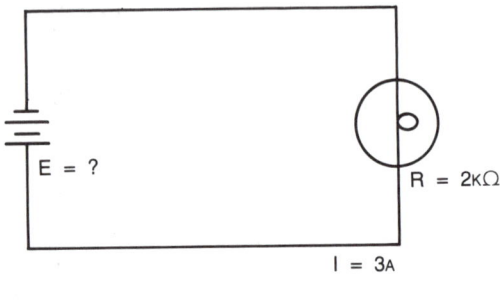

Fig. 11-3 When values are expressed in non-basic units, you must convert them to basic units before you substitute them into the formula.

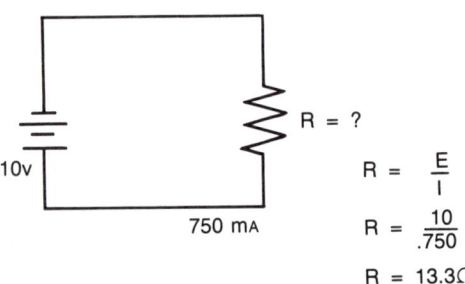

Fig. 11-4 Convert 750 mA to the basic unit, .75 A, before you plug it into the formula. One decimal place is usually accurate enough when your answer is not a whole number, unless you are working with very small values.

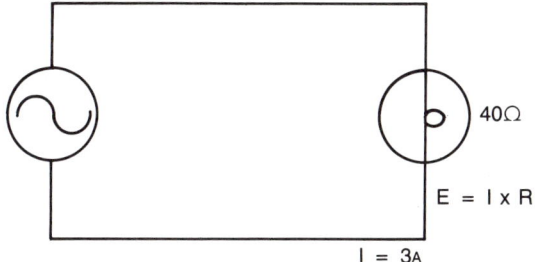

Fig. 11-5 Since voltage (E) is not given, it is the unknown value for which we solve.

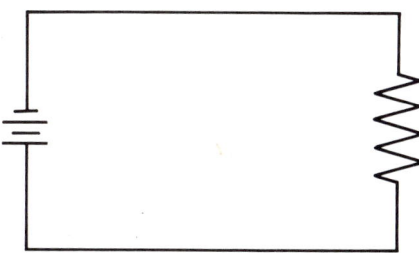

Fig. 11-7 What current flows if the resistor has 60 KΩ of resistance and the battery is 600 V? Remember to redraw the circuit on your paper and label the values.

Sample Problems

Complete each problem in Fig. 11-5 through 11-7 using the four steps discussed previously. When you have finished, look at Fig. 11-8 to check your answers. Figure 11-8 also contains comments about special features of some problems.

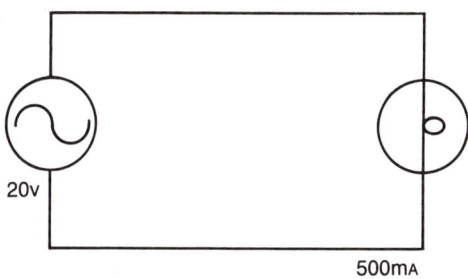

Fig. 11-6 Find the unknown resistance in this circuit. Choose the proper form of the equation, then substitute known values into it. Remember to convert all values to basic units.

Summary

Ohm's Law can be expressed mathematically as $E = I \times R$. When using this Law to solve problems dealing with the actual values found in circuits, follow four important steps. They may seem unnecessary, but they are important for building problem solving skills and will prevent many troubles when you work with more complex problems.

Before values are substituted into the Ohm's Law formulas, they must be converted to their basic unit forms. The next chapter will help you learn to use calculators so that you can solve problems in which the values do not always come out to nice, easy-to-handle numbers.

Problem	Answer	Comments
11-5	E = 120 V	
11-6	R = 40 Ω	500 mA = .5 A
11-7	I = .01 A	60 kΩ = 60,000 Ω. You should convert the final answer to I = 10 mA.

Fig. 11-8 Check your answers to the problems here. Your teacher may find some more problems for your practice.

Important Terms

logic
given
substitute
known
label
nonbasic units

Review Questions

1. Explain the relationship of E, I, and R.
2. List and explain the four steps to solving electronics problems.
3. Why are the four steps important even in simple problems?

Chapter 12
Using Calculators to Solve Problems

Calculators Simplify Computations

In the real world of electronics, the values you will actually find in circuits will frequently be large numbers with several decimal places. Many students have trouble working with such numbers and may easily become confused or lose count of zeros and decimal places. In the early days of electronics, engineers and students used **slide rules** to solve problems. See Fig. 12-1. Slide rules, however, did not account for **decimal places** and zeros. In fact, the slide rule may have done more harm than good because it does the multiplying or dividing for you, but you have to keep track of places in your head. Today engineers do not use slide rules. Modern electronic technology has created a far more accurate tool — the handheld electronic **calculator**. The calculator has more capability than some of the early computers which took up large amounts of space. Since you are studying about **technology**, it is only natural to use devices which represent the state of the art. Calculators make complex problems easy to solve for most students. The key is learning to use the calculator intelligently.

Operating the Calculator

Actually using calculators is very simple. Most of them work similarly, but some have unique steps. Read the instructions that accompany your calculator because they will give you all the details about your machine. If the instructions are lost, look on the bottom of the instrument. (Many companies print short instructions on their instruments.) If there are no instructions at all, a few simple experiments will help you learn to use the machine. Set up some easy problems which you can do in your head (like 2 × 3 and 8 ÷ 2), then try them on the

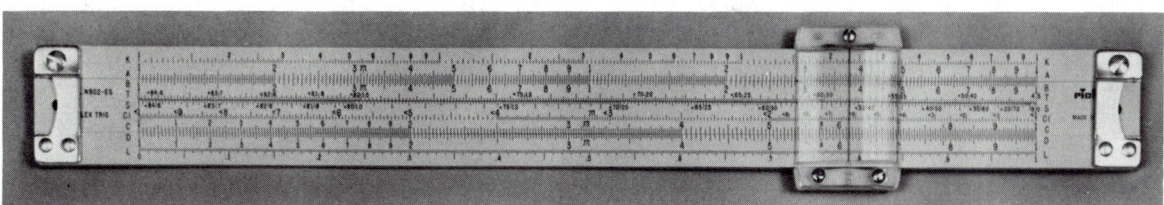

Fig. 12-1 Slide rules were a very common tool which electronics engineers and students used to help solve problems until the 1970s.

calculator. With most machines, to multiply you enter the first number, press the × key, enter the next number, and finally press the = key to display the answer. Some machines, however, require you to use another key labeled "enter" when you enter the numbers. These machines use a special type of logic called RPN (Reverse Polish Notation — so named because of the work of a Polish logician named Lukasiewicz). To multiply on RPN calculators, put in the first number and press the "enter" key, then you put in the second number and press the × key to display the answer. Some RPN machines do not even have an = key. Most calculators do not use RPN.

Dividing is a very similar process. Most machines require you to enter the dividend first, press the ÷ sign, enter the divisor, and press the = key last to display the quotient. If you are clearing a fraction, enter the numerator first, press the ÷ key, and then enter the denominator. A simple way to keep this straight in your mind is to say (under your breath) "8 divided by 2 equals," and enter "8 ÷ 2 =" as you say it.

Again, RPN calculators require you to push these keys: 8, "enter", 2, ÷.

Multiplying and dividing are not the only jobs the calculator can do. Many machines can find the **square** of a number just by pushing a key labeled x^2. (The square means to multiply that number by itself.) Another useful function is **square roots**, the opposite of squares. The most common label for a square root key is $\sqrt{}$, but some machines are different. Less expensive calculators may not have either square or square root functions. Figure 12-2 shows a couple of calculator keyboards with some special keys marked. Experiment with your machine using simple numbers until you are able to use it quickly and accurately.

When entering numbers like 6.0703, you would press six, decimal, zero, seven, zero, three. The machine will always assume there is a decimal at the end of your number if you do not tell it differently. If you forget to press the decimal key, the last entry would be understood by the machine as 60,703! If you are careful to enter zeros and decimal points when they are required, the machine will automatically keep track of places for you.

Fig. 12-2 Calculators which can find square roots by pressing a single key can be purchased for as little as $8.00.

Solving Problems

All calculators can do is complete arithmetic quickly. They cannot draw the circuit. They cannot select the formula. They cannot label the answers, and they cannot check the logic. You must do the four main steps yourself. Turn back to the problems you did in the last chapter, Fig. 11-5 through 11-7, and rework them with your calculator. This will be a good check of your ability to use the calculator.

Checking your logic is even more important with the calculator than it was when you did the arithmetic in your head. There are all sorts of clumsy errors that can be made. There are two safeguards you can

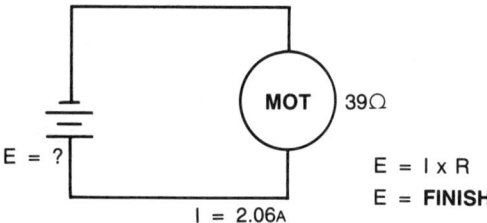

Fig. 12-3 A worker wants to know what voltage battery to put in the circuit above. What might happen if he or she forgets to enter the decimal in the current when solving the problem?

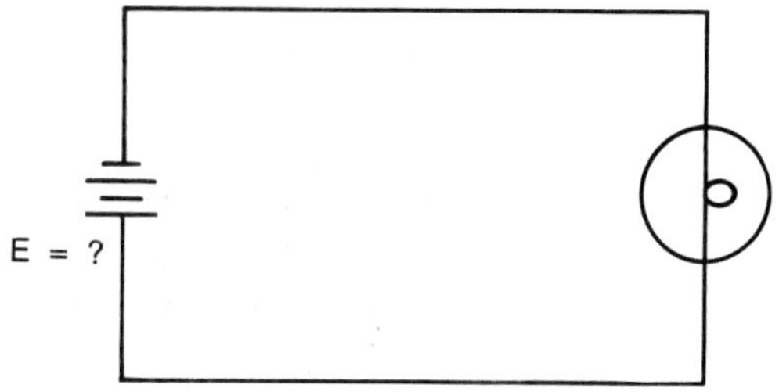

use to check your answers when you use a calculator. The first is to do the arithmetic twice. You can do the problem twice and still be faster than the old paper and pencil method. If you get the same answer the second time, you probably have used the machine correctly. The second safeguard is to look at the problem logically. Figure 12-3 shows an example problem. Work the problem, but intentionally enter a mistake — use 206 instead of 2.06 for the current. If you were on a job and you used this answer without looking at it logically to see if it were correct, you could cause serious damage to the circuit — possibly even a shock hazard or a fire! But saying to yourself "about 2 amps times about 40 ohms should be around 80 volts" would help you see that your answer of 8,034 V is wrong.

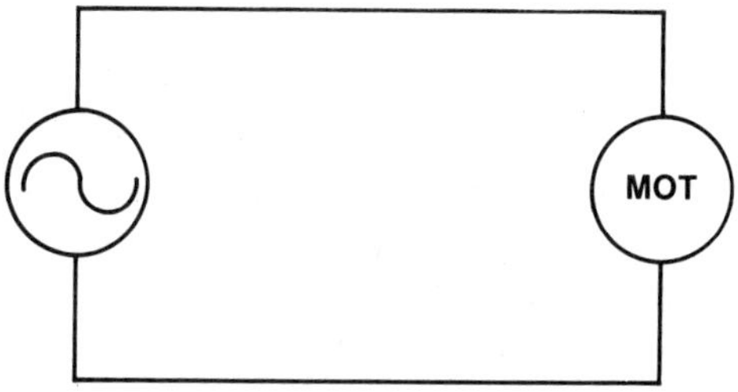

Fig. 12-5 What voltage battery would be needed to make our 20 Ω bulb draw exactly 300 mA?

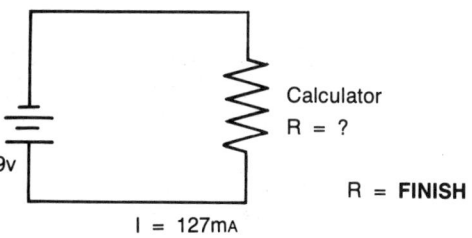

Fig. 12-4 A calculator draws 127 mA of current from a 9 V battery. How much resistance does it have?

Fig. 12-6 When the voltage output of the transformer to an electric train is at 9 V, the train engine runs slowly and draws 3 A. How much current will it draw when the transformer is turned up to 14 V? You cannot just start by using $I = \frac{E}{R}$, because you only know E. You must first find the resistance of the engine (motor), using 9 V and 3 A from the first part of the problem, and then plug that figure — which will not change for the life of the motor — into the $I = \frac{E}{R}$ formula.

Chapter 12 *Using Calculators to Solve Problems*

Calculators are supposed to help us do better work. They are tools just like power saws, but if either one makes a mistake, it's usually a bad one which can cause a lot of trouble.

Sample Problems

Finish working the problems in Fig. 12-4 through 12-6 with your calculator. Remember to use the four steps and to convert to basic units when necessary. The correct answers are found in Fig. 12-7. There are some additional problems at the end of this chapter for you to solve with your calculator.

Problem	Answer	Comments
12-4	R = 70.9 Ω	Remember to use basic units.
12-5	E = 6 V	If we want less current, we can lower the voltage.
12-6	R = 3 Ω I = 4.67 A	We find the resistance of the motor and then: solve for I using the new (higher) voltage and our 3 Ω (constant) resistance.

Fig. 12-7 Check your answers to the sample problems here. Rework those you missed to learn how to do them.

Summary

Calculators are tools which make the arithmetic part of solving problems easier. They permit you to solve more complex problems in a shorter time than by manual methods. You still must follow the four main steps to solving problems: draw circuit, select formula, label answer, and check logic. The next chapter will discuss the relationship between electricity and the ability to do work. This is called "power." It accounts for both voltage and current in a circuit.

Practice Problems

1. If a 47 kΩ resistor is carrying 32 mA of current, what voltage does the battery in the circuit produce?
2. A color television draws 1.3 A of current when it is connected to the 120 V line. How much resistance does it have when it is operating? (Remember that you can represent any load — even a whole television set — as a simple resistance when you draw the schematic of the circuit).
3. If a small black and white television was connected to the same 120 volt line and it only drew 986 mA, what would its resistance be? Think about this one logically first. You have the same voltage as in number 2, but lower current, should the resistance be lower or higher than before? Now solve the problem to see if you were right.
4. Find the current which flows through a 4 MΩ resistance when you apply a potential of 12.6 V.
5. A battery must be selected which will cause 52 mA to flow through a 13 kΩ load. What voltage must the battery produce?
6. If a lamp which is connected to a light dimmer draws 833 mA when the voltage is turned all the way up to 120 V, how much current will it draw when we turn the dimmer down to 80 V? This is a two part problem similar to the one in Fig. 12-6.

Important Terms

slide rule
decimal places
calculator
technology
square
square root

Review Questions

1. Why are calculators now used instead of slide rules for electronics problems?

2. What four steps to solving problems must you do yourself?

3. Why is checking your answer so important when you solve problems with a calculator? What are two ways you can check your work?

Pep Talk: Mathematics

Of all the courses you take in school, there are none more valuable to you in electronics than mathematics courses. Some degree of skill in math is required for any trade or professional job in electricity and electronics. Many jobs in electrical wiring and appliance repair require only a firm understanding of general mathematics. Engineering, however, demands a very strong background in math — study calculus as early as possible in school. You'll quickly find that an understanding of basic mathematic and algebraic concepts are essential to the enjoyment of electronics — even if electronics is just a hobby. If math is a difficult subject for you, it may be because you did not see how important it is in the real world — take a fresh look at math now that you are seeing important, practical uses of it in the exciting field of electronics.

Project: Motor Speed Control/Light Dimmer

Parts Lists:
- 1 wall plug (with ground preferred)
- 1 female receptacle (mate for above)
- 1 line cord
- 1 5 A fuse and holder (optional)
- 1 4.7 K ohm, ½ W resistor
- 1 50 K ohm potentiometer
- 1 NE-2 neon bulb
- 1 .1 mfd., 200 W.V. capacitor
- 1 SCR, at least 5 A, 200 V rating
- 1 PC board
- 1 suitable enclosure
- 1 control knob

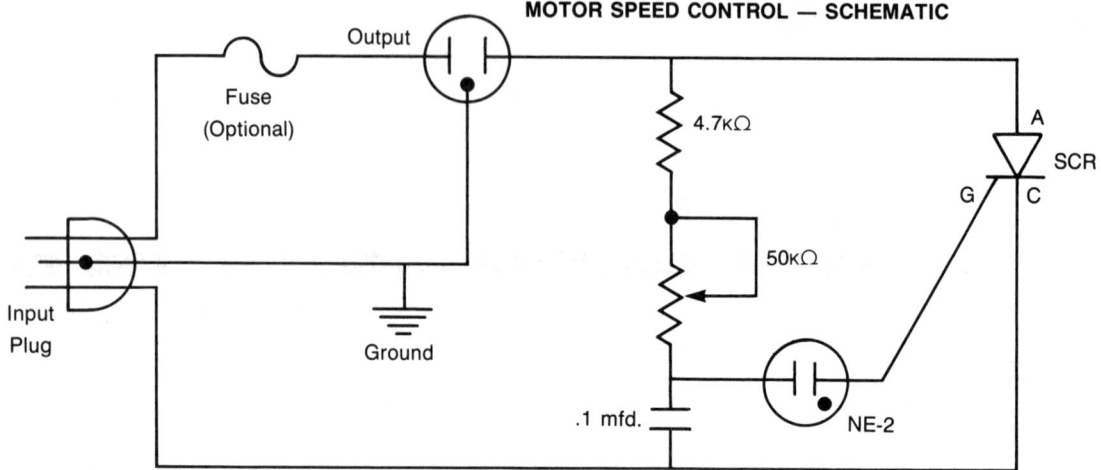

Fig. 12-8 This project should be fun to make and useful to have. When the control knob is turned, the speed of the motor (such as an electric drill) or brightness of a lamp may be controlled. Construct it carefully and be aware that it does use high voltage — do not plug it in until your teacher inspects it!

Notes:

The first three items (plug, receptacle, and cord) may be obtained by purchasing an extension cord and cutting it so as to place the SCR control box near one end as preferred. If desired, a TRIAC may be used instead of the SCR. The simple metal and wood enclosure shown in Fig. 12-9 may be used for this project (or many others), or a plastic or metal case may be purchased. If the case is metal, be careful not to allow any components to touch the case because the 120 volts of line current could be a hazard. Have the teacher inspect this project before you plug it in for testing. The job of the neon lamp is to trigger the SCR, but it may also double as an indicating lamp if it is mounted so that it may be seen from the outside of the enclosure. This could be done by simply forcing it into a grommet in the top of the case. An optional switch could also be added. Use fuses and heavy-duty cords if you plan to control power tools. Make sure that the motor is a universal type. Induction motors cannot be controlled with this project.

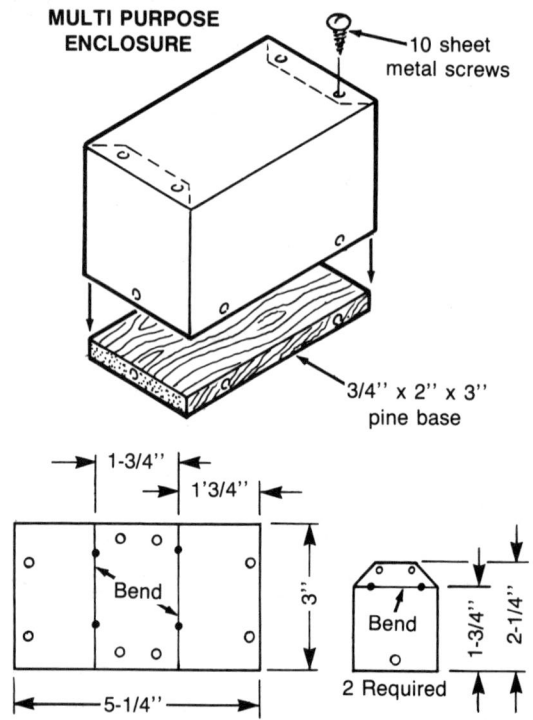

Fig. 12-9 This simple enclosure may be used for this project and many others. The metal parts may be made of any light sheet metal and joined with either sheet metal screws or pop-rivets. The wooden base could be a small scrap from the woodworking shop.

Electricity and Magnets at Work

Section 3

Chapter 13

Electrical Power

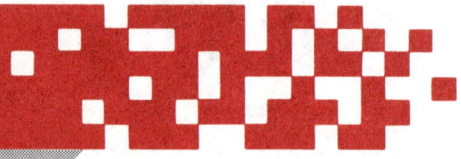

Mechanical Power and Electrical Power

Mechanical or physical power is quantified by **horsepower** units. Most engines and motors have a horsepower rating which tells how much work they can do when they are running in their maximum output conditions. To help you understand this rating, a small electric drill motor may be able to produce about ¼ hp, a lawn mower engine is typically between 2 and 4 hp, a moderately large motorcycle produces about 45 hp, and many automobiles produce well over 100 hp.

What does this have to do with electricity? Well, if you were using an electric motor, it would draw a certain amount of current at a specific voltage level in order to produce 1 horsepower. The unit you use to quantify electrical **power** is not the horsepower (which is a pretty large amount of power), it is the **watt**. It takes 746 watts to make 1 horsepower. A perfect motor would produce 1 horsepower when operated with 746 W of electrical power. But what is 1 watt? One watt is the amount of electrical power used when a circuit which is powered by a 1 volt battery draws 1 amp of current. If there were 2 amps flowing from a 1 volt source, then there would be a power consumption of 2 W. This relationship is very similar to Ohm's Law. This new relationship is called Watt's Law, named for Scottish inventor James Watt. Basically, Watt's Law says that the amount of electrical power in any circuit or device depends on how much voltage and current are *both* in the circuit. If either the voltage or the current is increased, the power goes up too. Watt's Law is mathematically stated as $P = I \times E$. See Fig. 13-1. The P is the symbol for power and the basic unit of power is the watt (W). The power or wattage rating of electrical devices tells you more than just how much power a motor would produce. It helps you tell the amount of light which bulbs will produce, how much heat resistance elements produce, how much sound a stereo can produce, and many other pieces of information.

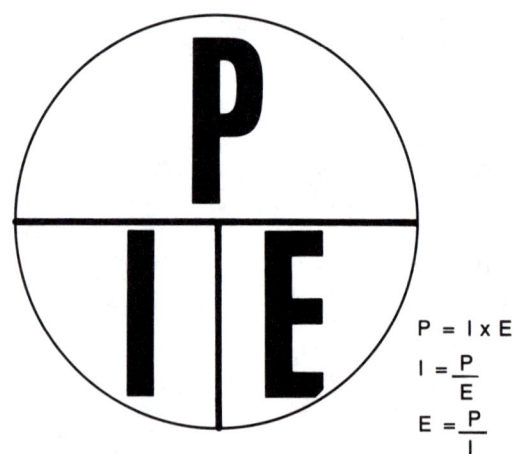

Fig. 13-1 This memory aid helps us remember Watt's Law. We can also imagine a big slice of "Pie" ($P = I \times E$).

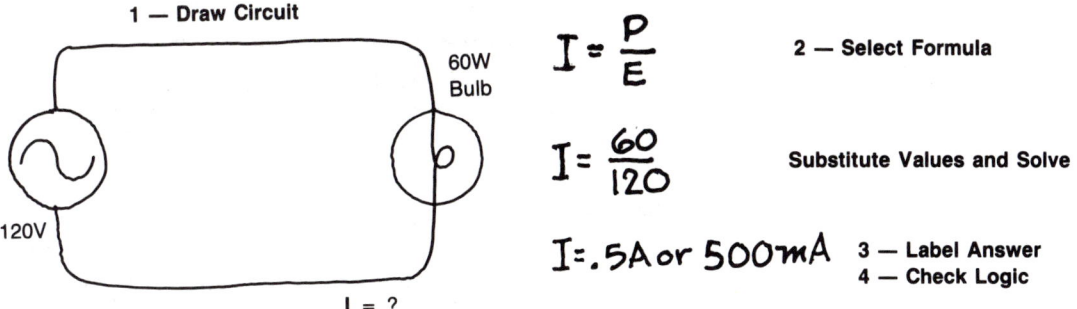

Fig. 13-2 If we removed this 60 W bulb and replaced it with a 100 W bulb, what would happen to current flow? Figure it out on a piece of paper by using Watt's Law.

Example Problems

Since you are pretty familiar with light bulbs and their ratings, let's use them as an example. Figure 13-2 shows the simple circuit of a lamp connected to the 120 V line. If the bulb has a rating of 60 W, then you can figure out, using Watt's Law, how much current it will draw. The same four steps you learned in the last two chapters work just as well for Watt's Law, so you should use them. As the figure shows, the current in the circuit is 500 mA. Now, what would happen to the current if you changed to a 100 W bulb? Think logically first. From your experience of changing light bulbs before, you probably already know that the 100 W bulb will burn brighter than the 60 W bulb. If it burns brighter, does more or less current flow? Now that you have thought about it, work the problem again using the 100 W bulb.

Wattage also tells you how much power is produced, just as it tells how much is used. If you had a perfect generator which did not waste any power, how many watts do you think it would produce if it were turned by a 1 horsepower engine? If the conversion from mechanical energy to electrical energy were perfect (had no losses), then a generator turned by 1 horsepower

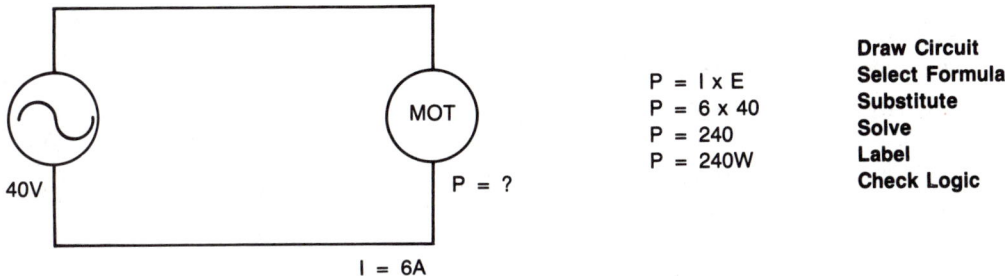

Fig. 13-3 Watt's Law problems are solved using the same steps as we used for Ohm's Law problems. Even the formulas are similar.

Chapter 13 *Electrical Power* 77

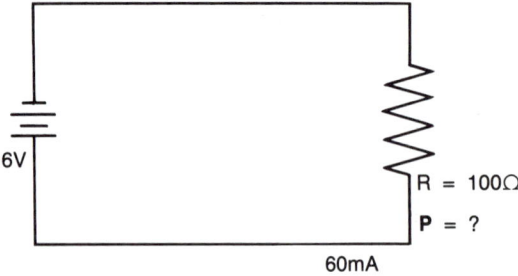

Fig. 13-4 Will a ½ W resistor be able to dissipate the power in this circuit safely? Use Watt's Law to find out. Even though the resistance of the resistor (100 Ω) is given, you do not need this to solve the problem — so ignore it.

would produce 746 W of electrical power. In other words, it would take about a 1 horsepower engine to turn the generator which produced the power to make seven 100 W bulbs burn. (A little power would be left over). Some of the power is lost due to resistance and other losses in the lines and other parts of the circuits.

If a generator produced electrical power at a voltage level of 40 V and you used that power to run a motor which drew 6 A of current, how much power would you be using? Figure 13-3 leads you through a step-by-step solution of this problem. Notice how similar Watt's Law problems are to Ohm's Law problems.

Wattage Rating of Resistors

You know how to read resistance values by means of the color code. Remember, the larger the physical size of a carbon resistor, the higher its wattage rating is. Figure 13-4 shows a circuit with a resistor. By using Watt's Law, you can find out how much power that resistor is **dissipating** (using up). Set up and solve a simple Watt's Law problem to determine whether a regular ½ watt resistor will be big enough here, or if you need to use a larger one.

Typical Levels of Power

Generally, since the voltage in the home is set at 120 V for most devices, the items which draw the most current are the ones that require the most power. Anything which produces heat uses a lot of power. A 100 W light bulb uses 100 W of power. A modern color television uses about the same amount — from 90 to 125 W — but an electric iron may use as many as 1100 W! Electric heaters, cooking ranges, and clothes dryers are all heavy power consumers. The electric power company charges its customers for the electricity they use by units called kilowatthours. Figure 13-5 shows a residential power meter. A **kilowatthour** means that you have used 1000 watts of power for one hour. This means that ironing for about an hour uses a little over 1 kWh. If you were listening to a stereo which only used 100 W of power, you could listen to it for 10 hours with the same amount of power which would be used by ironing for only one hour — 1

Fig. 13-5 When the aluminum disc in a kilowatt-hour meter turns fast, much power is being used.

kWh. Take a look at the rotation speed of the aluminum disc in the power meter at your home. You will notice that it turns much faster when heating devices are used than when they are not used. An electric cooking range can use as much as 12,000 W. Cooking a full meal on one of these ranges can consume power very quickly.

Equipment which has large motors in it tends to use a lot of power too. Air conditioners and refrigerators can be costly to operate.

Small devices which do not produce a lot of heat take very little power. When dealing with transistor radios, calculators, small battery operated toys, and the like, milliwatts (mW) frequently will be used as the unit of measure because they consume so little power.

Watt's Law and Ohm's Law Work Together

Some students ask, "Is it hard to keep straight when to use which law?" The answer is no, if you will approach problems logically and really think about what is known and what you want to find. Some books present several forms of the two laws and some special formulas to use in special cases. This will not be done here because it is better to develop a full understanding of what is happening in a circuit than it is to

To Solve For	If You Know	Use This Formula
E	I & R	$E = I \times R$
	I & P	$E = \dfrac{P}{I}$
	P & R	$E = \sqrt{P \times R}$ Rarely Used
I	E & R	$I = \dfrac{E}{R}$
	E & P	$I = \dfrac{P}{E}$
	P & R	$I = \sqrt{\dfrac{P}{R}}$ Rarely Used
R	E & I	$R = \dfrac{E}{I}$
	E & P	Find I first by $I = \dfrac{P}{E}$ and then use $R = \dfrac{E}{I}$
	I & P	Find E first by $E = \dfrac{P}{I}$ and then use $R = \dfrac{E}{I}$
P	E & I	$P = I \times E$
	E & R	Find I first by $I = \dfrac{E}{R}$ and then use $P = I \times E$
	I & R	Find E first by $E = I \times R$, then use $P = I \times E$

Fig. 13-6 Since almost all devices are rated by voltage with either wattage or current, the two formulas which have only R and P given (printed on gray background) would be very rarely used. Use this page as a reference if you need those formulas.

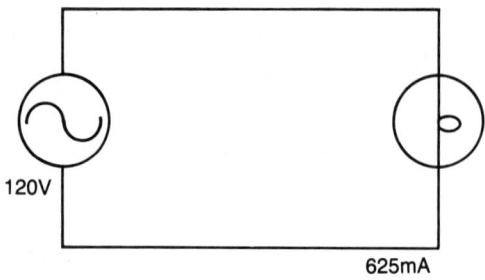

Fig. 13-7 Use Watt's Law to find what wattage bulb is in this circuit? Would a 60 W bulb draw more or less current? How much?

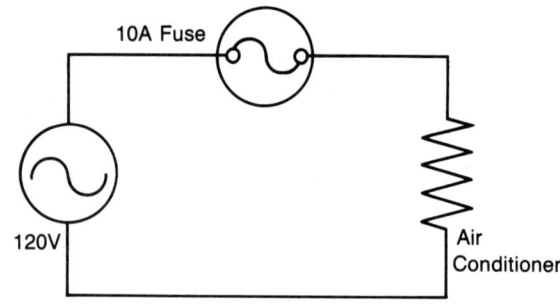

Fig. 13-9 If more than 10 A flows in this circuit, the fuse will burn out to stop the excess current from damaging anything. Use Watt's Law to decide if the fuse will blow or not when the 1440 W air conditioner is connected.

use a shortcut to the answer. After you have become very familiar with what is going on in the circuits, you may wish to look up these special-case formulas.

Basically, as shown in Fig. 13-6, Ohm's Law and Watt's Law use some of the same information, but each uses different parts of it and in different ways. The chart shows how you could solve for any value (E, I, R, or P) when you know any two of the other values. Sometimes, you may have to go through a two-step process, similar to what you did in the last problem in Chapter 12, but this is usually not necessary. Figures 13-7 through 13-9 are some sample problems for you to work. Remember to use the four problem-solving steps. Check your answers in Fig. 13-10.

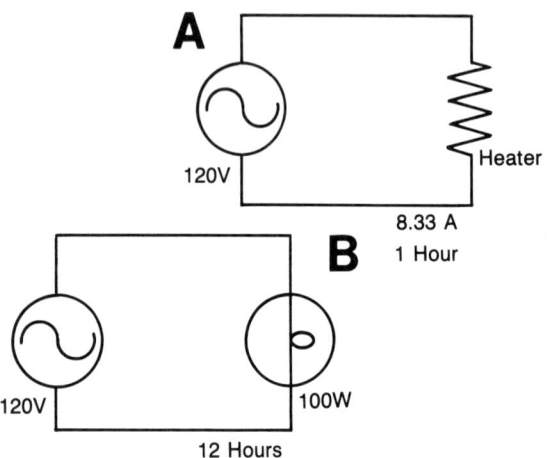

Fig. 13-8 Which will cost more, using the small heater (A) for one hour or the lamp (B) for 12 hours? Give the KWh figures for both.

Problems	Answers	Comments
13-7	P = 75 W I = 500 mA	first part with a 60 W bulb
13-8	A: 1 kWh B: 1.2 kWh B costs more.	Don't forget that it takes 1,000 W to make 1 kWh.
13-9	I = 12 A The fuse will blow.	

Fig. 13-10 Check your answers to the sample problems and then try the ones you missed again.

Summary

Power is the ability to do work such as cause motion (motors), make light, or produce heat. Mechanical power is measured in horsepower units and electrical power is measured in watts. It takes 746 W to equal one hp. Watt's Law ($P = I \times E$) is used to find how much power a circuit uses or how much a generator produces. The electric power company charges us for usage of electrical power by kilowatthour units (kWh). Heating devices are heavy power consumers. Upcoming chapters will deal with circuits which have more than one component or device in them.

Important Terms

horsepower
power
watt
dissipating
kilowatthour

Review Questions

1. If you use an electric drill that has a ¼ horsepower rating, about how many watts of electrical power *should* it consume? Why would it actually use a little more?

2. State Watt's Law in the form of a simple equation and explain what it means in your own words.

3. Which would cost more: Drying your hair for 10 minutes with a 650 W blow dryer or listening to a 50 W stereo for an hour?

4. If you had to find the power in a circuit, and all that you knew was the resistance of the circuit and the voltage of the source, could you find the power? If yes, outline the steps you would use. If no, tell why not.

History: James Watt

James Watt was born in Scotland in 1736. He became an instrument maker at the University of Glasgow in 1757. In 1763, while repairing a Newcomen steam engine, he became so interested in it that he went on to invent a better one. Watt worked with steam engines extensively and received his first patent in 1769. Among his innovations in steam engine design were the condenser, crank movements and planetary gears, double action, and governors — to mention only a few. But, in addition to making the first truly practical steam engines, James Watt also helped us to determine the relationship between work done, energy used, and electrical power. It is for this work that the watt and Watt's Law were named in honor of the man whose inventions made the industrial revolution possible.

Chapter 14

Series Circuits

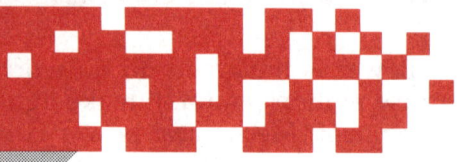

The Series Circuit

Two or more components can be wired into electrical circuits by **series circuits** and **parallel circuits**. Look at the examples in Fig. 14-1. Three lamps are used in these example circuits, but most other devices can be wired in either series or parallel, depending on circuit requirements.

The main difference between series and parallel circuits is that *all* the current must go through *all* the components in a series circuit. The current in parallel circuits divides at each junction. Thus, only part of it goes through each component. If you can trace a line from the power source, through every component in the entire circuit, and back to the power source without lifting your pencil or retracing any lines, the circuit is wired in series. This is exactly what electrons must do as they travel around a series circuit. See Fig. 14-2. If, while tracing the circuit, you find that you must retrace a line or lift your pencil, then you are working with a parallel circuit. Many people think of series circuits as "in-line" or "end to end" circuits because that is the way in which the components are actually wired. Similarly, parallel circuits

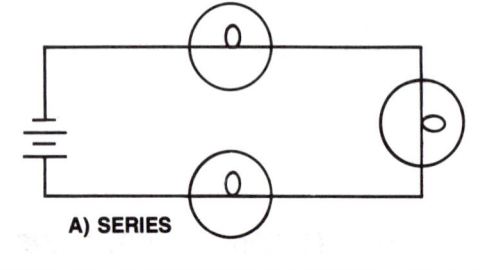

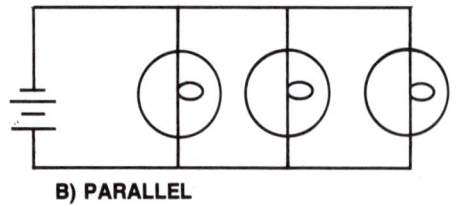

Fig. 14-1 There are two main ways in which three lamps could be wired together: A) The series circuit, which uses less wire, or B) The parallel circuit.

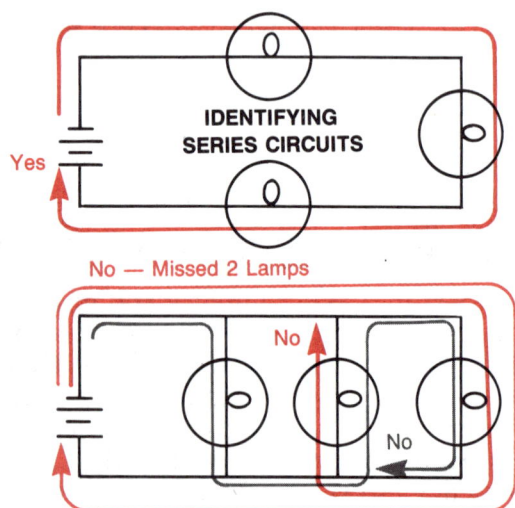

Fig. 14-2 The series circuit can be fully traced with one line which does not cross or skip but the parallel circuit cannot be.

82

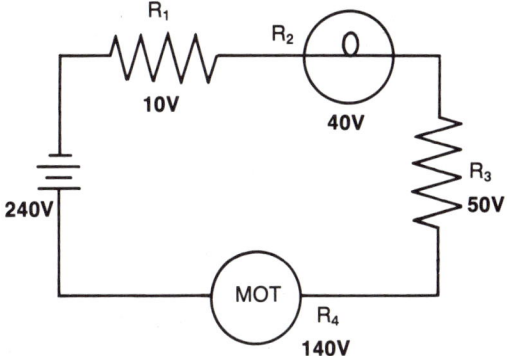

Fig. 14-3 The sum of the voltage drops around a series circuit loop equals the source voltage applied to that loop.

could be thought of as "side by side" circuits. Nearly all of the circuits which have been used in examples in earlier chapters have been simple series circuits.

Voltage Drops

When you studied voltage in Chapter 6, you learned several different names for EMF. One of the names is **drop** or potential drop. The meaning and importance of this term is especially clear when dealing with series circuits.

Figure 14-3 shows a series circuit which has four devices wired in series. Trace the circuit with your finger to prove that it is wired in series.

A voltage is labeled beside each device in the circuit. These labels show how much voltage is "dropped" by each component. Remember, each component in a series circuit will consume some of the voltage. When you add the drops together, your answer will be the same as the source voltage. Try it. You should get 240 V when you add the drops of loads R_1 through R_4 together.

Total Resistance

Just as you can add the drops of the individual loads to get the source voltage, you also can add the resistances of each load to get the **total resistance** of a series circuit. See Fig. 14-4 which has the resistance values of each load labeled.

This circuit has four loads, but some series circuits could have fewer or many more loads. It is best if you can state the relationship just discussed in a way that will work for all series circuits, not just for our one example. The way to do this is with a simple mathematical formula which is:

$$R_T = R_1 + R_2 + \ldots + R_n$$

The R_T means "the total resistance of the circuit." You already know what R_1 and R_2 mean. What about R_n? The "n" stands for the total number in the set. R_n is just a mathematical shorthand way of saying "the last resistance in the circuit." It does not matter how many loads there are in the circuit, this formula will work for any series circuit. The "..." means "and any other resistances there are between R_2 and R_n."

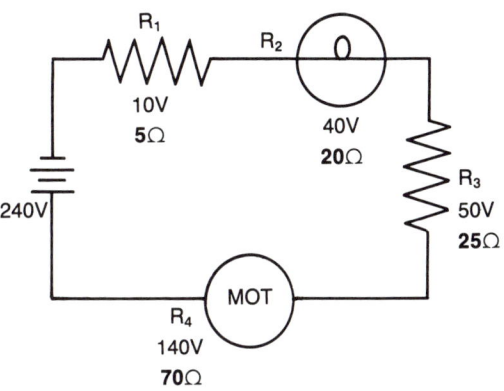

FINDING TOTAL RESISTANCE:
$R_T = R_1 + R_2 \ldots + R_n$
$R_T = 5 + 20 + 25 + 70$
$R_T = 120 \Omega$

Fig. 14-4 The resistances of the individual loads may also be added to find the total resistance.

Now that you understand the shorthand notation used in stating mathematical formulas of this type, you can see that voltage drop could be stated with the simple expression: $E_T = E_{R1} + E_{R2} + \ldots + E_{Rn}$. Use this formula to prove that the sum of the drops in a series resistance circuit equals the source voltage.

Ohm's Law Applications

These two formulas can be used to determine the total voltage and resistance of a series circuit. Once these two values are known, you can substitute them into the Ohm's Law equation for I to find the current of the circuit. For the example circuit, the current may be found by

$$I_T = \frac{E_T}{R_T}$$
$$I_T = \frac{240}{120}$$
$$I_T = 2 \text{ A}$$

Thus, the current which leaves the battery (the total current of the circuit) is 2 A. What would the current be in R_3? There are two ways to find this current. The first is to think logically. Since all the current which leaves the battery must go through all of the loads in a series circuit, the current must be the same in all parts of the circuit. If the electrons in a series circuit are moving at one point in the circuit, then they move at the same rate in all points of the circuit. So, since you know that the current leaving the battery in the example is 2 A, then there must be 2 A flowing through R_3, and, likewise, 2 A flowing through R_1, R_2, and R_4.

The other way to find the current of any particular component in any circuit is to substitute the known values of resistance and voltage for that single component into the Ohm's Law formula for current. Figure 14-5 shows the same circuit again with the currents figured out for all four loads by the Ohm's Law method. Would the currents of R_1, R_2, R_3, and R_4 add up to the total current of the circuit in this series circuit? No! Many students try to do this, but it is a very severe mistake in logic. Actually, if you wished to state the fact that the current is the same in all parts of a series circuit by means of a mathematical expression, you would write:

$$I_T = I_{R_1} = I_{R_2} = \ldots = I_{R_n}$$

Notice that there are = signs between all of the individual currents, not + signs as in the voltage and resistance formulas. If you can find the current of any one part of a series circuit, then you know the current of all parts of that circuit.

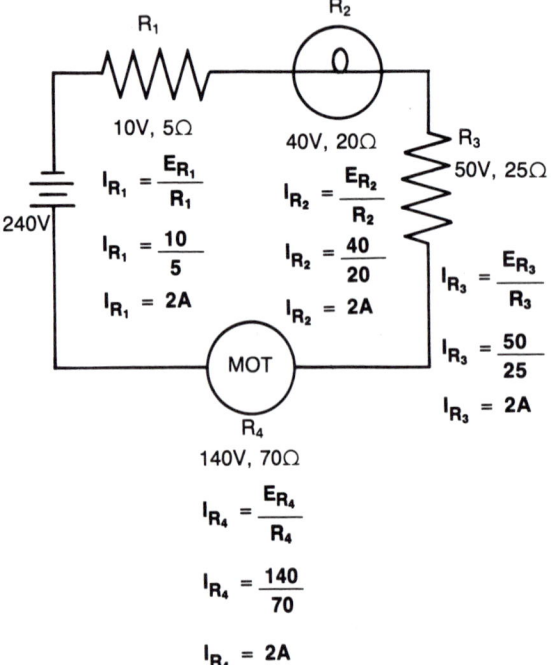

Fig. 14-5 The current could be found for each component separately. Notice that the current is the same in all points of a series circuit, so finding I_{R_1} tells us all that we need to know.

Statements for Series Circuits

Figure 14-6 shows the three mathematical expressions given in this chapter.

Series circuits have some advantages and disadvantages as compared to other types of circuits. What would happen if one part of our example circuit were to burn out (become open)? Since there would be an open in the only path electrons could take, electrons would stop moving in all parts of the circuit. Some Christmas tree lights are wired in series and one burned out bulb will prevent the whole string from lighting because of this problem. Another disadvantage is that each load in a series circuit will get a different amount of voltage, depending on its own resistance. Putting additional loads in the circuit without also increasing the voltage of the source could prevent some devices from getting enough voltage for proper operation. The main advantages of series circuits are: 1) simplicity of the design, and 2) they use less wire than parallel circuits do.

Finding Unknowns

Sometimes you may encounter a series circuit in which some values are known, but others are not known. The unknown values in such circuits can be found if you will apply logical reasoning and what you now know about series circuits along with Ohm's and Watt's Laws. Figure 14-7 is an example of a problem in which an unknown resistance must be found by using both formulas and logic. Since the old resistor burned up so badly that you cannot read its markings, you need to figure out what resistance and wattage rating the new one must have. The first step is to find out what voltage the resistor must drop. The voltage drop needed is 40 V. Now, all you need to know to find the unknown resistance value is the current which will flow through the resistor. Step 2 shows how you determine the current. Finally, in step 3, you use Ohm's Law to find the resistance value — using the values of E and I which you found in steps 1 and 2. Step 4 shows how to find the wattage rating for the resistor.

Summary Statements for Series Circuits

Voltage:	$E_T = E_{R_1} + E_{R_2} + \ldots + E_{R_n}$	The sum of the voltage drops in a series circuit equals the source voltage.
Resistance:	$R_T = R_1 + R_2 + \ldots + R_n$	The total resistance of a series circuit equals the sum of the individual resistances.
Current:	$I_T = I_{R_1} = I_{R_2} = \ldots = I_{R_n}$	The current is the same in all parts of a series circuit.

Fig. 14-6 These summary statements apply to all series resistance circuits. Use them for references.

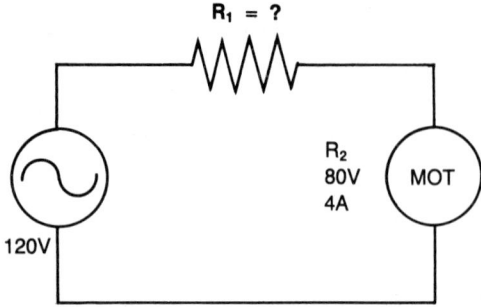

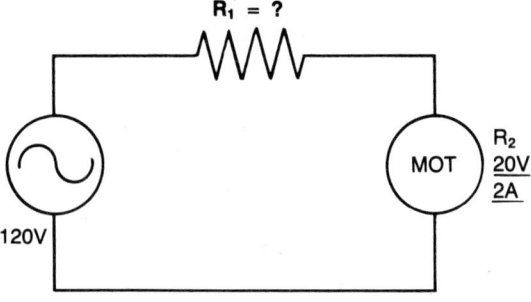

1) Find Voltage of R_1:

$E_T = E_{R_1} + E_{R_2} + \ldots + E_{R_n}$

$120 = E_{R_1} + 80$

$E_{R_1} = 40V$

2) Find Current of R_1:

$I_T = I_{R_1} = I_{R_2} = \ldots = I_{R_n}$

$I_T = I_{R_2} = 4\ A$

$I_{R_1} = 4A$ (and $I_T = 4A$ also)

3) Use I_{R_1} and E_{R_1} to find resistance of R_1:

$R_1 = \dfrac{E_{R_1}}{I_{R_1}}$

$R_1 = \dfrac{40}{4}$

$\boxed{R_1 = 10\ \Omega}$

4) Use I_{R_1} and E_{R_1} again to find the wattage rating:

$P_{R_1} = I_{R_1} \times E_{R_1}$

$P_{R_1} = 4 \times 40$

$\boxed{P_{R_1} = 160W}$

Fig. 14-7 By approaching the problem logically, you can find that you need a 10 Ω, 160 W resistor to replace the burned up R_1.

Fig. 14-8 This is the same problem as used in Fig. 14-7, but with different numbers. Solve to find the resistance and wattage ratings of R_1. Notice that the position of a load in the circuit does not affect the amount of voltage that it drops.

Fig. 14-9 You have found an old 5-tube radio that you would like a repair. The tube filaments are wired in series. One tube is missing and the schematic is partly destroyed. Since the first digit(s) of a tube identification number tell its filament voltage, what two numerals will complete the identification number of V_2? The 35W4 (V_1) drops 35 volts, the 12BE6 (V_4) drops 12 V, etc. What voltage must V_2 drop?

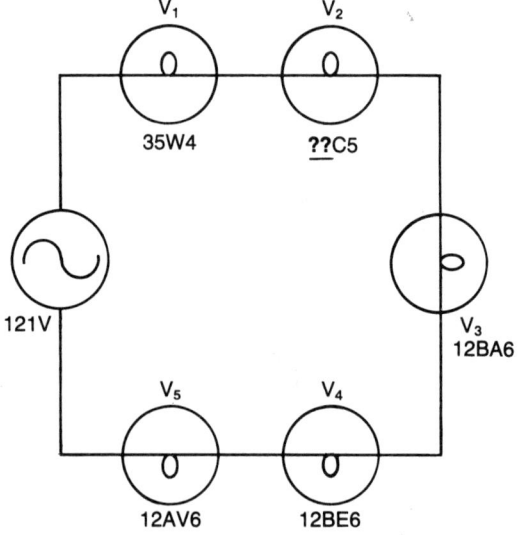

Note: "V_1" means Vacuum tube number one.

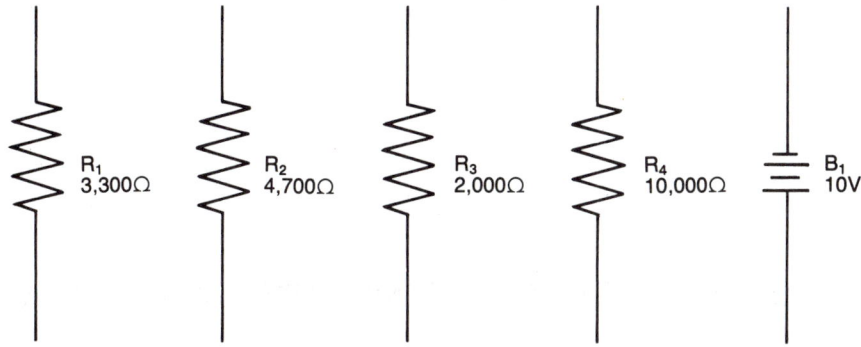

Fig. 14-10 How could these four resistors be combined to give a total resistance of 20 KΩ? Draw the circuit, using the 10 V source, and then figure out the circuit current and the voltage drop of R_3.

Additional problems for you to work are found in Fig. 14-8 through 14-10. Some of these problems are partially worked for you. You may check your answers in Fig. 14-11.

Summary

Series circuits are the simplest type of circuits, and it is very easy to find unknown values in them. The current is the same in all parts of a series circuit. If one component becomes open, all current in a series circuit stops flowing. The voltage drops of all of the loads in a series circuit may be added to find the source voltage. Likewise, the total resistance of a series circuit is equal to the sum of the resistances of all of the loads. Chapter 15 will deal with parallel circuits and you will see how different they are from series circuits.

Problems	Answers	Comments
14-8	R = 50 Ω P = 200 W	Use the same steps we used in 14-7
14-9	E_{V_2} = 50 V Tube type is 50C5	Use formula $E_T = E_{R_1} + E_{R_2} + \ldots + E_{R_n}$
14-10	I = .5 mA or 500 μA E_{R_3} = 1 V	Remember to convert to basic units for all formulas.

Fig. 14-11 Check your answers to the problems with this table.

Important Terms

series circuit
parallel circuit
drop
total resistance
E_{R_n}

Review Questions

1. How can you tell whether a circuit is wired in parallel or series?
2. Explain the relationship between E, I, and R in series circuits. How are total E, I, and R found?
3. What are the main advantages and disadvantages of series circuits?

Careers:
Hospital Technician

The professionals who repair and care for the technical equipment used in the health care field may be employed in hospitals or by the companies that sell and service the equipment. These people must be skilled, careful workers. If the television repairer makes a sloppy connection or a poor solder joint, the customer may be inconvenienced by missing a favorite program. But if the same defect appears in medical equipment it could cause serious injury, illness, or even death of a patient. Since dependability is so important in hospital equipment, the devices are produced using only the highest quality parts and materials. The equipment is very expensive because of its high quality and limited demand. Technicians who work on medical equipment are well paid. They need good manual skills, broad knowledge in general mechanics and shop practice, and specialized knowledge in both electronics and the use of the equipment they service.

Chapter 15
Parallel Circuits

The Parallel Circuit

Parallel circuits are different from series circuits in almost every way. Figure 15-1 shows a simple parallel circuit that uses the same four loads that were used in the series circuit examples of Chapter 14. Remember, one important feature of series circuits is that the current is the same throughout the circuit and all of the current must go through the whole circuit. This is not true in a parallel circuit. As the figure shows, only some of the current goes through R_1. Likewise, only some of the current goes through each load. All of the current which leaves the battery and goes to junction A does get back together at junction B for its return trip to the battery. The amount of current in each load is different, not the same as in a series circuit. In a **parallel circuit** there is more than one path for the current to take. Each of these separate paths is called a **branch**. In the example circuit, there are four branches. Some, but not all, of the current goes through each branch.

The amount of current which goes through a particular branch will depend on how much resistance that branch has. Branches that have a lot of resistance do not let as much current flow through them as do branches with less resistance. The motor in Fig. 15-1 (a load with much resistance) does not permit as much current to flow as R_1.

The position of the load does not make any difference. If R_4 had been first instead of last in position, it would still have passed the least current.

Voltage in Parallel Circuits

In series circuits, each load drops some of the voltage and the voltage of each load differs (unless all loads have the same resistance). Figure 15-2 shows that each

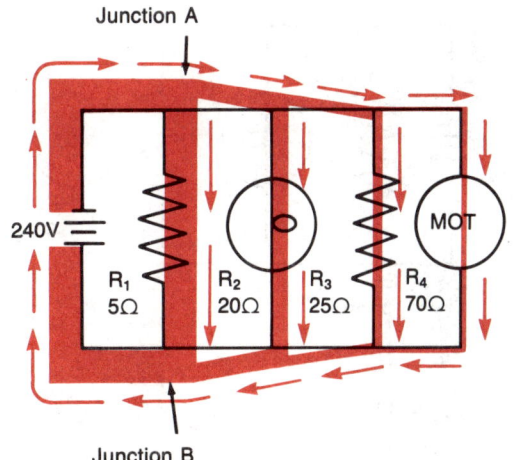

Fig. 15-1 Current divides at each junction of a parallel circuit and some, not all, of the current goes through each branch. The branch with the least resistance (R_1 in this case) carries the most current.

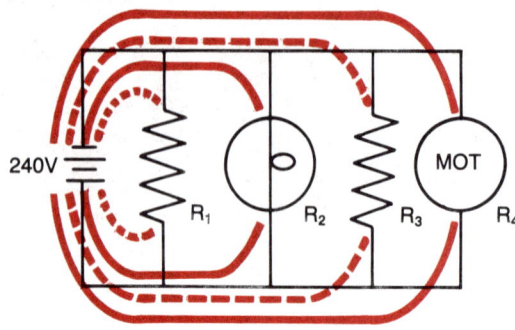

Fig. 15-2 The path of current for each branch is really separate. As far as R_2 is concerned, branches R_1, R_3, and R_4 do not even exist. Each branch is connected directly to the 240 V source.

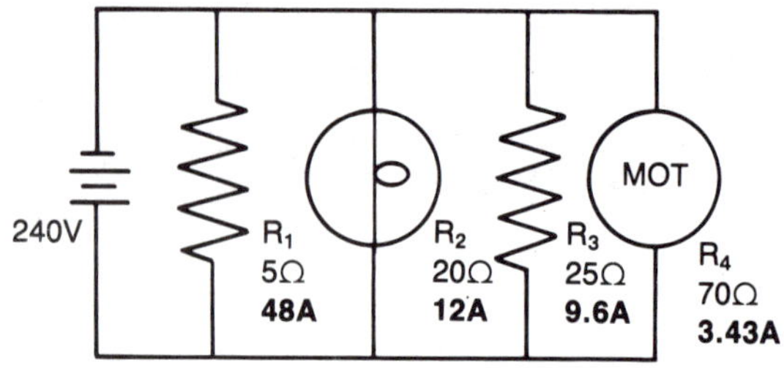

Fig. 15-3 The actual amount of current which each branch in our example circuit carries is labeled in bold. Find the total current of the circuit.

branch of a parallel circuit is actually connected to the battery directly. Therefore each branch gets its supply of voltage directly from the battery without anything in the path to block or drop voltage. Since branch R_1 is connected directly to the battery, it is easy to see that its voltage drop will be 240 V. What about branch R_3? To what voltage is it connected? It is easy to see that there is a wire going directly to the negative battery terminal from one end of R_3 and another one going directly to the positive terminal from the other end. The full 240 V is connected directly to branch R_3, so the voltage drop of R_3 is 240 V. Likewise, the voltages of R_2 and R_4 are also 240 V each. The voltage of each branch in a parallel circuit is equal to the source voltage:

$$E_T = E_{R_1} = E_{R_2} = \ldots = E_{R_n}$$

The total current of a parallel circuit equals the sum of the currents in the individual branches. Figure 15-3 makes this point clear. You could say that for parallel circuits:

$$I_T = I_{R_1} + I_{R_2} + \ldots + I_{R_n}$$

Use this formula to find the total current of the example parallel circuit.

$$I_{R_1} = \frac{E_{R_1}}{R_1} \qquad I_{R_2} = ? \qquad I_{R_3} = \frac{E_{R_3}}{R_3} \qquad I_{R_4} = \frac{E_{R_4}}{R_4}$$

$$I_{R_1} = \frac{240}{5} \qquad I_{R_3} = \frac{240}{25} \qquad I_{R_4} = \frac{240}{70}$$

$$I_{R_1} = 48A \qquad I_{R_1} = 9.6A \qquad I_{R_4} = 3.43A$$

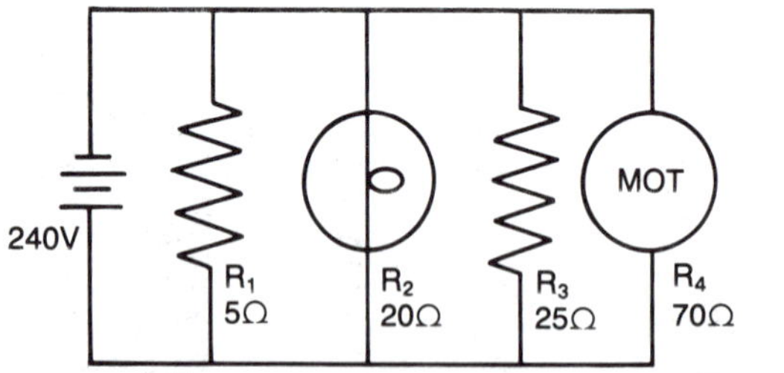

Fig. 15-4 Ohm's Law may be used to find the current of any single branch. Remember that the voltage is the same in all branches. Find the current of R_2 on a piece of paper.

Current in Each Branch

Now that you have the voltage of each branch and the resistance of each branch, you can use these values (and Ohm's Law) to find the current of each branch of the example parallel circuit. This is done in Fig. 15-4 for all branches except R_2. Use a piece of scratch paper to find the current of branch R_2 and then find the total current of the whole circuit by using the formula stated earlier in this chapter.

Total Resistance

When you wish to find the total resistance of series circuits, you just add the resistances of all the loads. In parallel circuits, however, this method will not work. To prove this point, add the resistances of the four branches of the example circuit. You should get 120 ohms. This cannot possibly be the total resistance of the parallel circuit because you can show by Ohm's Law that the total resistance is about 3.28 ohms.

$$R_T = \frac{E_T}{I_T}$$
$$R_T = \frac{240}{73}$$
$$R_T = 3.28 \; \Omega$$

Notice that this number is less than the value of the lowest value resistance in the whole circuit. This is the correct answer. What could account for this unexpected property? Look logically at the circuit for a moment. Remember, the reason the current divides in parallel circuits is because there are many different paths it can take. Each path (branch) will draw a certain amount of current from the source. The amount of current each branch draws will depend on the resistance of that branch. Since total resistance (R_T) of the circuit has to take into consideration how much total current (I_T) is permitted to flow from the total supply (E_T), the total resistance will always be lower than the value of the smallest resistor in a parallel circuit.

Adding a resistor in series circuits increases the total resistance and limits the current flow. Adding another resistance branch to the parallel circuit just provides another path for the current to flow. In other words, the added resistor draws some *more* current without having any effect on the current which the other two branches are already drawing. Thus, the total resistance has been decreased by adding another path.

How can you mathematically add resistances in parallel circuits together and get an answer that is lower than any of the numbers you are combining? There is a formula which allows you to do this. It uses reciprocals. When 1 is divided by any number (that is 1 is placed over top of that number), you get the **reciprocal** of that number. For example, the reciprocal of 4 is $\frac{1}{4}$. The formula for finding the total resistance of a parallel circuit is:

$$R_T = 1 \div \left(\frac{1}{R_1} + \frac{1}{R_2} + \ldots + \frac{1}{R_n} \right)$$

Notice that this formula is very similar to the one used for total resistance of series circuits except that it uses reciprocals and division into 1 as a final step. Figure 15-5 shows how to solve the example problem by using this formula. The first step (A) is to write the formula (as always). Next, substitute the known values into the formula (B). Then, compute the mathematical answer by working on everything inside the parentheses first. Before calculators were available, many people would find a common denominator and add the fractions. With a calculator it is much easier to change all the fractions to **decimals** (C) and then add the decimals (D). You do this by dividing the top number (1 in all cases in this formula) by the bottom number of each reciprocal. If you are familiar with the proper use of the **memory** function of your calculator, you may wish to add the decimals in the memory. However it will help you follow the problem if you will write the decimals on your paper anyway. After the decimals are added, the answer obtained is

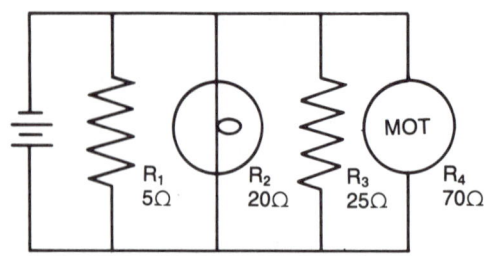

$R_T = 1 \div (\frac{1}{R_1} + \frac{1}{R_2} + \ldots + \frac{1}{R_n})$ A) State formula

$R_T = 1 \div (\frac{1}{5} + \frac{1}{20} + \frac{1}{25} + \frac{1}{70})$ B) Substitute

$R_T = 1 \div (.2 + .05 + .04 + .014)$ C) Find decimal-forms

$R_T = 1 \div (.304)$ D) Add decimals

$R_T = 3.29 \Omega$ E) Divide into 1 and label answer

Fig. 15-5 Notice that solving for R_T by this formula instead of by the Ohm's Law method results in a slight rounding difference in the hundredth's place, but we will still have the same result ($R_T = 3.3\ \Omega$) when we round to the tenth's place.

divided into 1 (E) to complete the operation. Last, the answer is labeled and looked at logically. Logic should tell you that the final answer (total resistance) should be smaller than any of the individual resistances in the circuit. Use the formula to solve for the total resistance of the circuit shown in Fig. 15-5 and then check to see if your answer is the same as you would obtain if you solved for R_T by using the Ohm's Law method:

$$R_T = \frac{E_T}{I_T}$$

Statements for Parallel Circuits

A set of summary statements for parallel circuits appears in Fig. 15-6. You may use these for quick reference and review. Take a moment now to compare these statements with those found in Fig. 14-6 to see just how different parallel and series circuits are.

The reason parallel circuits are so common is that all devices wired in parallel receive the same voltage. It takes a little

	Summary Statements for Parallel Circuits	
Voltage:	$E_T = E_{R_1} = E_{R_2} = \ldots = E_{R_n}$	The voltage is the same in each branch of a parallel circuit, and the voltage of each branch equals the source voltage applied.
Current:	$I_T = I_{R_1} + I_{R_2} + \ldots + I_{R_n}$	The total current of a parallel circuit is equal to the sum of the currents of the individual branches.
Resistance:	$R_T = 1 \div \left(\frac{1}{R_1} + \frac{1}{R_2} + \ldots + \frac{1}{R_n}\right)$	The total resistance of a parallel circuit is always less than the resistance of the branch with the least resistance.

Fig. 15-6 Note that parallel circuits are very different from series circuits. See Fig. 14-6.

lights and appliances receive the 120 V they require. This is true as long as you do not put so many loads in parallel that you try to draw more current than the source can deliver. **Overloading**, as this is called, will cause the voltage to go down, cause damage to the source, or cause a fuse or circuit breaker to blow.

Finding Unknowns

Just as with series circuits, you can use logic and the formulas to find unknown values in parallel circuits. The problems in Fig. 15-7 through 15-9 are for you to work. Some of these problems have some helpful hints included with them. The answers and some comments are provided in Fig. 15-10.

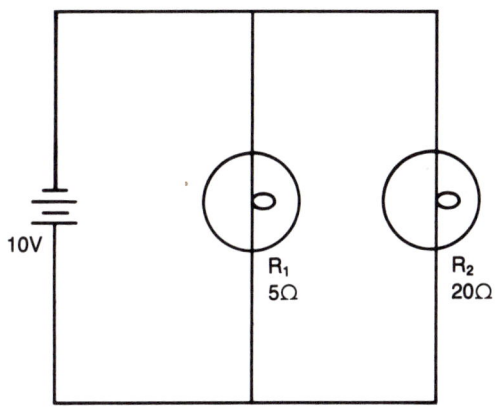

Fig. 15-7 What is the voltage of R_1? What is the voltage of R_2? Find the currents of R_1 and R_2.

more wire to connect things in parallel, but imagine the problems you would have if you had to increase the voltage of the supply to your home every time you turned on a light. Parallel wiring ensures that all

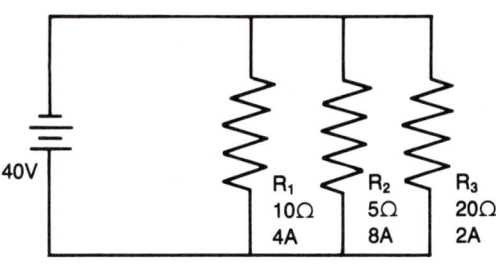

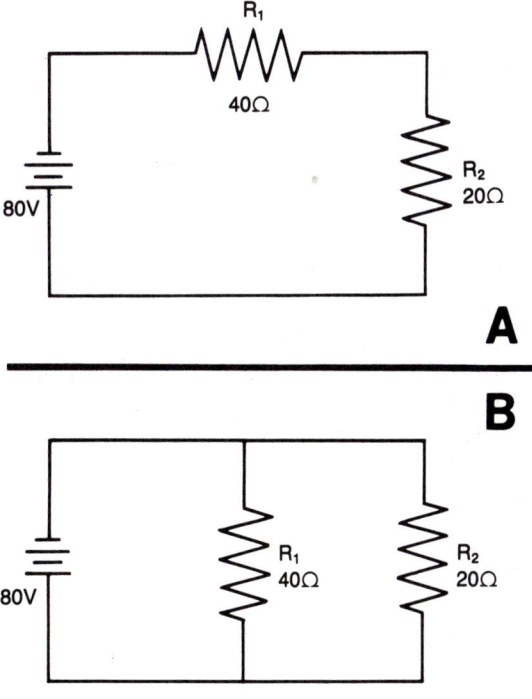

Fig. 15-8 Use the formula $R_T = 1 \div \left(\frac{1}{R_1} + \frac{1}{R_2} + \ldots + \frac{1}{R_n}\right)$ to find the total resistance of this circuit. Then use the value found to solve for the total current by the formula: $I_T = \frac{E_T}{R_T}$. Last, check your answer by solving for the total current by this formula: $I_T = I_{R_1} + I_{R_2} + \ldots + I_{R_n}$.

Fig. 15-9 Find the total resistances and total currents of both of these circuits. Both circuits use the same components and sources. Compare them.

Problems	Answers	Comments
15-7	$E_{R_1} = 10$ V $E_{R_2} = 10$ V $I_{R_1} = 2$ A $I_{R_2} = .5$ A or 500 mA	Voltage is the same in each branch. $I_{R_1} = \dfrac{E_{R_1}}{R_1}$
15-8	$R_T = 2.9\ \Omega$ $I_T = 13.8$ A $I_T = 14$ A	There is a small rounding error, but the answers agree.
15-9	A: $R_T = 60\ \Omega$ $I_T = 1.3$ A	Wiring in series increases R_T and lowers current flow.
	B: $R_T = 13.3\ \Omega$ $I_T = 6.0$ A	Wiring in parallel decreases R_T and allows more current to flow.

Fig. 15-10 Check your answers to the sample problems here.

Summary

Parallel circuits are especially useful because all branches of parallel circuits receive the same voltage. The current divides in parallel circuits so that each branch gets only the amount of current that its resistance will permit to flow. The current is often different in each branch of a parallel circuit. The total resistance of a parallel circuit is always lower than the resistance of any branch in the circuit because each additional branch gives current another place to flow. That is, each branch draws its own current. Ohm's and Watt's Laws are still useful with parallel circuits. In other regards, however, parallel circuits are almost exactly the opposite of series circuits. The next chapter discusses what happens when there are both series and parallel sections in the same circuit.

Important Terms

parallel circuit
branch

reciprocal
decimal

memory
overloading

Review Questions

1. Explain three ways in which parallel circuits are different from series circuits.

2. Why will adding another resistor to a series circuit decrease its current flow, but adding another resistance branch to a parallel circuit will *increase* the total current flow?

3. Why are houses wired in parallel instead of series? Do you think automobiles would be wired in series or parallel? Why?

4. If one bulb burns out in a circuit made up of four bulbs wired in parallel, will the other bulbs stop burning too? Explain your answer. What if the bulbs were wired in series?

Project: TV Bugger

Parts List:

1	10 K ohm resistor	R_1
1	4.7 K ohm resistor	R_2
1	1 K ohm resistor	R_3
1	Variable tuning capacitor	C_1
1	.001 microfarad capacitor	C_2
1	10 picofarad capacitor	C_3
1	spst switch (optional)	S_1
1	NPN transistor, medium power	Q_1
1	9 V battery with connector	B_1
1	coil: 9 turns of stiff wire, ½ inch length, 3/16 inch dia., with center tap.	L_1
1	small printed circuit board	

Notes:

Many of these parts may be salvaged from a discarded transistor radio, and the circuit can be built to fit into the case so that its identity is not apparent. The low power output of this circuit is intentional to avoid breaking FCC regulations. It should only be used in your own home to interfere with your own television set. Misuse or alterations to the circuit could get you into serious trouble.

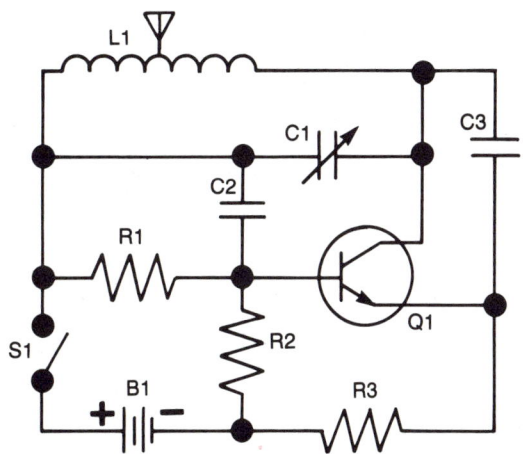

Fig. 15-11 The TV Bugger is a simple project to make and fun to use.

Chapter 16

Series-Parallel Circuits

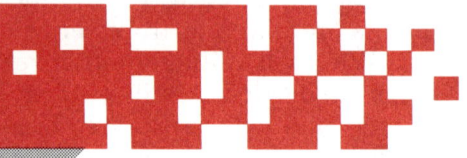

The Series-Parallel Circuit

When some of a circuit's components are wired parallel with each other, but others are wired in series, the circuit is a series-parallel type. Figure 16-1 shows a simple **series-parallel circuit**. Resistors R_2 and R_3 are obviously wired in parallel with each other, but they are not in parallel with R_1. Think of the combination of R_2 and R_3 as one resistance wired in series with R_1, then you can see what is happening in the circuit. R_2 and R_3 can be thought of as one resistance of value $R_{2\&3}$. This resistance ($R_{2\&3}$) is wired in series with R_1. What formulas and rules apply to this circuit? They all do!

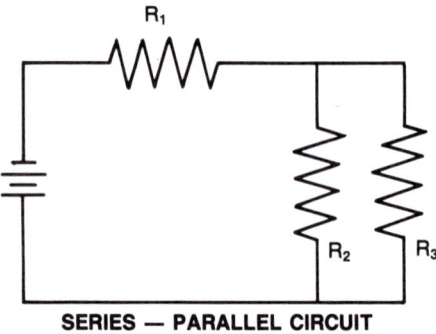

Fig. 16-1 A series-parallel circuit has some components wired in series and others in parallel.

Resistance in Series-Parallel Circuits

Let's think about total resistance first. The total resistance of the series circuit made up of R_1 in series with $R_{2\&3}$ is found by adding the resistance of R_1 to the resistance of $R_{2\&3}$. To find the resistance of $R_{2\&3}$ use the formula for R_T in parallel circuits with the real values of R_2 and R_3.

Figure 16-2 shows another series-parallel circuit with the values of components labeled. First, ask yourself which components are in series and which ones are in parallel? It looks as if R_1 is in series with the combination of everything else. You also should be able to see that the motor (R_4)

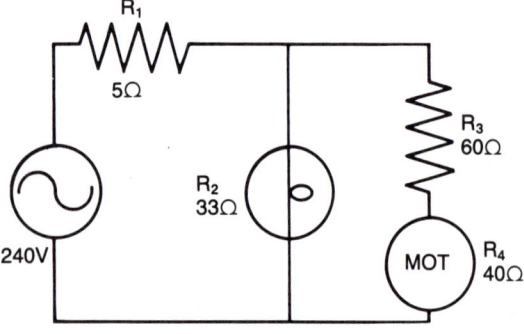

Fig. 16-2 This series-parallel circuit has four components with their values labeled.

96

and its resistor (R_3) are in series with each other. The combination of R_3 and R_4 makes up a **branch circuit** which is in parallel with the lamp (R_2). Figure 16-3 shows how each section can be thought of as a small part of the whole circuit. Start as far from the voltage source as possible.

To find the actual total resistance of this series-parallel circuit, use the proper equations on each section. First, find $R_{3\&4}$ by adding R_3 and R_4. Then, use this value in the parallel resistance formula to find the **combined resistance** of R_2, with $R_{3\&4}$ (that is $R_{2,3\&4}$). Finally, adding this value to the value of R_1 will result in the total resistance of the whole circuit. Notice that you have done no new mathematical operations, and there are no new formulas involved in this process. You merely use the formulas and rules in logical sequence that you already know.

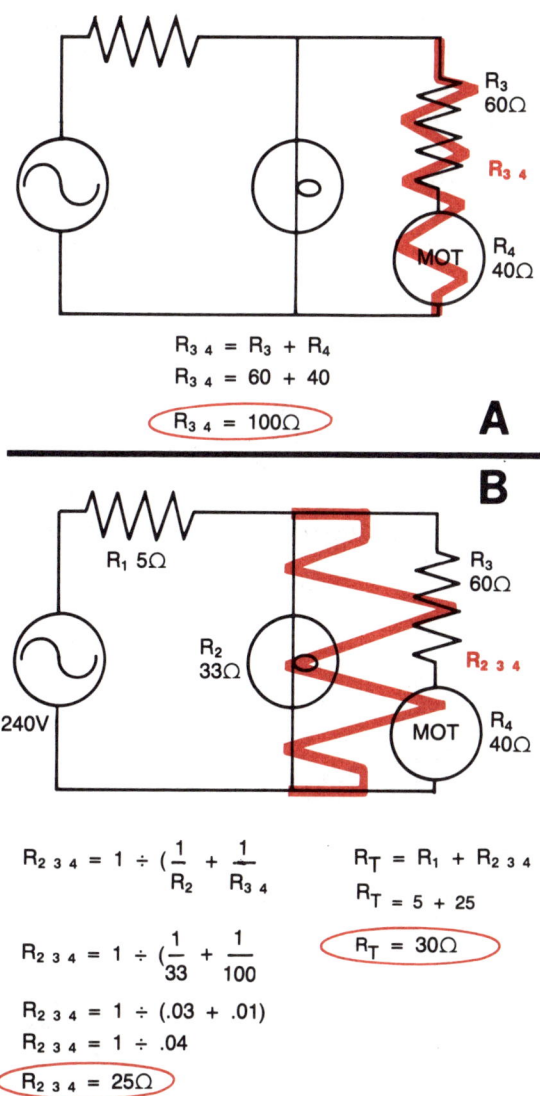

Fig. 16-3 A) R_3 and R_4 form a series combination. B) This series combination is wired in parallel with R_2 to make a bigger combination called $R_{2,3,\&4}$. Last, we add R_1 to $R_{2,3,\&4}$ to find R_T for the whole circuit.

Total Current

You can easily find the total current of any circuit, if you know the source voltage and the total resistance. Since you have just found the total resistance of the example circuit, and the source voltage was already labeled on the drawing, you can use Ohm's Law to solve for total circuit current (I_T).

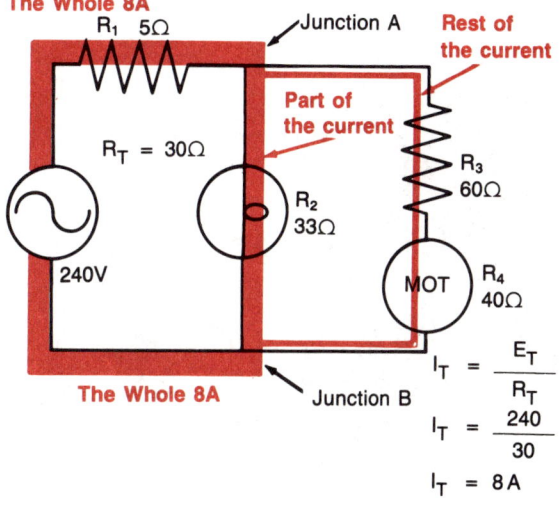

Fig. 16-4 The current divides in the parallel section of the circuit, but all of it must go through R_1.

The total current will be the value of current which leaves the source. All of this current will flow through R_1, but then it will divide at junction A and only part of it will go through R_2. The rest takes the path through R_3 and R_4. All of the current will get back together at junction B for the return to the source of EMF. See Fig. 16-4. The amount of current that flows through each branch will depend on the resistance of that branch. The branch with the least resistance carries the most current. You cannot find the exact value of current for each branch yet, because you do not know what voltage exists in them. Remember that you need two values to find a third one. In this case, to find I_{R_2} (the current of R_2) you would need to know the resistance of R_2 and the voltage of R_2. You know the resistance, but you have not yet found the voltage.

Voltages

What do you suppose will happen to the voltage in the example circuit? Will it all be dropped by R_1? Of course, it will not be. Some of the voltage will be dropped by R_1, and the rest will be used by the parallel section of the circuit. You can easily find how much of the voltage is dropped by R_1 if you know the current through R_1 and its resistance. You do know the resistance of R_1, and you just finished finding out how much current flows through it. Since it is in series with the rest of the circuit, the total circuit current flows through it. To find the voltage drop of R_1 use this value, I_T, as the current of R_1 in the formula:

$$E_{R_1} = I_{R_1} \times R_1$$

Once this value is known, you can easily see that the rest of the voltage from the source acts as the whole supply for the parallel section of the circuit ($R_{2,3\&4}$). Figure 16-5 shows that as far as the parallel section is concerned, the series part of the circuit does not even exist. The parallel portion acts as if it is all alone and connected to a source of 200 V. Thus, the voltage of R_2 is 200 V. The voltages of R_3 and R_4 will be values less than 200 V. The actual voltages of R_3 and R_4 depend on the resistance of each resistor.

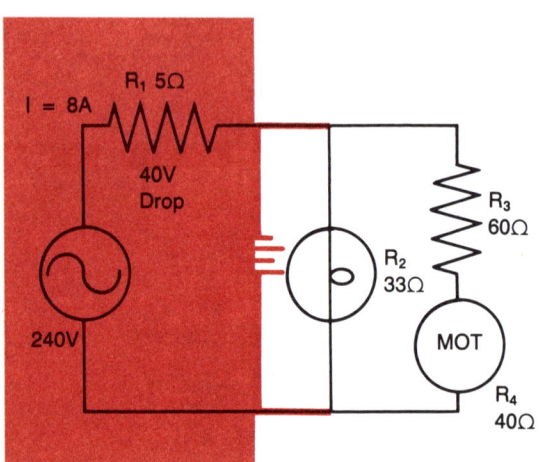

Fig. 16-5 The parallel branch only receives 200 V from the source because R_1 drops the other 40 V.

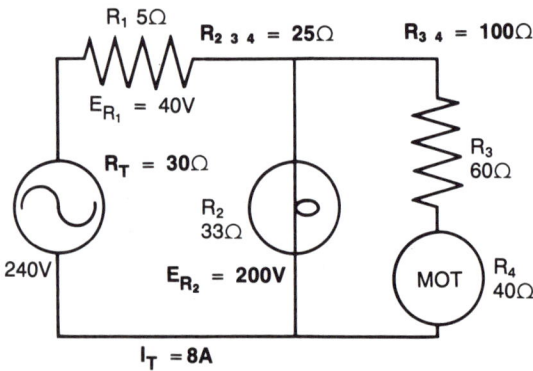

Fig. 16-6 Always label each new value as you find it. This will avoid confusion and make further work easier. The values we have found thus far are in bold.

Finding Unknown Values

You can find any unknown values in series-parallel circuits, if you use the formulas logically. Figure 16-6 shows the example circuit with all of the values labeled which you have found so far. If you wanted to find the current of R_2, you could now do this easily with the information which appears on the drawing. This points out an important secret to solving complex problems — as soon as a value is found, label it on your original schematic drawing. This simple, but crucial, step will keep you from becoming confused and using the wrong value in a formula. Now, solve for the current of R_2. Use the formula $I_{R_2} = \frac{E_{R_2}}{R_2}$.

You should get the answer 6 A. Next, use your own logic to find the voltage drop of R_4. The correct answer is 80 V. See if you can find this value.

Statements for Series — Parallel Circuits

The principles which apply to series-parallel circuits, in fact all circuits, can be summarized in two simple statements:
1. The sum of the voltage drops around a **series loop** equals the source voltage applied to that loop.
2. The current which leaves a junction is equal to the current which enters that junction.

These statements are known as Kirchoff's Laws. The first one is **Kirchoff's Voltage Law**. This law is most helpful when dealing with the series sections of a complex circuit, but it also applies to each individual branch (loop) of parallel circuits.

The second rule, **Kirchoff's Current Law**, is most useful with parallel portions of a circuit. Keep clear about what is happening in complex circuits. Look at Fig. 16-7 for a

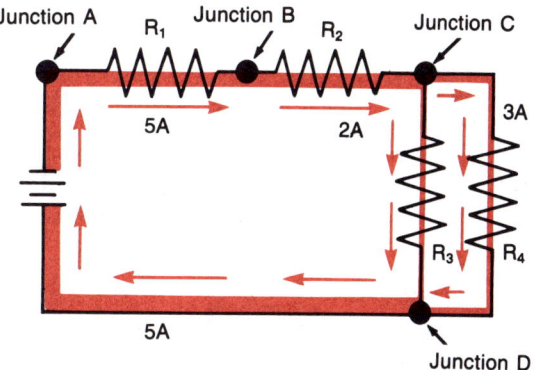

Fig. 16-7 The total current which leaves each junction is the same as the total amount of current which enters it.

moment. Whenever two or more components or devices are joined together in a circuit, the point at which they are connected together is called a **junction**. Thus, the point at which the battery is connected to R_1 could be called junction A.

Think for a moment about the current which leaves junction A to go around the rest of the circuit. Can there be any more or less current leaving junction A than the amount of current which comes into it from the battery? The answer is so obvious that it is almost silly, because there is nowhere else for the current to go. It goes in one side and out the other. The junction does not use or "eat" any of the current. Now consider junction B in the same way. Here again, it is obvious that what goes in must come out, and there is only one way it can go. Junction C is different though. Notice that the same amount of current enters it that had entered the other two junctions. However, when you look at the exit side of junction C, there are no current lines which are as large (have as much flow) as the entry side. To show that the Current Law works here, you must add the currents of the two exit lines together. When this is done, you will see that the current which leaves junction C is equal to the current which entered junction C. It just went in two different directions when it left. What about junction D? Two currents enter it. The sum of these two currents is 5A, and that is what must leave this junction to go back to the battery. Look back at Fig. 16-4 and apply what you now know about Kirchoff's Voltage and Current Laws to that circuit.

Sample Problems

Some sample problems for you to solve are provided in Fig. 16-8 and 16-9. Try to do

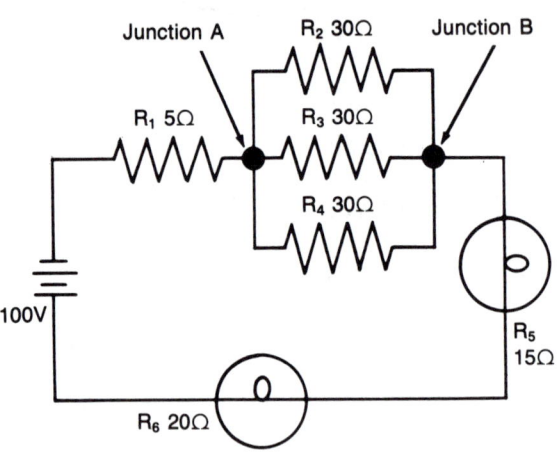

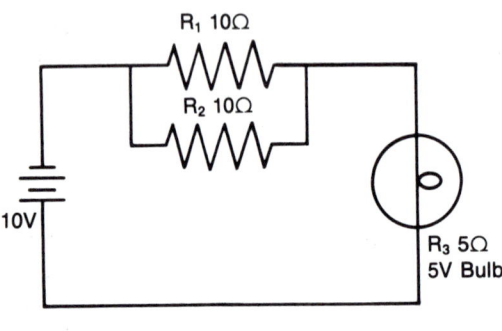

Fig. 16-9 Find R_T and I_T in this circuit. How much current enters junction B? How much current leaves it? What law applies here?

Fig. 16-8 A student was working an experiment which called for a 5 V, 5 Ω bulb to be powered by a 10 V battery. She decided that she needed to put 5 Ω of resistance in series with the bulb, but there were no 5 Ω resistors in her lab kit. Will the circuit she wired work? Find $R_{1\&2}$ to decide. Then find I_T, the current through the bulb.

Problems	Answers	Comments
16-8	$R_{1\&2} = 5\ \Omega$, yes	Use parallel resistance formula. Whenever two resistors *of the same value* are in parallel, their total resistance is half of the value of one of them.
	$I_T = 1\ A$	
16-9	$R_T = 50\ \Omega$	Find $R_{2,3,4}$ first, then just add all the others.
	$I_T = 2\ A$	
	2 A enters and 2 A leaves	Can be found by math or by logic.
	Kirchoff's Current Law	

Fig. 16-10 Answers to problems in Figs. 16-8 and 16-9. Correct those you did not solve and try them again.

each one, and then look logically at how Kirchoff's Voltage and Current Laws apply in each case. The answers and some comments about the problems appear in Fig. 16-10.

There are no new mathematical formulas needed to solve problems concerning series-parallel circuits. There are two laws which are helpful when working with series-parallel circuits. Kirchoff's Laws, as they are known, really say things which you have learned in Chapters 14 and 15. They just summarize those facts in a concise way.

Summary

Series-parallel circuits are nothing but combinations of series and parallel circuits.

Important Terms

series-parallel circuit
branch circuit
combined resistance
series loop
Kirchoff's Voltage Law
Kirchoff's Current Law
junction

Review Questions

1. Draw a simple series-parallel circuit with three resistances in it, and then use it to explain Kirchoff's Voltage and Current Laws in your own words. (If you really understand these Laws, you will not need to use any mathematical figures or formulas.)

2. Refer to Fig. 16-1 and explain how the total resistance of this circuit would be found if the values of the individual resistances were labeled.

3. Which parts of the circuit in Fig. 16-2 are series sections and which are parallel sections?

History: Gustav Robert Kirchoff

The work Georg S. Ohm did in developing his important law was further extended by the work of the German physicist, Gustav Kirchoff. Kirchoff is often credited as the first to show that electricity moves at the speed of light. Kirchoff's Laws of Current and Voltage are very important in electronics, but they may not have been his most important findings. He worked with Robert Bunsen, the chemist who invented the Bunsen burner. Together they developed the first spectroscope. This device, which uses the spectrum to analyze materials, has been very valuable in the study of outer space and metallurgy, the science of metals. Since Kirchoff died in 1887, many of his contemporaries probably had no idea how important some of his findings really were. Some of the findings were so advanced that their technological usefulness was not realized for more than 50 years. We remember the man who gave us new ways to look into the future when we use Kirchoff's Laws.

Chapter 17

Safety Devices

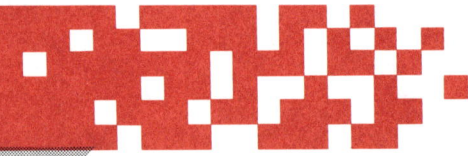

The Need for Safety Devices

Now that you have begun to learn about electricity and electronics, you have a responsibility to be on the lookout for sources of potential electrical hazard in your home and community. If you work on devices you have not learned to work on properly, fail to correct hazards, or intentionally bypass any safety devices, you are being **negligent**. Be very careful. Always do work of the highest quality and only that work which you know how to do.

Fuses

Many safety devices are only there to protect against events that could cause damage or injury. Some safety devices, such as fuses, destroy themselves as they do their jobs, but their cost is very low compared to the cost of repairing or replacing the things they protect. The simplest, but maybe the most important, of all safety devices is the common **fuse**. See Fig. 17-2. It is much cheaper to burn out a 25¢ fuse, for example, than it is to burn up a car when a short develops in the headlight circuit. A fuse is similar to the weakest link in a chain. If a load that is too heavy is placed on the circuit (or the chain), that weakest link will break down. Electrical wires are capable of handling a certain amount of current, depending on their diameter. If a wire is forced to carry more current than it can safely pass, great amounts of heat are produced (due to the friction of the electrons traveling in the wire). This excess heat can cause the insulation, surrounding materials, and the conductor to burst into flames. If this occurs behind a wall or in a wire which is near combustible floor or furniture materials, the

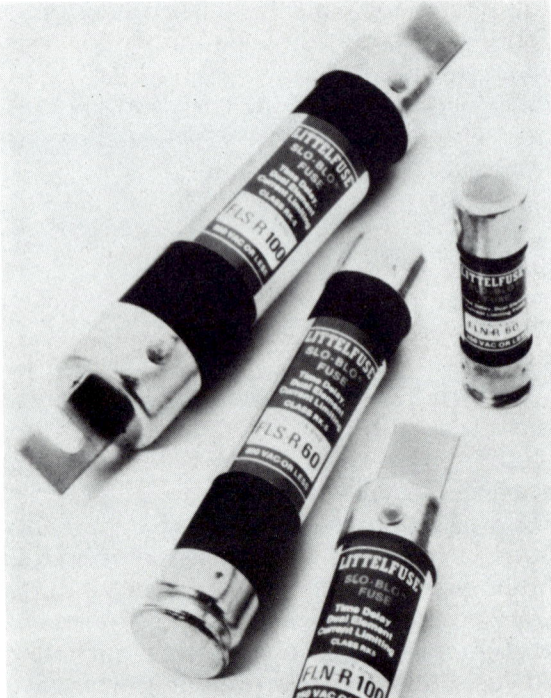

Fig. 17-1 Fuses come in many sizes and shapes for different types of circuits.

whole building may be destroyed by fire.

A fuse has a thin piece of wire in the circuit which will burn out first, whenever the current begins to get too high. This breaks the circuit and stops the current. This eliminates the danger. Figure 17-2 illustrates this point. The reason the fuse does not catch things on fire when it burns out is because it is safely enclosed in a glass or ceramic case. The case is able to withstand both the electrical potential and the heat produced as the wire burns out. Some people try to save a little money by bypassing the fuse — providing a path for current around the fuse. They put pennies under plug fuses or wire jumpers across other types of fuses and circuit breakers. This is very unwise. Many homes have burned to the ground because of such negligent acts. When fuses are bypassed, if a short or overload occurs, the wire will burn out somewhere else — under the floor for instance.

Fuses do not burn out frequently when their circuits operate under normal loads. A fuse will eventually burn out though, due to old age, even in lightly loaded circuits. Just because one fuse blows does not (by itself) indicate that there is a problem with the circuit. Try replacing the fuse with another one of the same amperage rating. Never use a higher rated fuse! If this new fuse cures the problem and the circuit operates normally, then the fuse either blew due to old age or there was a temporary surge or overload on the line.

When fuses continue to burn up, one after another in the same circuit, there is an **overload which must be corrected.**

Shorts are the most serious type of overload. Overloads also may be caused by placing too many loads (resistances) in parallel on the same circuit. To keep the fuse from blowing again and again due to this type of overload, some of the loads must be turned off or disconnected to get the current flow down.

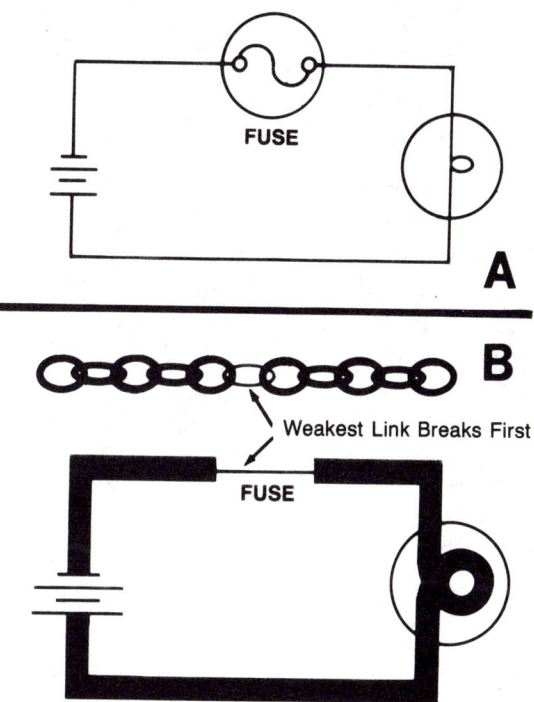

Fig. 17-2 A) The schematic symbol for a fuse may or may not include the circle which represents its enclosure. B) Just as the weakest link in a chain will break first, a fuse (which is the thinnest wire in a circuit) will burn out, and stop the current flow, before other wires burn due to overloads.

Circuit Breakers

Circuit breakers do the same job as fuses, except circuit breakers are not destroyed each time they do the job. When an overload occurs, a circuit breaker acts as an automatic switch which turns itself off. The power may be restored by simply resetting the circuit breaker. You can reset a circuit breaker by pushing a button or moving a lever to the fully off position and then back to the on position. Figure 17-3 shows the workings of a typical circuit breaker.

Most common circuit breakers operate

Fig. 17-3 Magnetic circuit breakers are used in many applications. They are more convenient than fuses.

either electromagnetically or by heat. In the magnetic type, when the current flow becomes great enough to create a certain strength magnetic field around the coil in the device, then the contacts are separated to "break" (open) the circuit. The thermal type has one of its contact points attached to a bimetallic strip. When current flow becomes great enough, heat is produced which causes the bimetallic strip to warp and open the points. Thermal circuit breakers reset themselves after they cool down.

Grounds and Insulators

Some circuits connected to the AC line should be grounded. To **ground** a circuit means to connect a special ground conductor from any metal enclosure which is around the circuit to the actual earth (ground).

Figure 17-4 shows how a typical circuit may be wired in the home. The black wire is the **hot** wire which brings current from the generator. The white wire, the **neutral** wire, is the current's return path in getting back to the other side of the generator. The other wire, the ground, is color-coded green. Usually, no current is flowing in this wire. Its purpose is to provide a way for current to get to the ground (and back to the generator) if there is a circuit problem or even a lightning strike. As long as everything is working correctly, the current flows only in the hot and neutral wires. The load is placed between the black and white wires. The frame or enclosure of both the generator and the load may be grounded. Most devices will work without a ground wire, but some of them are dangerous if their ground is removed. Simple devices such as lamps and radios usually do not have a ground wire. Electric drills and other shop tools do have ground connections in the plug. This is a three-pronged plug. The third prong is the ground. Some newer electric tools, however, are double insul-

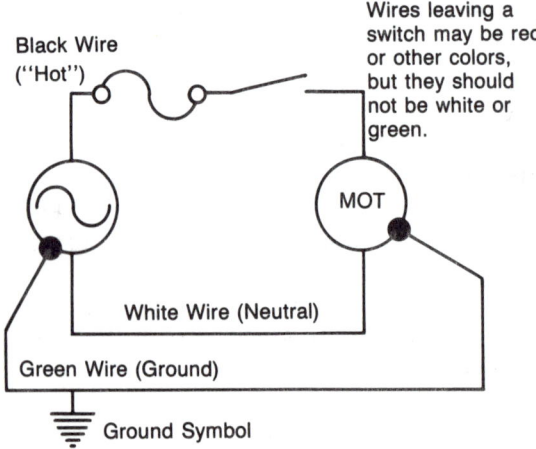

Fig. 17-4 The ground connections and green wire carry no current unless there is a problem of some type. Fuses and switches are placed on the "hot" side of the circuit because current may be able to use the ground wire to go around them if they were placed on the neutral line.

ated (that is, their motors have special insulated parts), and they do not need a separate ground connection. If a device has a ground, you should use it. Do not cut the ground off the plug if you do not have a three-hole socket. Use an adapter. Never remove the ground connection — it could be fatal!

You first read about insulating materials in Chapter 2. Insulators are used in places where you do not want electrons to flow. Wires are usually coated with insulating material to keep current from escaping them in case they contact other conductors or people. Electricians wear gloves made of insulating materials. Their tools are insulated, too. When you work on electrical devices, remember to check the insulation on the wires and replace the wires if they're in bad condition. Never leave wires uninsulated where they may be contacted. Grounds and insulation work together for your protection.

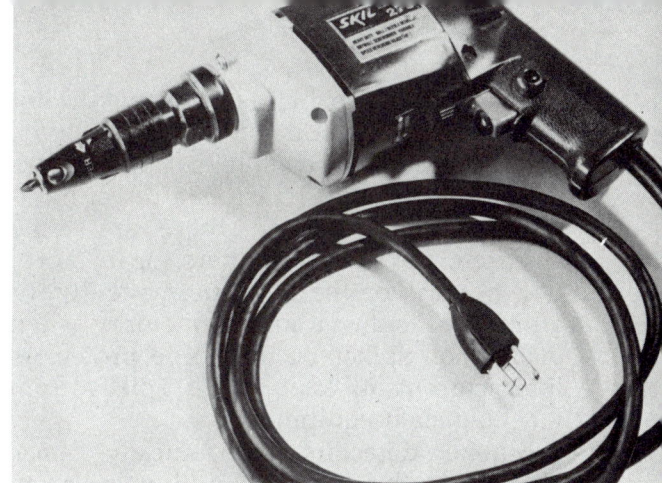

Fig. 17-5 Devices such as this electric tool are sometimes grounded by a third prong on their plugs. The ground plug is round — it should never be removed.

Ground Fault Interrupters

Fuses and circuit breakers only open the line when there is a major overload or a short. Sometimes, however, it is important to open a circuit to avoid electrical shock to a person or animal. A potential problem is that small amounts of current can leak in a defective device without causing an actual short. If you were to touch the "hot" side of the circuit, some current could travel through your body to ground. This would not be a short, but it could be even more dangerous! These problems can be safeguarded against with a **ground fault interrupter** (G.F.I.). Ground fault interrupters are used anywhere there is danger of people coming into contact with both water and the electrical system; for example, in the bathroom. They are also frequently used in laboratories, hospitals, outdoors, and where special dangers exist. Since they are relatively new, many older homes and buildings do not have them. Ground fault interrupters compare the current in both the hot and neutral lines, and they open the circuit whenever there is a difference in these two measured currents. A difference would mean that there was a defect or leakage in the circuit. This could be caused by a person contacting the line and providing a path for the missing current through his or her body.

Other Safety Devices

Modern technology has provided many other electrical and electronic safety devices. Two of the more common ones are a surge protector and a smoke detector. The **surge** or **spike protector** is used with computers and other sensitive electronic equipment. Sometimes, when heavy loads are connected to the line, or when they are removed, a burst of extra energy will be drawn from the generator. When this happens, the voltage may go up dramatically for a moment and then return to normal. You may have noticed your lights brighten and then dim — especially during an electrical storm. Surges like this do not hurt most electrical appliances (lights, motors, heaters, or even televisions) unless they are very great (such as a direct lightning strike). However, there are some electronic devices, computers for instance, which may be harmed permanently or which may suffer a loss of memory when even small spikes occur. Surge protectors limit current to keep these spikes from reaching such equipment.

Smoke detectors electronically sense changes in the air and sound an alarm to warn of smoke caused by fire. Some of these devices are even able to place emergency calls automatically or to operate safety and fire equipment such as a sprinkler system.

Summary

Safety devices protect people and property from damage and injury which could result when overloads or hazards occur. Fuses and circuit breakers do the same job — they stop current flow when there is a short or an overload. The only difference between them is that fuses are destroyed when they work the first time. Circuit breakers may be reset and used over and over again. Ground fault interrupters break both sides of the AC line, hot and neutral, whenever there is any leakage of current from the line. Fuses and circuit breakers usually only break the hot side of the line.

Important Terms

negligent
fuse
circuit breaker

ground fault interrupter
hot
neutral

ground
surge or spike

Review Questions

1. What is the difference between fuses and circuit breakers? What do they both do?
2. What are ground fault interrupters and how do they work?
3. What is the difference between the ground wire and the neutral wire?

Chapter 18

Magnetism

Permanent and Temporary Magnets

Magnetism is an invisible force that surrounds a magnet. It attracts iron and other materials which contain iron. There are two main types of magnets. One type is magnetic only when influenced by a magnetic field from another strong magnetic source. These magnets, referred to as **temporary magnets**, only pass along the magnetism of other sources. They have very little or no real magnetic force of their own. Some temporary magnets remain weakly magnetic after the outside magnetic force is removed, but others lose all of their magnetic properties immediately.

Permanent magnets remain magnetic after they are formed unless something happens to demagnetize them. Loadstone (also known as magnetite) is a natural magnet found in the earth. Other permanent magnets must be made from materials such as iron, high-carbon steel, or certain alloys. You might be asking yourself if permanent magnets are the only effective ones. No, they are not. Sometimes, in electrical devices, it is very important to have materials which may be quickly changed from being magnetic to not being magnetic — and back again. Permanent magnets cannot be quickly demagnetized and remagnetized, so they are very poor in these applications.

Magnetic Induction

A piece of lodestone (magnetite possessing polarity) or another permanent magnet may be used to produce another magnet from a piece of iron. Figure 18-1 shows that rubbing a bar of iron with a permanent magnet will magnetize it to make it a new permanent magnet. This process is called **magnetic induction**. The process of magnetic induction may be completed manually

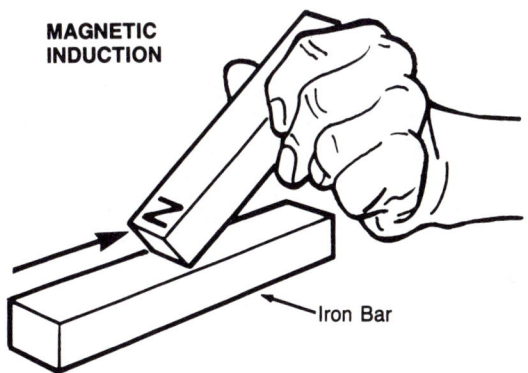

Fig. 18-1 To produce a new magnet by magnetic induction, rub a bar of ferrous (iron containing) metal with a strong permanent magnet. Lift at the end of each stroke and go back to start again at the other end so that you only rub in one direction.

or by using electric coils. This will be clearer after you read about electromagnetism in the next chapter.

The Nature of Magnets

The invisible force surrounding a magnet is a continuous force field. It is not lines of force as illustrated in most electricity texts for there is no way to illustrate an invisible force field. You can only imagine that the force exists in lines. Figure 18-2 shows the typical force field. The force field is actually a **flux field** and the imaginary lines are often called "lines of flux."

There is a flux field around the Earth and every magnet. Compasses used in navigation make use of this natural magnetic field around the Earth. The magnetic center of force located on one side of the Earth is called the North Pole. This is the side pictured as the top in maps. The South Pole is the other center of flux force. Since much of the early research on magnets dealt with their relationship to the Earth, the terms "north pole" and "south pole" are applied to all magnets today. The **poles** of magnets,

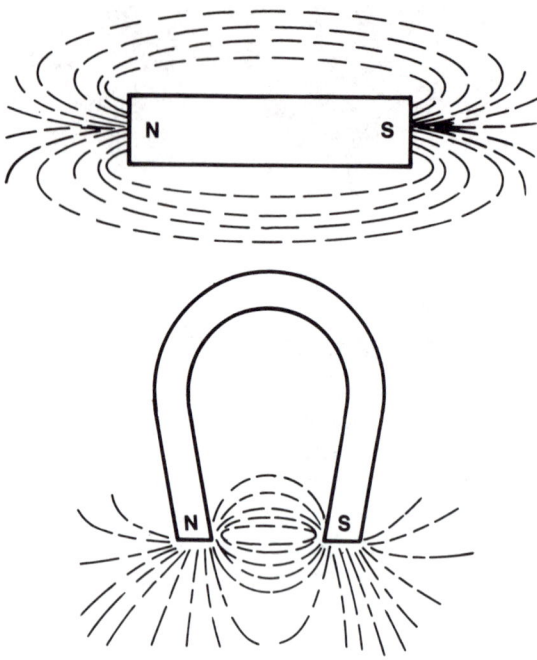

Fig. 18-2 Most illustrations, like this one, depict magnetic force as "lines of flux", but remember that flux is really a continuous force field. The force is greatest close to the magnet.

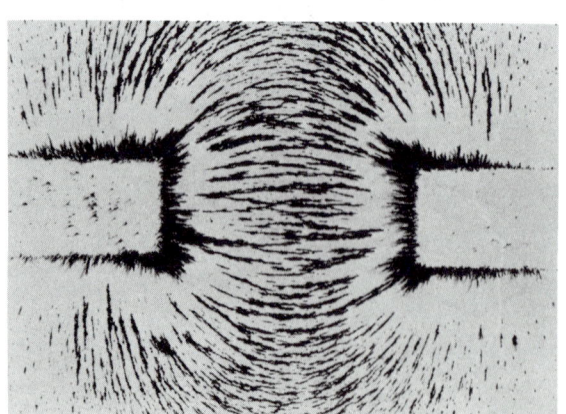 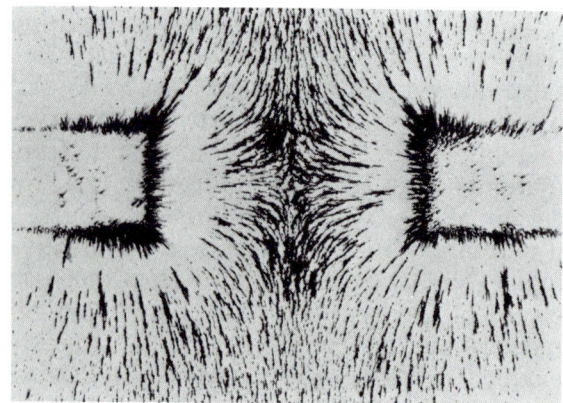

Fig. 18-3 These iron filings show the flux fields of pairs of magnets which are beneath a piece of paper on which the filings were scattered. Unlike poles attract as in A — and lines of filings seem to reach out to each other across the gap. When like poles (two north poles in B) are brought near each other, though, they push each other away — note the boundary of vertical lines midway between the magnets.

just as with the Earth, are the centers at which the lines of flux are most concentrated. The magnet is stronger at the poles than in the middle.

If two north poles are brought near to each other, they repel each other just as two negative charges should repel. See Fig. 18-3. Likewise, two south poles repel when placed together. But when a north pole is placed near a south pole, they both attract each other. Either pole will attract materials like iron or steel.

The poles of permanent magnets are also considered permanent. With a **temporary** magnet, however, it is possible to remove the magnetizing force and turn the temporary magnet around to make it have **reverse polarity**. This happens often in motors and other electrical devices.

Theory of Domains

Just what is a magnet and what gives it its force? The answer is not completely known, but a theory that stands up to many tests is the **Theory of Domains**. According to this theory, materials that can be made into magnets have many tiny crystal-like structures called **domains**. Each domain is made up of many atoms. Each domain has a small magnetic force of its own. When the material is not magnetized, the domains are haphazardly arranged — pointing in all directions — so that their tiny forces cancel each other. To make the material into a magnet, the domains need to be lined up so that their individual magnetic forces all help each other pull the same way. When *most* of the domains line up, the magnet becomes strong. When *all* of the domains line up in one direction, the magnet is **saturated**. It cannot be made any stronger regardless of how much more you try to magnetize it. See Fig. 18-4.

The ability of a material to be magnetized, even temporarily, and to transmit magnetic force is called **permeability**. Air has a permeability rating of about 1 which is very low. Many other materials have about the same low amount of permeability. Aluminum, copper, and water are examples. Other materials, however, have higher permeability ratings. Iron has a permeability of about 7000. Nickel and cobalt rate about 40 and 50, respectively. Permalloy, an alloy of other metals, has much greater permeability than even iron and is used to make very good magnets. Another very good material for making magnets is alnico, an alloy of aluminum, nickel, and cobalt. None of these three materials are very good as magnets indi-

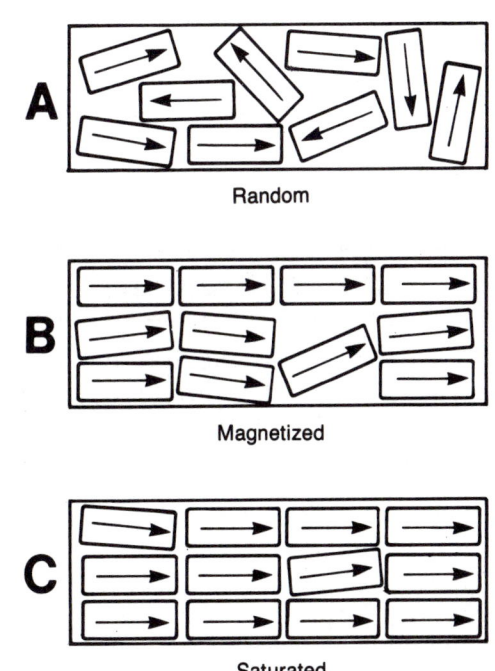

Fig. 18-4 A) Nonmagnetized steel has a random arrangement of domains. B) Fairly strong magnet — most domains are lined up in a similar direction. C) Saturated magnet — all domains aligned almost perfectly.

PROPER STORAGE OF MAGNETS

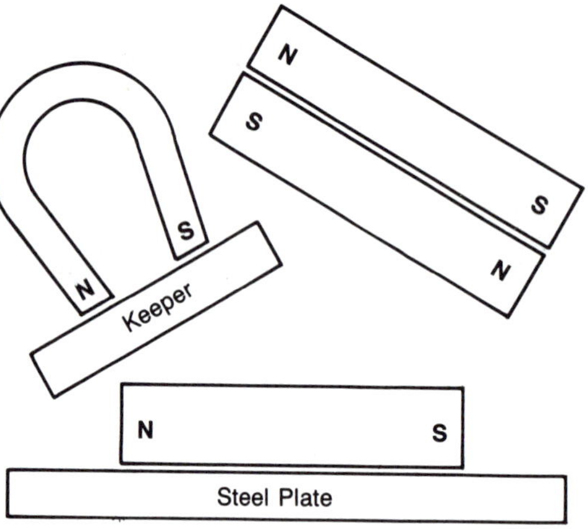

Fig. 18-5 Proper storage of magnets prolongs their lives. The worst way to store a magnet is in an unused and unprotected condition.

But what about using a magnet too long? Actually, using a magnet is better than not using it. When a magnet is clinging to a piece of metal, it temporarily makes that permeable piece of metal into a magnet. This new temporary magnet's flux pulls on the domains of the original magnet and helps to hold each one in place. In fact, if the permanent magnet receives a blow while it is holding onto this piece of metal, the domains may be drawn right back into place by the force of the temporary magnet. For this reason, horseshoe magnets should be stored with a "keeper" — a piece of permeable material attached to the poles. Bar magnets should be stored in pairs with the poles attracting (north to south) as in Fig. 18-5. If a pair is not available, sticking the magnet to a piece of mild steel is the next best protective measure. Magnets are more likely to be harmed when not in use than when they are used properly.

vidually. When alloyed together, however, they make a metal with very high permeability.

Care of Magnets

Many people mistakenly believe that magnets wear out with age or use. Actually, there are three ways a magnet may be harmed:
- Physical abuse.
- High electrical currents.
- High temperature.

When permanent magnets are hammered or dropped, their domains may become misarranged. This can also happen if high electric currents are passed through the magnet or through coils which create much greater magnetic forces near the permanent magnet. Finally, if permanent magnets are heated to very high temperatures, their domains will slip into a random arrangement.

Summary

Permanent and temporary magnets are both important, but for different uses. A magnet may be made by magnetic induction. Permeability refers to the ability of a material to accommodate magnetic flux, while saturation means that all of the domains are aligned and the magnet cannot be made any stronger. Magnets may be demagnetized by physical abuse, electric currents, or high temperature. They do not wear out with use. The next chapter will explain how magnets and electricity work together.

Important Terms

temporary magnet
permanent magnet
magnetic induction
flux field

poles
reverse polarity
Theory of Domains

domains
saturated
permeability

Review Questions

1. Explain the theory of domains.
2. What is the difference between temporary and permanent magnets? How is each one made?
3. How should you care for magnets? What can harm a magnet?
4. What are saturation and magnetic induction?

Project: Bike Burglar Alarm (or Scream Box)

Parts List

1	2.2 K ohm resistor	R_1
1	4.7 K ohm resistor	R_2
1	.05 microfarad capacitor	C_1
1	1000 to 4000 microfarad cap.	C_2
1	10 microfarad capacitor	C_3
1	Speaker, small	
1	PNP transistor (ECG 290A)	Q_1
1	LM 386 audio amplifier IC	Z_1
1	Audio output transformer (small)	T_1
1	9 volt battery with connector	B_1
1	Mercury switch (trigger)	S_1
1	DPDT slide switch (set)	S_2
1	Suitable enclosure	
1	Printed circuit board	

Notes

The value of C_2 determines how long the alarm will sound after the trigger switch is reopened. The larger the capacitor, the longer the sound. A 4000 microfarad capacitor will make the circuit sound for about 20 seconds. The sound will vary in pitch as the capacitor discharges, but will be loud enough to attract attention for the full discharge time. If used as a bike burglar alarm, the circuit should be hidden under the seat or in an accessory pack. The mercury switch should be positioned so that leaning the bike off its kickstand position will close the trigger switch. The "set" switch should be turned off when riding. There is some battery drain when the alarm is armed, so it should not be used overnight — it is mainly for short-term protection such as when you enter a store. A lock is also recommended. If you wish to build the circuit as a novelty device, construct it in a suitable box or enclosure and position the trigger switch so that anyone who picks up the box will get a screaming surprise that they cannot control. Put several "fake" switches alongside the real "set" switch so that your victims will fumble around for a few moments before they can turn the screaming sound off. You could even carefully wire the "fake" switches in parallel with the normal "set" switch so that only one pattern of switch positions would turn off the sound. If the switches are on the bottom, the mercury switch will be restarting the siren blast frequently while the victim fumbles to find the off switch.

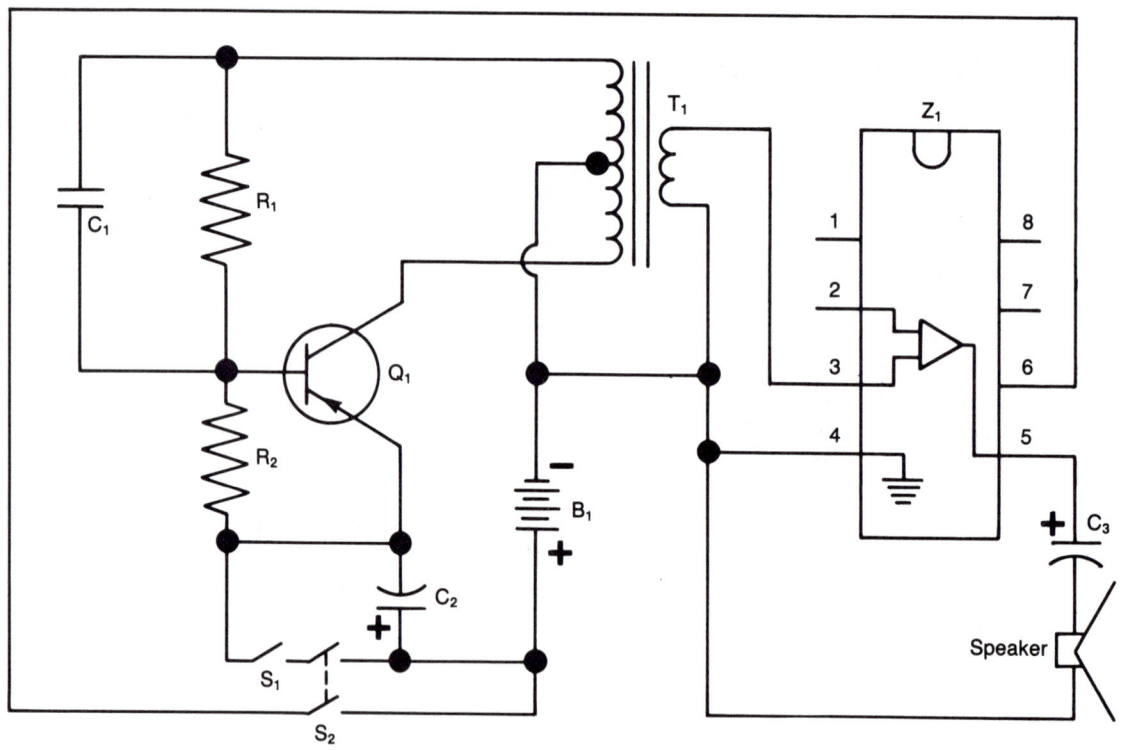

Fig. 18-6 This alarm circuit has a long delay before the siren stops after it is triggered. Closing S_2 arms the alarm. The alarm will sound when the trigger (S_1) is closed by tilting the mercury switch. The alarm will continue to sound until capacitor C_2 discharges — even if S_1 is opened again.

Chapter 19
Electromagnetism

What Are Electromagnets?

When an electric current flows through a wire, a magnetic field is created around that wire. The flux field it creates is an **electromagnet**. How do you tell the direction of the field created? The "left hand rule" for electromagnets says if you point the thumb of your left hand in the direction the current is flowing, your fingers will point in the direction of the magnetic field. This principle was studied by Christian Oersted in the early 1800s. Figure 19-1 shows another way to tell the direction of the field created.

The flux field around one current-carrying wire is quite small, but the field may be

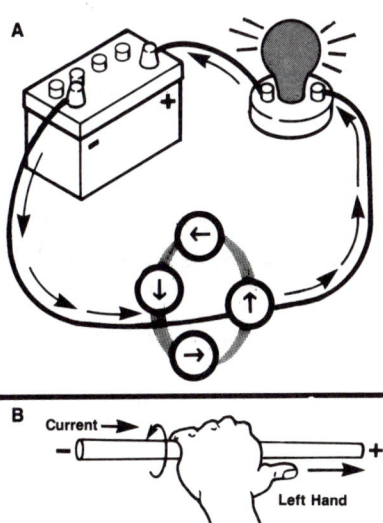

Fig. 19-1 In A, as a small magnetic compass is moved around a wire that is carrying a fairly high current, it will align itself with the magnetic field created by the current. If the battery is disconnected, then the compass will point toward the Earth's North Pole in all of these positions. Part B shows the left hand rule for finding the direction of the field when the current direction is known.

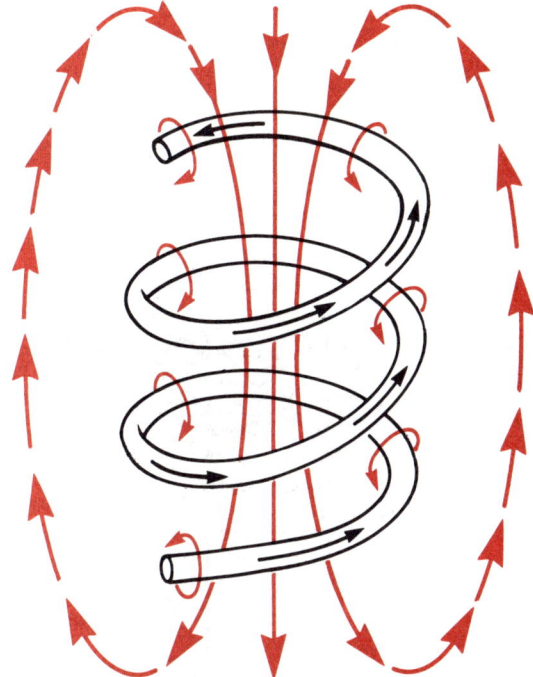

Fig. 19-2 Each loop of a coil has a small amount of magnetic pull when current flows through it. These small flux fields aid each other to make a larger, more powerful field around the whole coil.

strengthened by looping the wire into the shape of a **coil**. Figure 19-2 shows how the small fields of each loop of wire in a coil overlap and aid each other to make a much stronger field around the whole coil.

Strength of Electromagnets

The total strength of an electromagnet depends on three main things:
- Number of turns in the coil.
- Amount of current in the coil.
- Permeability of the core material.

The more **turns** in the coil of an electromagnet, the stronger it will be. Likewise, if a coil is being used as an electromagnet, the strength of the magnet may be increased by increasing the voltage of the power source. Since the resistance of the coil will not be changed, turning the voltage up will cause more current to flow. When the current flow increases, the magnet will get stronger.

The third factor which affects the strength of an electromagnet is permeability. This is not the permeability of the coil itself. It is the permeability of the **core** material. The core is whatever is inside the coil (what the coil is wrapped around). In the examples so far, the core material has just been air. Remember that air has a permeability of about 1. To produce a very weak electromagnetic field, air would be a possible choice for the core material. Air would be a very poor choice, however, for a strong magnet. If electromagnets with air and iron cores both have coils with the same number of turns and the same current flow which would be stronger? The more permeable iron core is much stronger. This is because the permeable iron core transfers the flux field better than the air core of the other coil. In fact, the materials which make the best cores for electromagnets are those materials that make good temporary magnets.

When materials are subjected to powerful magnetic flux fields, they actually change size and shape. Some materials shrink while others expand. This effect is called "magnetostriction." This effect is very small, but it has been demonstrated and could prove useful as technology broadens.

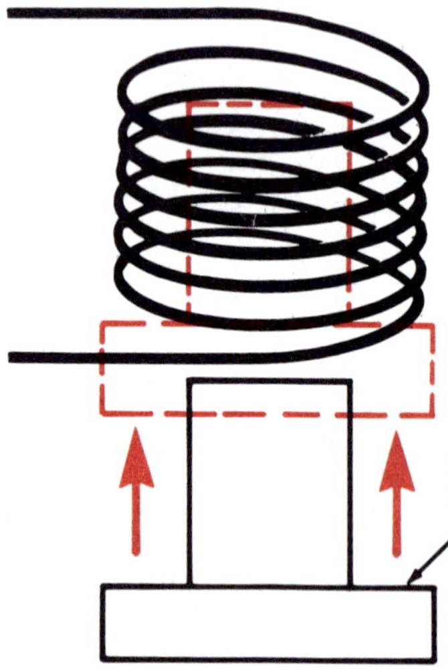

Electromagnets in Use

Two of the most important uses of electromagnetism are in transformers and motors (Chapters 30 and 54). Electromagnets are also used in common solenoids and relays.

A **solenoid** (Fig. 19-3), is an electromagnet with a movable core called a plunger. When the plunger is inside the center of the coil, the flux field has a path of low resistance to flow through. This resistance to accepting magnetic flux is called **reluctance**. Materials with great amounts of permeability have little reluctance. When the plunger is drawn out of the center of the coil, then

Fig. 19-3 When the coil of a solenoid is energized, it sucks the iron plunger into its center.

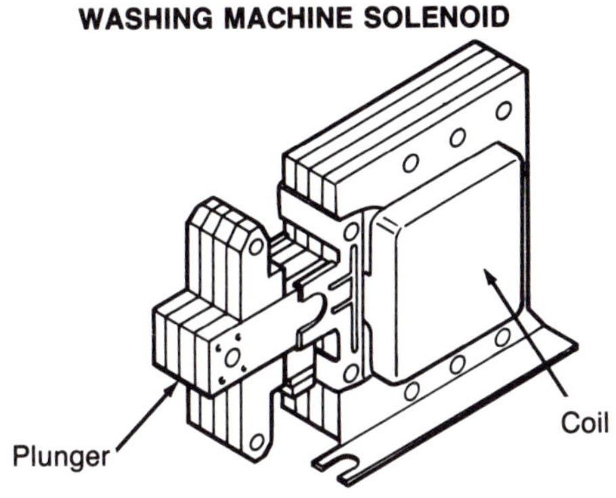

WASHING MACHINE SOLENOID

Fig. 19-4 Solenoids like this one convert electrical energy into linear (straight line) mechanical motion. Many washing machines use solenoids to change gears when told to do so by the timer.

The solenoid in Fig. 19-4 is used to change the gears in a washing machine. When the washer is in the wash cycle, the solenoid pulls a lever which engages the proper gear mechanism in the transmission of the machine. At all other times, a strong spring pulls the gear change lever to the neutral position. Another solenoid may be used to engage the proper gear for the spin cycle. The electromagnet in this solenoid is powerful enough that it would be difficult to force it open once closed. You would not want to get your finger caught in it! Solenoids are often used to open and close gas valves when electrical signals are sent from automated systems. They are also used to engage the starter motor gear in an automobile.

the core material is really air. Air has low permeability and, thus, forms a core with high reluctance. The coil attracts the iron plunger — actually seems to suck the plunger into its center — to reduce the reluctance and provide a low resistance path for its flux field. Usually, solenoids will have a spring which pulls the plunger out of the coil whenever the coil is not energized with current.

Relays

A special type of solenoid is the **relay**. Relays work in the same way as solenoids. That is, they close when the coil is energized, and they open when the coil's current is shut off. The main difference is the job that they perform. Instead of moving levers and changing gears, relays open and close electrical contact points. A relay and its schematic symbol are shown in Fig. 19-5. Relays are used to control a circuit auto-

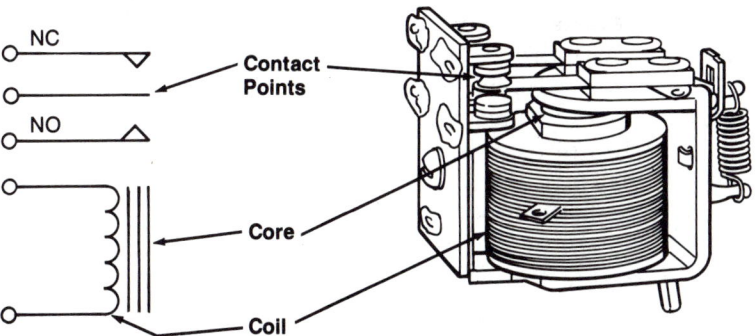

Fig. 19-5 Relays are really automatic switches. Some relays have more than one set of points. The NC points are "normally closed" and the No ones are "normally open". When the coil is not energized, the NC points are closed, but the NO points do not close until the coil is energized.

Chapter 19 *Electromagnetism*

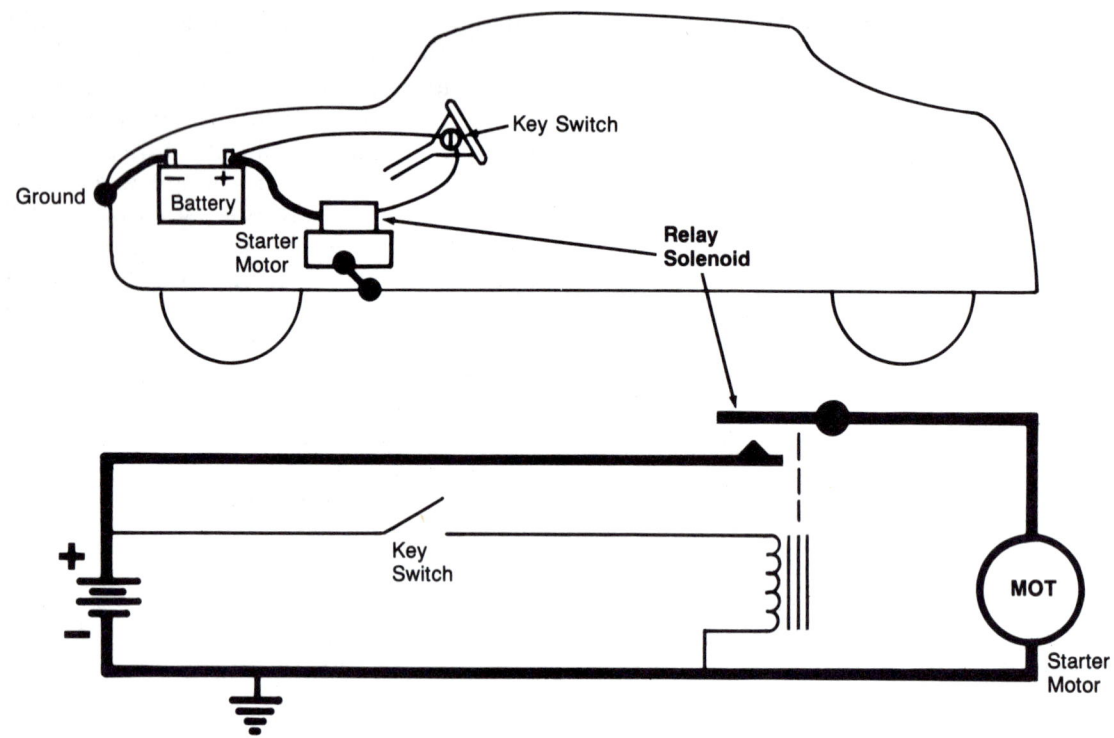

Fig. 19-6 Notice that large cables are needed to carry the current to the motor, but smaller wires work fine for the key switch and the coil of the relay because they only need a little current. The frame of the car is the ground (−) connection.

matically and to control a large current with a small one.

The relay in the starting circuit of an automobile is an example of the second use. In some cars this relay is actually a set of electrical contact points connected with the solenoid. In cars that do not use a solenoid to engage the gears, relay is a separate unit. Most mechanics and parts dealers call both types of relays "solenoids," though this is technically incorrect for one type. The relay (or solenoid) acts as a remote controlled switch with heavy-duty contact points which can carry the high current used to start the car's engine. It would be impractical to send this large amount of current all the way up to the key switch and then back to the motor. Besides, the key switch would have to be much larger to carry that amount of current. Instead of doing this, the circuit shown in Fig. 19-6 is used. Here, a small amount of current flows to the coil in the relay when the key switch is turned on. This causes the relay to close its contacts together. The relay's contacts are actually being used as the switch for the starter motor, so closing them allows high current to flow to the starter motor. When the key switch is turned back from the start position, the current to the coil stops, so a spring opens the contacts and removes the current from the starter motor.

The relay in a central air conditioning unit is another example (Fig. 19-7). The thermostat (temperature sensing unit) mounted inside the house operates on a 24 V circuit with fairly low current. The job of the thermostat, however, is to control the air conditioner's compressor and fans. The compressor requires 240 V and high currents to operate. The circuit uses a relay with a 24 V coil which can be energized by the low current circuit when the thermostat senses that the temperature is getting too high. When the coil is energized, it closes a set of points which permit current to flow in the high current (240 V) circuit to run the compressor. If there were no relay, then the thermostat would need to be very large, and a lot of heavy-duty cable would be required to carry the current into the house and back out. Thus, this relay does both jobs of relays: it controls one current (the large one) with another, and it makes automatic remote control possible.

Other Electromagnets

Even simple door bells, buzzers, and the ringers in telephones work on the principles of electromagnetism and solenoid action. Solenoids and relays are used in many applications in the home and in industry. Electromagnets are used in other applications as well. The picture tube in a television uses electromagnetic coils to produce a flux field which guides the electron beam to the proper point on the screen of the set for reproduction of each picture element. Chapter 57 will explain this use more fully.

Safety with Electromagnets

Electromagnets can be very powerful devices. Whenever you work around large coils, remember that they may have enough pulling force to crush parts of your body which get between them and metal parts they attract. In addition, do not wear rings, watches, or other jewelry around electric coils. You learned in Chapter 5 that when magnetic lines of flux pass through a wire, current is made to flow in that wire. If you were to get your ring near a powerful electromagnet, it is possible for the ring to act as a coil — a coil with one loop and almost no resistance at all! When there is little resistance, it does not take much EMF to make very high currents flow. When

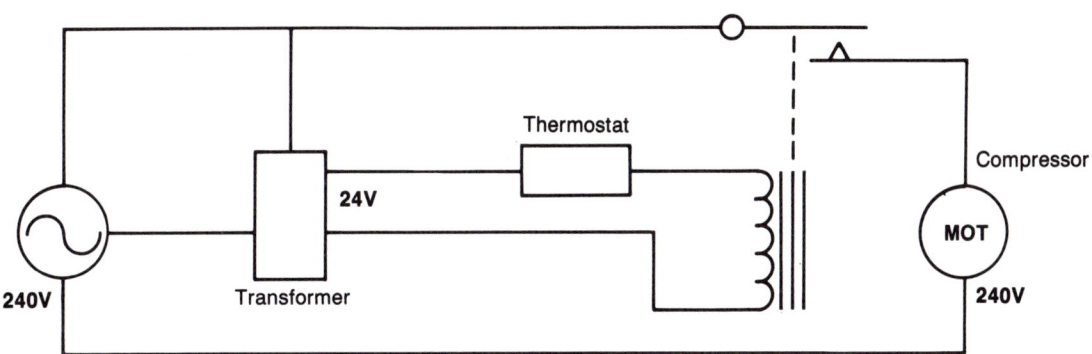

Fig. 19-7 The relay uses the 24 V from the transformer to control the 240 V current for the compressor. Only the 24 V current has to come into the house to the thermostat. This circuit is simplified for clarity.

high currents flow, heat is produced. In moments, the intense buildup of heat can burn your finger severely. Watches can burn your arm as well. Additionally, many watches may be harmed by the magnetism.

Summary

Electromagnets are used to make motors, transformers, solenoids, relays, and many other electrical and electronic devices. They operate on the principle that a magnetic field surrounds any wire which carries current. When wires are looped into coils, the field becomes stronger. The strength of an electromagnet depends on the number of turns in the coil, the amount of current in the coil, and the permeability of the core material. You should be careful when working around large coils; avoid wearing jewelry. The next chapter deals with another use of electromagnetism — using it to measure electric currents.

Important Terms

electromagnet
coil
turns
core
solenoid
reluctance
relay

Review Questions

1. Explain how an electromagnet works.
2. If you had a fairly weak electromagnet, name three things you could do to make it stronger.
3. What is the difference between a solenoid and a relay?
4. Why should a worker remove his/her ring when working on a large electric coil?

History: Heinrich Friedrich Emil Lenz

Lenz (lents) was a Russian theology student who lived from 1804 to 1865. He became interested in science and changed his career plans to become a physicist instead of a priest. He worked with electromagnetism and electromagnetic induction. A law that he discovered in 1834 is quite useful to engineers who design electrical equipment such as generators. Lenz's Law says that while electromagnetic forces are being used to induce new currents in a coil, forces that oppose those new currents are also being produced. This principle is called self-inductance, and you will learn more about it in Chapter 31.

Project: "Magnutic" Game

Parts List
- 1 Plastic spray can top, colorful
- 1 Strong magnet (or several buttons)
- 1 Wood or plastic base
- 30 8-32 chrome plated, steel nuts

Notes

Glue the magnet(s) under the center of the spray can top and glue the top to the base. The stronger the magnet is, the better the project will work. Make sure that the nuts (or other shapes) are of a permeable material such as steel. See if you can find the stacking patterns which allow the highest towers to be built. If larger magnets and enclosures are used, the size of the nuts can also be increased.

Fig. 19-8 This magnetic game can drive you nuts. It is simple and inexpensive to make, but it can provide hours of fun for all ages. Try to stack the nuts higher by using special patterns — you'll be surprised at the results!

Section 4
Meters and Measurements

Chapter 20

Meters

Uses of Meters

Many different types of meters are used to measure voltage, current, resistance, and other qualities of electric circuits. They are important for studying, designing, and repairing circuits. Some meters measure only one quality. Others have the ability to measure several different qualities. Engineers, service-people, students, and hobbyists work with meters in laboratories and shops.

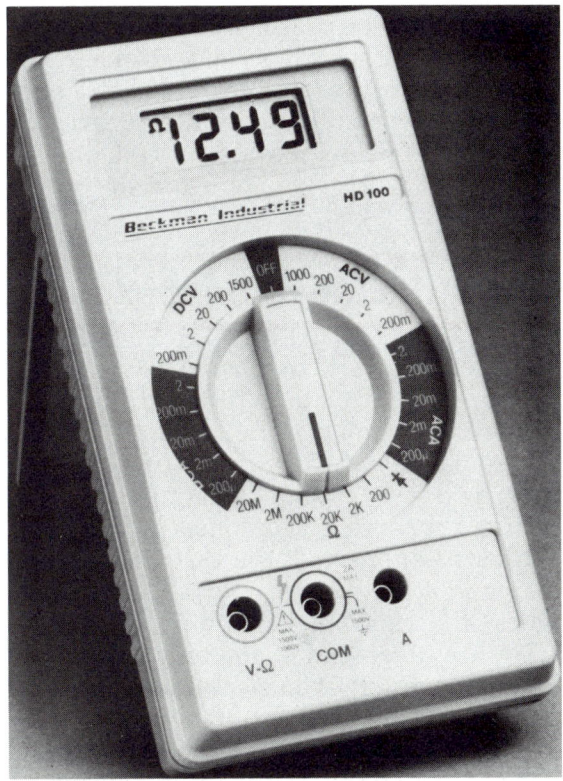

Fig. 20-1a Digital meters allow the user to read the measurement directly, just as if they were written on a piece of paper, but they are not very good for observing changing values.

Fig. 20-1b Analog meters allow you to follow changing values and to see trends of change.

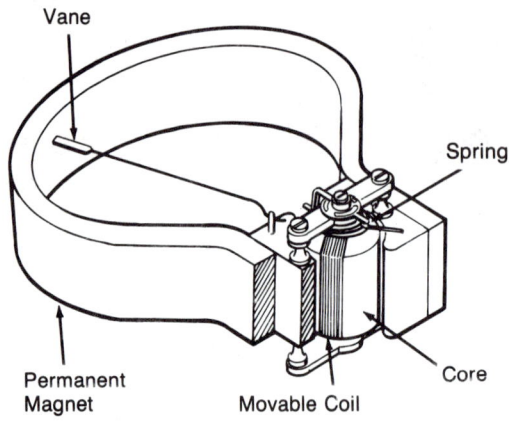

Fig. 20-2 The d'Arsenval meter movement is a very accurate analog unit which is used often.

Meter Readout Devices

A meter must have a method of communicating the measurements it makes to the operator. Some meters use a **digital readout** that shows the numbers which represent the value measured. These meters, which are discussed more fully in Chapter 25, are popular and accurate, but they have one major drawback. They tell the value being measured at any given moment, but they are very difficult to interpret when the value is changing. If you want to watch a value rise or fall, these meters are not a good choice because they will be flashing new values so quickly you cannot follow them.

The more traditional meter, with a moving pointer and a meter scale, is much better when you need to observe a trend or measure a changing value. This type of meter is an **analog** device. This chapter explains how these analog meter movements work.

The d'Arsenval Meter Movement

One of the best and most popular mechanical meter movements works on the same general principle as one invented by Arsene d'Arsenval. This device is called the **d'Arsenval meter movement**. Refer to Fig. 20-2 as you read the following discussion.

The d'Arsenval movement has several main parts. First, it has a vane, (the pointer which indicates what the measurement is). The pointer is connected to a small, rectangular coil that moves freely around a core. The coil, which is very light, rotates between two bearings. A spring keeps the coil positioned so that the vane is at the 0 point when no measurement is being taken. Last, a strong permanent magnet is positioned so that it shields the coil from other magnetic influences, and it provides a strong flux field around the coil and its core.

How the Movement Works

Current flows through the coil when a measurement is made. When current flows in the coil, it makes the coil into a small electromagnet which is free to turn because it is held only by the bearings and the spring. When the coil becomes magnetized, it tries to align with the flux field of the permanent magnet. If only a little current flows, the coil will make a weak magnet and only turn a small amount. Remember that the strength of an electromagnet partly depends on the amount of current flowing in the coil. Therefore, if more current is forced through the coil, the magnetic field it produces will increase in strength, and it will turn farther. The spring is always trying to hold the coil back (so the pointer will aim at 0), but the coil will turn more and more as the current gets higher. In other words, as the current increases, the magnetic flux field of the coil will become strong enough to overcome the effect of the spring. The coil will turn to align its field with the field of the permanent magnet. The vane will point in whatever direction the coil turns it. If the meter scale (face) behind the pointer has been properly drawn up, the pointer can be made to indicate exactly the value of current flowing in the circuit.

Accuracy of Meter Movements

The accuracy of the movement (how well it tells the true value) depends on several things. One of the most important factors is the quality of its parts. The bearings must be of very high quality to allow free movement of the coil. The coil should be very light and have enough turns of wire in it to create the flux needed to turn the whole assembly. The springs must have the right amount of tension, and they must provide good, even resistance to the turning of the coil. The permanent magnet should be capable of providing a strong, uniform flux field with which the coil can turn to align itself. The vane and all other moving parts should be as light, yet rigid, as possible. Lastly, the markings on the meter scale must be very accurately **calibrated**. The word calibrate has two related meanings: It means to print the markings on the scale when the meter is produced, and it also means to adjust an existing meter to properly register measurements.

There Must Be Current

The d'Arsenval movement (and other similar ones) are used to measure many different qualities and values. They can be used in meters which measure voltage, current, resistance, temperature, motor speed, pressure, wattage, and many other qualities. Other than current, none of these qualities can make a magnetic flux field around the little coil. Heat could be applied all day to the wires leading to the coil, and it would probably not move at all. Remember that current must pass through the coil to make it electromagnetic, and it will not turn until it is electromagnetized. If you can convert that heat to electric current flow, then the meter movement can measure the current. If you are careful to make the current increase with rises in temperature, and you mark the meter scale carefully in degrees instead of amps, then you can use the meter to measure heat. Figure 20-3 shows a simplified circuit for measuring heat. The movement itself only measures current, but it can be used in circuits to measure other things. To use these analog

SIMPLIFIED PYROMETER CIRCUIT

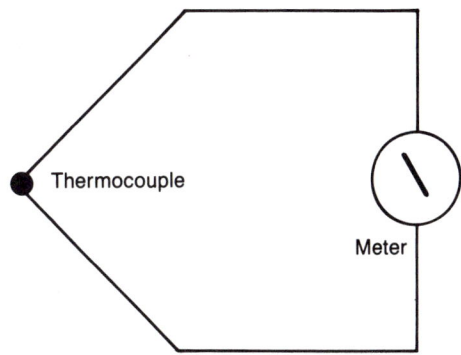

Fig. 20-3 When the thermocouple is heated, it produces a current. The current will make the meter vane move in proportion to the amount of heat. The scale should be marked to read degrees instead of current.

movements to measure qualities other than current, current must flow through the meter movement in proportion to the amount of the quality you want to measure. The meter movement cannot measure a quality such as the amount of light by itself. However, if you connect a photovoltaic cell to it, it could measure the current output of the cell! Circuits that allow you to use current-measuring meter movements to measure other qualities are called **auxiliary circuits**. Upcoming chapters will deal with several auxiliary circuits in detail.

Other Types of Movements

Sometimes, meter movements will use the same principle the d'Arsenval movement uses, but in a slightly different way. In some movements, the permanent magnet is mounted so that it may turn (with the vane attached to it) while the coil is held stationary. In this case, the magnet is small and light but the coil may be larger and should produce a stronger flux field.

Two other types of meter movements are the plunger-vane movement, which uses solenoid action, and the hot wire movement, which uses heat produced by the current to change the length of a wire that moves the vane. These movements are somewhat rare. All of these movements **require auxiliary circuits to convert the quality being measured to current. They all can only measure current flow.**

Reading Meters

Analog meters are a little harder to read than digital meters because they do not just print out the exact numbers for you. You already know how to read some analog meters. The speedometer in an automobile is an analog meter. If the vane points directly at 40, you know that the speed is 40 mph. What would the speed be if the vane pointed midway between 40 and 50? The speed would be 45 mph. Many speedometers have a small mark between each numbered calibration. If the vane pointed halfway between the 40 and the mark for 45, you would read it as about 42.5 mph. Here you would have a little more chance of error, but the error would be quite small. Figure 20-4 shows the scale of a voltmeter which reads 120 V. Where should the vane point if the value went down to 98 volts? What if it went up to 125 V? Be sure to check the markings on the scale (or on switches that select the function of meters which have several ranges) so that you will know what units each calibration represents.

One error which can occur in reading meters is called **parallax error**. When you are sitting in the passenger's seat of a car,

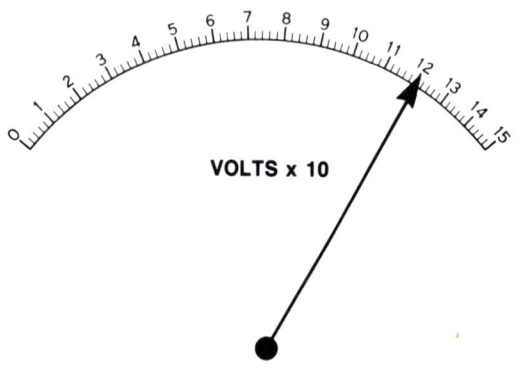

Fig. 20-4 The marking on the scale says the meter reads in "volts × 10". That means that the "1" stands for 10 V and the "15" stands for 150 V. The meter now reads 120 V. Where should it point for 98 V and 125 V? Make up a few more readings to use for more practice.

and you look over at the speedometer, you see a speed which is higher or lower than the speed the driver sees. This is because the pointer is not flat on the scale, and you are looking at it from an different angle. A parallax error also will happen when you read an electrical measurement meter unless you make certain you are directly in front of the scale each time you read a measurement.

Summary

Meters are used to measure many qualities and values in electronics. They come in two main types: analog and digital. Most analog meters use the d'Arsenval movement (or some variation of it) which works on the principle of electromagnetism. Because current must be sent through the coil to make the electromagnet work, these **meters require auxiliary circuits to convert** any quality being measured into an electric current. Analog meters are easy to read if you remember to check the markings and switches carefully and avoid parallax errors. The next few chapters will describe the auxiliary circuits needed to use d'Arsenval movements to measure electrical values.

Important Terms

digital readout
analog
d'Arsenval meter movement
calibrated
auxiliary circuits
parallax error

Review Questions

1. Explain how the d'Arsenval movement works.

2. List the main parts of the d'Arsenval movement and tell what each one does.

3. What is parallax error? How do you avoid it?

4. What is the only value the d'Arsenval movement can measure directly?

Chapter 21

Current Meters

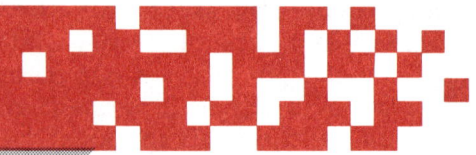

Meters for Measuring Current

Since the basic meter movement measures current, it is very easy to make a current meter using this movement. The meter includes both the meter movement and the auxiliary circuits connected to it — usually enclosed in the same case. If the movement measures current, why do you need any auxiliary circuits at all to make a meter for current? This is a good question. You can make a complete current meter by just attaching two test leads (wires with probes) to a d'Arsenval movement. See Fig. 21-1. Such a meter would work fine and be very accurate, but it could not measure very high currents. Figure 21-2 shows how a current meter is connected into a circuit to measure the circuit's current flow. Notice that all of the current in the circuit must go through the meter. That is, the meter is in series with the rest of the circuit. The simple current meter we designed in Fig. 21-1 will work in this circuit only if the current flow in the circuit is quite small. This is because the tiny coil in the meter movement has to carry all of the whole circuit's current. Auxiliary circuits, though, could make it possible for the movement to work on part of the current while the rest takes a bypass. The auxiliary circuit provides the bypass, so all of the current will go through the meter, but only part of it will go through the movement.

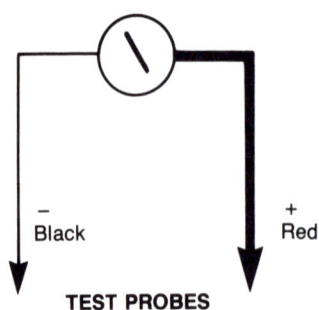

Fig. 21-1 Merely attaching test leads with probes to a meter movement makes a very simple current meter.

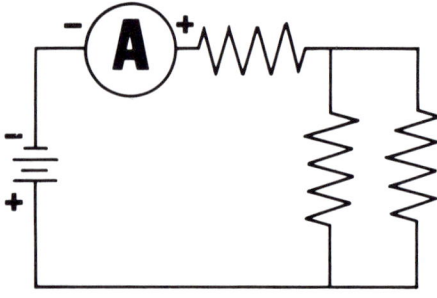

Fig. 21-2 The symbol for a complete meter is a circle with the measurement range identified. This circuit has an *ammeter* (for measuring Amps). Ammeters are placed in the circuit in series.

Meter Movement Sensitivity

As the current flowing through a meter movement's coil increases, the vane of the movement moves farther and farther upscale (usually to the right). Once a certain current is reached, the vane will point at the extreme upscale marking. There, a small pin will prevent the vane from moving farther. When this happens, the movement is **pinned** (or "pegged"). The best term to use for such a reading is **full scale deflection**. This means there is *exactly* enough current to make the vane point at the highest marking, and no more. When the meter is truly pinned, you do not know if there is just the right amount of current for full scale deflection or if there may be considerably more than enough — causing the movement to strain against the pin in an effort to move even farther. The exact amount of current which is required to make a meter movement show an exact full scale deflection is called its **sensi-tivity** rating. The sensitivity of a typical d'Arsenval movement may be as low as 50 or 100 microamps (µA), but this value depends on the number of turns in the moving coil and the strength of the permanent magnet and springs. The true sensitivity of any particular movement may be found experimentally. The importance of meter movement sensitivity is that, since the sensitivity rating is the current required for full scale deflection, a movement may not be used to measure any higher currents than its sensitivity rating. The exception would be if some of that current were sent through an auxiliary circuit. This, then, is the answer: auxiliary circuits in current meters make it possible to measure higher currents than the movement can handle on its own.

Auxiliary Circuits

By using auxiliary circuits, you could use a movement with a sensitivity rating of 50 µA to measure much higher currents, even currents of over 100 A! There are different types of auxiliary circuits for current meters.

Single Shunt Meters

If your only purpose for a meter is to measure currents up to 1 A and you wish to use a 50 µA movement to make this meter, then you could use a circuit like the one in Fig. 21-3. The resistor is placed in parallel with the movement so that it provides a bypass for some of the current to take around the movement. Notice that the dotted lines show that both the movement and the parallel resistor are inside the meter case. Resistors placed in parallel with a movement are called **shunts**. The shunt just gives some of the current another path to take in its trip through the meter, without going through the actual movement.

When the test probes are connected in series into a circuit, all of that circuit's

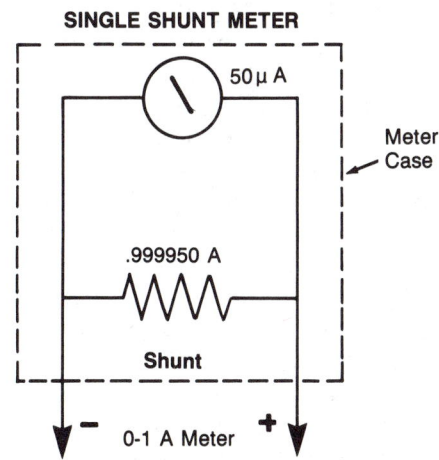

Fig. 21-3 The shunt provides a bypass for current to take through the meter, but around the movement. Current will increase and decrease proportionally in both the movement and the shunt when measurements are made.

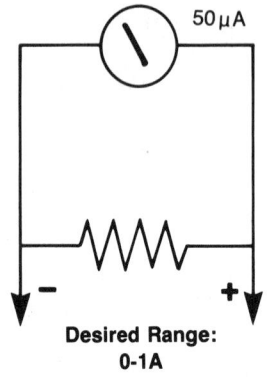

1.000000 Meter Range Desired, I_t
$-.000050$ Movement Sensitivity, I_m
$.999950$ Shunt Current, I_s

$R_m = 100\Omega$ (internal resistance of the movement)
To find Shunt Value (R_s), use this formula:

$$R_s = R_m \times \frac{I_m}{I_s}$$

$$R_s = 100 \times \frac{.00005}{.99995}$$

$R_s = 100 \times .00005$
$R_s = .005\Omega$

Fig. 21-4 When the sensitivity and the internal resistance of a movement are known, you can figure out the values of shunts needed to make meters of selected ranges.

current will go through the test probes and through the meter. Only part of the circuit's current, however, will go through the movement. If the circuit has a small current, then a little bit will go through the movement and the rest will go through the shunt. If the circuit has a high current, then a higher amount of current will go through both the movement and the shunt. Figure 21-4 shows how to find the proper value of shunt resistor to use in making a meter with a specific range from a given movement.

Multirange Meters

The single shunt meter only has one **range**, which means that it can measure up to the limit of its range (1 A in the case of the example). You could use this meter to measure any currents up to 1 A. If you wanted to measure a very small current, such as 20 µA, the vane would barely move at all. You would not be able to see the change well enough to read the measurement. Likewise, if you wanted to measure a current of 5 A, you still could not use the single shunt meter of the example. It is possible, though, to make a meter which can measure many different ranges of current. Such a meter would be called a **multishunt meter**.

Figure 21-5 shows a meter which would be capable of measuring any current from very low values up to 10 A in five ranges.

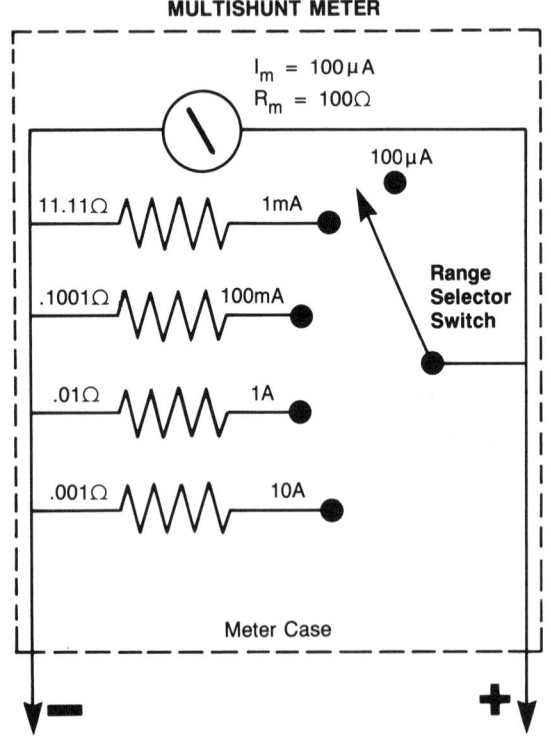

Fig. 21-5 The range selector switch chooses which shunt is used. The smallest value shunt is used for the 10 A range. The 100 µA range has no shunt (essentially, a shunt of infinite ohms resistance).

With this meter, you could always select the best range to get a deflection near the center of the scale. The extreme ends of the scale are not quite as accurate as the center region. You should begin by using the highest range and then turn back one range at a time until the vane is near or above a midscale deflection, then make the reading.

The switch selects which of the four shunts is placed in parallel with the movement for each range. No shunt is used for the lowest range, so the limit of that range is the sensitivity value of the meter movement. The value of shunt resistance for each range may be found in the same way as the value needed for a single shunt meter.

Alternating Current Meters

All of the meters and movements discussed so far will only work with DC current. If you try to measure an AC current with a d'Arsenval movement, you will have a real problem. Since AC currents change direction rapidly (120 times per second for house current) the magnetic field of the coil in the movement would be changing direction rapidly also. That means that the coil and vane assembly would be trying to move first one way and then the other way 120 times per second. Even if it could do this, your eyes would not be able to see it do so. You could not possibly read the meter. Actually, the movement cannot follow this demand for rapid movement and it will just sit still and oscillate (shake violently). If the AC current continues, the meter movement may be badly damaged.

There are meters which can measure AC currents. What these meters do is change the AC current to DC current in their auxillary circuits so that only DC is fed to the movement. An AC ammeter (current measuring meter) is shown in Fig. 21-6. The part which changes the AC to DC is the **diode**. The AC current which tries to flow through the meter movement looks like A in Fig. 21-7. The diode will only allow current to pass through in one direction, so half of the current is cut off (the pulses which are trying to flow backward through the diode). Thus, the current which actually flows through the movement looks like B in Fig. 21-7.

Notice that this current still pulsates, but

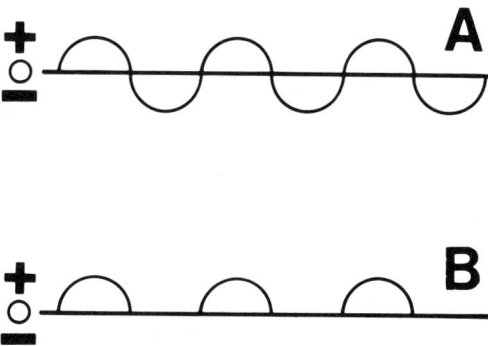

Fig. 21-7 An AC current appears in A. This current tries to flow through a meter, the diode only allows current in one direction, so the current through the meter would look as it does in B.

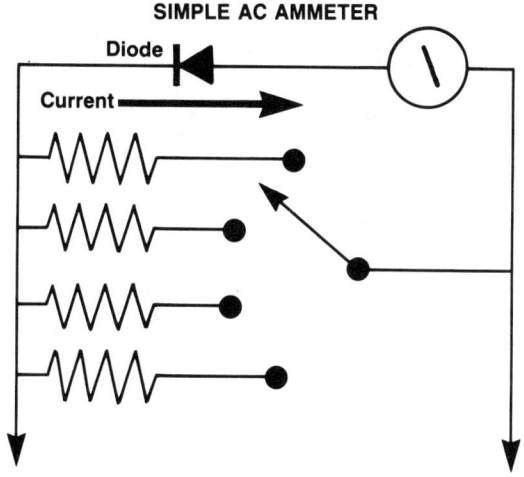

Fig. 21-6 The diode only lets half of the AC current pass, so pulsating DC goes to the movement and AC goes through the shunts.

it flows in only one direction. The + and — are never reversed. When this pulsating DC current is fed to the movement, it will still vibrate a small amount, but it never tries to travel back and forth between two extreme positive and negative readings as it would with AC current. The small amount of vibration which does occur is too slight and rapid for our eyes to detect, so the meter appears to hold still.

The shunt resistors are capable of carrying AC current with no problems. All of the current which flows through the meter, except the small amount that actually goes through the movement, is regular AC current.

Measuring Total Circuit Current

Figure 21-8 shows the proper way to connect an ammeter into a circuit. The circuit must be broken so that the meter may be inserted in series with the circuit load. In this arrangement, all of the current which flows through the circuit will have to travel through the meter. This rule only applies to current measurements, voltage and resistance measurements are made differently. If you were to become confused and connect an ammeter in parallel with the source, as shown in Fig. 21-9, what would happen?

Remember that Ohm's Law tells that the amount of current which will flow in a circuit depends on the amounts of voltage and resistance. Remember also that, in parallel circuits, each new branch provides another path for current to flow in, and each branch is connected directly to the full source voltage. This means that the meter which is incorrectly connected makes a new branch with far less than 1 ohm of resistance connected to a 100 V source. If you used 1 ohm as a rounded figure for easy calculation, you can easily see that the current flow through the meter would be 100 A! You can mathematically show that the current in the circuit should be 50 mA, so you are not getting a true reading when you connect the meter in parallel. What is

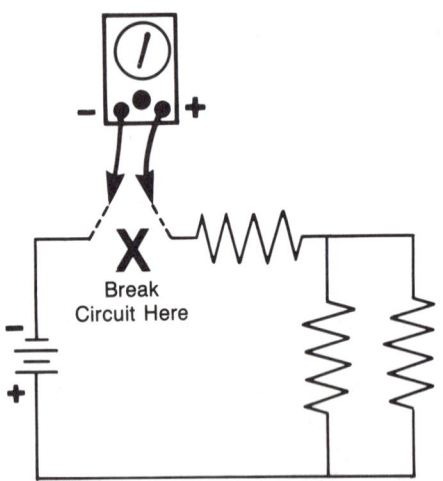

Fig. 21-8 The circuit must be broken so that the ammeter may be inserted in series into the line. Check the polarity carefully.

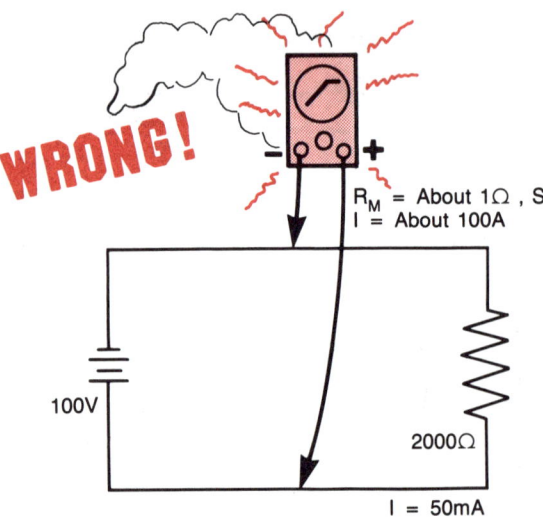

Fig. 21-9 Never connect an ammeter in parallel with the source or a load!

even worse, the high current which would flow through the meter would ruin it. Be very careful to always connect the meter in series when you wish to measure current.

Another important point is that the ammeter must be connected into the circuit in the proper polarity. In other words, the positive test lead (usually red) must be connected to the + side of the circuit and the negative (black) test lead will be connected to the other unconnected point in the circuit. Refer back to Fig. 21-8. Always connect the positive lead to the side of the circuit which is nearest to the + battery terminal. If the meter is connected backward, it will try to display the current in reverse (downscale), and it may be damaged badly if there is high current. Ammeters for AC currents do not have any polarity and may be connected either way.

Current Measurement Troubles

The fact that current meters must be connected in series limits their usefulness

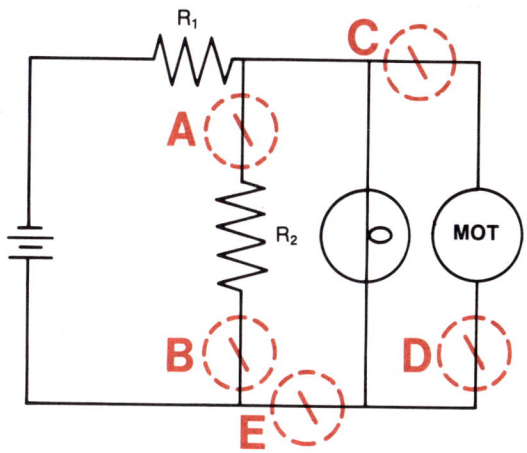

Fig. 21-11 Meters placed at A or B would measure only the current of R_2. Meters in locations C or D would only measure the motor current. What currents would a meter at E measure? It would measure the combined currents of the lamp and the motor. Where else in this circuit could that same measurement be obtained?

as test instruments. This is because the technician would have to break a connection and then reconnect it after the measurement was completed. One way around this is to just measure the voltage with a voltmeter (which does not require breaking the circuit) and then to calculate the current by Ohm's Law. When fairly large AC currents are to be measured, a clamp-on ammeter may be used. These meters, however, will not work for DC currents or for very small currents.

To use a clamp-on AC ammeter, a lever is squeezed which opens the clamp so the wire to be measured may be slipped into the open area inside the clamp. As the current flows through the wire, the magnetic flux field that it sets up around the wire is monitored by the clamp-on meter. As the amount of current increases, the flux strength goes up, too. The meter works on the same principle as a transformer (covered in Chapter 30).

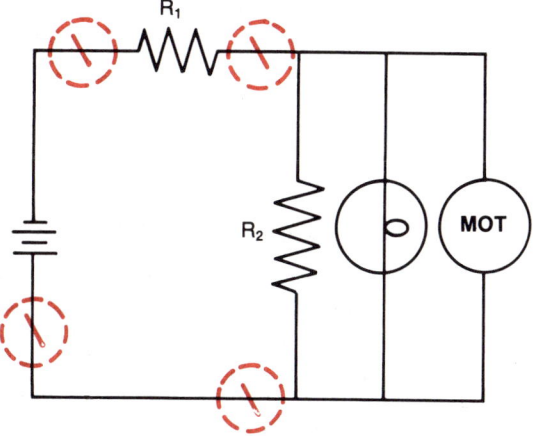

Fig. 21-10 Here are four places in which this circuit could be broken to measure the total circuit current.

Chapter 21   *Current Meters*

Measuring Partial Circuit Currents

Figure 21-10 shows four places in the circuit that you could place an ammeter to measure all of the current in the circuit. What would you do to measure only the current which went through R_2 or the motor? The meter must be placed in a series location in the circuit that will only allow the current which goes through that particular device to pass through the meter. Placements for meters to measure the current through specific devices and branches are shown in Fig. 21-11. Remember to connect meters in proper polarity.

Summary

Ammeters are used for measuring current. The auxiliary circuits of ammeters use shunt resistors to allow current measurements that are higher than would be possible with the movement alone. Sensitivity refers to the current required for full scale deflection. Ammeters are connected in series in the circuit. They must be connected in proper polarity. Ammeters for AC currents have diodes to convert the current to pulsating DC. They do not have polarity.

Important Terms

pinned
full scale deflection
sensitivity
shunt
range
multishunt meter
diode

Review Questions

1. Draw a series-parallel circuit with several components and devices, and then mark the proper places to measure the total current and the currents through a few chosen devices.
2. If you had an ammeter which could measure up to 1 A and you wished to make it capable of measuring higher currents, what would you have to do to it?
3. Why must ammeters be connected in series?

Careers: Physicist

Physics is an important area of science. It includes (among other things) the study of atomic structure, motion, laws of nature, mathematics, and electronics. Many of the advances in electronics were first imagined and calculated by physicists. Many of the great scientists who have contributed so much to electronics were physicists. A physicist may be a researcher and teaching professor at a university or may work for the research department of a major corporation or governmental branch. To become a physicist, you would need a very strong background in mathematics and science. Your electronics training also will be helpful. Advanced college or university training is required for upper level jobs in the area of physics.

Project: Easy Use Multitester

Parts List

1	Tri-color LED (Tri-state)	D_1
1	3 V battery with holder	B_1
1	150 ohm resistor	R_1
1	220 ohm resistor	R_2
2	330 ohm resistors	R_3 & R_9
4	100 ohm resistors	R_4, R_6, R_7 & R_8
1	33 ohm resistor	R_5
1	3.9 K ohm resistor, 5 watt	R_{10}
1	5 K ohm resistor, 5 watt	R_{11}
1	10 position rotary switch	S_1
2	Test leads with probes or clips	
1	Suitable enclosure	
1	Small grommet or bezel for LED	

Notes

This is a very accurate and simple to use multitester which can measure voltages as low as 3 V or as high as 240 V. The "tristate" LED will glow green if the voltage is of correct DC polarity (according to the placement of the probes) or red if the polarity is reversed. The LED will glow yellow if the measured voltage is AC. Continuity checks cause the LED to glow green. To use the multitester for voltage measurements, set the switch to the highest setting (240 V) and connect the probes to the source you wish to measure. Then, slowly rotate the switch to each voltage setting (in turn) until the LED indicates a voltage. When the LED glows with normal brilliance, the setting of the switch indicates the voltage. The color of the LED indicates the type of voltage. Continuity checks should only be made on unpowered circuits. If using salvaged parts and the 10 position rotary switch is not available, a smaller one may be used by eliminating the least useful positions. However, if this approach is taken, **do not** omit the resistors because their resistance in series is added to obtain the proper resistance for any higher voltage measurements.

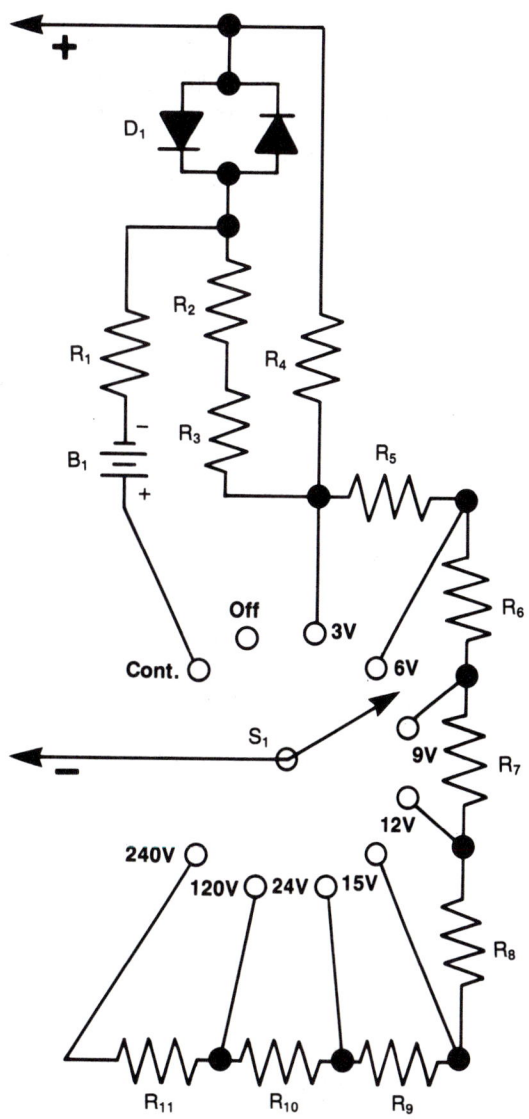

Fig. 21-12 An inexpensive multitester may be made by using a tri-state LED instead of a meter movement to indicate the value measured.

Chapter 22

Voltage Meters

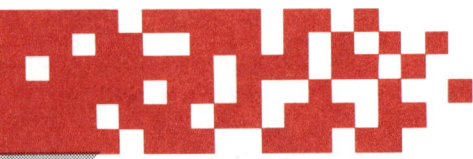

Voltage Measuring Meters

The simplest form of current meter discussed in the last chapter was a d'Arsenval movement with two test leads attached. Though this meter would work for current, it would not work for measuring voltage. Figure 22-1 will help you to see why. Since **voltmeters** are connected in parallel with the source or drop they measure, if the sensitivity rating of the meter movement is only 100 microamps or so, then even very low voltages would pin the meter. With an internal resistance of 200 ohms and a sensitivity of 100 μA, the meter movement would register a full scale deflection with only .02 V. A potential of even a tenth of a volt would peg the meter violently! Since resistance of the meter is very low, connecting it in parallel with a source could draw so much current that it would actually change the circuit's voltage by pulling it down. This effect is called **loading**. Therefore, all voltage meters, even the very simplest ones, need some added resistance in their auxiliary circuits.

Auxiliary Circuits

The auxiliary circuit for a simple, single range voltmeter is shown in Fig. 22-2.

Fig. 22-1 The voltage needed for full scale deflection is found by Ohm's Law:
$E = I_m \times R_m$
$E = .0001 \times 200$
$E = .02$ V   Thus, the movement (by itself) is useful as a voltmeter only in *extremely* low voltage circuits.

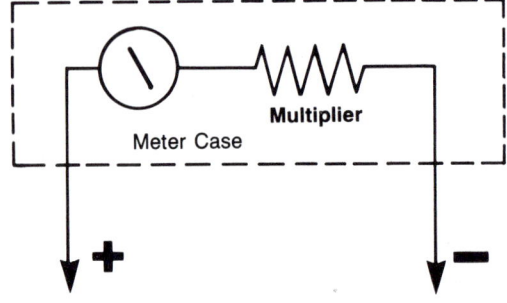

Fig. 22-2 Series connected *multiplier* resistors limit the current in voltmeters.

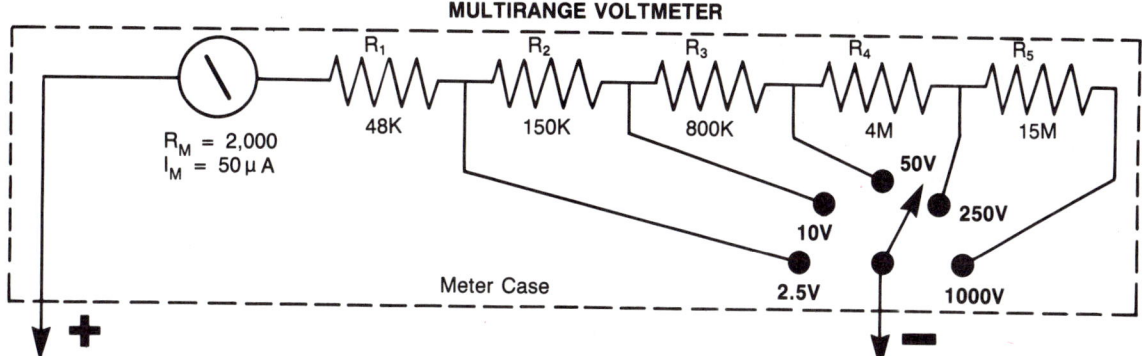

Fig. 22-3 When the 1 V range is selected, only one multiplier is used, but when the 50 V range is selected, the current must pass through three multipliers: R_1, R_2, and R_3. This simplified circuit uses the actual values of a very popular meter.

The resistors in ammeters, connected in parallel, are called shunts. The resistors in voltmeters, however, are connected in series with the movement and are called **multipliers**. The multiplier resistor limits the amount of current that the meter draws, so it helps to minimize the loading effects of uisng the meter in a working circuit. It also prevents too much current from flowing through the movement.

Multirange Voltmeters

The circuit for a multirange voltmeter is very similar to the one for a multirange ammeter. The main difference is that the resistors must be in parallel with the movement in the ammeter, but in series in the voltmeter. Figure 22-3 shows a five range voltmeter. Notice that there is at least some extra resistance (besides the internal resistance of the movement) for even the lowest voltage range. Notice also that this meter uses the combined resistances of two or more multipliers for all ranges except the lowest one. The current for the movement must flow through each multiplier selected.

Figure 22-4 shows how to find multiplier values used in making a voltmeter for measuring a certain range when the movement sensitivity and internal resistance are both known.

Alternating Current Voltmeters

Just as with ammeters, you can make an AC voltmeter by using a diode to **rectify** the current. Rectify means to change AC to DC. A simple AC voltmeter is shown in Fig. 22-5. Some AC voltmeters and ammeters use more than one diode. You will learn more about these sophisticated rectification circuits in the Chapter 41 discussion on power supplies. For now, accept the fact that smoother forms of DC can be produced from AC if more than one diode is used.

Measuring Voltages

In the previous chapter, it was said that ammeters were not very handy test instruments because you have to break the circuit to insert them into it in series. Voltmeters are the easiest to use of all meters for power-on measurements. You do not have to turn the power off. In fact, to do so would make it impossible to measure voltage. You do not have to disconnect or break the circuit either. A voltmeter is connected in parallel with the source of voltage or the voltage drop which is being measured.

SIMPLE AC VOLTMETER

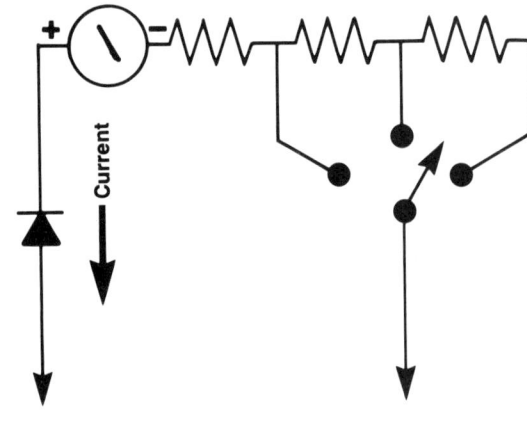

Fig. 22-5 A diode may be used to make a simple AC voltmeter.

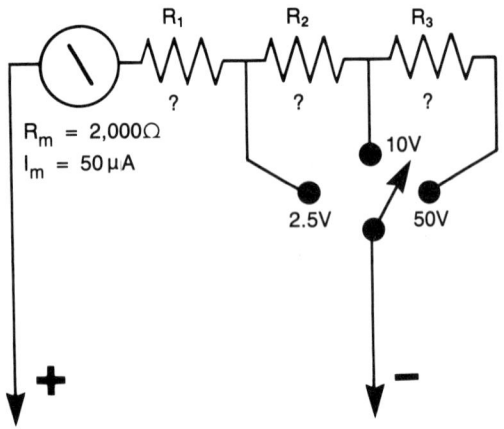

To find the proper multiplier for the 2.5V range, we use,

$$R_{mult.} = \frac{E_{range}}{I_m} - R_m$$

$$R_{mult.} = \frac{2.5}{.00005} - 2000$$

$$R_{mult.} = 50,000 - 2000$$

$$R_{mult.} = 48,000 \Omega \text{ OR } 48K\Omega$$

Then, to find the needed value of R_2, you include the value of R_1 as part of the movement's resistance. Complete this calculation on scratch paper, and then check your answer by looking back at Fig. 22-3.

$$R_{mult.} = \frac{E_{range}}{I_m} - R^{m+r}$$

$$R_{mult.} = \frac{10}{.00005} - (2,000 + 48,000)$$

$$R_{mult.} = ? \quad \text{(Finish this one)}$$

Can you find the value of R_3 on your own?

Fig. 22-4 Finding multiplier values.

MEASURING SOURCE VOLTAGE

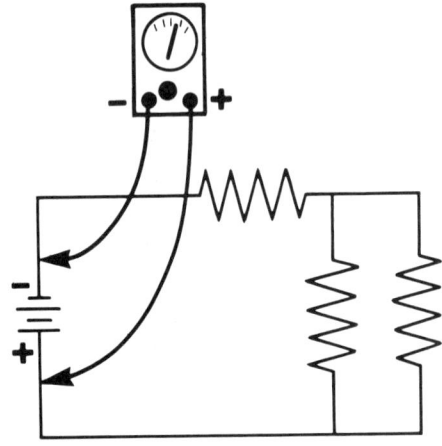

Fig. 22-6 Carefully check the polarity, and then connect the meter in parallel with (or, as some people would say, "across") the source. Note that the meter is also in parallel with the rest of the whole circuit, so its reading would be the total voltage drop of the circuit, as well as the source voltage. This is in keeping with Kirchoff's Voltage Law.

Figure 22-6 illustrates the proper way to connect a voltmeter to read the voltage of the source in a circuit. This same connection method would be used to read the total voltage dropped by this whole circuit! The polarity must be carefully checked so that the meter may be connected correctly. Again, as with ammeters, the red (positive) lead is connected to the plus side of the circuit and the black (−) lead goes to the negative line. Like AC ammeters, AC voltmeters have no polarity.

To measure a single voltage drop of a particular device or component, the positive lead is connected to the end of the device nearest to the positive terminal of the source. The negative lead is connected to the other end of the device. This arrangement puts the voltmeter in parallel with only that one device so that it measures only that one drop. See Fig. 22-7. Sometimes, you may need to measure the combined drops of several components. In this case, connect the voltmeter in parallel with the combination of drops as shown in Fig. 22-8.

Never break the circuit and place a voltmeter in series with any part of the circuit. This adds the high resistance of the voltmeter in series into the circuit. This also limits current flow in the circuit so that the circuit would probably not work correctly, and any reading which registered on the meter would be useless anyway. With multirange meters, begin the measurement of any unknown value by using the highest range on the meter, and then turn back one range at a time until a readable measurement is obtained.

Since voltmeters are so easy to operate, they are the most often used test instruments. Learn to use the voltmeter capably. If you want to find the current in a circuit, it is often easier to measure the voltage, read the resistance by color code, and then calculate the current by Ohm's Law than it is to disconnect the circuit just to use an ammeter.

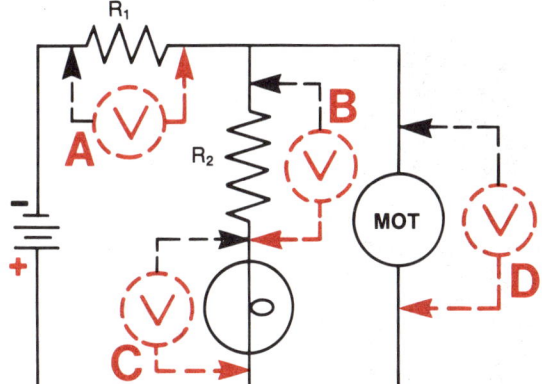

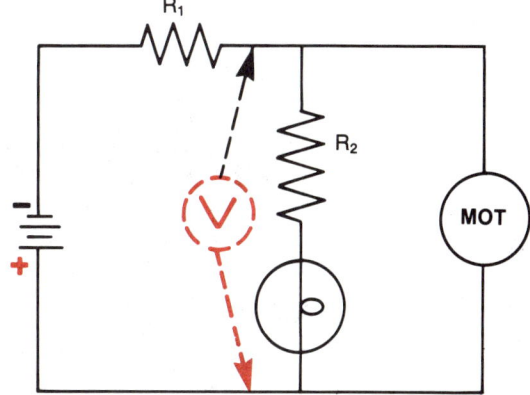

Fig. 22-7 Placing the voltmeter in parallel with R_1 (location A) will measure the drop of R_1 alone. Location C (across the lamp) measures only the drop of the lamp. What do locations B and D measure?

Fig. 22-8 Connecting the voltmeter as shown here will measure the voltage dropped by the whole series-parallel network of R_2, the lamp, and the motor. It will only omit the drop of R_1.

Chapter 22 *Voltage Meters*

Electronic Voltmeters

If there is not much resistance in the auxiliary circuits of a voltmeter, the operating characteristics of the circuit being measured may be seriously affected by using the meter. Remember, since you connect a voltmeter in parallel, you are actually forming another parallel circuit branch which will draw more current from the source. This is called circuit loading. Some circuits are so sensitively balanced that the loading caused by using the simple voltmeters in this chapter would greatly upset their operation. These meters are fine for many circuits, however. In those cases where loading is a serious problem, special electronic auxiliary circuits which use transistors or vacuum tubes may be needed to prevent circuit loading. Two of these special meters, the transistor voltmeter (TVM) and the vacuum tube voltmeter (VTVM) are discussed in Chapter 25.

Summary

Voltmeters are easier to use for circuit testing because they are connected in parallel. The resistors in voltmeters are called multipliers, and they are in series with the movement. Circuit loading can be a problem when using voltmeters, so they should have considerable resistance. (Remember that current meters frequently have less than one ohm of resistance.) AC voltmeters are available. The circuits for voltmeters are similar to the ones for ammeters except that the parallel shunts are replaced with series multipliers. Chapter 23 deals with meters for measuring resistance — ohmmeters.

Important Terms

voltmeter   multipliers
loading   rectify

Review Questions

1. What is the basic difference in the construction of ammeters and voltmeters with several ranges?
2. What is the difference in the procedure for using voltmeters as opposed to ammeters?
3. Why do voltmeters need some resistance in addition to the internal (coil) resistance of the movement even for very low ranges?
4. Why are voltmeters used more often than ammeters for circuit testing in the field?

Chapter 23

Resistance Meters

Meters for Measuring Resistance

Like voltage and current, resistance may also be measured with electrical meters. Meters used to measure resistance are called **ohmmeters**. Ohmmeters have some important features which make them different from either voltmeters or ammeters. First, voltmeters and ammeters are powered by the circuits they measure. Ohmmeters, however, do not measure voltage or current. Instead, they measure resistance. Since resistance alone cannot make electrons flow through the coil in the movement, there must be some power source for the meter. Consider what would happen if you connected a meter movement to a resistor with no power source. Nothing in this circuit makes current flow in the coil of the movement, so it will not move at all — regardless of how much or how little resistance the resistor has.

You want to make a circuit which will cause the meter to move a different amount when there is a change in the resistance of the component or circuit being measured. Figure 23-1 shows the simplest ohmmeter powered with a small battery.

Zero Adjust Control

There is another part in this meter that you have not seen in voltmeters and ammeters. This part is the rheostat (R_2), labeled "zero adjust." A potentiometer could be used in place of the rheostat.

As the battery ages, it weakens. If there were no way to adjust for this, then the meter would show different readings for the same resistance when measured with new batteries or old batteries in the meter. Such a meter would be of little value. The **zero adjust control** allows you to lower the resistance of the auxiliary circuit when the battery gets old and weak, but then to add more resistance when the battery is changed. The proper way to adjust this control is discussed later in this chapter.

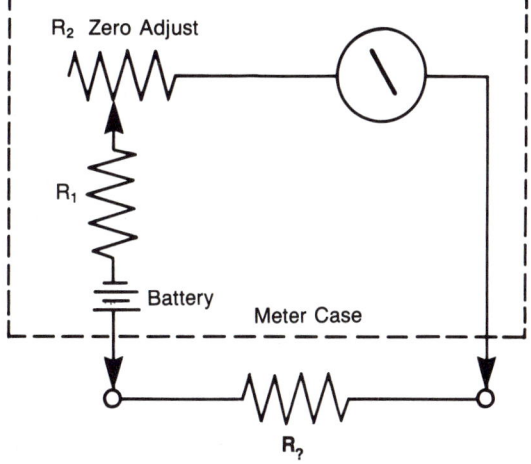

Fig. 23-1 The power source for an ohmmeter is usually a battery. The fixed resistor (R_1) and the zero adjust control (R_2) help calibrate the meter.

Chapter 23 *Resistance Meters*

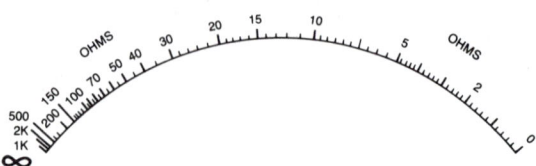

SERIES OHMMETER SCALE

Fig. 23-2 In the series ohmmeter, the scale reads backwards — from right to left. Notice also that the values crowd together at the high end of the scale. That is, the scale is not linear. Refer back to Fig. 20-4 to see a linear scale.

How Ohmmeters Work

To measure the value of an unknown resistance, connect that component or circuit between the test probes. The battery would force electrons to flow through the component being measured ($R_?$), then through the meter movement, R_1 and R_2, and back to the battery. If the unknown resistance is a small value, a lot of current will flow and the meter movement will make a large deflection. If the component has high resistance, less current will flow and the movement will deflect less. If the unknown resistance is infinite (as in an open circuit or component), then the meter movement will not move at all. If the component is shorted (zero resistance), then the movement will show a full-scale deflection. Notice that the movement deflects the most when there is lowest resistance and the least when there is infinite resistance. For this reason, the scales of many ohmmeters have 0 on the right end and infinity on the left end. See Fig. 23-2. This is the opposite of the way that you generally expect to find things because our language reads from left to right. However it makes sense when you think about how the meter actually works. The fixed resistor in the meter's auxiliary circuit (R_1) ensures there is some resistance in the circuit to prevent pegging the meter. Since the fixed resistor, zero adjust rheostat, battery, and $R_?$ are all in series with the meter movement, this particular device is a **series ohmmeter**.

Calibrating Ohmmeters

Before an ohmmeter can be used to make a resistance measurement, it is important to set the zero adjust control to allow for the condition of the battery. Calibrating the meter, as it is called, must be done before any measurement is made. Multirange ohmmeters should be recalibrated whenever the range is changed. To properly calibrate an ohmmeter, touch the two test probes together — this is called "shorting the leads" — and observe the meter. If the meter is properly calibrated, it should register a full scale deflection (a reading of 0). Shorting the leads is the same as measuring an unknown resistance of 0 ohms. If the vane points exactly at the 0 mark on the meter face, then the meter is properly calibrated. If the vane points slightly past 0 or just a little short of 0, then the 0 adjust control must be turned to move the vane to an exact 0 reading.

Shunt Ohmmeters

There are ohmmeters which place the unknown resistance in parallel with the meter movement. These devices are called **shunt ohmmeters**. They have some advantages and disadvantages when compared to series ohmmeters. Figure 23-3 shows the circuit of a simple shunt ohmmeter. The important feature is that the movement is in parallel with (shunted by) the unknown resistance. Look also at Fig. 23-4 to see the scale for a shunt ohmmeter. Notice that, unlike the series ohmmeter, this scale reads from left to right (0 to infinity). Study the circuit to see why this is so.

When a high resistance component is being measured, the total circuit resistance is high, so little current is drawn from the battery or through R_1 and R_2. As Ohm's Law would predict, when the current flow through $R_{1\&2}$ is low, the voltage drop of the pair is also low. That means that both the meter movement and the unknown resistance ($R_?$) receive fairly high amounts of the battery's voltage. This potential will cause many electrons to flow through the movement and produce a large deflection. On the other hand, if $R_?$ is a low value, the total circuit resistance will be lower. So, the current through $R_{1\&2}$ will be higher when $R_?$ is a low value and more voltage will be dropped by the auxiliary circuit ($R_{1\&2}$). Thus, with a low value of $R_?$, the movement will not carry as much current and will not deflect as much. The values of R_1 and R_2 are set so that when $R_?$ is infinite (or too high for the meter to measure), there is just enough current flow in the auxiliary circuit itself to cause exact full scale deflection. If the leads are shorted, all of the current will take the easy path from R_1 back to the battery through the shorted leads. No current will be available to go through the movement and it will not move. The same thing would happen if a shorted component were measured. Thus, the meter scale reads from left to right with 0 on the left.

The main advantage of shunt meters is that they can measure very low resistances. There is, however, a serious disadvantage. Since there is always a path for current to travel, even when no measurement is being made, there must be a switch in the battery circuit. If you forget to turn this switch off when the meter is not in use, the battery will be wastefully discharged in a short time. Many ohmmeters have both series and shunt sections to allow both low and high resistance measurements. Multirange ohmmeters are also available.

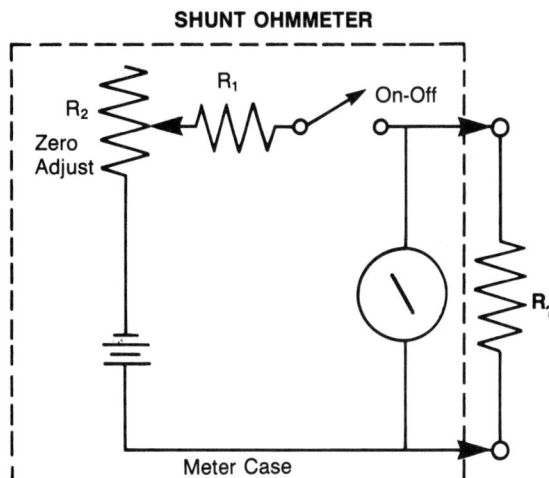

Fig. 23-3 In the shunt ohmmeter, the component being measured ($R_?$) is placed in parallel with the movement.

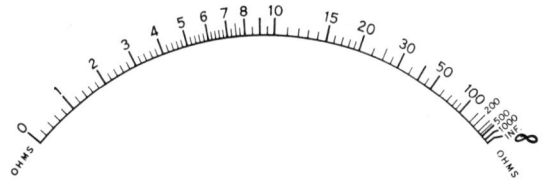

Fig. 23-4 The scale of a shunt ohmmeter is similar to a series ohmmeter scale, except that it reads from left to right.

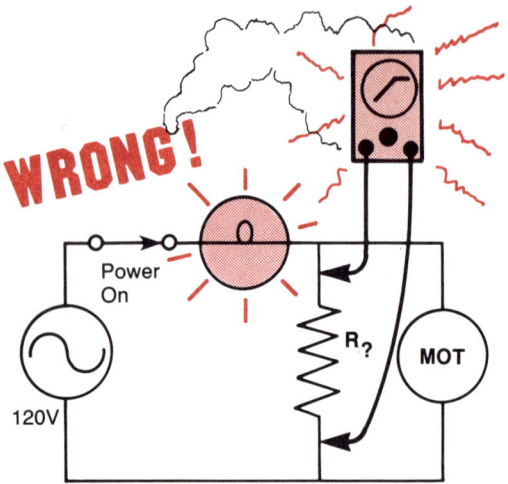

Fig. 23-5 Never connect an ohmmeter into a circuit which has the power on! The circuit's voltage and current will make the measurement incorrect and could damage the meter severely. There is even danger of personal injury to the user and damage to the circuit under test. Disconnect the circuit's power source first!

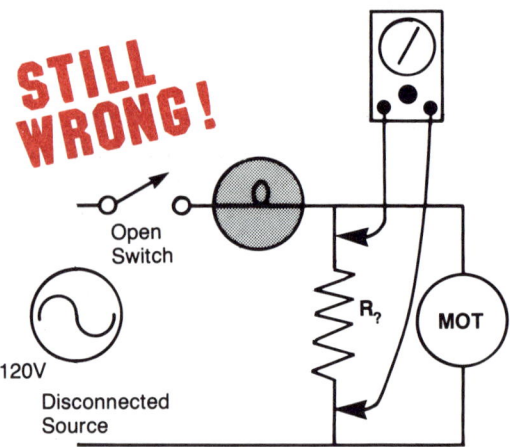

Fig. 23-6 Here, the circuit's power has been properly removed, but the meter is measuring the parallel resistance of $R_?$ and the motor, instead of $R_?$ alone. The reading will be lower than it should be.

Using Ohmmeters

Since ohmmeters have their own power source (usually a battery), they need no power from the circuit being measured. In fact, any extra power obtained from a circuit would disturb the measurement and make it useless. If the external EMF is more than a few volts, it is likely to ruin the meter. See Fig. 23-5. So, the first thing to do when using an ohmmeter, is always disconnect the circuit and turn the power off. If a single component which is not in a circuit is to be measured, just touch the leads from the component with the test probes and read the meter. When the component is in a circuit, though, this method will not work. Even with the power off, if you connect an ohmmeter as in Fig. 23-6, the current used to measure the resistance of $R_?$ will also have a path to travel through the motor. Therefore, more current will be drawn than should be drawn and the movement will be tricked

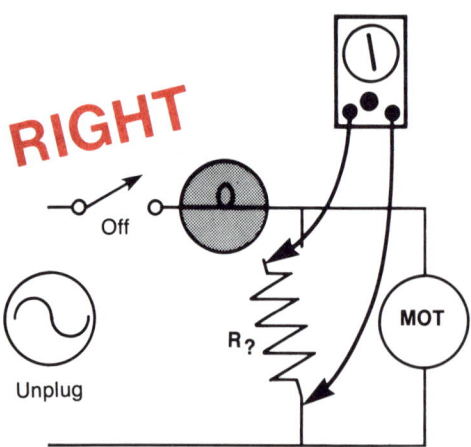

Fig. 23-7 The true value of $R_?$ may now be measured safely. The power is disconnected and so is one lead of $R_?$.

into thinking that there is less resistance that that of $R_?$. Actually, the movement will be correct. You have added another parallel branch to the circuit (the motor), and thus reduced its total resistance. The proper way to measure $R_?$ is to disconnect one of its leads from the circuit as shown in Fig. 23-7. Now, the meter's current can only go through $R_?$ because there is no other complete path back to the meter. So, two things must be done before in-circuit resistance measurements may be made:
1. Turn off and disconnect power.
2. Disconnect one lead of the component being measured.

Summary

Two types of ohmmeters are used to measure resistance. Series ohmmeters read backwards; they do not need a power switch. Shunt meters read left to right and can measure smaller values of resistance, but their batteries discharge if they are not turned off when not in use. Ohmmeters have their own power sources and may be damaged if the power to the circuit they are measuring is not disconnected. Resistance measurements may be confused and made inaccurate by the home circuit of a component to be measured, so one lead of the component must be disconnected when the measurement is made. The ohmmeter must also be calibrated by the zero adjust control before accurate measurements may be made. The next two chapters discuss meters that can measure more than one electrical quality, like the ones you use in your class.

Important Terms

ohmmeter   series ohmmeter   shunt ohmmeter
zero adjust control

Review Questions

1. Why do ohmmeters use a built-in power supply instead of depending on the power of the circuit they are measuring?
2. Explain how to make a resistance measurement of a component in a circuit.
3. Why does the scale of a series ohmmeter read backward?

Chapter 24

Multimeters

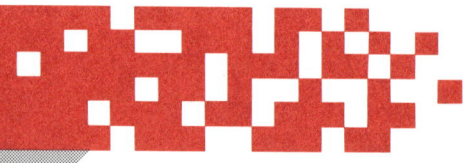

Three Meters in One

People who work with electricity or electronics make many different types of electrical measurements. They could use separate ohmmeters, voltmeters, and ammeters for each measurement, but this would be expensive and awkward. Since the most expensive single part of any meter is the movement, it makes sense to use one movement with several auxiliary circuits. This is exactly what is done in the multimeter. A **multimeter** has one meter movement with a complex auxiliary circuit that enables it to measure voltage, current, and resistance. A good multimeter is the first instrument you should buy if you wish to collect your own set of electrical equipment.

Uses of Multimeters

Most multimeters can measure voltage in several ranges from a couple of volts up to 1000 or 5000 volts AC or DC. Multimeters can also measure resistance. Some multimeters use a shunt (parallel) ohmmeter circuit for a "low ohms" range and a series ohmmeter circuit for other ranges such as "R × 1," "R × 100," "R × 10K," etc. see Fig. 24-1. The more ranges, the easier it is to make accurate measurements.

A multimeter can test current, too. Most multimeters are limited to only a few DC current ranges. In many multimeters, the highest current ranges are 1 A or lower. Ranges of 500 mA are very common. Ranges of up to 2 A or 10 A are found less often. For this reason, the current measuring section of a multimeter is usually called a "milliammeter." AC current ranges are

Fig. 24-1 Most multimeters have three or more resistance ranges.

generally not found in multimeters.

Most professionals use the initials **VOM** instead of the word "multimeter." There is some confusion, however, about the meaning of these initials. The multimeter is most often used to measure voltage or resistance because breaking the circuit for a power-on current measurement is so much trouble. In fact, some very inexpensive multimeters do not even have current ranges. Since they are used mostly for measuring voltage and resistance, many people think the initials VOM stand for "volt-ohm meter," but this is not true. VOM really means **volt-ohm-milliammeter**, and a true multimeter does have all three capabilities.

Advantages and Disadvantages of Multimeters

It is certainly less expensive to buy one multimeter than to buy separate volt, ohm, and milliamp meters. It is also easier to carry around one meter than to carry three separate ones. Likewise, it is less trouble to turn the range selector knob on a VOM than it is to hunt for and connect another meter each time you want to change from one type of measurement to another. High quality VOMs also have attachments that are used to convert them into transistor testers, thermometers, and even clamp-on AC ammeters.

One disadvantage is that if you want to measure both the voltage and the current in a circuit at the same time, you will need two separate meters. Some inexpensive VOMs are not very accurate instruments, and they tend to load circuits excessively. These meters are fine for crude electrical work, but do not use them in delicate electronic circuits. If you want to puchase a multimeter, buy a high-quality instrument.

Reading the Multimeter

Reading the multimeter requires you to observe two things. First, the range and/or function selector knob(s) must be set properly. Then the actual reading may be made from the meter scale. See again Fig. 24-1. It shows the front of a typical VOM. Notice that there are three knobs, seven holes, and several scales on this meter. The knob on the left selects whether AC or DC ranges are to be used, and it also can reverse the polarity of DC current. The right knob is the zero adjust control for the ohmmeter section. The large knob is the main range/function selector. Reading clockwise from the lower left (about 7 o'clock) position, you

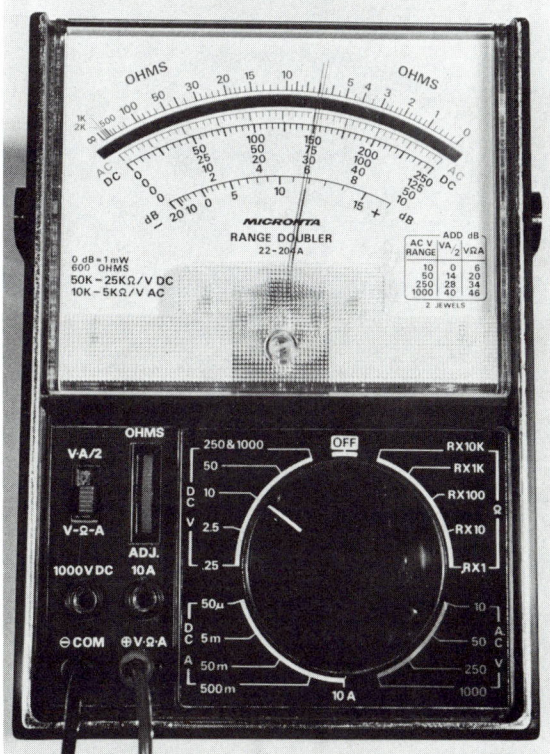

Fig. 24-2 We read the 10 V scale because the 10 V range has been selected by the main function knob.

find "off," seven voltage ranges, two low-power ohms ranges, four normal ohms ranges, and four current ranges. The two most important holes are the ones in the lower left corner labeled "common" (for the black test lead) and "+" (for red lead). These are used for most measurements. The other holes are jacks for special functions which are not used often. Inexpensive multimeters with only eight ranges that are selected by plugging the test leads into different holes are also available. Such meters are less expensive, but not as convenient to use.

Making Voltage Measurements

Figure 24-2 shows how to read the meter when it is set for a voltage measurement. First check the range setting of the selector knob. The 10 V range has been selected. Now, look at the meter face and decide which scale to read for the measurement. Since you want the 10 V DC range (as selected by the function knobs), look to find a scale which is labeled to let you know that it is a DC voltage scale. Some meters may say "volts" on this scale, this one just says "DC." Here your choices of DC voltage scales are 10, 50, and 250. The function knob is set to 10 V, so it is easy to see that you must read from the 10 V scale. The reading is 6 V. If the function knob had been set for the 100 V setting, you would read the same scale, but the "10" at the high end of the scale would really represent 100 instead of 10. Therefore the measurement would be 60 V. If the knob had been in the 250 V position, instead of the 10 or 100 V settings, you would read the scale marked 250 at the highest end. Then the measured value would have been 150 V (see the top voltage scale). What would the reading have been if the pointer was in this same position, but the function knob was set for 500 V? Again, find the proper scale ("50" in this case) and then read it. You should get 300 V for your answer.

Try to read the meter in Fig. 24-3. Make sure you have checked which scale to read by looking at the setting of the function knob. The answer is 3.4 V. The large mark midway between 2 and 4 on the lower voltage scale represents 3. The reading is two small marks past this large (3) mark. Since there are five small spaces between the large ones, each small mark stands for 1/5 of a large one. If a large mark on this scale stands for 1, then a small mark here stands for 0.2. The measurement is two small marks beyond the 3, so the meter would be read as 3.4 V. If the function knob were in the 100 V position, the reading would be 34 V. What would the reading be if the selector were set for 250 V? Read the top scale. The reading would be 85 V. On this range, the large mark between 50 and

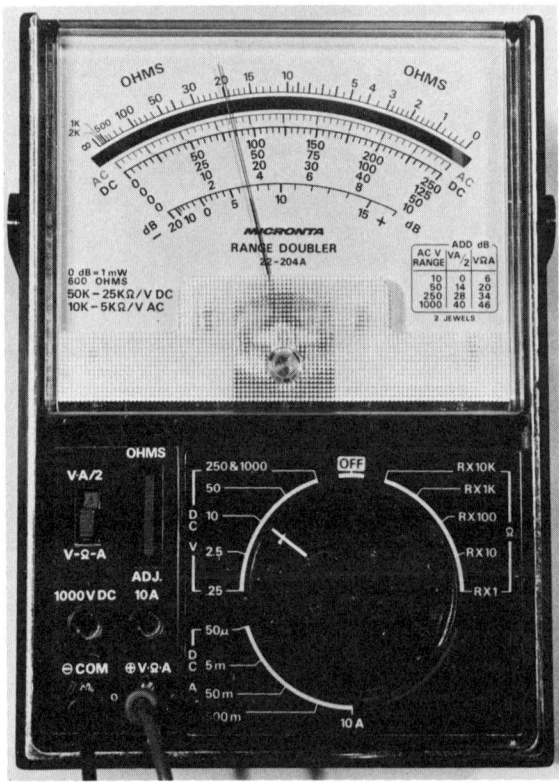

Fig. 24-3 The divisions between numbers change in value depending on the scale and range with which they are used.

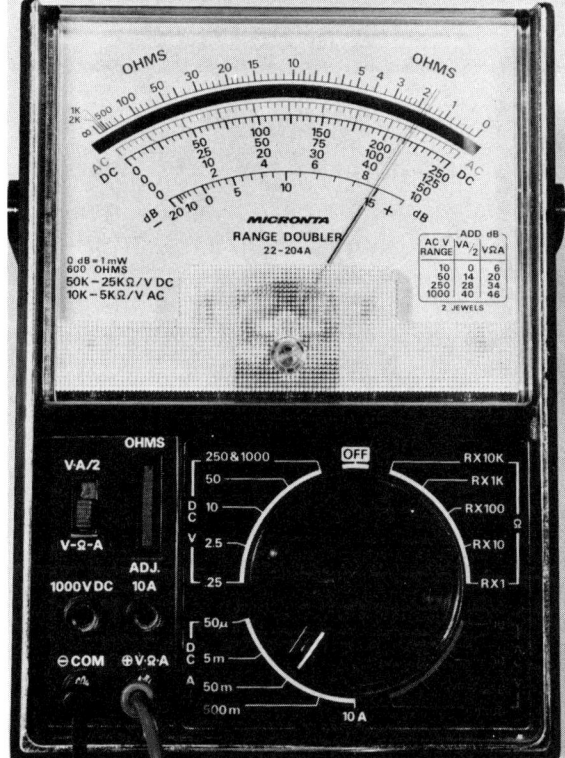

Fig. 24-4 Reading current measurements is much the same as reading voltage measurements.

Making Resistance Measurements

For this discussion, the range marked "R x 100" will be used and the very top scale which goes with it. Figure 24-5 shows a resistance measurement. Try to read it on your own. Always remember to disconnect the power from the circuit and loosen one lead of the component being tested when you use the ohmmeter section of a VOM. Also, always check the zero adjust setting whenever you change ranges.

100 stands for 75. Make up a few more examples to try on your own. The AC scale (below the DC scale) reads the same way, it is just slightly offset from the DC scale.

Making Current Measurements

If the meter is inserted in series into a circuit, the current functions can be used. Figure 24-4 shows a possible current measurement. On this meter, the same scale which was used for voltage (DC) is used for DC current. Some meters have separated scales. The function knob is turned to the 50 mA range, so the measurement is 43 mA. If the function selector were set on 500 mA, what would the reading be?

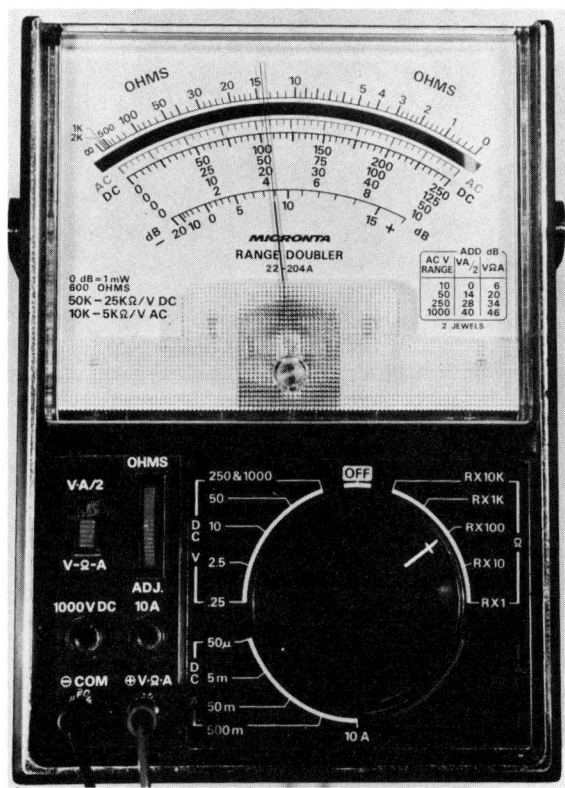

Fig. 24-5 Check the range selector and then multiply the actual reading by the multiplier selected (100 in this case). Did you get 1400 Ω for your final answer?

Chapter 24 *Multimeters*   147

Special Features

Many companies produce multimeters in a wide price range. Figure 24-6 shows a **mirrored scale** which is found on the more expensive meters. The mirror helps to avoid parallax errors. When you are directly in front of the scale as you read the meter, the reflection of the pointer in the mirror will be perfectly hidden by the actual pointer which is in front of it. If your head is off to either side, however, you will be able to see both the pointer and its reflection in the mirror. So, if you can see two pointers in the mirrored section of the scale, move your head to avoid parallax error. Some more expensive meters are equipped with overload protection circuit breakers. Generally, the more expensive a multimeter is, the more accurate it will be and the more features it will have. However, a very inexpensive meter will work fine for most home applications.

Summary

Multimeters, or VOMs, are widely used in the field of electricity. They measure all three of the most important qualities in DC electrical circuits in addition to AC voltages. VOMs vary in price, features, and accuracy. To read a VOM, you must observe the positions of the range selector knob and the meter pointer. You also must read from the proper scale. The next chapter concerns electronic meters, which are similar to multimeters.

Fig. 24-6 If the pointer does not cover its own reflection in the mirror, then your head is not directly in front of the meter and a parallax error is being made.

Important Terms

multimeter   volt-ohm-milliammeter
VOM   mirror-scale

Review Questions

1. What can you measure with a typical multimeter?

2. Since there are usually several scales on a VOM, how can you tell which scale to read when you make a measurement?

3. Why do most multimeters have a battery in their auxillary circuits?

Careers: Instrumentation

Instrumentation includes many job titles, but they all require understanding of basic measuring techniques and instruments. In electronics, instrumentation involves all sorts of electrical and electronic measurement devices. It could be the operation and repair of the equipment or it could require observing the equipment to ensure certain conditions are maintained. For instance, you could be involved in checking the instruments that control the flow of liquids for a major pumping station. Specialized training beyond high school is required for many jobs in instrumentation. The length of the training period varies with the job title and responsibilities. Mathematics, science, electronics, and technology education courses, combined with the ability to communicate are helpful in preparing for a career in instrumentation.

Chapter 25

Electronic Meters

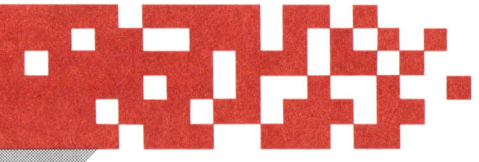

Types of Electronic Meters

To be considered an "electronic" meter, an instrument must use either transistors or vacuum tubes as key part of its **auxiliary** circuits. One very popular electronic meter is the vacuum tube voltmeter, or **VTVM**. Another common electronic meter is the transistor voltmeter, **TVM**. Both the VTVM and the TVM are analog meters which use d'Arsenval type movements. The digital multimeter, **DMM** uses a digital readout instead of an electromagnetic movement.

Fig. 25-1 The VTVM was the first truly electronic meter to gain widespread use. It is still a very important test instrument. The digital multimeter looks quite different than the VTVM and TVM because of its digital readout.

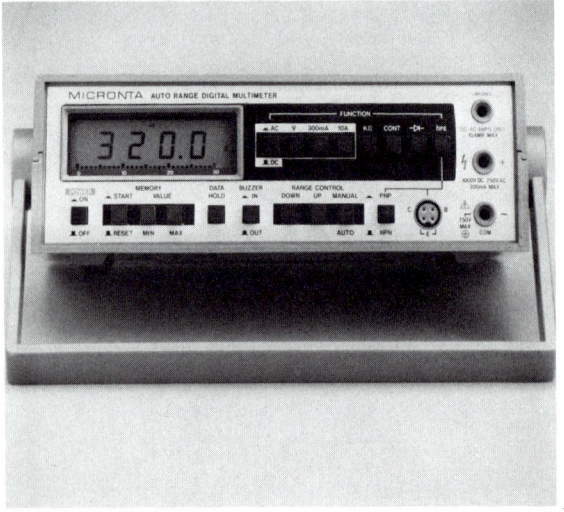

Vacuum Tube Voltmeters

For several decades, the vacuum tube voltmeter was the most important electronic test instrument. It is especially good for voltage measurements in sensitive electronic circuits which may be adversely affected by the loading effects of a regular multimeter.

Voltage is not all that this meter can measure, though. Most VTVMs have six or more ranges for voltage and resistance. The voltage ranges include AC and DC functions. In fact, most VTVMs are calibrated to measure AC voltages in more than one way. Most VTVMs do not have current ranges, but they have more capabilities in other ways than a typical multimeter (VOM).

The biggest disadvantage of a VTVM is that it requires a 120 V external power source. Its auxiliary circuits consist of the usual array of resistors and switches found in any multimeter, but with some important additions. The additional components make up a vacuum tube amplifier circuit which makes it possible to measure extremly low and extremely high values safely and without overloading the circuit under test. Therefore, the VTVM is a much better test instrument for a laboratory or shop than a VOM, but the VOM is better to carry in a toolbox if you go to people's homes to repair their washing machines. The VTVM is physically more fragile — it cannot withstand rough treatment and bouncing around in a truck. A VOM is less fragile than a VTVM. It is the choice of most electricians and appliance service-people. The VTVM is more likely to withstand electrical abuse, however. This is because the amplifier circuits protect the meter movement from damage when values which are slightly over the range of the meter are measured. VOMs do not have this electronic protection.

The VTVM is not as popular today, due in part to increased use of digital readouts in other types of more rugged meters and to the fact that a 120 V power source is needed. Another factor that has influenced the VTVM's decline in popularity is its relatively high cost. Still, the VTVM is one of the most accurate and versatile analog test instruments available, and no well-equipped electronics shop is complete without one.

Transistor Voltmeters

It would appear that everything said about VTVMs should be true of TVMs except that they carry their own power source. True, a small battery contained in the instrument case is the only power source required for a TVM. True, many of the same measurements made with VTVMs may be made with TVMs. Likewise, a TVM is an analog device. But, there are important differences between these two instruments. One advantage of the TVM is that it is more rugged than a VTVM. The TVM is usually smaller and lighter, and (with its internal power supply) is much more portable than a VTVM. There is one advantage, however, that a VTVM has which a standard TVM does not have. A VTVM does not load circuits heavily when it is used to test low voltages in circuits which have low circuit resistance. This is especially important in electronic servicing. A standard TVM uses transistors in the input section of its auxillary circuits, and transistors used here act something like low values of resistance. They can draw enough current from the circuit being tested to make it stop functioning correctly. Therefore, the earlier standard TVMs had little advantage for electronics work over regular VOMs. Now there are transistorized multimeters which use a special type of input transistor called a **field effect transistor** (FET). When FETs are used in the input circuits, they act like very high resistances and prevent circuit loading. Instruments with these input circuits, **FET multimeters**, have many of the advantages of the VTVM, the standard TVM, and the VOM combined, with few of the disadvantages, except high cost.

Digital Multimeters

If circuits like those in TVMs and FET voltmeters are used to drive a digital readout instead of an electromagnetic, analog meter movement, the result is a *digital multimeter* (DMM). Currently, these are the most popular instruments on the market. DMMs come in several models. Some are designed for shop use. Others are more suited for work in the field. See Fig. 25-2. Size and durability are the main differences in the two types, though some bench models are exceptionally accurate. DMMs are so popular because they are so easy to read. Figure 25-3 shows a standard VOM and a DMM. Which one is easier to read? Many of the newer DMMs feature "auto ranging." This means that no selector knob is needed because circuits inside the meter set the range automatically. Some DMMs use **light emitting diode (LED)** readouts while others use a **liquid crystal display (LCD).** The LED type is usually red, and it can be difficult to see in bright light. It is excellent when low light levels exist and at night. The LCD type is nearly the opposite — the black digits can be read fairly well even when there is some glare, but a small

Fig. 25-2 Though this DMM may be carried around, it is a highly accurate meter.

Fig. 25-3 The DMM is easier and quicker to read than an analog meter. You do need to check the range and function selectors, though, to know what the reading means.

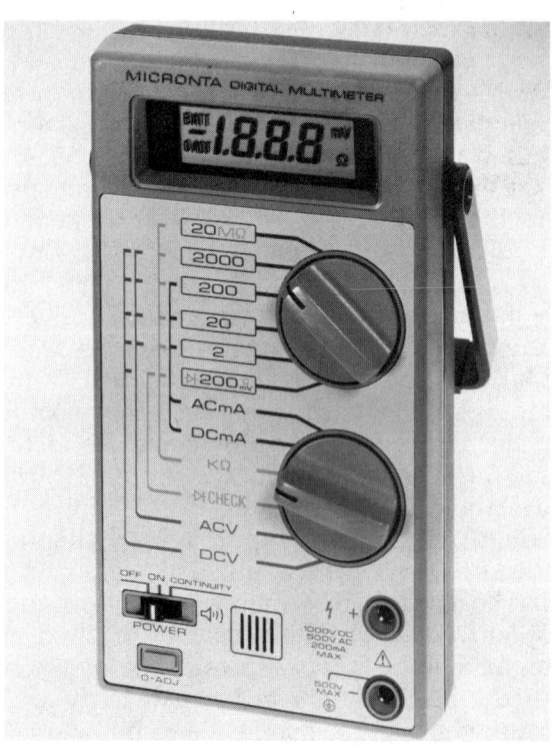
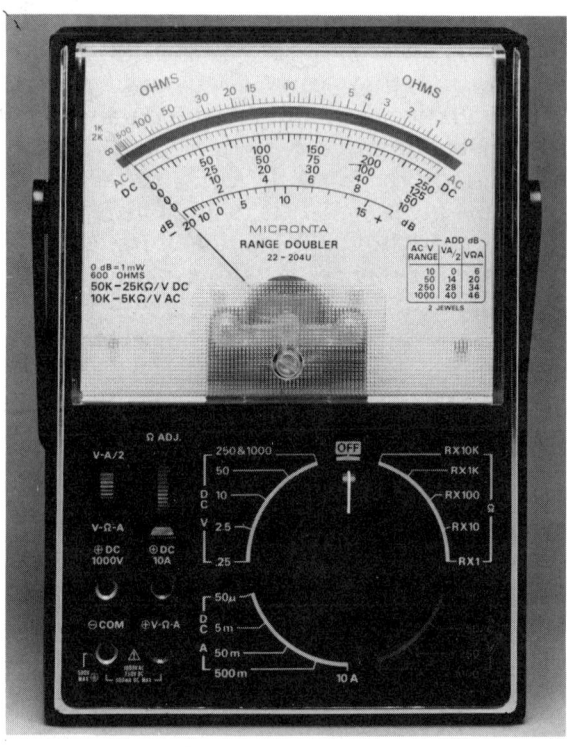

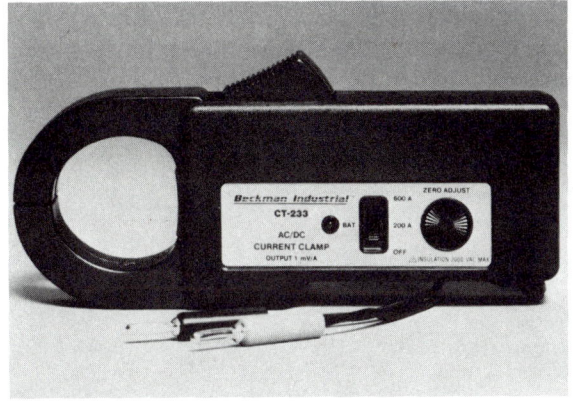 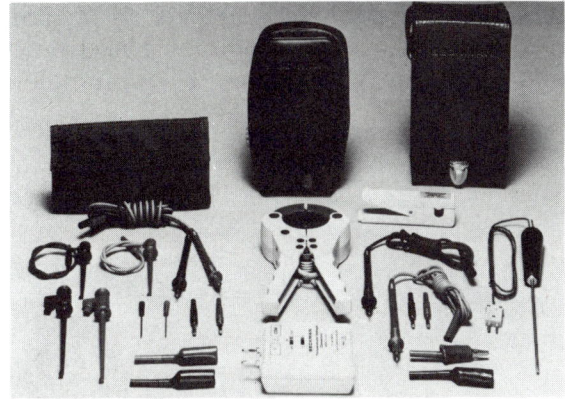

Fig. 25-4 These are only two examples of the many accessories which are available for modern meters.

light may be needed when it gets dark. The LED uses a lot more current to operate than the LCD. It will discharge the meter's batteries much sooner.

A disadvantage of DMMs is that they blink when the quality being measured changes. Expensive DMMs are pretty stable. They do not blink much when the voltage being measured is fairly constant, but even they flash continuously when the voltage is rising or falling. Inexpensive DMMs blink almost all the time, regardless of what the quality being measured does. For this reason, a cheap DMM is a very poor investment. It is no fun trying to read a set of numbers which constantly flashes from one value to another.

Special Features and Accessories

Like multimeters, some electronic meters can be fitted with accessories and attachments for special jobs. One of the most commonly used accessories is a high voltage probe. This device increases the range of a meter to about 50 kV (50,000 volts). One use of this probe is checking the voltage of a television picture tube.

Another probe measures high frequencies of AC current. This is an **RF (radio frequency) probe**. AC currents which change direction very rapidly, such as radio frequency AC currents, cannot be measured correctly by the auxillary circuits in many meters. The RF probe is needed to make it possible to measure them.

Temperature measuring, transistor testing, and clamp-on attachments are also available for many meters. See Fig. 25-4. These accessories are also made for the newer DMMs. It is usually best to use accessories made by the same company that produced the meter you are using. This ensures that the equipment is well matched for accurate results. It is generally unwise to use cheap accessories with an expensive meter.

Summary

Electronic meters are similar to multimeters in many ways, and they have some of the same functions. The main differences are in the auxillary circuits. Electronic meters use tubes or transistors in amplifier circuits which change their operating characteristics. In terms of use and appearance most electronic meters are very similar to a standard VOM. Only the DMM is unique in appearance due to its digital readout. The most common electronic meters are the VTVM, the FET multimeter (a special type of TVM), and the DMM. Each of these instruments has its own special capabilities and limitations.

Important Terms

VTVM   field effect transistor   liquid crystal display (LCD)
TVM   FET multimeter   RF (radio frequency) probe
DMM   light emitting diode (LED)

Review Questions

1. What are the advantages and disadvantages of a VTVM? Why should you learn about these meters when DMMs are so accurate and so inexpensive today?

2. What was the biggest limitation of the early standard TVMs?

3. Why is the DMM so popular now?

Project: Multimeter from Surplus Parts

Parts List

2	1N34A diode	D_1 & D_2
1	SPST Push Button Switch	S_2
1	Meter movement	M_1
5	Resistors	R_1 through R_5
1	Potentiometer	R_6
2	Test probes	
2	Banana jacks (or similar)	
1	Banana plug (to mate above)	
1	Battery with holder	B_1
1	Rotory switch	S_1
1	Suitable enclosure	

Notes:

Many parts of this project may be obtained by salvaging an old surplus meter. The original prototype was constructed from a discarded radiation meter. The unit had a good case, meter movement, range switch, battery holder, and potentiometer. Part of the design of this project is your job. After you find the surplus parts, you will need to perform portions of experiments 21, 22, 23, and 24 with the actual meter movement you will use in the project. This must be done to determine what values of resistors are needed to make the VOM from the movement you have found. Depending upon what markings are on your meter movement when you find it, you may decide to use it as is or make new markings on its face. If you are unable to find some parts, make substitutions or delete certain functions from your design. Be very careful with the **current** ranges, because it would be easy to make a mistake which could ruin the meter movement. Switch (S_2) is for safety, it should only be pressed when making current measurements. Then, it should be pressed only after carefully checking to be sure the meter is correctly connected into the circuit to be measured.

MULTIMETER FROM SURPLUS PARTS

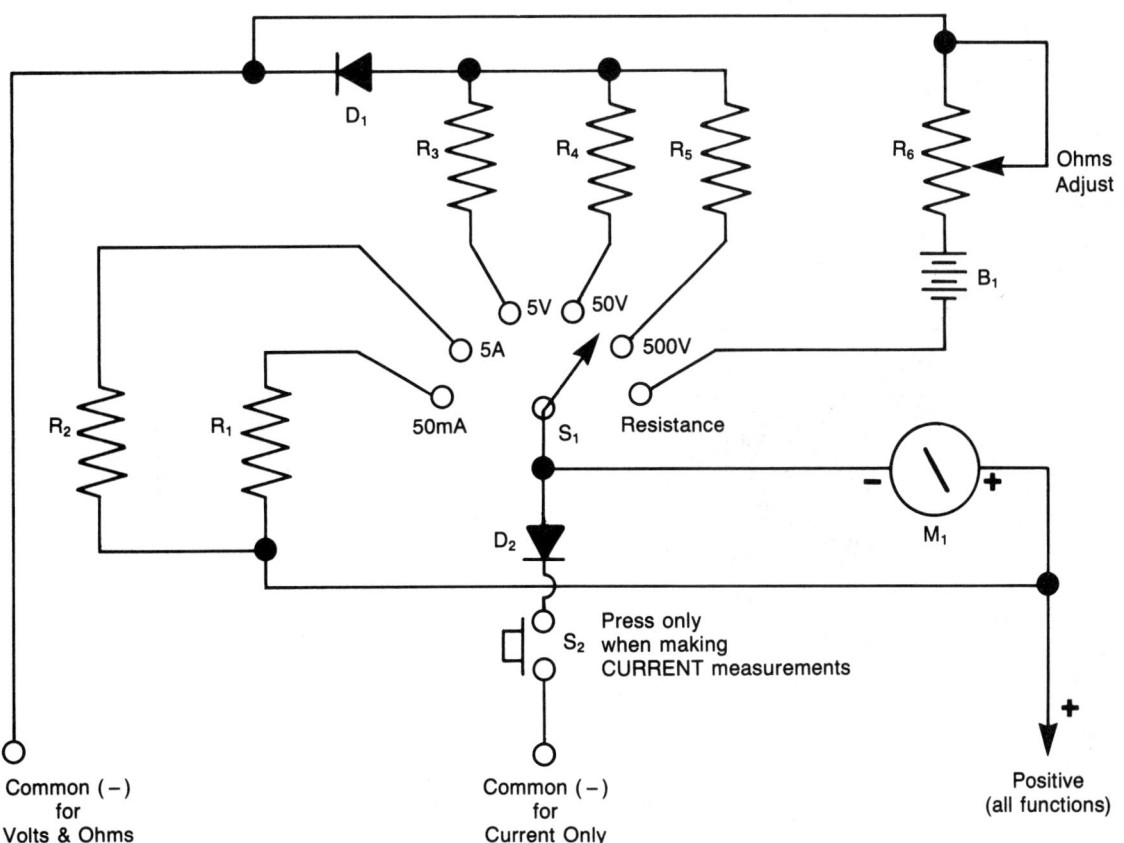

Section 5: Direct and Alternating Currents

Chapter 26

Batteries

Types of Batteries

There are two main groups of batteries (or cells). **Primary cells** make up batteries which are used once and then thrown away. The other group, **secondary cells**, may be recharged when they begin to lose power. A standard flashlight battery is really a **dry cell** which is in the primary group. An automotive battery is made up of secondary cells. There are several types of cells and batteries within these two main groups.

Dry Cells

Dry cells are a very common type of inexpensive, light duty power source (Fig. 26-1). Dry cells produce about 1.5 V of EMF each. When they are new, they produce slightly more than this value. As they age, they will produce less voltage. This loss of voltage will occur whether the cell is used or not, although heavy use weakens cells quickly. As the figure shows, the chemicals are contained in a zinc can. The dry cell has both a negative and positive pole. The can is actually the negative pole of the cell. A carbon rod in the center of the cell is the positive terminal. The chemicals, or **electrolyte** in the cell are a pastelike mixture of ammonium chloride, carbon, and manganese dioxide.

The voltage output of a cell depends mostly on the chemicals and materials used to produce it. A large dry cell has only 1.5 V just as a very small dry cell does. That is if the large cell really is only one cell. Large batteries are sometimes made by putting two or more dry cells into a single container and connecting them in series. The voltage of such a battery is found by multiplying the number of cells by the voltage of a single cell.

Though the voltage of a single cell does not depend on the physical size of the cell, its current producing ability (current capacity) does depend somewhat on size. If they were used in the same circuit, a regular D size dry cell would last much longer than a smaller AA (penlight) cell of the same voltage.

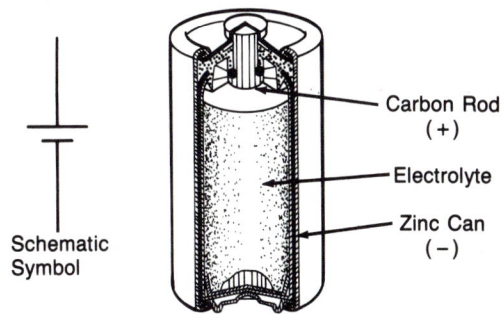

Fig. 26-1 Dry cells are a dependable source of portable power.

Very high temperatures cause the chemical reactions to speed up and can make the battery expand. This results in more rapid discharge of the battery. Very cold temperatures will not permanently damage dry cells, but they do slow down the chemical reactions, so there is usually a lower output from a cold dry cell. Room temperature, about 70°F (21°C), is the best operating range for dry cells. They store very well at lower temperatures.

Other Primary Cells

Dry cells are not the only cells that cannot be recharged. **Alkaline cells** have become very popular in recent years. They are used in most of the same applications as regular dry cells, but they last much longer. Figure 26-2 shows the construction of an alkaline cell. It is similar to the regular dry cell, but the chemicals are different. Alkaline cells are preferred for use in clocks, smoke detectors, and other devices because they last up to 10 times as long as regular carbon-zinc dry cells. Alkaline batteries come in a variety of sizes, similar to regular dry cells, and each cell produces about 1.5 V.

The **mercury cell** is usually quite expensive. It is used in applications that require extremely long life in a very small package. Watches, hearing aids, light meters in cameras, and many medical devices use mercury cells for dependability. Mercury cells are not often used in devices which draw high currents that would discharge them quickly. Mercury cells produce about 1.35 V each.

Silver oxide cells are similar to mercury cells. They are used in some of the same applications, but their output is 1.5 V. Another miniature cell, the lithium cell, is also similar to these.

Much progress has been made in developing improved primary cells. When primary cells discharge and when they remain unused for long periods of time, liquid byproducts of the chemical reactions inside them may leak out and corrode contacting metal parts. This used to be a very frequent and troublesome problem. Now, however, batteries have better cases and other technical improvements which reduce these problems. There was a time when leaving a "dead" battery in a flashlight or radio for even a very short time would ruin the device. Damage can still result from long periods of neglect, but high-quality batteries do not leak for some time after they have lost their power.

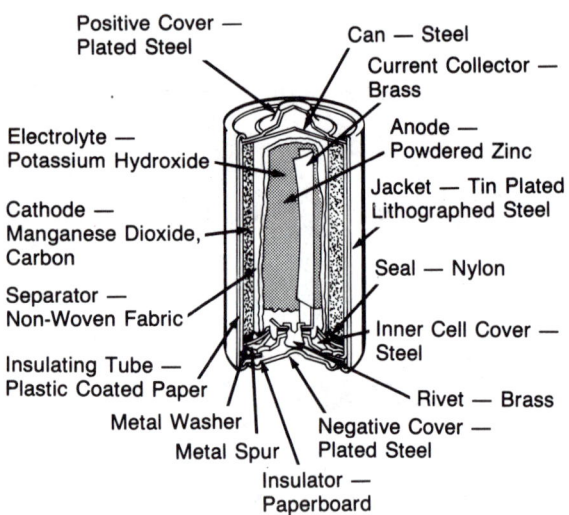

Fig. 26-2 Alkaline cells last much longer than zinc-carbon dry cells.

Fig. 26-4 A) Shows a fully charged cell with good, strong electrolyte. B) As the cell discharges through the load, some of the sulfuric acid from the electrolyte is used to create lead sulfate on the plates, and the electrolyte becomes weaker. C) A dead cell has very little sulfuric acid in its electrolyte and the plates are heavily "sulfated". D) When reverse polarity current is used to recharge the battery, the sulfation leaves the plates (if it is not too heavy), and the electrolyte becomes stronger again. This whole chemical process is expressed in the equation: Spongy lead + Lead Peroxide + Sulfuric Acid $\rightleftarrows$ Lead Sulfate + Water

Terminal Post
Plate Strap Casting
Negative Plate Group
Separator
Element
Positive Plate Group

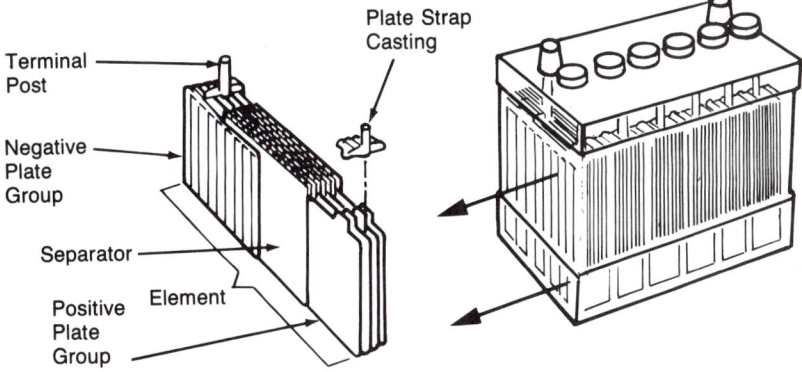

Fig. 26-3 A six-cell lead acid battery and a close up of one of its cells.

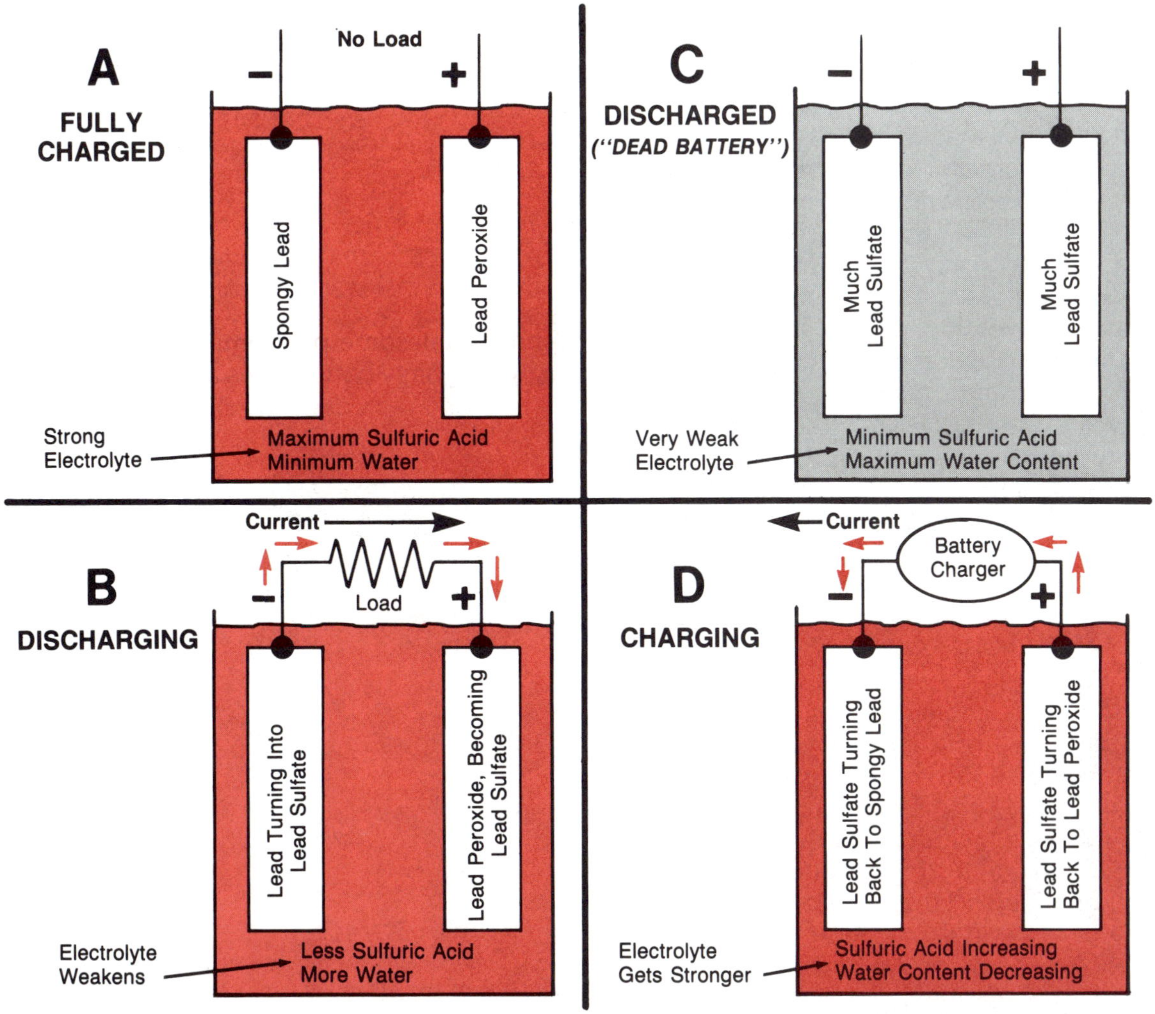

A
FULLY CHARGED
No Load
−
+
Spongy Lead
Lead Peroxide
Strong Electrolyte
Maximum Sulfuric Acid
Minimum Water
C
DISCHARGED
("DEAD BATTERY")
Much Lead Sulfate
Much Lead Sulfate
Very Weak Electrolyte
Minimum Sulfuric Acid
Maximum Water Content
B
DISCHARGING
Current
Load
Lead Turning Into Lead Sulfate
Lead Peroxide, Becoming Lead Sulfate
Electrolyte Weakens
Less Sulfuric Acid
More Water
D
CHARGING
Current
Battery Charger
Lead Sulfate Turning Back To Spongy Lead
Lead Sulfate Turning Back To Lead Peroxide
Electrolyte Gets Stronger
Sulfuric Acid Increasing
Water Content Decreasing

Secondary Cells

The storage battery in an automobile is a **secondary cell**. When it is first installed and filled with chemicals, it is **charged** by sending an electric current into it from a battery charger. If this step is not taken, the chemical reaction would produce a very weak (and short-lived) battery. Secondary cells actually store energy better than produce it. The charger pumps current into the battery. The battery stores this potential in the form of chemical energy. When the driver turns the key to start the car, the chemical energy is converted back to electrical EMF — to force electrons through the circuit. The battery uses one chemical reaction to store the energy and then uses the reverse chemical reaction to convert it back to electricity. When the automobile's engine is running, its charging circuit sends current back into the battery to recharge it so it will be ready for the next start. This cycle is repeated many times before the battery can no longer "hold" a charge.

The construction of an automobile storage battery is shown in Fig. 26-3. The plates are made of lead (−) and lead peroxide (+). The electrolyte between the plates is a mixture of sulfuric acid and distilled water. Acutally, the name **lead-acid battery** best describes an automobile battery. The battery shown in Fig. 26-3 has six cells. Each cell of a lead-acid battery produces about 2 V, so this is a 12 V battery. The caps on the top allow the user to replace the distilled water which slowly evaporates during use. Some maintenance-free batteries do not have caps because of technical improvements and venting designs which limit excess water evaporation. Allowing lead-acid batteries to get too low on liquids ruins them very quickly. Another way to damage a storage battery is to leave it in an uncharged condition for a long period of time. Figure 26-4 shows what actually takes place chemically when the battery charges and discharges.

The sulfuric acid is much heavier than regular water. Its specific gravity is higher. When the electrolyte is strong (contains a lot of sulfuric acid), its specific gravity is high. The specific gravity of the electrolyte may be measured with a **hydrometer**. (See Fig. 26-5.) If the specific gravity is low, the battery is discharged. Measuring the specific gravity of the electrolyte tells more about the condition of the battery than merely measuring its voltage output. Even a very weak battery will measure almost full potential voltage under a no-load condition, but the electrolyte test would show low specific gravity.

Safety with Storage Batteries

When a car battery is allowed to discharge too much to start the car's engine, another car's battery may be wired in parallel with it for a short time to start the engine. As the engine runs, its charging circuit will recharge the battery (unless it is so damaged that it will not hold a charge).

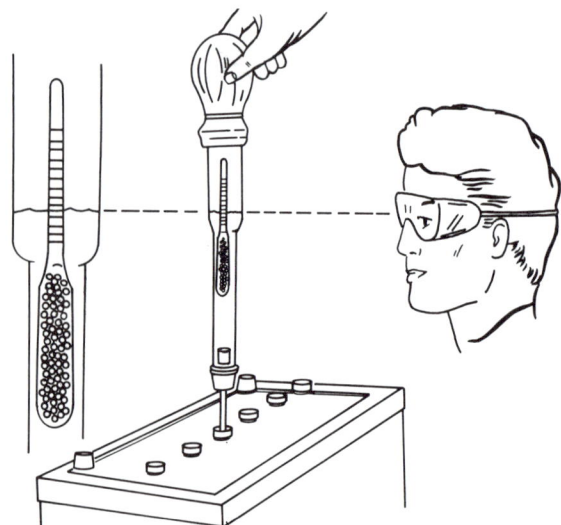

Fig. 26-5 The hydrometer is used to test the specific gravity of the electrolyte in a storage battery. Wear goggles to protect your eyes from splashing acid.

This process, called **jump starting** a car, can be done safely, but there are some important precautions which must be taken. As batteries charge and discharge, the chemical reactions inside them result in the giving off of some vapors (gases). They should only be charged in well ventilated areas. Do not breath the vapors or allow the vapors to concentrate near your eyes. The vapors have an even greater danger. They are very explosive! Even a small spark could cause a battery to explode by igniting these fumes. For this reason, battery chargers and loads must be connected carefully, and eye protection must be worn. When jump starting a car, a fairly large spark is often produced when the last connection is made between the two cars' circuits. This spark can easily cause one of the batteries to explode. Therefore, steps must be taken to ensure that the spark occurs far away from the batteries and their fumes. Likewise, great care must be taken to be sure that the batteries are not connected incorrectly. This not only damages the batteries, but the electrical systems of the cars as well.

The proper way to jump start a car is shown in Fig. 26-6. Only high-quality, well insulated, heavy duty cables should be used and eye protection should be worn. Do not lean over the batteries. Disconnect the cables in exactly the reverse order for connecting them. The cables should be disconnected as soon as the disabled engine is running.

Charging Batteries

A neglected battery should be charged by an external battery charger, because the car's charging system is not designed to recharge dead batteries. The car system's job is to keep a good battery freshly charged. Batteries should be charged slowly, with low currents at a voltage that is only a little higher than their own. Charging slowly is called **trickle charging**. It extends battery life greatly. Connect the charger to the battery before the power is turned on to prevent making dangerous sparks near the terminals.

Other Secondary Cells

New technology is being applied to secondary cells, too. Small lead-acid batter-

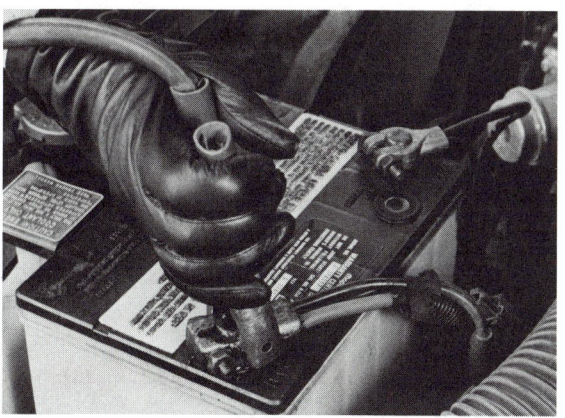 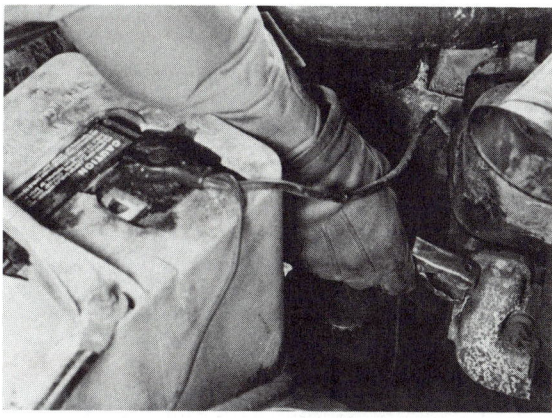

Fig. 26-6 First connect the end of one cable to the positive terminal of the booster battery. Connect the other end of that cable to the positive terminal of the dead battery. Next connect one end of the second cable to the negative terminal of the booster battery. Finally connect the other end of the second cable to the engine block of the car with the dead battery. Do not attach it to the negative terminal of the dead battery.

ies are available for use in portable equipment and emergency lighting systems. The "gel cell," which uses a gelled electrolyte, may be used in places which would cause normal lead-acid cells to leak. Nickel-zinc batteries do the job of lead-acid batteries with about half their weight.

The **nickel-cadmium cell** ("Ni-Cad" for short) looks like a standard dry cell, comes in the same popular sizes as dry cells, and may be used to replace them in many applications. Its advantage is that as a secondary cell, it can be recharged. Each cell produces about 1.25 V. That's lower than a dry cell but enough for most devices.

Fuel Cells

A new type of high technology battery is the **fuel cell**. Fuel cells have been used experimentally for a number of years. They are used in the space shuttle program. Like batteries, they produce electrical power by means of chemical reactions. With a fuel cell, the chemicals must be fed to the cell continuously as it is used. Hydrogen and oxygen have been used successfully. Other chemicals can be used as well, but they are not as efficient. As the cell operates, the natural byproduct of its chemical reaction is water. The water produced is drained off. Fuel cells are very heavy for the amount of power they produce. This poor weight-to-power ratio and the danger of using hydrogen (which is very explosive) remain as the major technical problems in the use of fuel cells. As energy becomes less and less plentiful, research on new technologies, such as fuel cells, will expand greatly.

Summary

Primary and secondary cells are available in a variety of types, shapes, and sizes. The voltage, uses, and operation of cells depends mostly on the chemicals and materials used to produce them. Larger batteries usually last longer and can deliver more current, but the size of a cell has little to do with its voltage output. Only secondary cells are rechargeable.

Important Terms

primary cell
secondary cell
dry cell
electrolyte
alkaline cell
mercury cell
silver oxide cell
charged
lead-acid battery
hydrometer
jump starting
trickle charging
nickel-cadmium cell
fuel cell

Review Questions

1. List some types of primary cells, and tell their uses.
2. Explain, how a storage battery works.
3. Explain how to properly (and safely) jump start a car.
4. Where are nickel-cadmium cells used, and what is their advantage over dry cells?

Chapter 27
Generators and Alternators

Motion and Magnetic Fields

Batteries, or chemical reaction, are one important source of electrical energy. Another important source of EMF is magnetism. Magnetism is the principle on which generators and alternators work. Recall from earlier chapters the relationship between magnets and electricity. You should be familiar with these relationships before you study this chapter.

As Chapter 5 explained, you can make current flow by causing magnetic flux fields to move in such a way that they cut through a coil (you would draw them as lines of flux). There are two ways to do this. You can either hold the coil still and move the magnet past it, or you can hold the magnet still and move the coil. Either way, you will make "lines of flux" pass through or "cut" the coil. The quotation marks are used here to remind you that the flux field is really continuous. Likewise, the coil is not really "cut" (as in sliced). The word "cut" is merely used to mean that the coil is influenced by the field. The traditional way to state this principle is to say that current is induced (created) in a coil when the coil is cut by moving lines of magnetic flux. When the flux field stops moving, the current stops flowing. This is true even if the magnet is touching the coil! This is so because the motion of the magnetic field or the coil is needed to cause the motion of the electrons. Actually, holding a magnet close to a coil, but not moving, would tend to freeze electrons in a still position rather than cause them to move.

To create electric current by magnetic action, there are three requirements:
1. There must be a coil and magnetic field near each other.
2. There must be motion of either the coil or the magnetic field.

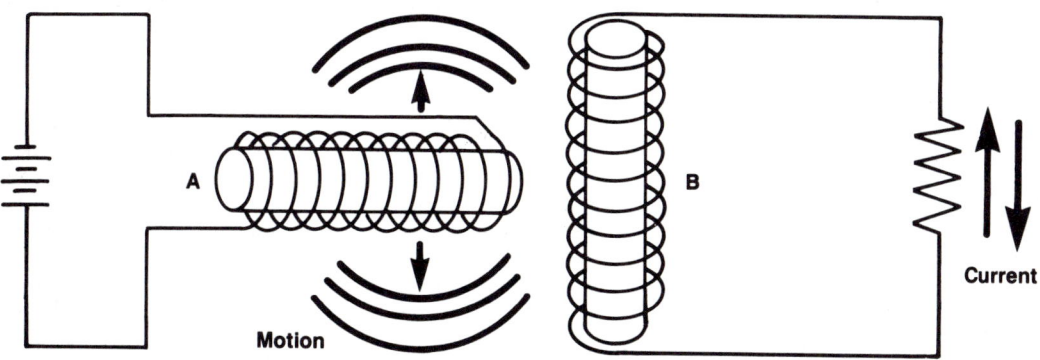

Fig. 27-1 Coil A is connected to a battery to make it an electromagnet. As coil A is moved up and down, its magnetic field induces a new current (AC) in coil B.

3. The coil must be connected to a complete circuit.

How the Magnetic Field is Produced

One way to make a magnetic field is to use a permanent magnet. It is also possible to make a magnetic field by means of an electromagnet. Remember that you studied this in Chapter 19.

Figure 27-1 shows how this could be done. Just as with permanent magnets, when electromagnets are used to produce the flux field, there must be motion of either the field or the coil into which the new current is to be induced. If not, there will be no current flow. It is also important for you to notice that there is no electrical connection needed between the source of EMF for the electromagnet's coil and the coil you are trying to make a new current in. Indeed, it is usually best if they are completely insulated from each other electrically.

Motion

Though it would be possible to make generating equipment that depended on magnets moving back and forth, it is far easier and more common to use rotational (turning) motion. There are two main groups of devices which produce electrical current by using the magnetic and induction principles discussed sofar. These two groups are *generators* and *alternators*. The difference between these groups is that **alternators** produce AC current. DC **generators** produce DC current. Some alternators, such as those in automobiles, have diodes and other devices that change their AC current into DC before it is used. Initially, however, they produce AC. Within these two main groups (alternators and generators) there are also several types. The types differ as to which parts (coils or magnets) move, whether they use permanent magnets or electromagnets, and how the coils are connected electrically.

Alternators

The simplest generating device, and the easiest to understand, is a **permanent magnet alternator**. Figure 27-2 shows how this simple generating device works. As the

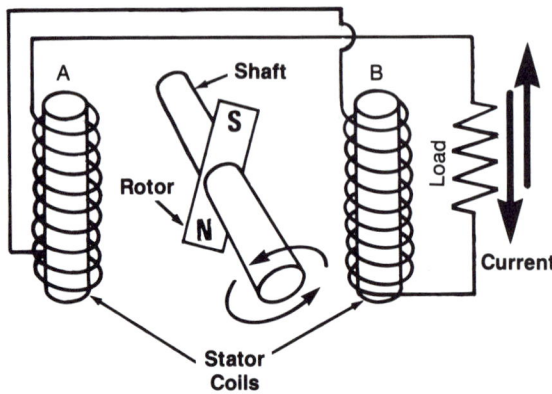

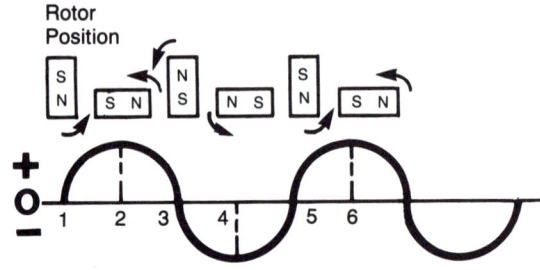

Fig. 27-2 As the waveform shows, there is no current (O) when the magnet is straight up and down between the coils (point 1). When the south pole of the magnetic rotor is nearest to coil A, we get a current peak as at point 2. The amount of current flow falls off as the rotor continues to turn — pulling the south pole away from coil A. At position 3, the magnet is vertical again, but upside down. Position 4 shows the current flow when the north pole of the rotor gets closest to coil A (south is now at B). The cycle repeats over and over again as the rotor turns, so we get AC current through the load.

magnetic poles of the **rotor** turn past the coils in the **stators**, electrical currents are induced into the coils. Since each coil will be first influenced by a north pole and then a south pole of the rotor, the current will be going back and forth. Another way to say this is to say that it alternates. Thus, it is an alternating current (AC).

It is more common to use a rotor made up of coils instead of one made of permanent magnets (Fig. 27-3). When this is done, the coils in the rotor are fed current from an outside source (such as the battery, in the case of an automobile alternator). These rotor coils are not the ones in which new currents are produced. They merely act as electromagnets to create the moving magnetic fields. In other words, the rotor coils just replace the permanent magnets. The currents are still induced in the stator (outside) coils, just as they were in the permanent magnet alternator. The current which powers the electromagnetic coils of the rotor gets to them through the slip rings and brushes on the rotor shaft. In very large alternators, such as the ones used to supply the current for homes and factories, this current for the rotor coils is supplied by a separate DC generator called an **exciter**.

Often the exciter is located directly above or beside the main alternator so that it may be turned by the same power shaft. Figure 27-4 shows an automobile alternator. Why go to the trouble of making AC current in an alternator and then using a diode circuit to change it to DC as is done in automobiles? The reason is the alternator produces more current when the engine idles. Thus, alternators charge the battery more effectively than DC generators do. Older cars used to have many more charging system problems because they used DC generators.

DC Generators

It is impractical to make a DC generator using permanent magnets, but this device will be used to explain the basic principle of DC generation. Then it will be explained how usual DC generators differ. Figure 27-5 shows a permanent magnet DC generating machine for discussion. The first difference that you will notice between the alternator and the DC generator is that the coils the new currents are induced into are the parts

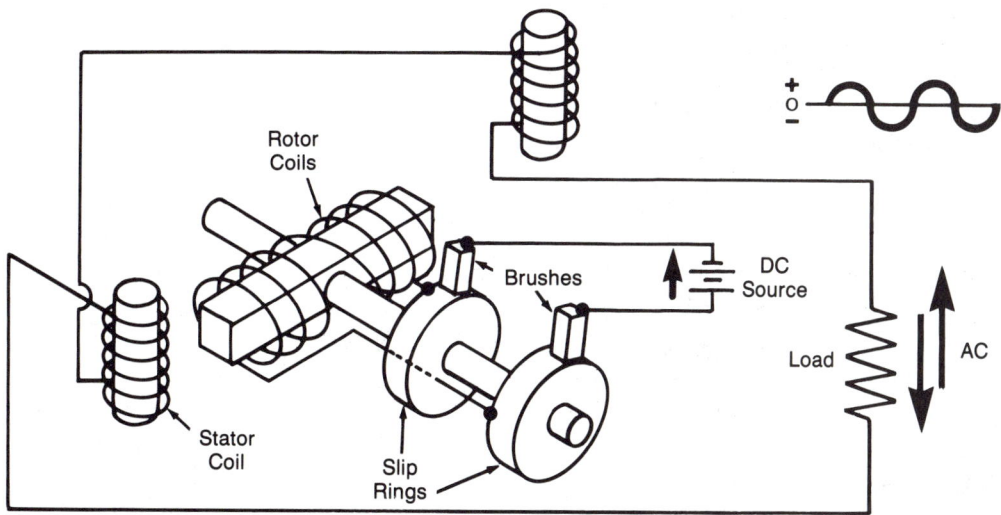

Fig. 27-3 The battery in this alternator excites the coils of the rotor. It works the same as the permanent magnet alternator, except that the rotor has electromagents.

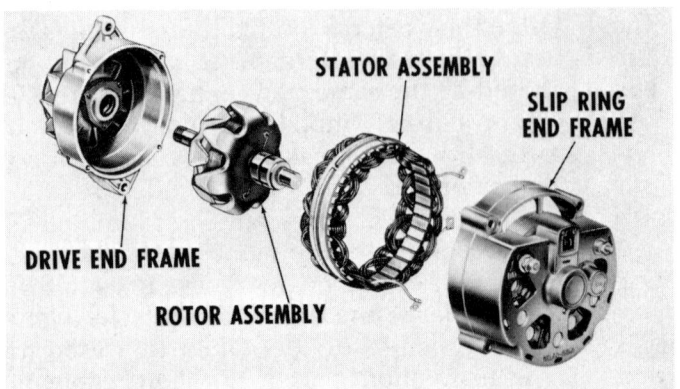

Fig. 27-4 There are several sets of rotors and stator coils for more power in this automobile alternator. The brushes and diodes are in the slip ring end frame.

that move in the DC generator. These moving coils make up the armature windings. Remember that the still (stator) coils of the alternator were the ones developing the new currents. In DC generators, the magnetic field stays still and the coils move through it. Another big difference is that this device needs brushes even though the field is made by a permanent magnet. The brushes in the alternator were used to get exciter current into the rotor windings to energize the electromagnetic coils. These generator brushes, however, ride on a new part, called the **commutator**. Their job is to get the new currents produced out of the generator. The reason that the commutator is needed, instead of regular slip rings as in the alternator, is that the commutator changes the current from AC to DC. The current in the **armature** windings (coils) goes back and forth, just as it would in an alternator. The commutator changes the direction of the current leaving the generator each time the polarity reverses. The

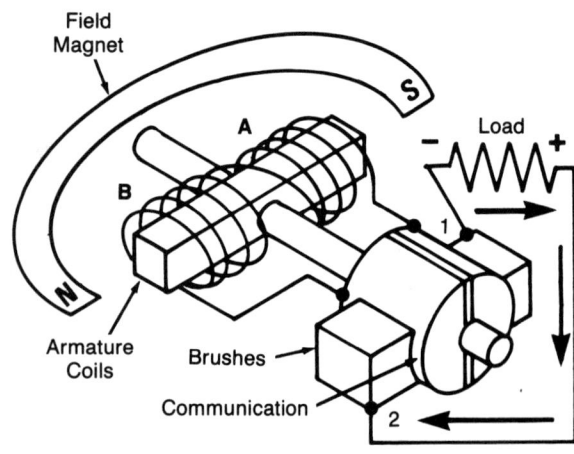

Fig. 27-5 In this position, the current is flowing from the shaft end out to the outer end of coil A and then out through brush 1. When the armature turns further, coil A will be influenced by the north pole of the field. Then, the current in coil A will be going from the end toward the shaft (backward). But, the commutator will also have turned, so the current in the load will still go the same way as it did before. The commutator reverses the current direction for the load.

Current in coils = ⁓⁓⁓ actually AC.
Current in load = ⌒⌒⌒⌒ which is pulsating DC.

simple drawing used here shows a generator with only two poles in the field and two coils on the armature. Practical generators would actually have several of each and separate sections on the commutator for each set of armature windings.

When permanent magnets are not used to produce the stationary fields in DC generators, field coils are used. The field coils may be wired in the generator in several ways. The **independently excited field generator** (Fig. 27-6) uses a separate battery to supply power for the field coils. The resistance in the circuit regulates the output of the generator by varying the strength of the field. The **shunt, series**, and **compound generators**, Fig. 27-7, do not use separate batteries to power their field coils. In these designs, the field coils are powered by using part of the current produced by the generator. Each of these methods of winding the field coils has special technical advantages and disadvantages. The shunt generator produces a

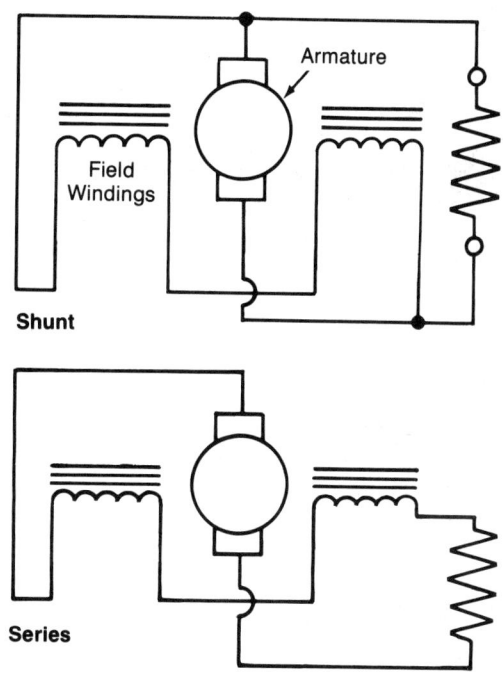

Fig. 27-7 Shunt and series DC generators are shown here. Compound generators (not shown) have some field coils in series and some in parallel with the armature. When no outside exciter source is used, the generator begins the first power cycle by using the small amount of magnetism left in the poles from the previous time it was used.

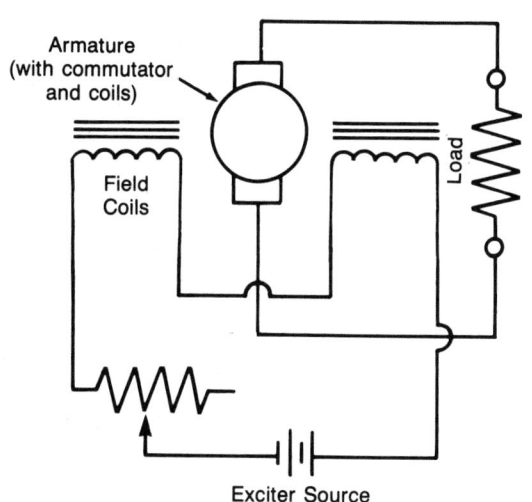

Fig. 27-6 The battery in an independently excited generator merely energizes the electromagnetic field coils.

fairly constant value of voltage. The series design is rarely used. Compound generators may be produced to change slightly in output as their loads vary. In a power station generator (Fig. 27-8), the large bottom section is the alternator that produces the AC current used in homes and factories. The smaller section on the top is the exciter generator that is actually a DC generator similar to the ones just discussed. The only job of this DC generator is to supply the power for the electromagnetic coils in the alternator.

Chapter 27 *Generators and Alternators*

Fig. 27-8 A power station generator. Notice how small the man seems beside the machines.

Summary

Generators and alternators use magnetic induction to cause electrons to flow. Alternators are really special types of generators that produce AC currents. The word "generator" can be used for AC or DC types, but it is generally used it to identify DC machines. Either permanent magnets or electromagnetic coils may be used to produce the magnetic field. Coils are used more often in larger, more powerful devices. Either the field or the coils in which the new current is to be induced may be moved. Only DC generators will have a commutator. Its job is to change the AC in the coils to DC before it leaves the generator. Chapter 28 will explore AC currents more closely.

Important Terms

alternator	stator	independently excited
generator	exciter	field generators
permanent magnet	commutator	shunt generator
alternator	armature	series generator
rotor		compound generator

Review Questions

1. Explain how a permanent magnet rotor alternator works.
2. Why do some alternators need an exciter?
3. What is the difference between slip rings and a commutator? Where is each one used? Why?
4. Why are alternators used in modern automobiles? How do we get DC current from them to charge the battery?

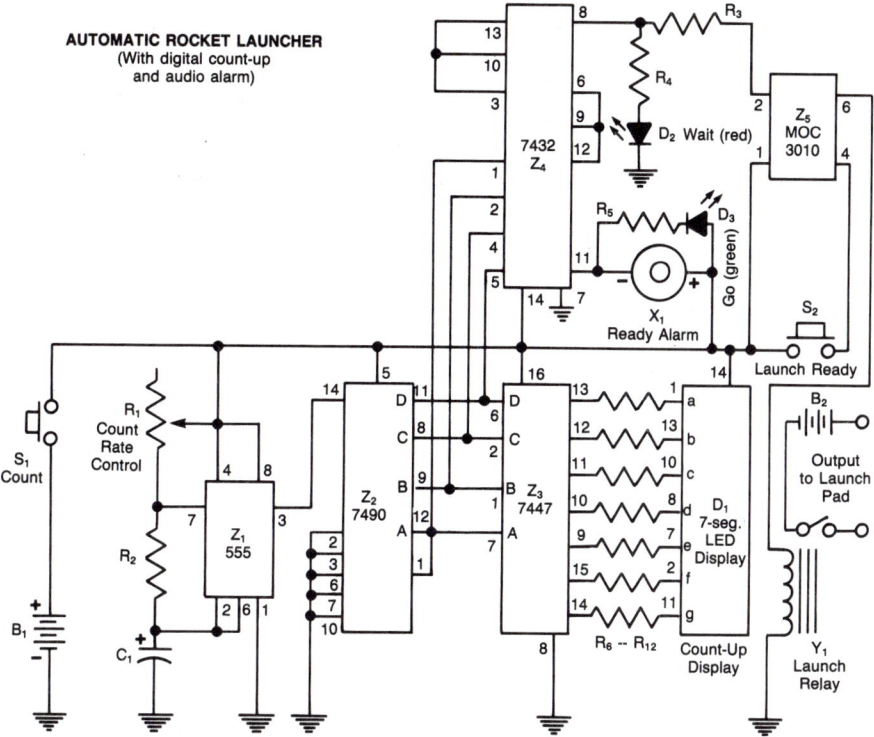

Fig. 27-9 The automatic rocket launcher with digital count-up and audio alarm.

Project: Automatic Rocket Launcher

(with digital count-up and audio alarm)
Parts List:

1	4.5 to 6 V battery (control)	B_1
1	6V lantern battery (launch)	B_2
2	SPST push button switches	S_1 & S_2
1	1 Meg. ohm potentiometer	R_1
1	1 K ohm resistor	R_2
10	330 ohm resistors	R_3 through R_{12}
1	Piezo Buzzer (273–065)	X_1
1	10 microfarad capacitor	C_1
1	Relay, 5V low current coil, with 2 Amp contacts (275–244)	Y_1
1	7-Segment LED display, common A	D_1
1	Red LED	D_2
1	Green LED	D_3
1	555 IC	Z_1
1	7490 IC	Z_2
1	7447 IC	Z_3
1	7432 IC	Z_4
1	MOC3010 (optocoupler, 276–134)	Z_5
2	microgator clips for launch pad	
1	PC board	
1	suitable enclosure (numbers given are Radio Shack)	

Notes:

This circuit can be adapted for many uses. The operation is simple. Pressing S_1 (Count) will start the display counting from 1 to 0 (10). The red LED will remain lit while counting is in progress. When the display reaches 10 (0), the red LED will go off, and the green LED will glow as the alarm sounds. If the Launch Ready switch (S_2) is off, no launch will be made. If the Launch Ready switch is pressed at the time the display reaches 0, the relay will close and power the launcher. A launch may be aborted by releasing either switch. The rate of the count-up may be adjusted with R_1. The wires leading to the launch pad from the control box should be at least 50 feet long and shelter should be available. If the rocket fails to launch, release both switches, disconnect the Launch Power Battery (B_2), and wait several minutes before cautiously approaching the rocket for troubleshooting. Do not attempt to troubleshoot the circuit with a "live" rocket on the pad.

Chapter 28

AC Currents

What Is AC Current?

AC current is **alternating current**. It alternates, flowing in one direction first, then flowing in the opposite direction. Figure 28-1 will help you to understand this principle. The current is induced into the coils of the alternator by changing of poles. As the north pole goes past the coil, it induces a current in one direction. When the south pole passes, it will induce current going the exact opposite way. Many devices, such as simple resistors, heating elements, lights, some relays and electromagnets, and even some motors work no matter which way the electrons in the circuits are flowing. All they require is movement of electrons. Such components and devices will work equally well with DC or AC current. Some components, however, require DC current, which moves in only one direction. Still other devices must have AC current. Transformers, discussed in Chapter 30, need AC.

The Sine Wave

Figure 28-2 shows a graph of the voltage produced by an alternator. This graph is a **sine wave**. The sine wave gives much information about the AC voltage in a circuit. First, you can see the **polarity** (positive or

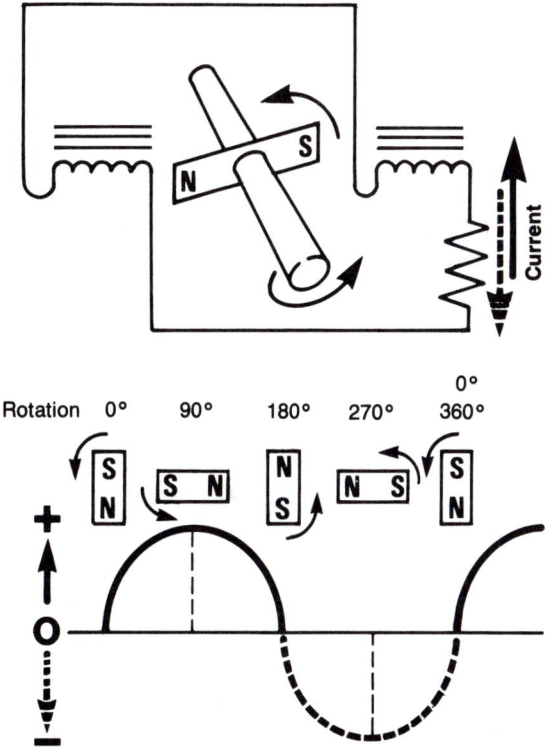

Fig. 28-1 The AC cycle begins at 0° when no current is produced. Maximum positive (+) voltage occurs when the alternator rotates to the 90° position. At 270°, the maximum negative voltage is produced.

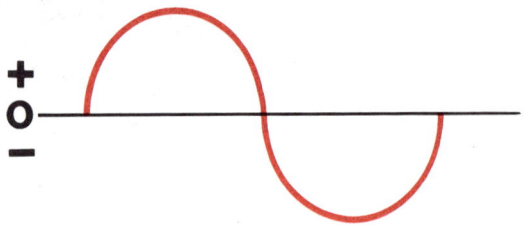

Fig. 28-2 The colored line shown here is the sine wave of one cycle of AC.

current each time they turn. Half of one cycle, one positive or one negative hump, is called an **alternation**.

The current that is available from the outlets in your home is produced by an alternator which completes 60 whole cycles every second. The term **hertz** (Hz) means "cycles per second." So the current in your home is 60 Hz AC current. If 60 Hz means that there are 60 complete cycles per second, how many alternations are there every second? If you said 120, you are correct. There would be 60 positive alternations and 60 more negative alternations making up the 60 complete hertz (cycles).

Frequency

The number of complete cycles which occur in a second is the **frequency** of the current. The higher the frequency, the more cycles there will be in each second. See Fig. 28-3. The AC currents that make sound in the speakers of a stereo system may have frequencies as low as 20 Hz for low bass notes or up to 20,000 Hz (20 KHz) for high musical notes. The electrical currents in the antenna circuit of an FM radio may be as high as 108 MHz in

negative charge) of the voltage at any given time. The humps of the wave that are above the center line (0) are positive (+). The parts of the waveform beneath the 0 line are negative (−) pulses. If you graph the voltage output of the AC alternator for several rotations, you get a series of several of these + and − humps. Each pair of humps (one + and one −) makes up a full *cycle*. A **cycle** is one revolution of the alternator when there is only one set of magnetic poles. Most alternators actually have several sets of poles, so they can produce more than one complete cycle of

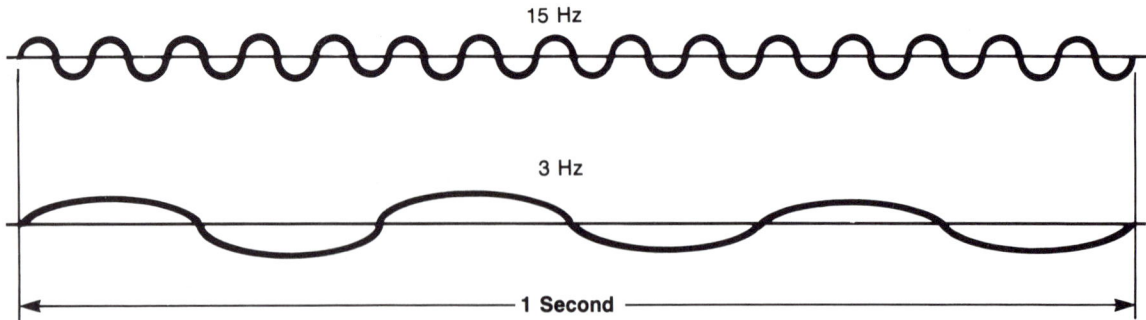

Fig. 28-3 The top waveform (15 Hz) is a much higher frequency than the lower one (3 Hz). The word "frequency" refers to how frequently the current flow changes directions.

frequency. That is, they could change directions 216 million times every second! Some devices which use AC currents demand certain frequency ranges for proper operation. To view and measure frequency ranges, an **oscilloscope** is used. If we were observing an AC current on an oscilloscope, we could make the humps get closer together or further apart by turning the alternator faster or slower. By changing the speed of the alternator, we change the number of cycles per second (hertz), or the frequency of the current.

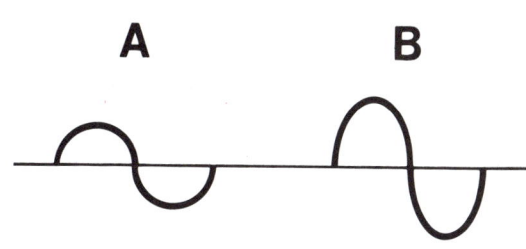

Fig. 28-4 Though waveforms A and B are both for currents of the same frequency, the voltage of B is about twice as great as the voltage of A. This fact may be shown by its amplitude (height).

Amplitude

Frequency is not the only important thing about AC currents. The height of each hump, that is its **amplitude**, tells you how much voltage (or current, in some cases) there is. On the oscilloscope, a 10 V AC waveform will be twice as tall as a 5 V waveform, even if they are both the same frequency. Frequency is a quality that is different from voltage. If you wish to change frequency, you change the speed of the alternator (cycles per second). If you want to change the voltage (amplitude of the waveform), you must increase or decrease the strength of the magnetic field in the alternator. You measure AC voltages, on an oscilloscope, by the amplitude of their waveforms as shown in Fig. 28-4.

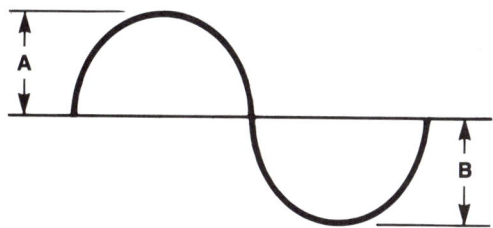

Fig. 28-5 Measuring *either* A or B will tell the peak voltage.

Four Types of AC Voltage Measurements

There are four ways to consider AC voltages. There are advantages to each of these methods.

Peak Voltage

Figure 28-5 shows the simplest way to think about measuring AC voltage. This type of measurement is known as **peak voltage measurement** because it measures the amplitude of the AC waveform at its peak. Peak measurements are useful when it is crucially important that the actual voltage in a circuit or device does not exceed a certain limit for even an instant. They do not measure the levels of voltage which exist beneath the peaks. Peak voltage may be measured on the top or the bottom half of the wave.

Peak-to-Peak Voltage

Both the positive and negative alternations can be included in the measurement by using the amplitude of the two together as in Fig. 28-6. If your meter were calibrated to measure peak-to-peak voltage and you wished to find the peak value, divide the actual measurement by 2. Likewise, if the peak value is known and you wished to find the peak-to-peak value of the sine wave, multiply the peak by 2.

Average Voltage

Peak and peak-to-peak voltages only tell what the values of voltage are at their maximum points of the sine wave. Sometimes it is important to know the **average voltage**. This is the value you would get if you measured the voltage at many different points around one half of the sine wave and then averaged them. This would be impractical because it would be hard to get enough separate values to make it easy to compute. Another way to find average voltage is to multiply the peak voltage value by a constant value. A **constant value** (or constant) is a number which mathematically works in a certain formula for all possible values. The value π (pi), which equals about 3.14, is a well known constant for working with circles. In fact, it is very useful in electronics too. The constant value used to find average voltage is .637. To find the average voltage of a circuit, multiply the peak value by .637.

Root-Mean-Square Voltage

Actually, the most useful measurement of AC voltage (except measurement by oscilloscope) is the one made by most meters. The meter measures AC voltage in terms of how much DC voltage it would take to have the same effect in a circuit. If you use AC voltage to operate a device that works equally well with AC or DC, say a heating element, then you can compare this amount of "effective AC voltage" to the DC voltage in a meaningful way. This "effective voltage" could be found by measuring the voltage in many equally spaced points around one cycle of the

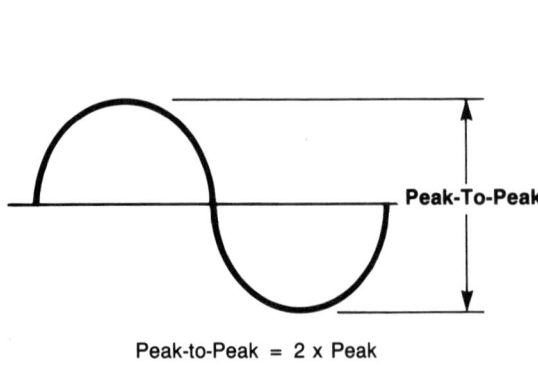

Fig. 28-6 The measurement of the total amplitude from the peak of the top alternation to the peak of the bottom alternation is called peak-to-peak measurement.

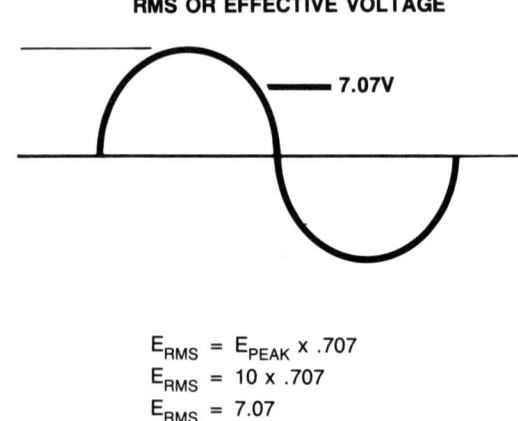

Fig. 28-7 The effective voltage of an AC sine wave is called the RMS voltage. RMS voltage is found by multiplying the peak value by .707 — another constant.

sine wave, squaring all the values found, averaging all the squared values, and finally, taking the square root of the average. This whole process, and the value it gives you, is called the **root-mean-square** voltage, or simply RMS voltage. Again, this calculation is not difficult to understand, but it is burdensome to complete, and nobody does it this way in practice.

Another constant value has been found for RMS voltage. The constant .707 may be multiplied by the peak value to find RMS voltage. See Fig. 28-7. If you look at many high quality AC voltmeters, you will see that they have separate scales calibrated to measure in RMS and peak-to-peak voltages. Using these values, and the constants just given, you should be able to find any needed values.

Phase

If two AC waveforms were being compared to each other, you might find that they were the exact same frequency and that the graph of one fell exactly on the graph of the other one. By using a special "dual trace" oscilloscope, you could display both waveforms together. If they are exactly together, they are not only the same frequency and amplitude, they are also "in phase." If they were the same frequency, but one started a little later than the other one, then they would be "out of phase." Figure 28-8 shows in phase and out of phase waveforms.

Shape

So far, only AC currents in the form of sine waves have been discussed. Some AC currents are not sine waves. Their waveforms may look like a series of triangles, sawteeth, squared pulses, or even erratic pulses with no identifiable pattern. The constant values given earlier in this Chapter only work with sine waves.

Summary

When an AC sine wave is examined, it can tell you much about the voltage. The frequency can be found by the number of alternations which occur each second. The peak voltage is measured by the amplitude of one alternation of the wave. The constant numbers 2, .637 and .707 may be multiplied by the peak value to find peak-to-peak, average, and RMS voltages (in that order). Phase has to do with whether or not the humps of two waveforms occur at exactly the same time. Chapter 29 will explain how to use an oscilloscope to display waveforms so that these measurements may be made.

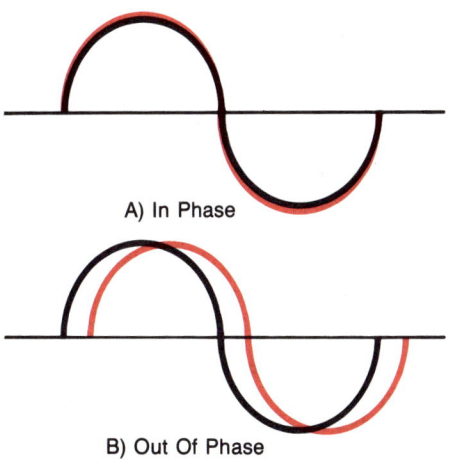

Fig. 28-8 In both A and B, the two waveforms are the same frequency and amplitude, but in B they are not in phase with each other because the orange wave peaks a little later than the black one.

Important Terms

alternating current
sine wave
polarity
cycle
alternation
hertz
frequency
oscilloscope
amplitude
peak voltage measurement
peak-to-peak voltage
average voltage
constant value
RMS voltage
phase

Review Questions

1. In many other countries, the standard AC current in homes is 50 Hz instead of the 60 Hz that we use in the United States. How many alternations would there be in one second with a 50 Hz current?

2. If you were measuring the voltage in a circuit and you found that the peak value was 169 V, what would the peak-to-peak, average, and RMS values be? (Round your answers to the nearest whole volt.)

3. Why is the RMS value so important?

4. What is the difference between frequency and phase? (Drawings may help you explain this difference.)

Chapter 29
Oscilloscopes

What Is an Oscilloscope?

In the last chapter you learned that **oscilloscopes could be used to show an AC waveform**. Figure 29-1 shows an oscilloscope. It looks like a small television set, and it is a lot like a television set in many ways. The screen of the oscilloscope is a **cathode ray tube** (CRT) very similar to the one in a TV. See Fig. 29-2. When the oscilloscope is used to measure a voltage, a beam of electrons is fired at the screen from an electrode gun (called the cathode) in the neck of the CRT.

If the amplitude of the voltage being measured goes up or down, the beam is directed up or down with it. This is accomplished by creating high voltage static charges in the electrostatic deflection plates. These plates are located so they will cause the beam to bend in any direction desired.

When the voltage becomes more positive, the beam is deflected upward. Since the beam is made up of electrons (negative charges), it tries to get away from the negatively charged lower plate and nearer to the positively charged upper plate. Likewise, when the voltage goes

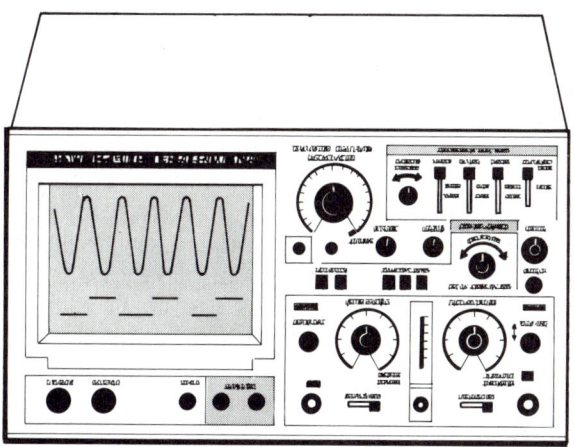

Fig. 29-1 An oscilloscope may be used to display waveforms.

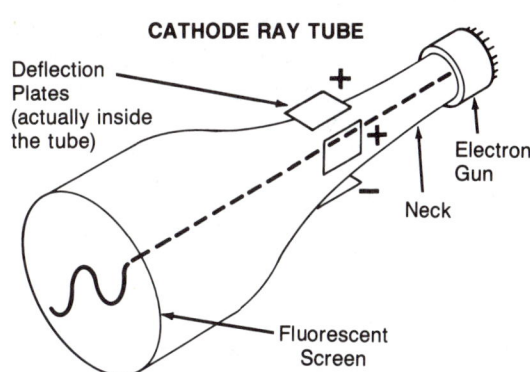

Fig. 29-2 An electron beam in the CRT traces the waveform on the screen. The beam is deflected (bent) by the electrostatic charges in the plates.

negative, the beam is bent downward. The plates on the sides make the beam sweep from side to side of the screen.

As the electrons from the beam hit the screen, they cause it to glow. The screen will continue to glow for a short time, so it will appear that it stays lit because the next sweep of the beam passes by before the light from the first sweep fades out. If you could draw a sine wave with an ink that disappeared quickly, you would have to go back and redraw it over and over again very quickly in order to keep it visible for an extended viewing time. This is exactly what the oscilloscope does — it makes one sweep, or trace, and then it goes back to the beginning and traces it again very quickly. This process is repeated many times each second.

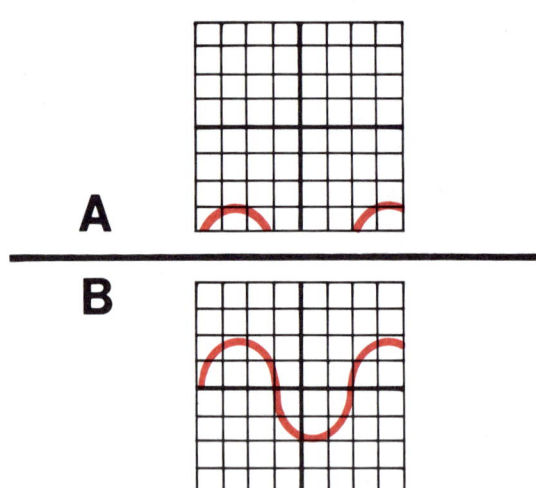

Fig. 29-3 Both A and B show the same waveform, but in B the horizontal and vertical position controls have been adjusted to make the waveform align with the graticule markings.

Controls on the Oscilloscope

Some of the controls on the oscilloscope are similar to those on a television set. The first two controls are the power switch and the intensity control. On many "scopes," these will be controlled by the same knob. The intensity control makes the image brighter or dimmer just like the brightness control on a TV. A focus control makes the image clear. Position controls, usually vertical and horizontal, center the image or trace on the screen so that the image may be analyzed. Vertical controls move the image up and down. Horizontal controls move it from side to side. Sometimes it is helpful to put the trace near the bottom of the screen, or to align it with a marking on the screen which acts like the calibration marks on the face of a meter. These markings on the face of the screen are called the *graticule*. Figure 29-3 shows how the position controls could be used to align the trace with the markings on the graticule. Notice that the position controls do not change the size of the image — they only move it around the screen. Oscilloscopes also have controls that affect the size of the waveform. These controls are like the range selectors on multimeters.

Vertical Amplitude Controls

Oscilloscopes, like meters, need range selectors to measure values higher or lower than the natural capabilities of the CRT. They also have auxiliary circuits. The *vertical range* selector usually has two knobs. Sometimes the knobs are together, or there may be a row of switches in place of one of them. Either way, one of the vertical controls will be a switching-type control that selects various ranges like the range selector on a multimeter. This switch-type vertical control is a "coarse" control because turning it one position makes a great difference in the height of the waveform displayed.

The other vertical knob is a "fine" control. It is usually an analog device, such as a potentiometer, which is capable of making small changes in the height of the waveform. By using both controls, it is possible to display a wide range of voltages on one scope. The fine control is used on some scopes to calibrate the scope — that is, to adjust it so that measurements may be made accurately. When this is done, a special reference voltage is measured, and the fine vertical control is adjusted until the proper waveform is displayed with the correct height on the screen. The graticule is used to measure the height of the waveform. The reference voltage source is usually built into the scope. It may be measured by touching the probe to a calibration terminal on the scope, or by selecting a special position marked "calibrate" or "cal" on the range selector switch. Many newer (and more expensive) scopes use a different system of calibration. These instruments are called **calibrated scopes**. They have a position marked "cal" on the fine vertical adjustment. When the fine control is set for this "cal" position, the range switch actually tells the value of each mark on the graticule without making a fine adjustment before each measurement (Fig. 29-4). Calibrated scopes are easier and quicker to use.

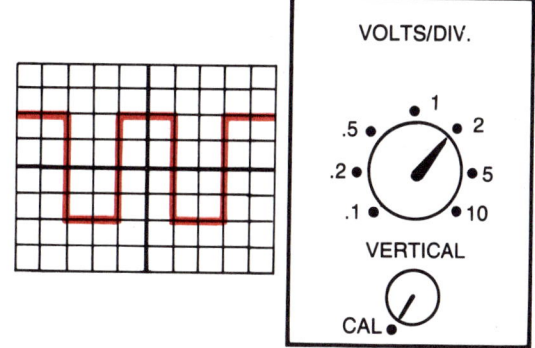

Fig. 29-4 This calibrated scope is set so that each vertical division on the graticule means 2 volts. What is the peak-to-peak voltage of the squarewave that is displayed? The answer is 8 V.

Horizontal Controls

The vertical size of the waveform tells its voltage. The horizontal movement of the electron beam which writes the waveform on the screen is controlled by time. In some scopes, the beam moves across the screen continuously at a set rate (as selected by the horizontal range selector control). These are "continuous sweep scopes." Sometimes it is helpful to have the sweep start precisely at the time a pulse of the observed current begins. These are called **triggered scopes**.

Triggered scopes are generally more expensive than nontriggered (continous sweep) scopes. As with the vertical controls, there are both coarse and fine controls for horizontal sweep. In nontriggered scopes, the coarse control selects a range of frequencies within which to measure (say, 15 to 150 Hz or 1500 to 15K Hz). The fine control on these scopes is used to make the waveform expand or contract for clear viewing. There is no way to tell the frequency of the waveform on continuous sweep scopes by merely looking at the basic waveform. The advantage of triggered scopes is that their horizontal controls work a lot like the vertical controls of calibrated scopes. That is, when the fine control is set to a special position labeled "cal," then you know exactly what the coarse control setting means in terms of time. With the scope in the calibrated mode, as shown in Fig. 29-5, the setting of the coarse control tells how long it will take for the beam to sweep across each division of the graticule. This means that the user can mathematically figure out the exact frequency of a waveform by counting the number of peaks in a given number of markings. Also, it is usually easier to get a triggered scope to

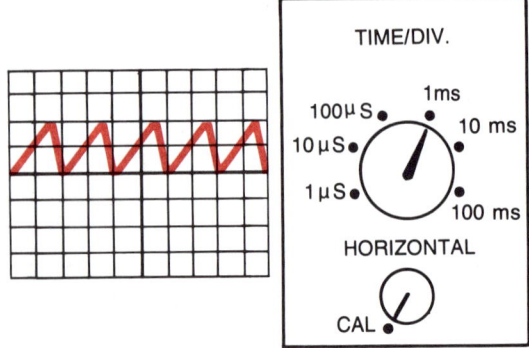

Fig. 29-5 This triggered scope is set to sweep across each division of the graticule (horizontally) in one millisecond. Since each cycle of the sawtooth wave displayed requires two divisions, it takes 2 ms. to complete each cycle. Thus, 2 ms. is the "period" of the wave. The frequency may be found by dividing the period into one. Our frequency here is 500 Hz.

hold the waveform steady for easy viewing. Some scopes have many more controls than the ones mentioned here. Your teacher will explain the specific controls used on the scopes in your school laboratory.

Special Features and Uses

Oscilloscopes are used in many applications in electronics. Some scopes are capable of displaying two waveforms at the same time so that they may be compared with each other. We call these instruments **dual trace scopes**.

When used with special circuits, it is possible to display special patterns on the scope to help determine frequency and phase relationships. These figures are called **lissajous patterns** (pronounced lee-sazhew.) They may be displayed on even very inexpensive noncalibrated, non-triggered scopes. The patterns are created when one AC waveform is connected to the vertical input, and another waveform is connected to the horizontal input. The patterns, Fig. 29-6, are produced as the electron beam tries to move across the screen at a rate set by one of the inputs, and up and down at a rate set by the other input. The number of vertical or horizontal loops is counted to determine the proportion of the two frequencies. If the exact frequency of one of the inputs is known, then the frequency of the other input may be found mathematically from the information given in the patterns. The patterns also shift when phase relationships change.

Oscilloscopes are not limited only to use as electronic measurement instruments. An EKG machine measures the electrical impulses that control the human heart. Scopes also are used in automobile engine analyzers and a variety of other equipment. The ability that a scope has to draw a picture of the actual shape of an AC waveform makes it a very valuable instrument. Some scopes can be used to measure DC voltages, too. A

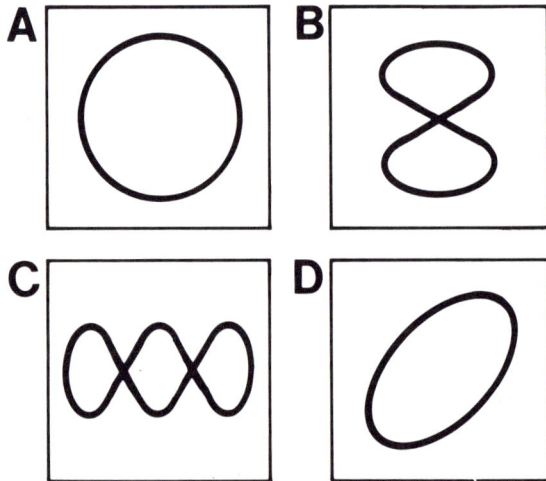

Fig. 29-6 The meanings of these lissajous patterns are: A) Both inputs are the same frequency. B) Horizontal frequency is twice as great as vertical frequency. C) Vertical frequency is 3 times as great as horizontal frequency. D) Both are the same frequency, but they are out of phase with each other.

true DC voltage would be displayed on such a scope as a flat line above or below the zero (or reference) line. The distance that the voltage line is away from the zero line tells its voltage (amplitude), and the direction (up or down) tells its polarity.

Summary

The oscilloscope is a very important instrument, especially for measuring AC waveforms. There are different types of scopes. The controls differ among scopes, but there are some similarities in most scopes. Calibrated scopes are easier to use because they do not require adjustment before each measurement. Triggered scopes allow the user to figure out the frequency of the waveform mathematically, while users of nontriggered scopes must display a lissajous pattern and then figure the frequency. Chapter 30 discusses transformers, and upcoming chapters deal with other devices mainly used in AC circuits. The oscilloscope will be very useful for you as you study these devices.

Important Terms

cathode ray tube
graticule
calibrated scope
triggered scope
dual trace scope
lissajous pattern

Review Questions

1. Explain the function of the intensity, focus, and position controls.
2. What do the coarse and fine vertical controls do on a noncalibrated scope? How do the same controls differ on a calibrated scope?
3. What is the biggest advantage of a triggered scope?

Chapter 30

Transformers

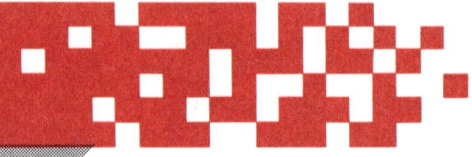

Uses of Transformers

Now that you understand the basics of AC current and the general use of oscilloscopes, you are ready to learn about some special devices and components. Let's begin with transformers. Transformers are very important because they are able to change the voltage and current in AC circuits. Many more complicated pieces of equipment that you will study in upcoming chapters will have transformers in the power supply sections of their circuits.

Have you ever played with a toy electric train? The device that controls the speed of the train is called a transformer. **Transformers** change the voltage in AC circuits. In the electric train set the transformer changes AC from 120 V (that comes from the wall outlet) to about 6 to 15 V (needed to run the train). Actually, this is a special type of transformer because it is variable. Most transformers are not variable, but they do change the voltage in AC circuits. Transformers do not work in DC circuits.

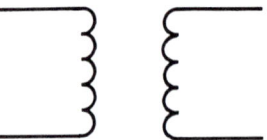

Fig. 30-1 A simple transformer is merely two coils which are near to each other.

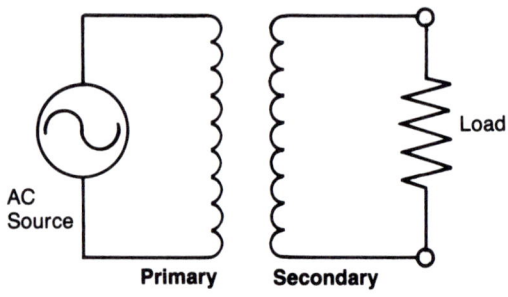

Fig. 30-2 The coil connected to the source is the primary. Secondary coils deliver power to loads.

How Transformers Work

Figure 30-1 shows the schematic diagram of a simple transformer. A transformer is two coils of wire close enough to each other for the magnetic field of one coil to affect the other coil. A circuit with a transformer in it is shown in Fig. 30-2. Here, there is an alternator connected to the first coil and a load (resistance) connected to the second coil. The first coil (the one that gets power from the source) is called the **primary**. The second coil, which delivers current to the rest of the circuit, is the **secondary**. When the alternator turns, it produces AC current that flows through the primary. This AC current makes the primary an electromagnet with rapidly reversing polarity. The magnetic field around the primary coil will be changing directions with each alternation of the AC current.

The magnetic flux field that it produces surrounds and "cuts" the secondary coil, but no current can pass between the two. Since this field is always moving, it is able to induce AC current into the secondary coil. See Fig. 30-3.

The AC current in the primary creates a reversing electromagnetic field around the coils. This magnetic field induces *new* currents into the secondary. The new current is AC because it changes directions at the same rate as the original current from the alternator. None of the alternator's current — or the current of the primary — goes to the secondary. Transformers use electromagnetic induction to transfer the power from the primary coil to the secondary coil.

Transformers will not work on DC because DC current would make a "still" magnetic field. Remember that the field must be moving to induce current in a coil. Refer back to Chapters 5 and 27 if this is not clear.

Step-Up and Step-Down Transformers

The coil with the most windings will have the highest voltage. The transformer shown in Fig. 30-4A is a **step-up transformer.** As the schematic symbol shows, the secondary coil is larger than the primary coil. In this case, this means the primary coil's low voltage will be transformed into a higher

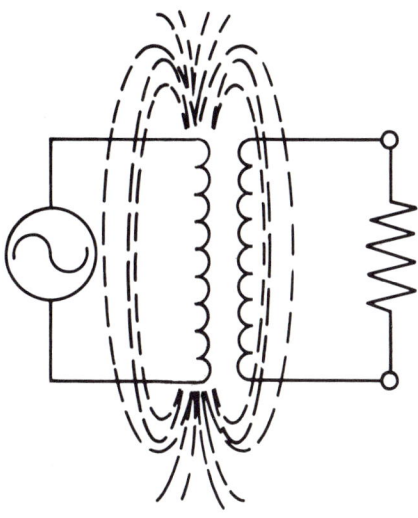

Fig. 30-3 The alternating magnetic field produced by the primary cuts the secondary coil and induces current in it.

voltage on the secondary side. If a transformer has a larger primary coil and smaller secondary, as in Fig. 30-4B, it is a **step-down** type. An electric train uses a step-down transformer. If both the number of turns in the two coils and the voltage of one of the coils is known, you can figure out the voltage of the other coil mathematically. The formula to use is:

$$\frac{E_p}{E_s} = \frac{T_p}{T_s}$$

This is called the **turns ratio**. E_p and E_s stand for the voltages of the primary and secondary, while T_p and T_s represent the number of turns (loops) in the two coils. Figure 30-5 shows a transformer with the turns ratio worked out.

You can identify step-up and step-down transformers on the basis of what they do to voltage. Transformers do step voltage up or down, but they cannot change the power in the circuit! The power that comes out of a transformer can be no greater than the power that goes into it. A simplified formula may be used to illustrate what happens to the voltage and current in a

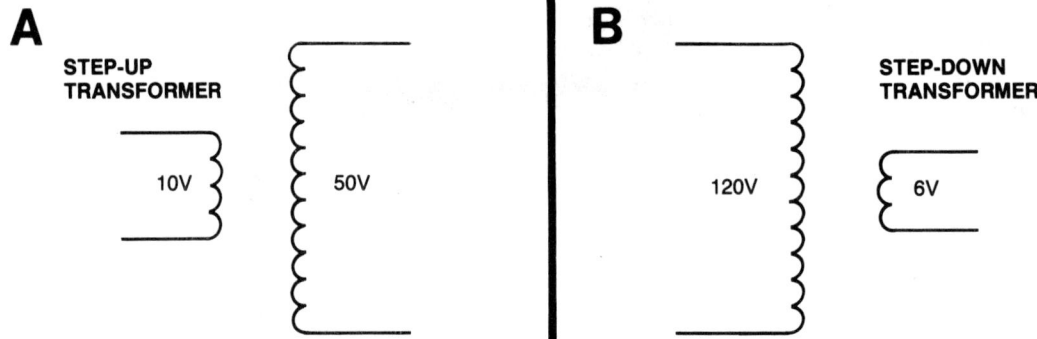

Fig. 30-4 The coil with the most turns (loops) of wire has the highest voltage (A). If the primary is larger than the secondary, then the transformer is a step-down type. (B).

transformer. The formula is:

$$I_p \times E_p = I_s \times E_s$$

Notice that each side of the equation is really the expression for power in one coil of the transformer. If the voltage is stepped down what must happen to the current capability? It will go up. See Fig. 30-6. An engineer who was really designing a transformer would need to include multiplying the power of the secondary by an efficiency factor — a percentage figure which tells how well the device actually transfers power. You may find the efficiency factors of certain transformers in your laboratory work experimentally.

100V | 25V
? Turns ?
1000 Turns

TURNS RATIO:

$$\frac{E_p}{E_s} = \frac{T_p}{T_s}$$

$$\frac{100}{25} = \frac{1000}{T_s}$$

$$100 \, T_s = 25{,}000 \quad \text{(Cross Multiply)}$$

$$T_s = 250 \text{ Turns}$$

Fig. 30-5 The turns ratio formula may be used to find any of the four values when the other three are known. Here, an engineer wishes to know how many turns the secondary of this transformer will need to give an output of 25 V. If the transformer in Fig. 30-4A has 1000 turns in its secondary, how many turns are needed in its primary? The same formula will work.

Schematic symbols that show three lines between the coils indicate an iron core.

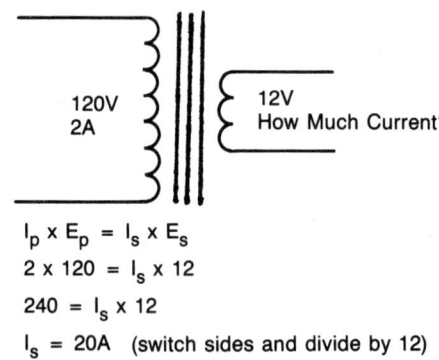

120V
2A

12V
How Much Current?

$$I_p \times E_p = I_s \times E_s$$

$$2 \times 120 = I_s \times 12$$

$$240 = I_s \times 12$$

$$I_s = 20A \quad \text{(switch sides and divide by 12)}$$

Fig. 30-6 Since the power in both sides of the transformer must be about the same, we can *estimate* any of the four values in this equation if the other three are known.

Transformer Losses

The power you get out of a transformer will always be a little lower than the power you put into it.

There are three reasons for this loss of power: *copper losses, eddy current losses*, and *hysteresis losses*.

Copper Losses

When you studied resistance, you learned that all materials have some resistance. Even the copper wires used to wind the coils in transformers will have some small amount of resistance. This resistance will cause some of the transformer's power to be lost in the form of heat. Making sure that the wire used to wind the coils is large enough in diameter helps to limit these losses.

Eddy Currents

Between the coils in a transformer is a core. The core may be air, iron, or some other materials. The iron core helps to concentrate and conduct the flux field, so there is greater transfer of power from primary to secondary with low frequency AC. There are problems, though, with iron cores. When the reversing electromagnetic field of the primary coil tries to induce new AC current into the secondary coil, it also will induce new, unwanted currents into the metal of the core! These little currents are called **eddy currents**. They reduce power transfer because they disturb the magnetic actions of the domains in the core. They create little magnetic fields of their own which try to induce currents opposing the wanted currents. In addition, they cause heat in the core — unneeded heat is usually a sign of power loss. When a solid core is used, these eddy currents can become quite large and have great effect on the power transfer. You can reduce the effect of eddy currents by simply reducing their size. This is done by using several layers of core material laminated together with insulating coatings between them. Since this keeps each eddy current small, it lowers the power loss. When higher frequencies of AC current are used, the core may need to be made of powdered material to further decrease these losses.

Hysteresis

Each time that the AC current in the primary alternates, the magnetic domains in the core must turn around and face the opposite way. Turning around like this is not an easy job. Heat is built up. This is why special metal alloys, such as silicon steel are used to make cores. They reduce the friction. The higher the frequency, the greater the **hysteresis** loss becomes. For this reason, many high frequency transformers and coils use air cores instead of metal ones.

These same three losses occur in any device that uses coils and AC currents. They account for loss of power in motors, transformers, generators, and other equipment.

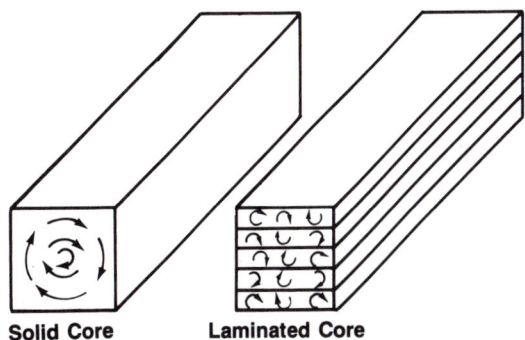

Fig. 30-7 Eddy currents are unwanted currents in the core material. Laminated cores keep the eddy currents small so these cores waste less power.

Types of Transformers

Figure 30-8 shows some special types of transformers. Each of these has unique uses, and design features. Center tapped transformers, used in many power supplies, make it possible to divide the secondary's voltage in half. Auto transformers are the only ones that really do have electrical connection between the primary and secondary. Multiple secondaries are found in power transformers in most television sets.

Summary

Transformers use electromagnetic induction to transfer power from a primary coil to a secondary coil. The voltage may be stepped up or down, but the opposite thing will happen to the current because the power delivered will be about the same as the power put in. The coil with the most turns will have the greatest voltage. Copper losses, hysteresis, and eddy currents are causes of power loss which affect all AC devices with coils. There are different types of transformers for special applications. The next chapter will discuss the uses of single coils in AC circuits.

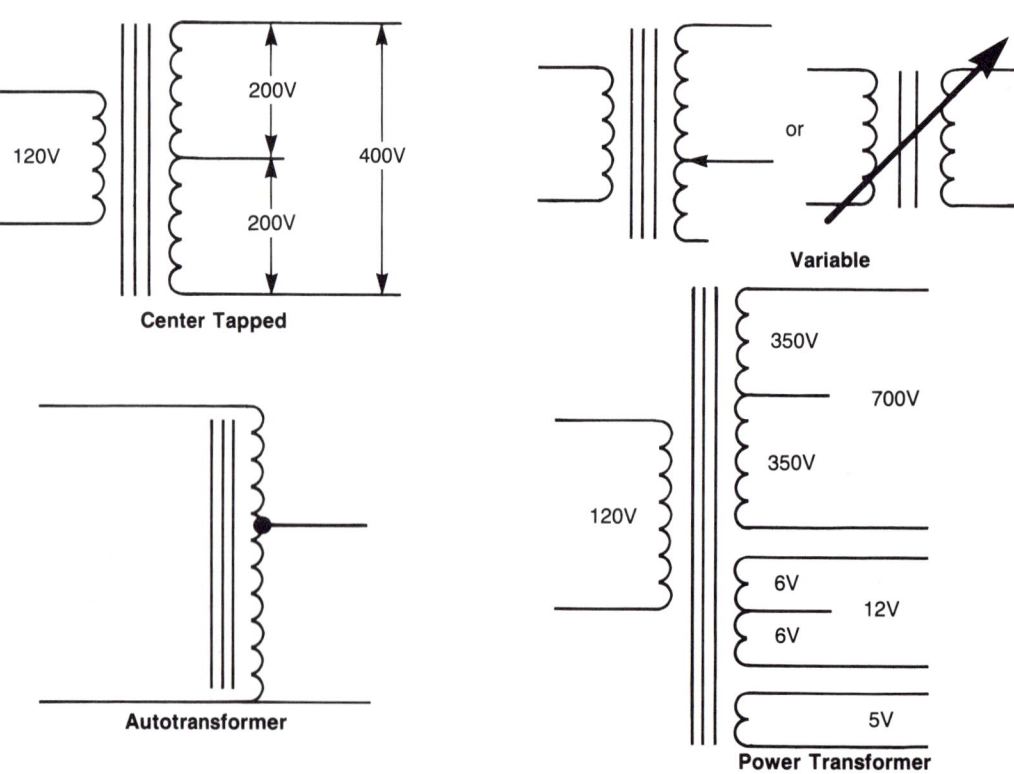

Fig. 30-8 Transformers are available in many types and sizes.

Important Terms

transformer
primary
secondary
step-up transformer
step-down transformer
turns ratio
eddy currents
hysteresis

Review Questions

1. How does the power get from the primary to the secondary in a transformer, if there is no electrical connection between them?

2. Will the current in the secondary be higher or lower than the primary current in a step-up transformer? What about the voltage? Why?

3. What causes each of the three transformer losses? How may each loss be reduced?

4. Why don't transformers work with DC current?

Project: Three-Channel Adjustable Color Organ

Parts List

3	5 K-ohm potentiometers	R_1, R_2, & R_3
1	1 K-ohm resistor	R_4
1	2.7 K-ohm resistor	R_5
2	2.2 K-ohm resistors	R_6 & R_7
3	SCR, 4 A, 200 V (SK 3570)	SCR_1, $_2$, & $_3$
1	120 V AC line cord	P_1
1	SPDT switch	S_1
3	0.1 mfd. capacitors	C_1, C_3, & C_4
1	0.5 mfd. capacitor	C_2
1	audio transformer (8–1000 ohm)	T_1
3	120 V AC colored light loads, see notes on drawing, DO NOT exceed 150 W with any single load. Select colors and placements which give a good effect as music changes.	
1	cabinet with translucent front assorted wires, PCB, and hardware as required.	

Notes

Potentiometer, R_1, adjusts the gain of the whole color organ, while the other two controls adjust the high and low frequency responses individually. If the light loads are carefully selected and positioned, a very professional unit can be produced with this simple circuit. Basically, the larger lights will respond to booming bass sounds, while the tiny midget lamps will twinkle with the high frequency sounds. The unit will seem more professional and interesting if the lamps are arranged randomly with the larger lamps near the back and the smaller ones in front. A unique effect can be added by placing the larger bulbs behind a slowly rotating disk with holes in it. The holes will allow the light to pass to different areas of the front panel as the disk rotates, giving the effect of motion. The disk may be powered with a small electric motor. Use your imagination to add other special touches to your color organ.

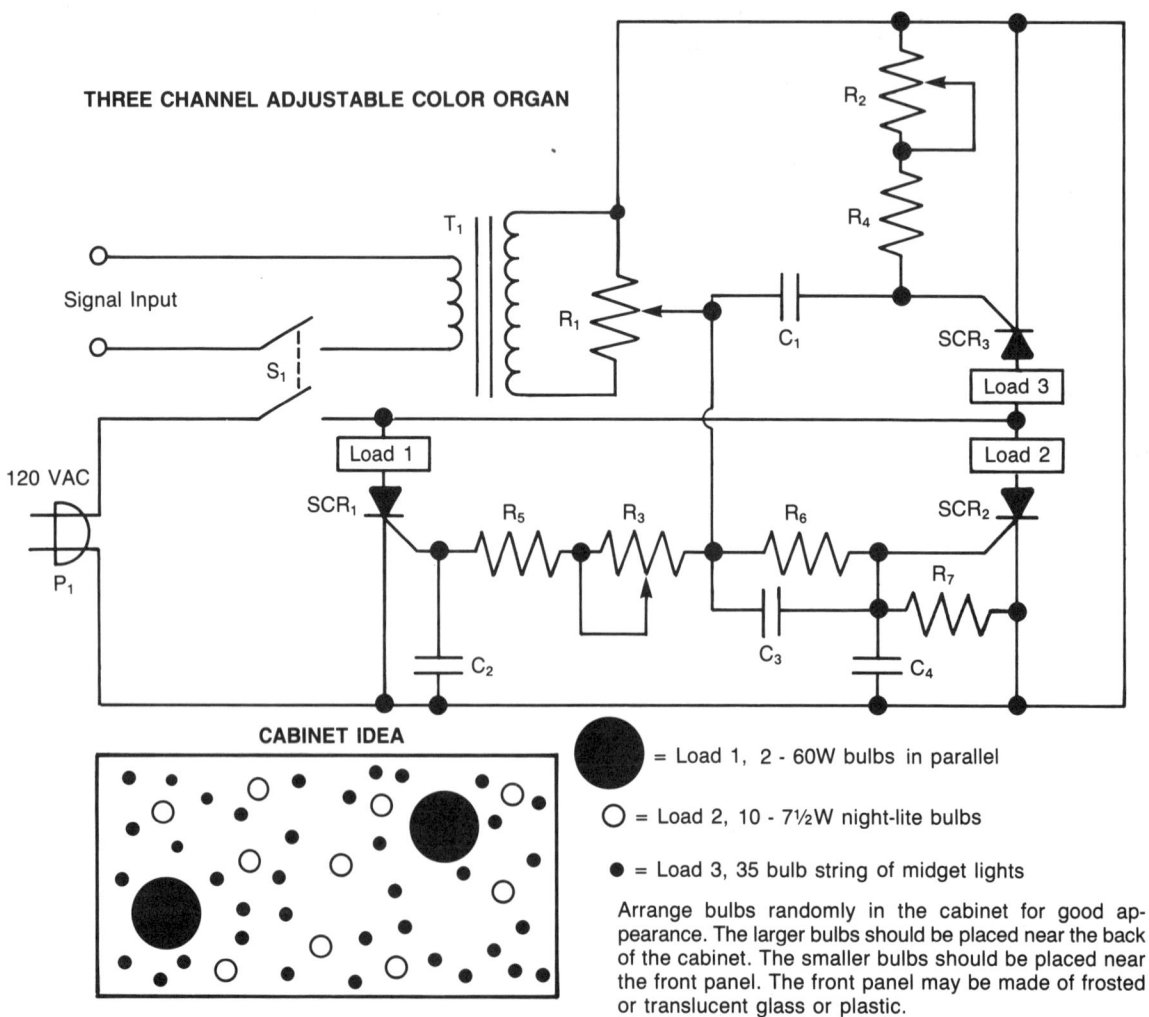

Fig. 30-9 The three channel adjustable color organ is an interesting project which is fun to build and use.

Chapter 31
Inductance

Coils

Coils have many uses in electronic circuits other than making magnetic fields. Let's see what happens when current passes through a coil, and why they are so important in electronics.

DC Currents in Coils

Figure 31-1 shows a simple coil in a DC circuit. The coil is labeled L-1 for inductor number 1. When the switch is open, no current flows in the circuit. This is the starting point. When you first close the switch, current begins to flow through the entire circuit, including the **inductor** (coil). The current passing through the coil creates a magnetic field around the coil, just as if you were trying to make an electromagnet. You know this would happen because a magnetic field builds up around any wire that is carrying current. Remember, though, that the opposite is also true. Any time a wire or coil is cut by a moving magnetic field, a current is produced in the wire.

When the switch is first turned on, not all of the possible (potential) current will flow right away. It takes a little time to build up to full current flow. During this building period, the current flow rate and the strength of the magnetic field around the coil are both changing (growing) quickly. This means the magnetic field is moving around and cutting the wires of the coil. It acts the same way as a permanent magnet would if it were moved quickly up to the coil.

If this occurred, a surge of current would be produced in the coil. Likewise, the building motion of the magnetic field will be cutting the very coil which is producing it.

The current induced in the coil only lasts as long as it takes to build the field to full magnetic strength. It stops because no current is produced when the field does not change in size, and the field remains mo-

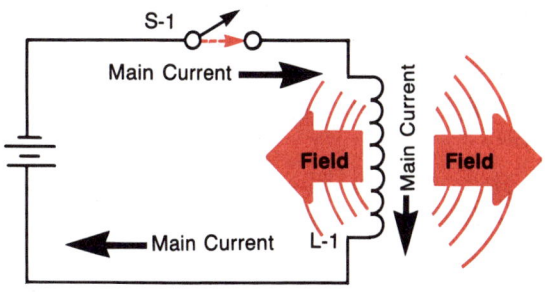

Fig. 31-1 When the switch is first closed, current flows in the coil (L-1) and builds a magnetic field around it.

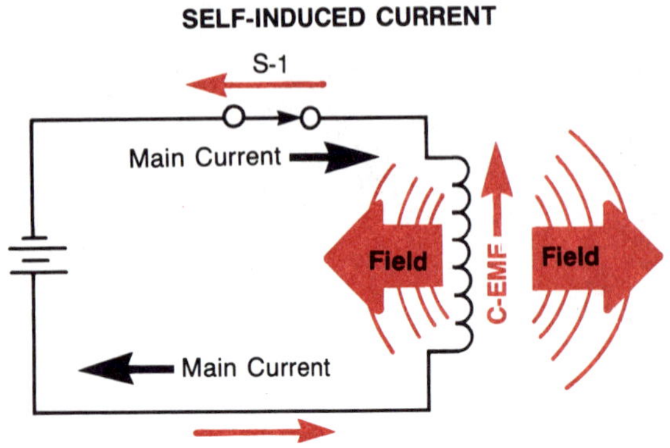

SELF-INDUCED CURRENT

Fig. 31-2 The movement of the building field self-induces a new, backward current into the coil. This smaller, reverse current is the counter-EMF (C-EMF).

tionless after it reaches full strength. This momentary current the coil makes is called **self-induced current** or **inductance**.

What is really surprising, is that this self-induced current goes in the *opposite* direction to the original current that created the field in the first place. In other words, when you first close the switch, a magnetic field begins to build up around the coil. While it builds, it creates a new reverse current in the coil that tries to stop the main (original) current from flowing. The main current is stronger than the self-induced current, so it eventually overcomes this added opposition to flow. Because of this, it takes a little more time than if there was no coil in the circuit. Because the self-induced current tries to go against the main current, it is called **counter-EMF**. See Fig. 31-2.

Counter-Counter-EMF

The self-induced current produced when you first turn the switch on tries to stop the main current from flowing. After the main current has reached full flow rate, you may later wish to turn the switch off to stop current flow. Figure 31-3 shows what will happen when you do this. While the main current was flowing at full rate, the magnetic field remained at full strength around the coil because it was supported by the main current flow. As soon as you open (turn off) the switch, the main current stops flowing. This means the field, left with no support, collapses. While it was supported by the main current, it did not "move" or change in strength. Therefore it did not induce any currents into the coil. But when you turn the switch off and the field begins to collapse, it moves again. This time it is moving in the opposite direction from its earlier building motion. The collapsing field will induce current into the coil, just as the building field did. This time the current will flow in the *same direction* as the main current was flowing! This means that it is going against the direction of the counter-EMF. Therefore, this current induced by the collapsing magnetic field is called a **counter-counter-EMF**. This self-induced counter-counter-EMF will try to keep the main current flowing for a little while after the switch is opened. It takes about the same amount of time for the current to fully stop after being turned off as it did for it to build to full flow rate after being turned on. This is because of the effects of the two self-induced currents.

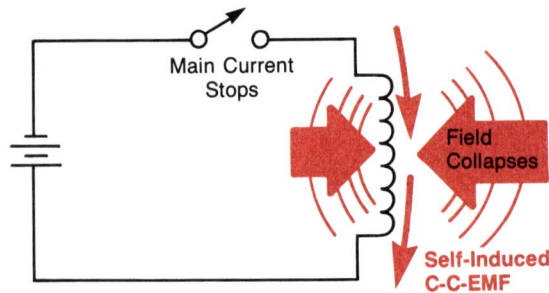

Fig. 31-3 When the main current stops, the field collapses and induces a counter-counter-EMF into the coil in the *same direction* as the main current had traveled.

Notice that the coil opposed any attempts to change the flow rate of the DC current. When there was no current flowing in the circuit and you tried to start current flow by turning on the switch, counter-EMF was self-induced into the coil to resist the flow of the main current. Likewise, when full current was flowing and you wished to stop it by turning the switch off, then counter-counter-EMF was self-induced. This helped the main current keep going for a little while, even after the switch was opened.

AC and DC Currents

In DC circuits, the production of counter-EMFs is not very important because the counter-EMF only lasts for a very short time. The main current can overcome it relatively quickly. However, in AC circuits the production of counter-EMFs is important. Consider what you already know about counter-EMF. Counter-EMF is only produced when the field is building. Counter-counter-EMF is only produced when the field is collapsing. The field only builds and collapses when the main current flow rate is changing. In DC circuits you have counter-EMF when you first close the switch and counter-counter-EMF when you first open it. Neither one affects the circuit at all during continuous operation or non-use. Once the circuit is turned on and the main DC current overcomes the initial counter-EMF, the main current may flow for hours without ever producing even one little pulse of counter-EMF.

If the main current is an AC current, it is always coming on, reversing, going off, coming on in the opposite direction, reversing again, going off again, and then coming on again over and over very rapidly. Figure 31-4 shows how counter-EMF and then counter-counter-EMF are produced repeatedly the entire time an AC current tries to flow through a coil. Since counter-EMF opposes the main current, this rapid production of new counter- and counter-counter = EMFs will cause a great amount of opposition to the flow of AC currents.

In other words, DC currents flow through coils with very little opposition, because the resistance of the coil is very low. AC currents are met by a new type of opposition to current flow which is quite large. This special opposition to AC currents is called "inductive reactance" and it is the topic of Chapter 33.

Uses of Coils

Because of the different ways coils affect AC and DC currents, it is said that coils "pass DC and block (oppose) AC." Figure 31-5 shows one use of an inductor. In this circuit, the coil is being fed a pulsating DC

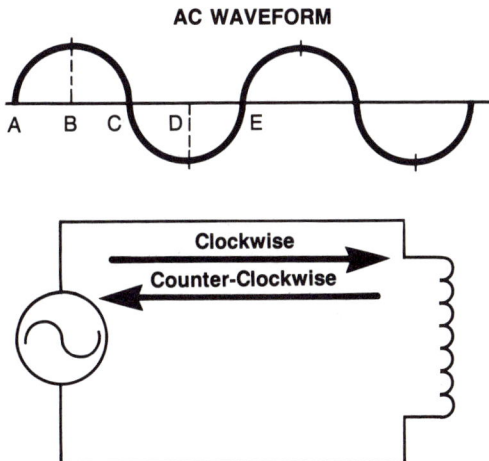

Fig. 31-4 During the time period from A to B, the main current flows clockwise, but C-EMF tries to go counterclockwise. From B to C the main current still flows clockwise, but it is falling in intensity, so it creates C-C-EMF that tries to keep the main current at a high level. From C to D the main current has reversed and is flowing counterclockwise, but a new C-EMF opposes it in a clockwise direction. What happens from D to E? These patterns repeat and repeat very rapidly.

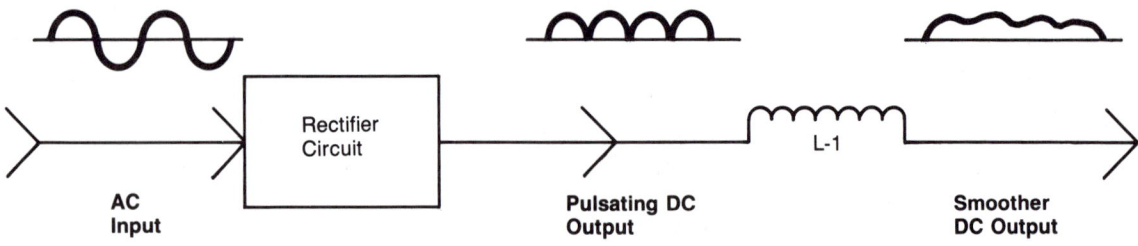

Fig. 31-5 The choke, L-1, smoothes out the peaks and valleys of the pulsating DC by opposing these great changes in current flow rate. The result is a nearly smooth DC output with some ripple.

current from the rectifier circuit. The rectifier changes an AC current to DC, but it does not make it perfectly smooth. As this pulsating DC current tries to go through the coil, it is always changing in intensity. Therefore the magnetic field it creates is also changing rapidly. As the field grows and collapses, it creates counter-EMF and counter-counter-EMF which oppose the changes in current flow. Therefore, the current which the coil actually allows to flow is smoothed out. The coil will let the DC part of the current flow with little resistance, but the changing part of the current (the little peaks and valleys) will not be permitted to flow through the coil. The coil will treat the peaks and valleys as though they were AC. In fact, the coil cannot tell that they are not AC, so it puts up strong opposition to their flow. In this use, the coil is most often called a **choke** because it chokes out the AC component (part) of the current but passes the DC component.

Amount of Inductance

Inductance is measured in units called a **henry** (H). The henry is related to other units, such as volts and amps, in the following way: A 1-henry inductor will create a counter-EMF of 1 volt when the current flowing through it changes at a rate of 1 amp per second. Since the henry is a fairly large basic value, most inductors are labeled in millihenrys (mH) or microhenrys (µH).

Four factors determine the inductance of a coil. These factors are:

1. Number of turns — A coil with many turns will have more inductance than a coil with fewer turns.

2. Diameter of the coil — A coil with large-diameter loops will have more inductance than a coil with smaller diameter loops. This is because more wire is needed to wind the coil.

3. Length of the coil — If the loops are placed closely together, the magnetic fields they produce add strength to each other. Widely spaced turns do not affect each other greatly. Therefore, a longer coil will have less inductance than a shorter coil.

4. Permeability of the core — A coil wound around a highly permeable core, such as a powdered iron compound, has higher inductance than the same type of coil wound around an air core.

A variable inductance may be made with a core that can be moved in and out of the coil. The more of the powdered iron "slug" that is inside the coil, the higher the inductance. When the slug is pulled farther out of the coil, the inductor acts as if it had

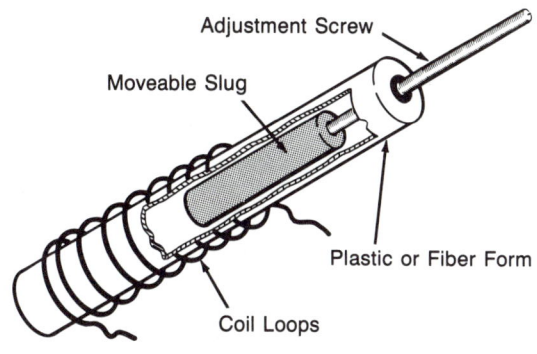

Fig. 31-6 The slug may be adjusted by turning it so that it moves in and out of the coil by screw threads.

an air core, so the inductance goes down. Variable coils, such as the one in Fig. 31-6, are used in many circuits that select the channel or station in radios and televisions.

When two or more inductors are placed in circuits in series with each other, the total inductance is found by adding the values of all inductors, just as with resistance. The total inductance of parallel inductive circuits is found by a formula just like the one for parallel resistances. The total value is always lower than the value of the smallest inductance in the circuit. See Fig. 31-7. The coils must be magnetically isolated from each other for these formulas to work. If they are not, more complex formulas which include "mutual inductance" as a factor must be used.

Unwanted Inductance

There is some inductance in every wire and connection in all circuits. The wires themselves act as inductances. This is not much of a problem with DC and low frequency AC currents, because there is not a lot of inductance and these currents can pass through with little opposition. With higher frequency AC currents in which the current changes direction very quickly, even a little extra inductance can be enough to prevent the current from flowing. Loose connections and cold or poor solder joints add much unwanted inductance to circuits. They often cause student projects to fail. This is especially the case when higher frequencies are involved. Even the length of wires can become a problem with high frequency circuits. Longer wires have more inductance. Avoid leaving a lot of extra wire length laying around in the projects you build.

Summary

Inductance, measured in henrys, is the quality of a coil which causes it to create self-induced counter-EMF when current

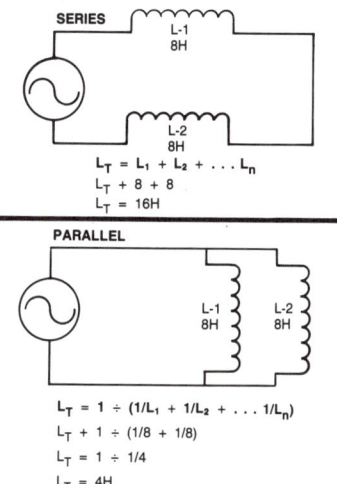

Fig. 31-7 The formulas for finding total inductance are the same as those for finding total resistance. *Note: These formulas only work if the coils are far enough away from each other so that their magnetic fields do not overlap.*

rate increases and counter-counter-EMF when current rate decreases. These self-induced currents oppose any changes in circuit current flow rate. Therefore coils allow DC current to flow easily, except for a little delay when the flow first starts and a similar delay in stopping when the switch is opened. However, coils create much opposition to the flow of AC currents because AC is always trying to change in rate. The four factors affecting inductance are number of turns, diameter, length of coil, and permeability of the core. All wires and components have some inductance, and unwanted inductance affects high frequency AC circuits a great deal. The next two chapters deal more with the affects of inductance in DC and AC circuits.

Important Terms

inductor
self-induced current
inductance
counter-EMF
counter-counter-EMF
choke
henry

Review Questions

1. Explain how counter-EMF and counter-counter-EMF are developed in coils when DC currents are turned on and off.

2. Why is no counter-EMF created when DC current is flowing at a steady rate through a coil?

3. Why do coils seem to pass DC but oppose the flow of AC currents greatly?

4. What is unwanted inductance, and what are its causes?

Chapter 32

RL Time Constant

Time Constant of RL Circuits

The counter-EMF produced when a DC current is first passed through a coil only lasts a short time. The length of time the counter-EMF lasts is determined by the amount of resistance in the circuit with the coil and the value (in henrys) of the coil itself. Figure 32-1 shows a circuit that has both a resistor and a coil. The graph in Fig. 32-2 helps you determine the duration of the counter-EMF and the percentage of potential (full) main current permitted to flow at any moment during the time current flow is building.

Notice that the graph is divided into five sections. Each vertical line marks the end of a period of time known as a **time constant**.

A time constant is the time it takes for the current to build up to 63.2 percent of its eventual full flow rate. The length of time that it takes for one time constant is determined by the formula: $T = \dfrac{L}{R}$. In this formula, T means "the length of time in seconds for one time constant," R is the amount of resistance in ohms, and L is the amount of inductance in henrys. The letter L is used for the inductor because I has already been used for **current** in other formulas.

To determine the length (duration) of *one* time constant, substitute known values of L (inductance) and R (resistance) into the formula and then divide to find T (the duration of *one* time constant expressed in seconds). With 6 henrys of inductance, and

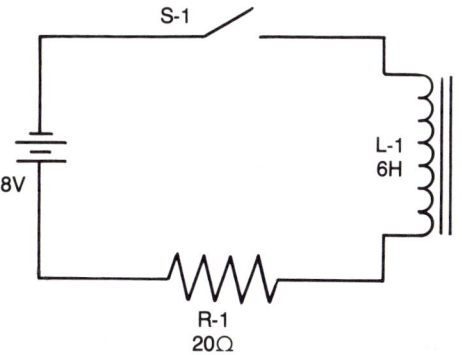

Fig. 32-1 Some circuits have both resistance and inductance.

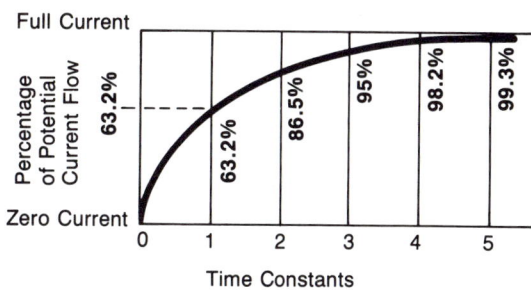

Fig. 32-2 This graph shows how the current slowly builds as counter-EMF is overcome. Notice that five time constants are needed to build nearly full current.

20 ohms of resistance, each time constant will be 0.3 seconds long.

The actual length of time included in one time constant depends on the particular circuit. A time constant is not a standard unit for measure like an inch or a gallon. It is more like a slice of pie. If you cut a pie into five equal slices, the size of the slice depends upon the size of the original pie. In the same way, the time constant may be thought of as a "⅕-sized slice" of the total time for full current flow to be reached. Most circuits will have different time constants. Yet, no matter what value of T you find for one time constant, it will always take five time constants for the current to build to nearly full value (99.3% of full potential current flow). For all practical purposes, this is full current and that is why the graph shows five time constants.

Using Time Constants

The length of one time constant may be found for any RL circuit by using the preceding formula. An **RL circuit** is one that has resistance and inductance together. Now, look again at the graph in Fig. 32-2. What the graph shows is the percentage of full main current flow at the end of any time constant. It is known that each time constant will provide enough time for the current to build up to 63.2 percent of its eventual full flow rate. So, at the end of one time constant after the switch is closed (regardless of how many seconds one time constant actually lasts for a given circuit), there will be about 63 percent of the potential main current flowing in the circuit.

At the end of the second time constant, 63 percent of the remaining difference between full and present current flow will be built up. The colored area of Fig. 32-3 shows this remaining portion of the potential current. The figure shows how to find new points on the graph. For instance, say you find the new point B (86.5 percent). If you then take 63 percent of what remains between it and full current, you can find the percentage of current flow at the end of three time constants. The main thing to

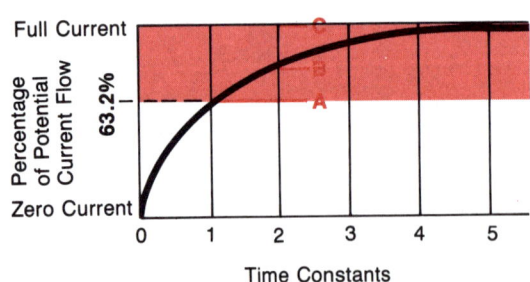

Find the space from A to C by subtraction:

C = 1.000 (100%)
-A = -.632
―――――――――――
       .368 = space from A to C

Find 63.2% of this remaining space:

   .368
x .632
―――――
   .233

Add this value to the height of point A to find point B:

A = .632
  + .233
―――――
B = .865

Plot new point B on the graph as 86.5%

Fig. 32-3 This figure shows how the graph was developed. In each time constant, 63 percent of the remaining potential current is developed. Here, we find point B which is the current percentage at the end of the second time constant.

remember is how to use the graph. What percentage of the potential full current will flow at the end of four time constants? If you said 98.2 percent, you are correct.

You can never reach actual full potential current flow. The curved line in the graph will always approach the imaginary 100 percent level, but it will always be a little below it. This is true no matter how many time constants you continue to calculate. For this reason, it is necessary to assume that some point must be treated as full current flow. Since five time constants allow current to build up to 99.3 percent of the full potential value, and this is in every practical way almost the actual full value, it is said that "full current flows after five time constants." In the circuit you have been studying, one time constant is 0.3 second long. Therefore, it takes 1½ seconds for full current to flow in this circuit — 5 × 0.3 = 1.5 seconds. The same process may be used to find how long it takes for full current to build up in any RL circuit. The time it takes for it to build to full current is **delay**.

Figure 32-4 allows you to find the time for full current to flow in another circuit. Follow the steps you have just read.

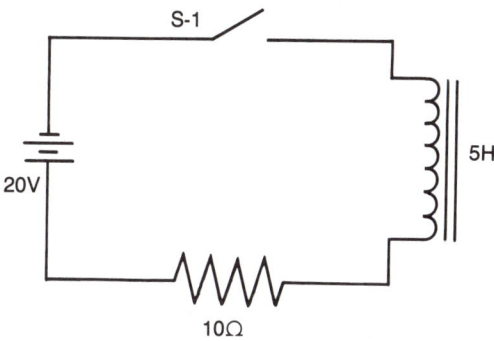

1) Find the Length of one Time Constant: **T = ?**

2) Look up the number of Time Constants required for full current on the graph (Fig. 33-2): __?__ Time Constants

3) Multiply the answers to steps 1 and 2:

Time for Full Current = T × number of time constants

Fig. 32-4 The time needed to reach full current flow may be found easily.

Instantaneous Current Rate

In some situations, a circuit may do its work with less than full current. Therefore it is sometimes helpful to be able to tell how much current is actually flowing at any given time during the building-up period. The actual instantaneous current flow rate for the circuit may be found for any instant during the building period by finding the percentage on the graph and then multiplying it by the eventual full current flow. Look again at our circuit in Fig. 32-1. By using Ohm's Law ($I = \frac{E}{R}$), you can find what current will flow once the building period is over. For example, you wish to know the actual momentary current flow rate of the circuit after 0.9 seconds. The graph shows that at 0.9 seconds — the end of three time constants — 95 percent of the eventual full current will be flowing in the circuit. This figure may then be used with the predicted full current flow rate (I = 400 mA) to find the momentary current (380 mA). See Fig. 32-5.

To find the momentary current at an instant which does not exactly end at a time constant, you must estimate the distance on the graph to find the percentage. For example, after exactly 1 second the current will be a little over 95 percent but not quite 98.2 percent. It will be closer to 95 percent.

PREDICTING MOMENTARY CURRENTS

1) Find eventual full current value (Ohm's Law)

$$I_T = \frac{E_T}{R_T}$$

$$I = \frac{8}{20}$$

$$I = .4A \text{ or } 400mA$$

2) Use 95% (found from the graph) to find the actual instantaneous current flow rate by multiplication:

$$.95 \times .4 = .380A \text{ or } 380mA$$

Fig. 32-5 Sometimes it is important to find the momentary current flow rates during the building period. For the circuit in Fig. 32-1, at the end of 0.9 seconds, there will be 380 mA of current flowing.

Uses for RL Circuit Time Delay

A possible circuit which uses the RL time delay is shown in Fig. 32-7. This circuit could be used for monitoring the flow of parts on a factory conveyor belt. As long as parts are flowing at the proper (normal) rate, the "normal" lamp will burn or blink at a desired rate to indicate that operation is correct. If the belt stopped or a part became jammed and blocked the path, then the "alarm" lamp will also light. The alarm light would usually not build up enough

An estimate of 96 percent would be close enough for most practical situations. See Fig. 32-6.

Try to find the momentary current flow rate of the circuit in Fig. 32-4 at the end of exactly 1 second. Did you get an answer of 1.73 A? If not, carefully follow the steps outlined and try again. When you get the right answer, find the momentary current at the end of exactly 1.5 seconds.

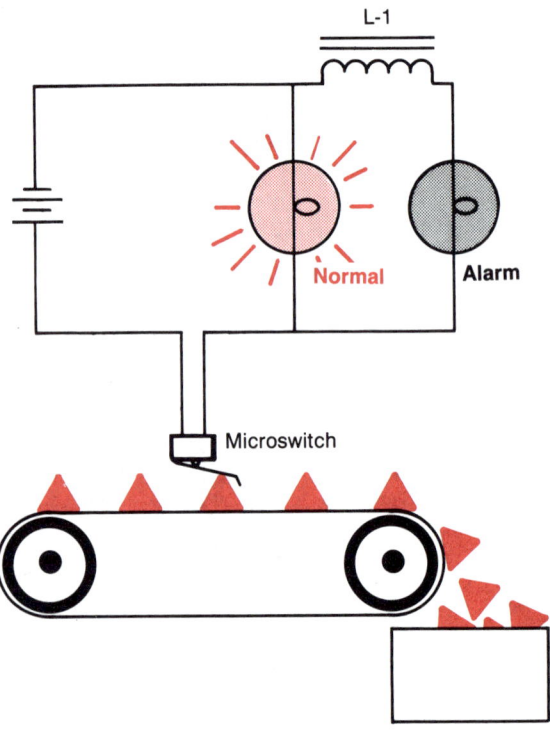

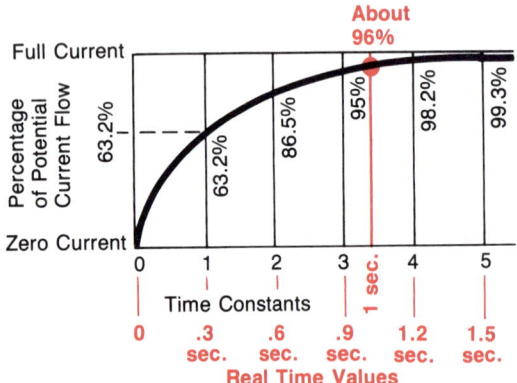

Fig. 32-6 Finding the instantaneous current at the end of exactly one second for the circuit in Fig. 32-1. You must read between the lines of our graph to estimate the values.

Fig. 32-7 In this possible circuit, mass produced parts moving along a conveyor belt are monitored by a microswitch. The alarm lamp will only light if a back-up causes the switch to be closed long enough for coil (L-1) to pass nearly full current.

current to glow. This is because the coil (L-1) would cause a time delay so that any part triggering the current to flow would have passed by before current reached a level high enough to make the alarm lamp burn. The length of the time constant required could be determined by using the formulas in this chapter.

The circuit could be expanded so that it sounded an alarm or even operated a relay which would stop the production line. In actuality, using coils for this type of circuit would be rare because this can be done with capacitors. They are generally cheaper to use.

Current Lag in RL Circuits

When the switch in the circuit of Fig. 32-8 closes, the **full voltage** (pressure to push electrons) is applied to terminals A and B immediately. The lamp, however, will not burn right away due to the time delay. This proves that the current does not flow immediately because the lamp will burn whenever enough current flows.

Why is this so? What does that voltage do? How can the pressure be pushing at full value but so few electrons actually make it through the lamp? The answer is that a large part of it is used to build the field around the coil. Until the field is built up, the coil seems to be a fairly high opposition to the flow of the DC current. Building the field only requires a very short time. As the field builds, the amount of opposition to the current flow it produces goes down dramatically until it is almost nonexistent after five time constants.

When you first close the switch, the coil acts as a very high resistance and the lamp acts as a very low one. You only have 6 V to work with, so at first most of it is dropped by the coil. (Remember Ohm's Law, the higher value resistance will drop the most voltage.) At the same time, there is high total circuit resistance during the first time constant, so total circuit current will be very low. It would not be high enough to light the lamp.

After a few time constants, the field will be built almost to full size, and the coil will now seem to have almost no resistance. At this time, the lamp will be the higher value of resistance and the total circuit resistance (coil and lamp together) will be fairly low. Again, Ohm's Law would tell you two things.

1. More current will flow in the circuit because you have decreased the total circuit resistance.
2. The lamp will now cause a larger voltage drop than the coil because it is now the higher value of resistance.

When you first closed the switch, the voltage was all there and available with no delay, but the current did not flow through the circuit until a little later (Fig. 32-8). For this reason, it is said the the current lags the voltage in an RL circuit. That is, the current is slower to build up to full value than the voltage. The voltage was at full value immediately. The current came lagging along behind it. Notice also that current continues for a short time after the switch is opened and the voltage is taken

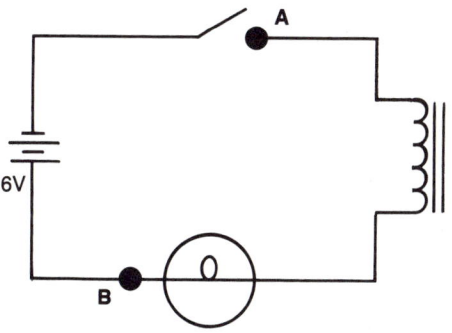

Fig. 32-8 The instant that the switch closes, full voltage is available between points A and B — only the current is slow to start.

away. The amount of lag and the importance of this factor are dealt with in Chapter 33 because they are even more important in AC circuits. Just remember for now that current lags voltage in RL circuits.

Summary

Circuits with both resistance and inductance (coils) are known as RL circuits.

When DC current is used in RL circuits, the voltage is available immediately after the switch is closed. However, a time delay occurs before full current flow is reached. The length of the time delay is determined by the values of resistance and inductance in the circuit. The time required for full current flow is divided into time constants on a standard graph which may be used in solving problems to determine instantaneous current flow. Chapter 33 will apply these concepts in AC circuits.

Important Terms

time constant
RL circuit
delay
full voltage
current lag

Review Questions

1. Explain why the current and voltage do not both reach full value as soon as the switch is closed in a DC RL circuit.
2. What is a time constant and how can you tell the length of the time delay?
3. Which lags in an RL circuit (voltage or current) and why is this so?

History: Joseph Henry

Joseph Henry was born to a poor family in Albany, New York, in 1797. At 13, he was apprenticed to a watchmaker. According to tradition, he gained his first desire to study in science from a book titled *Lectures on Experimental Philosophy* which he found while chasing a rabbit under an old church building. Henry discovered that electromagnets could be made strong by insulating the wires so that the coils did not short each other out. His wife's silk petticoat was sacrificed for one of these early experiments.

He succeeded in making electromagnets able to lift more than a ton (2000 lbs.) by 1831. (Nine pounds had been the previous record!).

He also invented the relay and was one of the discoverers of self-inductance. He became the first secretary of the Smithsonian Institution.

In 1878, Joseph Henry died. He did not patent much of his own work — preferring to remain a "pure" scientist. Several others became wealthy using principles he discovered. We honored Henry by naming the unit of inductance for him.

Project: Pulsed Alarm Circuit

Parts List

1	10 K-ohm resistor	R_1
1	33 ohm resistor	R_2
1	220 ohm resistor	R_3
1	0.1 microfarad capacitor	C_1
1	500 microfarad capacitor	C_2
1	10 microfarad capacitor	C_3
1	audio output transformer	T_1
1	PNP transistor (SK 3114A, or ECG 290A)	Q_1
1	microswitch or mercury switch	S_1
1	SPST switch	S_2
1	3 V battery with holder	B_1
1	9 V battery with connector	B_2
1	LM 383 audio amp. IC (8-watt)	Z_1
1	speaker (2 to 16 ohm)	SP_1

Suitable enclosure, wires, PC board, and hardware as required.

Notes

This circuit can produce either a continuous or pulsed alarm sound. With an 8-watt IC (the LM 383), the sound can be quite loud, if the circuit is properly loaded. Switch S_1 is the trigger for the alarm. S_2 selects the type of output. Values of capacitors and some resistors may be changed to vary the tone and the length of pulses. Try experimenting a little on your own. Experiment with different speakers also to see the effects of different loads on the amplifier IC. You should breadboard the circuit first and make any desired design changes, before you make a PC board for the final circuit.

Fig. 32-9 The pulsed alarm circuit may be used in applications that require loud, attention-getting sound.

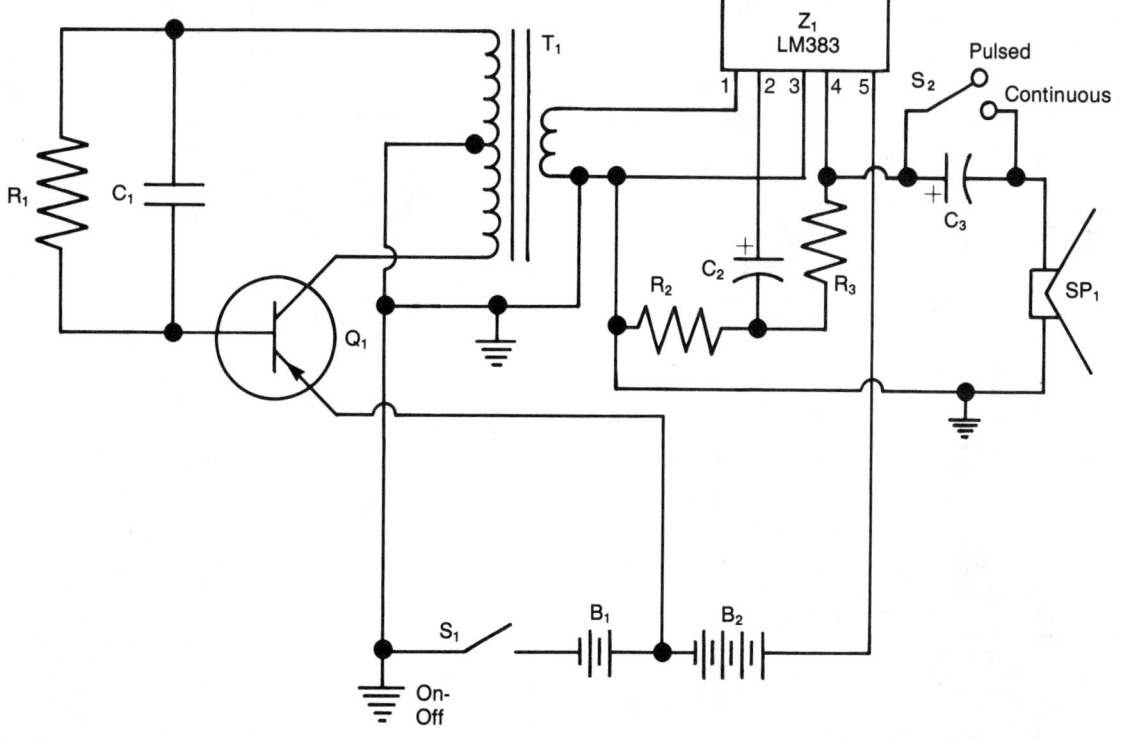

Reactances and Resonance in AC Circuits

Section 6

Chapter 33

Inductive Reactance

A New Type of Resistance to Current Flow

Coils act very differently in AC circuits than they do in DC circuits. For this chapter, it is important that you understand the concepts in chapters 28 (AC Currents), 31 (Inductance), and 32 (RL Time Constant). It would be a good idea to review those chapter summaries and study questions.

An AC power source is shown in Fig. 33-1. The AC current in this case is 60 Hertz AC (60 cycles per second). That means it changes directions 120 times every second.

The AC sine wave of this current is shown in Fig. 33-2. This full cycle is completed 60 times each second, so one cycle requires 1/60th of a second, as shown. During the first fourth of the cycle (point A to B on the graph), the voltage goes from off to maximum positive value. This requires 1/240th of a second and would be very much like turning the switch on in a DC circuit.

Fig. 33-1 This circuit has a 60 Hz AC power source.

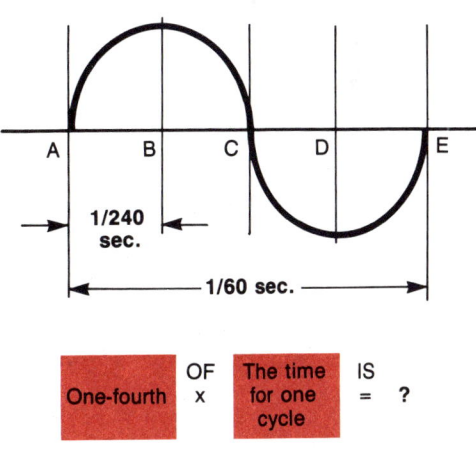

Fig. 33-2 During the first 1/240th of a second the current builds rapidly, then the current changes directions repeatedly.

Figure 33-3 shows what would happen if you apply the graph of time constants to the AC circuit. The new point for the "fully on" condition (positive) would be placed at the 1/240-second point on the graph (roughly where the colored mark is drawn). At this point, the percentage of current flowing (read from the left edge of the graph) would be very, very small. Almost no current would be able to flow. According to Ohm's Law for current $\left(I = \frac{E}{R}\right)$, the total circuit resistance must be very high because 10 volts of pressure is only able to push a few electrons through the circuit. Indeed, the coil is acting like an extremely high value resistance!

After the first quarter cycle, the current reverses. Then a totally new graph (beginning with another first time constant) would apply, so you are always going to be operating in the first 1/240th of the graph. Therefore this graph is not used with AC circuits in actual practice. You will never even come close to reaching the end of the first time constant and the current will always be very low. If the current is low, but there is plenty of voltage, then the opposition to current flow must be very high.

Opposition to current flow has been called resistance in all of the circuits that you have studied so far. There are, however, other types of opposition to current flow. The special opposition to current in this circuit is only active when you use an AC power source. If the AC generator were removed and a DC battery were used, then high current would flow (about 500 mA) after ½ second (five time constants).

Resistance affects both DC and AC circuits the same. The 20 ohm resistor (R-1 in both circuits) has the same resistance (20 ohms) in both circuits. The coil, however, has almost no resistance in the DC circuit after the first half second, but a very great amount of something like "resistance" in the AC circuit at all times. The special quality that acts like a type of "resistance" to only AC is called **inductive reactance**. Inductive reactance is caused by the continuous changing of directions that takes place with AC currents. It does not apply at all to DC circuits.

Factors Affecting Inductive Reactance

The amount of inductive reactance in a circuit depends on two factors.
1) If the value (measured in henrys) of the coil in the circuit is high, there will be more inductive reactance.
2) If the AC current were very low in frequency, it would take a long time for each fourth cycle.

If, for example, you had a frequency of 3 Hz (3 cycles per second), then the first change in direction would occur after 1/12th second. See Fig. 33-4. According to the graph in Fig. 33-3, this would be very near the first time constant, so the current would at least be able to flow at about 50 percent

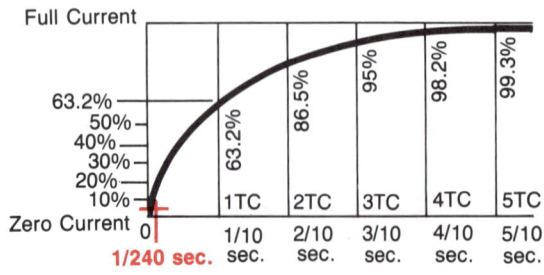

Almost no current flows in the first 1/240 second.

Fig. 33-3 In the first 1/240th second, the current flow is almost zero and the opposition to current flow is very large.

before the first change in direction. This is a lot more current than used for the 60 Hz source. Briefly, the higher the frequency, the higher the inductive reactance. Lower frequencies produce less inductive reactance and permit higher current flow.

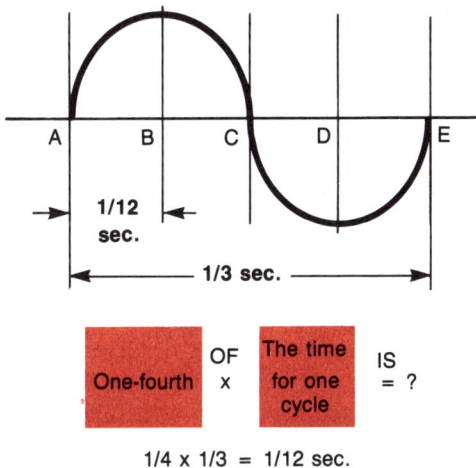

Fig. 33-4   A very low frequency AC sine wave. Here the first direction change does not occur until the ¹/₁₂ second point.

other types of reactance and other letters are used to identify the X when needed. These will be studied later. The number 2, located at the right side of the formula, is a constant in the formula. That means that it will be the same every time you use the formula. Regardless of how much the other values may change from one problem (circuit) to the next, you will always use the number 2. Next, you have another constant called **Pi**. Pi is the Greek letter π and it is pronounced "pie." The value of Pi is approximately ²²⁄₇ or 3.1416. You may use either the decimal or fraction form, whichever is easiest for you to use at the time. If you are using a calculator, there may already be a key labeled π on the keyboard. Pressing that key will enter an even more accurate decimal value of Pi for you.

The third symbol on the right side of the formula is f. When solving problems to find inductive reactance, you must enter the frequency of the circuit in the place marked by the f. The basic unit for frequency, the Hertz, is used.

The final factor is the inductance, L. Again, the basic unit is to be used. For inductance, use the value of the inductor in henrys (H). Solving a problem is very

Finding Inductive Reactance

As with all other qualities in electronics, exact values for inductive reactance may be found by using a mathematical formula. Fig. 33-5 shows the formula for finding inductive reactance. The symbol for reactance is X. **Reactance** is any type of opposition to current flow which is caused by something other than simple resistance. The little L in X_L tells you that you are concerned with "the reactance of an inductor" or inductive reactance. There are

INDUCTIVE REACTANCE

X_L = 2 π f L   Formula to use.

X_L = 2 x 3.14 x 60 x 2   Substitute real values.

X_L = 753.6 or 754 Ω   Multiply all values in any order and label answer.

Fig. 33-5   Finding the amount of inductive reactance (X_L) for the circuit of Fig. 33-1.

simple. Fig. 33-5 shows how to find the inductive reactance of the circuit from Fig. 33-1. The solution is very simple. Enter the real values, multiply, and label the answer. Since the inductive reactance is a form of opposition to current flow (similar to resistance) use Ω (omega, just as with normal resistance) to label the answer and the value of inductive reactance is measured in ohms.

Effects of Changes in f or L

Earlier in this chapter, it was explained that increasing either frequency or inductance would make the opposition to current flow (inductive reactance) increase and would limit current. You can now prove this mathematically. First, it was illustrated that lower frequency AC current would have less inductive reactance and permit more current to flow (review Figs. 33-3 and 33-4).

Prove this with real numbers by finding the actual inductive reactance of the circuit but using a frequency of 3 Hz as in the example. You should get an answer of 37.7 ohms. If not, try again. The inductive reactance was much lower with the lower frequency just as predicted in the earlier example, so more AC current would be allowed to flow. Now work another problem using the same circuit, but with a frequency of 200 Hz. Do you expect to get a higher or lower value of inductive reactance? Prove it.

What value of inductive reactance would you expect to find for the circuit that had a DC battery for a power source? When you try to substitute real values into the formula, what will you use for frequency (f)? Since DC current has no frequency (because it does not change directions), the value of f is 0. When you work the problem, you should get an answer of 0 ohms.

This is correct because there is no inductive reactance in a DC circuit. Inductive reactance only applies to AC circuits. The coil only causes high opposition to current flow during the first few time constants in a DC circuit. In AC it causes a lot of opposition all of the time.

Now demonstrate that the amount of inductance also affects the value of inductive reactance. Do this by increasing the value of the inductor in Fig. 33-1 from 2 H to 10 H and solve the problem (using a frequency of 60 Hz as shown). Compare your answer here to the one found in Fig. 33-5 which used 2 H and 60 Hz.

Inductive Reactances in Series and Parallel

When two or more resistors are placed in a circuit in series, you can find their total combined resistance by adding. This is also true for inductive reactances in series. Fig. 33-6 shows a circuit with two coils and

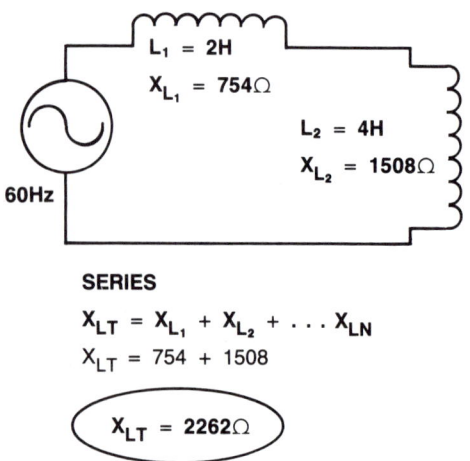

Fig. 33-6 Individual inductive reactances may be added to find the total inductive reactance of series circuits.

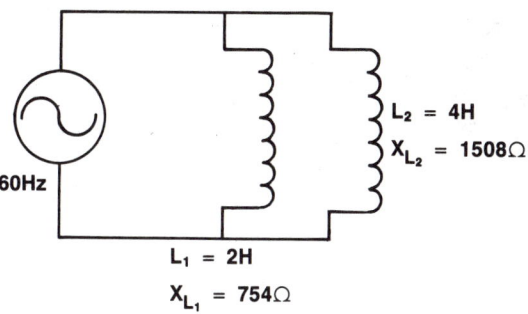

PARALLEL

$X_{LT} = 1 \div (1/X_{L_1} + 1/X_{L_2} + \ldots 1/X_{LN})$
$X_{LT} = 1 \div (1/754 + 1/1508)$
$X_{LT} = 1 \div (.001326 + .000663)$
$X_{LT} = 1 \div (.001989)$

$X_{LT} = 502.7\,\Omega$

Fig. 33-7 We use reciprocal addition to find total inductive reactance of coils in parallel. Remember that the total reactance must be less than the value of the lowest value coil. The simplest method of solution when using a calculator is shown — other methods may be used.

the formula for finding their total inductive reactance.

To find the total inductive reactance of two or more coils which are in parallel in a circuit, use a formula like the one for parallel resistances. The formula is shown in Fig. 33-7.

Summary

Coils produce a special opposition to current flow in AC circuits known as inductive reactance. There is no inductive reactance in DC circuits. The amount of inductive reactance in a circuit depends on the frequency of the AC current and the value of the coil. When coils are placed in circuits in series and parallel, their combined inductive reactances may be found just as series and parallel resistances are found. The next chapter discusses something which is almost the exact opposite of inductance. It is called capacitance and it is produced by capacitors.

Important Terms

inductive reactance
reactance

constant

Pi (π)

Review Questions

1. How is inductive reactance similar to resistance? How is it different?
2. What factors determine inductive reactance?
3. Why is there no inductive reactance in DC circuits?
4. If 4 H coils were placed into two circuits, one powered by a 60 Hz AC generator and the other powered by a 1000 Hz AC generator, which circuit would be most affected by adding the coil? Prove your answer mathematically.

Chapter 34

Capacitance

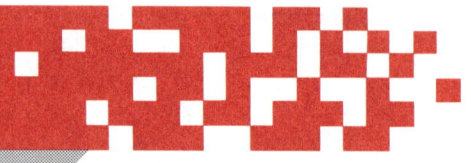

Capacitors

A **capacitor** is a very simple device, but its function in electronic circuits is very important. Figure 34-1 shows a simple capacitor. You can even make one like it. The **plates** are made of conducting material, such as metal or foil, and the space between them is called the **dielectric**. Dielectrics may be made of many insulating materials such as air, paper, or ceramics.

How Capacitors Function

The example circuit in Fig. 34-2 will help you see what capacitors can do. When the circuit is first assembled, before any current is allowed to flow, the capacitor would have no definite positive or negative charge because there would be about the same number of electrons on both plates (A). At B, switch no. 1 closes and the battery pumps out electrons that try to flow

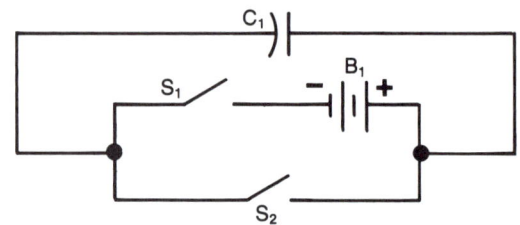

A) Both S_1 and S_2 are open (before charging the capacitor)

B) Switch S_1 closed, the capacitor charges. Electrons flow until capacitor voltage is the same as the battery voltage, then stop.

C) Both S_1 and S_2 open again, but the capacitor is fully charged. Capacitor holds the charge.

D) Switch S_2 is closed, but S_1 is left open. Capacitor discharges until charges are equal again.

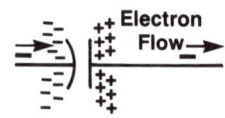

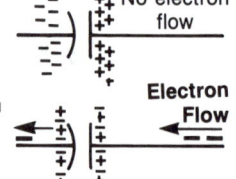

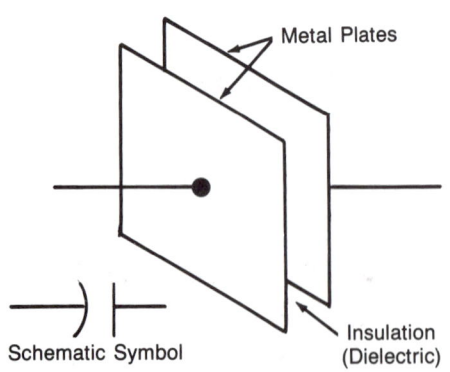

Fig. 34-1 A capacitor may be made by separating two conductor plates by an insulating material (dielectric). In this drawing, air is used for the insulator.

Fig. 34-2 This simple circuit illustrates how capacitors can store a charge.

through the capacitor. If there was no capacitor (and the circuit were open), the battery would be unable to pump electrons out because they would have no place to go. They simply would not leave the negative battery terminal. But the capacitor plays a trick on them. The plates of the capacitor are so large and so near to each other that they have two effects. First, they provide a place to store the positive and negative charges. Second, the electrons pumped onto the negative plate have a tension (like a static charge) which makes them "want" to jump through the insulation to get to the large, positively charged plate on the other side. Since the battery that is pumping them will not let them back up, the electrons are tricked into believing that they will be able to complete their trip. When they get to the dielectric (insulation), however, they are not able to go through it, so they wait. That is how the capacitor is **charged**.

If you again open both switches, the electrons that have been forced to "wait" on the negative plate will just stay there (C). You may wonder why they do not just drift away from the plate. The answer is that they are strongly attracted by the positive charge on the other plate. They are attracted to the plate by a force somewhat like magnetism (static charge) trying to pull them through the dielectric. Remember that you can have voltage without current. The "charge" is not moving, so the capacitor stores a voltage charge, not current.

If you now close switch 2 (as in D), there will be no pressure pushing electrons onto either plate or preventing them from flowing backward, but there will be a path for them to follow to get to the positively charged plate. So, some electrons will flow from the negative plate, through S-2, and to the positive plate. This will continue until there is no difference in the amount of electrons on the two plates. That is, until the capacitor is again **discharged**.

Discharging Capacitors

This cycle could be repeated as often as you wish. After the capacitor is charged and both switches are opened (C), the capacitor could hold its charge for a long time. Theoretically, a perfect capacitor would hold the charge forever, but most capacitors do allow some of the charge to "leak" off after a while. It is important to remember the capacitor's ability to store a charge because you could become seriously injured by touching the terminals of a charged capacitor. For this reason, you should assume that all capacitors are charged. They should be discharged before they are handled. Figure 34-3 shows one way to discharge a capacitor. Your teacher will show you other effective methods to use in different situations. Sometimes the charge in capacitors may be hundreds or even thousands of volts, so be very careful!

Capacitors in AC and DC Circuits

The ability to receive and store a charge is very useful in many circuits. Capacitors actually do quite different things in DC and AC circuits. In DC circuits they will allow current to flow only until they become fully charged, however since no electrons can actually go through the dielectric, no current flows after the capacitor is charged. In AC circuits, however, the current is rapidly changing directions. Figure 34-4 shows what happens when AC voltages are applied to a capacitor. When the AC generator pushes the electrons through the circuit in a clockwise direction (alternations A and C), the plate on the left side becomes

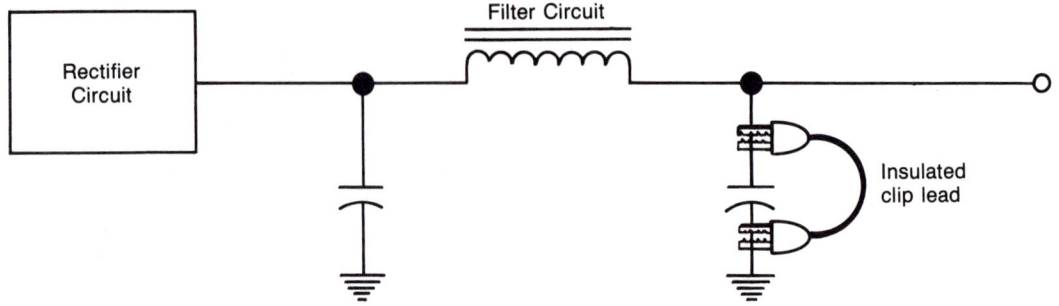

Fig. 34-3 Using an insulated clip lead to discharge a capacitor by "shorting" across its terminals.

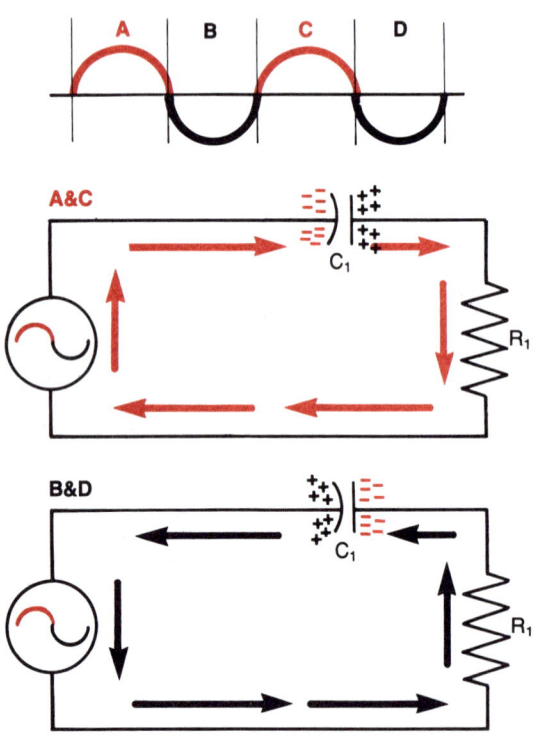

Fig. 34-4 During alternations A and C, the electrons flow through the circuit clockwise to charge the plate on the left side of the capacitor. They flow counter-clockwise in B and D, charging the other plate.

charged. Electrons flow downward through the resistor during these alternations. Electrons just build up on the plate.

During alternations B and D, the electrons are pumped back around the circuit counter-clockwise to be stored on the right plate. Again, no electrons actually flow through the dielectric, but the resistor now has current flowing through it in an upward direction. So, the resistor had electrons flowing through it during all of the alternations of AC. Or you can say that the circuit passed AC current, yet none of it went through the capacitor! Instead, current goes around the capacitor from one side to the other.

For this reason, it is better to say that capacitors appear to pass AC current, but they block DC. Frequently people will shorten this by saying that capacitors pass AC and block DC. This sounds as if the current is going through the dielectric. Be sure you understand that it does not! If current could go through the dielectric, then DC current could "pass," too. The DC is allowed to flow for a very short time to charge the capacitor and then it is blocked or stopped. Chapter 35 will explore this concept further.

Amount of Capacitance

The amount of **capacitance** (capacity) of a capacitor is measured in **farads**. The farad refers to how many electrons may be stored on the plates of a capacitor if it is charged by 1 volt of pressure. If a capacitor can store 1 coulomb of electrons on its plates when it is charged by only 1 volt, you would say that it is a 1-farad capacitor. (Chapter 7 explains what a coulomb is).

A farad is such a large value of capacitance that you probably will never use 1-farad capacitors. Smaller units are used to label and measure capacitors. The units used most often are the **microfarad** and the **picofarad**. These units are explained in Fig. 34-5. The farad must be used in formulas and problems.

Factors Affecting Capacitance

Three factors determine the capacitance of a capacitor:
- Area of the plates
- Spacing between the plates
- Type of dielectric used.

Area of the Plates

It stands to reason that since the purpose of the plates is to store the charges, larger plates will make "better" capacitors. This is basically true. The larger the plates, the more capacitance a capacitor will have. Often, large plates of thin foil are rolled up to form a tubular capacitor.

Spacing Between the Plates

The closer the plates are to each other, the more effect the positive charge on one plate will have on the electrons attracted to the other plate. Plates which are separated a great distance have little effect on each other, just as magnets do not pull as hard when they are held greater distances apart. Thus, to make a higher capacity capacitor, make the space between the plates smaller.

Type of Dielectric Used

Some materials make stronger dielectrics than others. A vacuum is considered a "basic" dielectric. That is, other dielectric materials are compared to a vacuum. A vacuum is assigned a number, called a **dielectric constant**, of 1. The dielectric constant of air (1.0006) is so close to that of a vacuum that it is considered to have a value

Units of Capacitance		
Microfarad:	1 µf (or 1 mfd.)	= .000,001 farad
Nanofarad:	1 nf	= .000,000,001 farad
Picofarad:	1 pf (or 1 µµfd.)	= .000,000,000,001 farad

Fig. 34-5 The microfarad, nanofarad, and picofarad are generally used in practical circuits because the basic unit (farad) is too large, but the farad is used in solving problems.

of 1. Other dielectric materials and their dielectric constants are listed in Fig. 34-6. The higher the dielectric constant, the higher value capacitor the material will make (assuming all the samples are the same thickness).

Types of Capacitors

Several types of capacitors are available. The capacitor to use in a circuit depends on several considerations. Obviously, the capacity of the capacitor is the first consideration. However, the physical size of the capacitor also may be important. A ceramic capacitor may be a better choice than a paper one if a small size is needed, even though the needed value is available in both types. Frequency is another consideration. At higher frequencies, mica and ceramic capacitors are generally better than paper ones.

Working Voltage

Another consideration in the selection of a capacitor is its working voltage. Capacitors have one rating for their capacitance and a second rating called **working voltage** (W.V.) which is the highest voltage at which the capacitor may be used safely. The working voltage is determined by the thickness of the dielectric and the **dielectric strength** of the materials from which it is made. (This is slightly different from dielectric constant discussed earlier in this chapter. The dielectric strengths of some common materials is shown in Fig. 34-7. If a capacitor was made from paper and the dielectric was 2 mils in thickness, then it should not be used in a circuit more than 2,500 V because the current may burn through it. Another name for working voltage is "breakdown voltage." For safety's sake, select capacitors that have a higher working voltage than the circuit requires.

Variable Capacitors

Fig. 34-8 shows a variable capacitor and its schematic symbol. The way it works is that one set of plates is able to move closer to and further from the other set of plates. When you change the spacing between the plates in this manner, you also change the capacitance of the capacitor. Remember,

Dielectric Constants

Material	Dielectric Constant
Air or vacuum	1
Paper	2–6
Oil	2–5
Mica	3–8
Glass	8
Ceramics	80–1200

Fig. 34-6 The dielectric constant tells how suitable materials are for use in the dielectrics of capacitors. The specific values actually depend on the exact chemical makeup of the sample used.

Dielectric Strength

Material	Strength in Volts per mil of thickness*
Air or partial vacuum	20
Paper	1,250
Oil	375
Mica	600–1,500
Glass	200–300
Ceramics	100–1,200

* a mil is 1/1000th of an inch

Fig. 34-7 The dielectric strength and the dielectric's thickness together determine the working voltage of a capacitor.

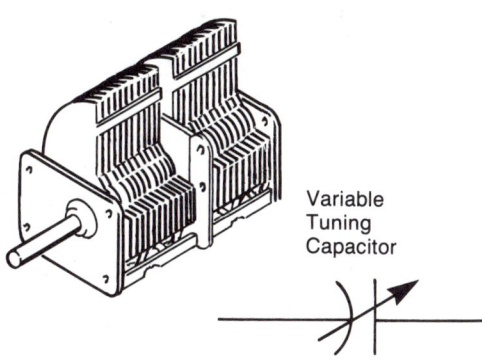

Fig. 34-8 Turning the shaft on a variable capacitor changes the spacing of the plates.

spacing is one of the factors affecting capacitance. The further the plates are from each other, the less capacitance you will have. Variable capacitors are often used to select the station in radios. Smaller variable capacitors, which are set by technicians when equipment is serviced or aligned, are often called "trimmers."

Electrolytic Capacitors

Electrolytics are special capacitors because of their unique manufacturing process that creates a very, very thin film of dielectric oxide. The advantage is that, since the dielectric is so thin, a lot of capacitance can be packed into a small container. However, the oxide film is very fragile electrically and it only works in one direction. That means that the film will break down (sometimes even cause an explosion) if voltage is applied to it backward.

Most capacitors are two-way devices which may be used in both AC and DC circuits. Electrolytics must be used in DC situations only. Also they must be put into the circuit correctly, they are polarized devices. Since they are polarized, they will be marked + and − to let you know how to correctly install them into circuits.

Two electrolytic capacitors may be connected back-to-back to allow their use in AC circuits, but their capacitance is reduced when this is done. It would be safest for you to follow the guidelines provided by the designers of circuits and to not make substitutions with electrolytic capacitors.

Additionally, the working voltage is more important with electrolytics than with most other types of capacitors. It is usually safe to use a capacitor with a much higher working voltage than is really needed, but this is not true with electrolytics. If an electrolytic capacitor is used in a circuit much below its working voltage rating for a long time or allowed to sit unused for a long time, it is possible for it to "reform" itself to the lower voltage. It could even damage the capacitor and cause a hazard. Therefore, electrolytics should always be used in circuits which have voltages between 75 percent and 100 percent of their actual working voltage rating. Figure 34-9 shows the schematic symbol for a can capacitor with four electrolytics.

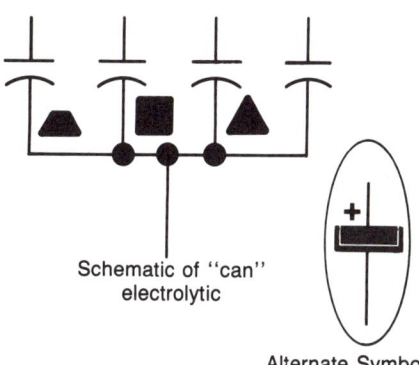

Fig. 34-9 Electrolytic capacitors give more capacitance in smaller packages. The schematic symbol for a "can" type, multiple capacitor is shown. The little symbols ▲, ■, and ▲ are actually printed beside the terminals they identify.

Capacitors in Series and Parallel

When capacitors are placed in parallel with each other in a circuit, in effect you make their plates larger. See Fig. 34-10. Larger plates would mean that the capacitance would be greater for the pair than for either one separately. To find the value of capacitors in parallel, simply add their values. Notice that this is the exact opposite of what happens with both resistors and coils in parallel.

Figure 34-11 shows how to find the total capacitance when capacitors are placed in circuits in series with each other. The area of the plates is not being increased this time. The distance between the outermost plates is increased. You must add reciprocals to find your answer, and the total capacitance will always be lower than the capacitance of the smallest capacitor in the circuit. This is the opposite of both resistors and coils.

Working Voltages in Series and Parallel

The working voltage depends on the thickness of the dielectric and has little to do with the plate area. Therefore when capacitors are in parallel, since the spacing does not change, the working voltage for the pair will be the same as that of the lowest working voltage rated capacitor. For Fig. 34-10, this rating will be 10 V because the 9 µfd. capacitor will burn out if any more than 10 V is applied to the pair.

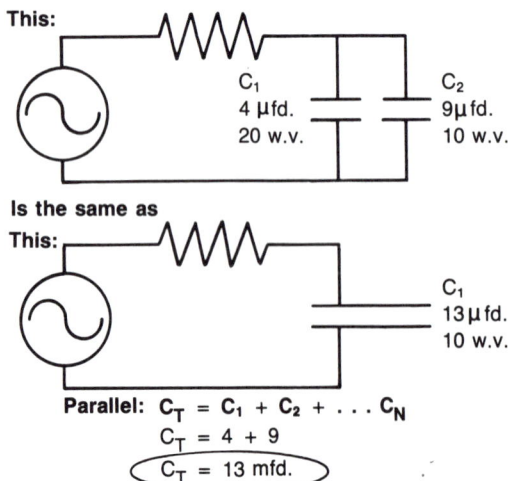

Fig. 34-10 Finding the total capacitance of capacitors in parallel.

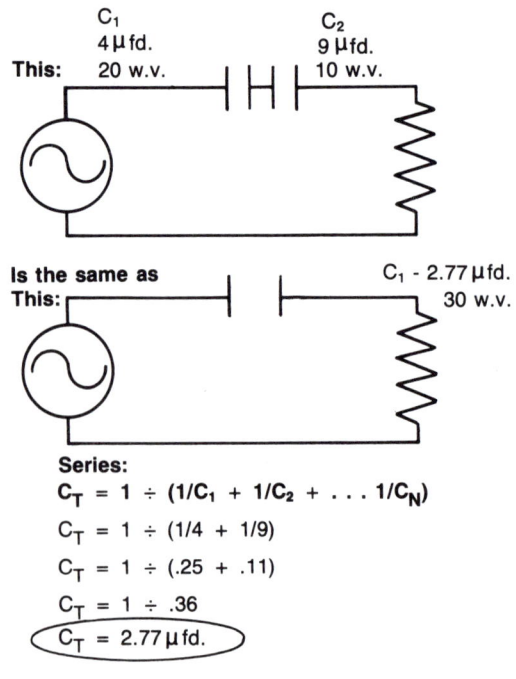

Fig. 34-11 Finding the total capacitance of capacitors in series. Notice that capacitors in series use a formula like the one for coils and resistors when they are in parallel!

When capacitors are placed in series, their total working voltage is found by adding their individual working voltage ratings. Since series connection is like widening the distance between the plates, the pair of capacitors can handle more voltage than either one could alone. So, for Fig. 34-11, the total working voltage rating for the pair would be 30 V.

Summary

Capacitors block DC and appear to pass AC. Capacitors are able to store a charge. They should be handled carefully and discharged before they are touched. Capacitance measured in micro-, nano-, and picofarads, depends on the size of the plates, the spacing between the plates, and the dielectric constant. There are many types and sizes of capacitors, including electrolytics and variable types. Capacitors may be connected in series or parallel to change their capacitance. They are almost the exact opposite of coils. Chapter 35 will explore more about capacitors and now they differ from coils.

Important Terms

capacitor
plates
dielectric
charged
discharged
capacitance
farad
microfarad
picofarad
dielectric constant
working voltage
dielectric strength

Review Questions

1. What do capacitors do when a DC voltage is applied?

2. What do capacitors do when an AC current is applied?

3. What determines the capacitance of a capacitor? How do variable capacitors work?

4. How are capacitors different from coils?

5. What is the difference between dielectric constant and dielectric strength? What does each mean?

Chapter 35

RC Time Constant

Charging Time

In the last chapter, you learned that the capacitor can hold the charge after the voltage source is removed. Charging of capacitors does not occur instantly. A little time is required for the capacitor to build up to full charge. This is the **charging time**. When the switch closes, electrons will be pumped around clockwise to charge the top plate of the capacitor. When they begin to travel, there will be no extra electrons on the plate (both plates will have about the same neutral charge), and so there will be plenty of space to store the extra electrons. As the plate fills with extra electrons, a longer period of time will be required to force additional ones onto it.

A graph (Fig. 35-1) of the rate the capacitor is charged would show a high rate of charge at first and a slower rate as the capacitor's plates become more fully charged. Does the shape of the curve look familiar? It should. It is the same shape as the one for time constants of inductors (coils).

This graph used with coils told you how long it took the coil to allow full current to flow through the circuit. The shape may be the same with capacitors, but the meaning of the graph is different. When applied to capacitance, this graph tells you the percentage of full voltage charge the capacitor will have at the end of each time constant. At the end of five time constants, the capacitor will have 99.3 percent of its potential full charge and you would say it is fully charged.

It takes just as long to discharge a capacitor (through the circuit) as it does to charge it. (The graph of the charging rate also applies to the discharge rate, except that the curve will be turned upside-down.)

Finding the RC Time Constant

Regarding coils, the formula for finding the duration of one time constant required you to divide. The formula was $T = \frac{L}{R}$. It has been said that capacitors and inductors are almost exactly the opposite. To find the duration of one time constant for **RC circuits** (resistive-capacitive circuits), you perform the mathematical operation which is the opposite of division — multiplication.

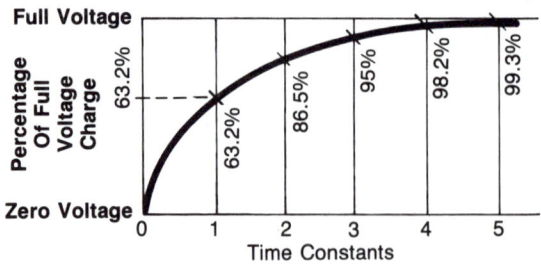

Fig. 35-1 Graph of the time required to charge the capacitor in a DC, RC circuit.

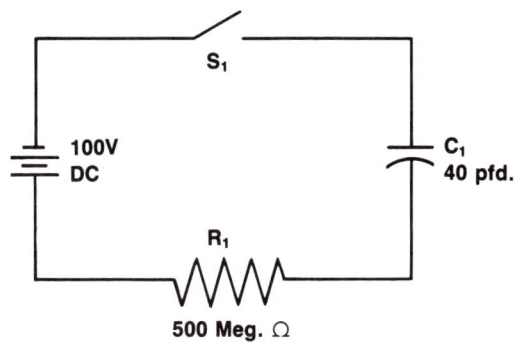

1) How long will it take for C_1 to charge to the full 100V? Remember to change to basic units
2) How long will it take for C_1 to charge to 63V?
3) What would the voltage charge of C_1 be after it had charged for exactly .06 seconds?

Fig. 35-2 Work the problems carefully and check your answers here. Correct any parts of the problems which you miss. Answers: 1) .1 sec., 2) .02 sec., 3) 95 V.

The formula for finding the duration of one time constant for resistive-capacitive circuits is $T = RC$. This formula is: The time for one time constant (in seconds) is equal to the value of the resistance (in ohms) multiplied by the value of the capacitor (in farads). Remember to always use basic units. Consider a circuit with a 50 kΩ resistor in a series with a 2µf capacitor:

```
T = RC                  Convert to Basic
                        Units and substitute
T = 50,000 x .000,002   values into the
T = .100,000 SEC.       formula.

    OR  1/10 SEC.       Multiply and then
                        label your answer.
```

Use the formula to solve the problem in Fig. 35-2. If you do not remember how to use the graph to find instantaneous values, refer to Chapter 32. Feel free to try additional problems for practice.

All circuits have some resistance, but what if you had a capacitor in a DC circuit with absolutely no resistance? How long would it take to charge or discharge? The value for R would be 0, and 0 times any number is 0, so your time constant would be 0, also. This means that the capacitor would charge or discharge instantly. In fact, when there is very little resistance in the circuit, charging and discharging both take place very quickly. Increasing the value of either R or C will make the capacitor take longer to charge and discharge.

A Use for the RC Time Constant

The RC time constant is used in many circuits for timing events and changing the shapes of waveforms. Stereo circuits that make the bass or treble louder are based on this concept. A very simple tone control circuit is shown in Fig. 35-3. Since it takes time for the capacitor to charge, it is possible to send the fast-acting high frequency (treble) tones to ground through a resistor-capacitor circuit. That is, you can cut some of them out (or "waste" them) instead of amplifying them. This would make the music have louder sounding bass. The bass sounds would not be affected as much because they are slower acting (lower in frequency), and they would last much longer than needed to charge and discharge the capacitor. Therefore instead of going to ground, these tones would go on to the rest of the amplifier circuits. The potentiometer is used to adjust how much of the treble you wish to pass to ground.

Voltage Lags Current

Another way in which capacitors are the opposite of inductors is that the voltage lags the current with capacitors. When you

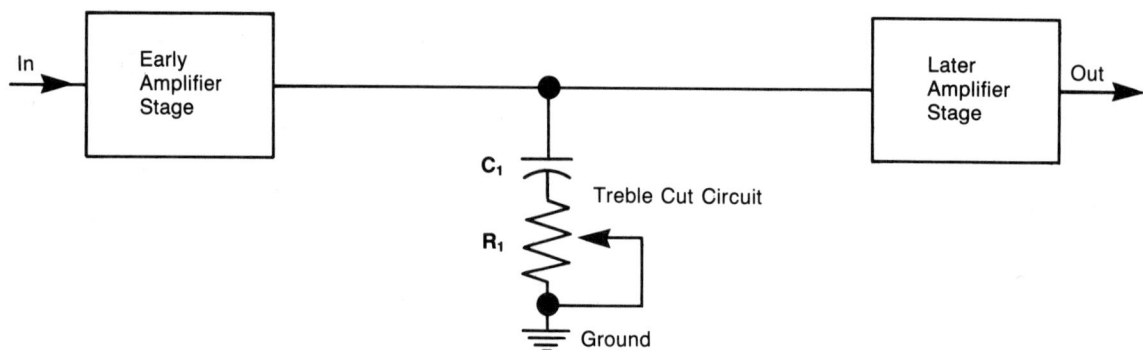

Fig. 35-3 A simple "treble cut" tone control circuit which will pass high frequencies to ground but let the lower bass frequencies go on for further amplification.

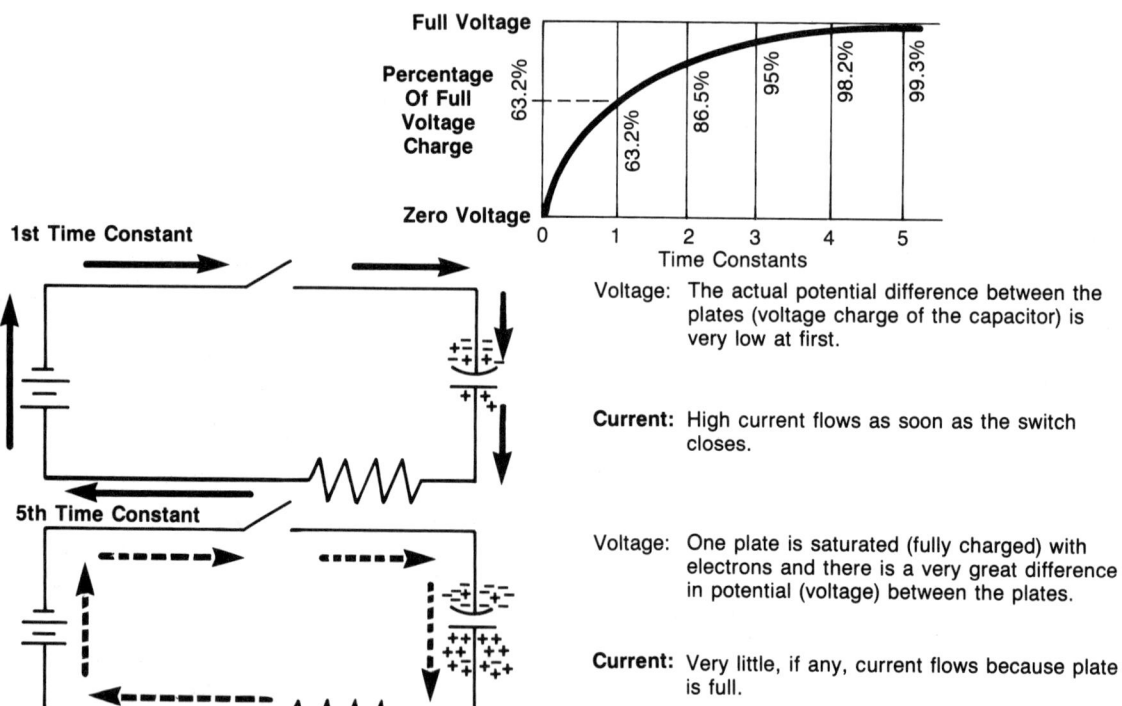

Fig. 35-4 Immediately after the switch is turned on, current flows at a high rate, but the capacitor has almost no voltage charge. The voltage charge catches up later. In other words, voltage LAGS current in capacitive circuits.

studied inductors, you read the current would always be slower to build up than the voltage. This is because the pressure to push electrons is there immediately after the switch is turned on, but some time is required for the flowing electrons to overcome the effects of counter-EMF. With capacitors, however, you have a different situation. Figure 35-4 illustrates what happens in a DC RC circuit. Notice you are considering the actual voltage charge which is existing in the capacitor. The current to make the charge grow must come first, so the voltage must lag the current in capacitive circuits. As you saw earlier, this is the exact opposite of inductance.

Opposition to Circuit Voltage Changes

Still another way capacitors are the opposites of inductors concerns what they oppose. Inductors opposed any changes in circuit current flow. When current was building, the counter-EMF of the coil tried to stop it from building. When it was decreasing, the counter-counter-EMF tried to keep it going. Capacitors have two jobs:
1) Block DC, but appear to pass AC.
2) Oppose changes in circuit voltage.

This section deals with keeping circuit voltage steady. When peaks of the pulsating DC from the rectifier circuit get to the capacitor, some of that pressure will be used to charge up the capacitor (Fig. 35-5). Therefore, only part of the original pressure will be left to go on to the circuit output. After the peak passes, a valley comes along. By the time the valley gets to the capacitor, it will be freshly charged with the extra pressure from the last peak. The capacitor will act like a small power source (similar to a battery) and pump its extra electrons back into the circuit to fill in the valley.

The capacitor will not "know" it, but it draws an imaginary line for the "typical" voltage of the circuit. Any time the actual circuit voltage is higher than this imaginary line, the capacitor will act like a "load" to use up some of the extra voltage. Any time the actual circuit voltage is below the imaginary line, the capacitor will give that stored extra voltage back to the circuit to fill

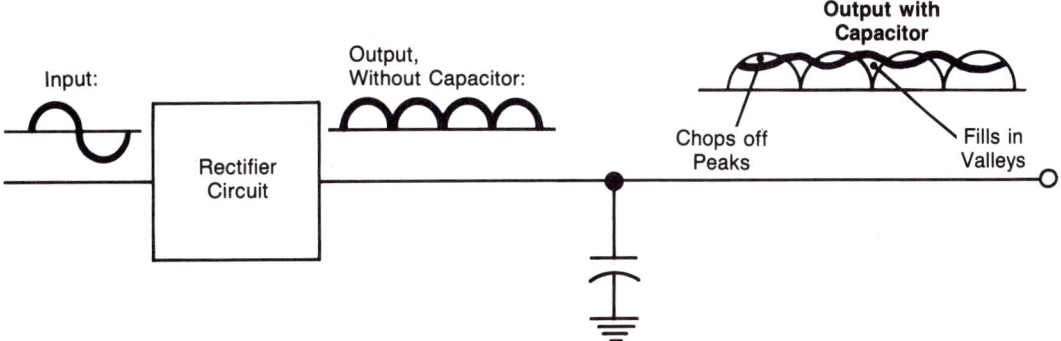

Fig. 35-5 The capacitor opposes the rapid changes in circuit voltage by soaking up the extra electrons pumped out during the peaks and then giving them back during the valleys. The orange line in the final graph is the circuit output. It could be made much smoother by adding a choke and another capacitor to the circuit.

in the valleys. So, the voltage is not allowed to go below a certain standard nor to go above it without overcoming the action of the capacitor. The capacitor "opposes all changes in circuit voltage." In this sense, capacitors act like road graders that cut off the tops of hills and push the dirt from them into valleys to make the road flat.

Summary

Capacitors have two main jobs: blocking DC while appearing to pass AC, and keeping circuit voltage steady by storing charges and releasing them when the voltage goes too low. When capacitors are used with resistors in DC circuits, the time required for them to charge and discharge may be found by a graph method similar to the one used for percentage of current flow in RL circuits. However, since capacitors are the opposite of inductors, you multiply to find their RC time constant and the voltage will lag the current in RC circuits. The next chapter will explain the special resistance capacitors offer in AC circuits and describe additional differences between capacitors and inductors.

Important Terms

charging time
RC circuits

Review Questions

1. What is the difference between the graph shown in this chapter and the one used with inductive circuits?
2. What are the two jobs of capacitors? Explain each of these jobs.
3. Explain two ways in which capacitors and inductors are the opposite of each other.

History: Michael Faraday

An English blacksmith's son who was apprenticed to a bookbinder at age 14 became an outstanding scientist in chemistry and electricity. Michael Faraday worked with electrolysis and magnetism. One of his most important contributions was the invention of the transformer.

Faraday was not well educated in mathematics. This may have limited possible advances in his own work. Working in the early to middle 1800s, some of his more important discoveries were not even understood until Maxwell (about 1870) and Einstein (1915) used mathematics to apply and predict broad theories from the principles Faraday had discovered. He was probably the first to demonstrate that electrical current involved the movement of particles. He also did important work with magnetic fields, including the invention of an early generator. He died in 1867. The unit for capacitance, the farad, commemorates his accomplishments.

Project: Bargraph Lie Detector

Parts List

1	10-Element LED Bargraph display	D_1
1	LM3914 LED Bar Display Driver IC	Z_1
1	2.2 microfarad capacitor	C_1
1	1.2 K-ohm resistor	R_1
1	3.3 K-ohm resistor	R_2
1	100 K-ohm potentiometer	R_3
1	15 K-ohm resistor	R_4
1	47 K-ohm resistor	R_5
1	10 K-ohm resistor	R_6
1	1 k-ohm resistor	R_7
1	470 ohm resistor	R_8
1	ECG 289A NPN transistor	Q_1
1	ECG 290A PNP transistor	Q_2
1	9 V battery with connector	B_1
1	SPST switch	S_1
2	pennies soldered to long wires for test probes	
	Suitable enclosure, wires, PC board, and hardware as required	

Notes

This circuit gives a bargraph reading of a person's skin resistance. Skin resistance tends to drop when a person is "sweating-out" a situation, such as when they tell a lie. Professional lie detectors used by police departments are much more accurate than this simple circuit, because they also measure other factors such as heart rate and breathing rate. This circuit should be sensitive enough, however, to make it fun to use.

The probes are made by soldering pennies to long wires and then attaching them to the back of the client's hand with a large rubber band. They also may be held in the fingers, but this makes it easier for the person to alter the test by squeezing the probes at different pressures.

After the probes are attached, adjust the potentiometer so that the reading is near the center of the graph. Position the circuit so that the person cannot see the readings shown by the bargraph display. Then, ask some "neutral" (that is, nonthreatening) questions which the person can answer truthfully without fear of embarrassment. All questions should be stated very directly so that the person can answer with a simple yes or no. If they have to make a statement of several words, the meter may register a false reading due to body motions. Examples of neutral questions are:

- "Is your name Alice?"
- "Are you wearing a red shirt?"
- "Do you like ice cream?"

Make minor adjustments to the potentiometer while you ask the series of neutral questions to calibrate the meter. Then, without telling the person that you are actually beginning the test, ask more serious questions that will make him or her sweat a little. Do respect your friend's privacy in personal and highly embarrassing matters. Major drops in skin resistance could mean a statement is false. Do not try to depend upon this simple circuit for any serious detective work, and keep the fun in good taste.

BAR GRAPH LIE DETECTOR

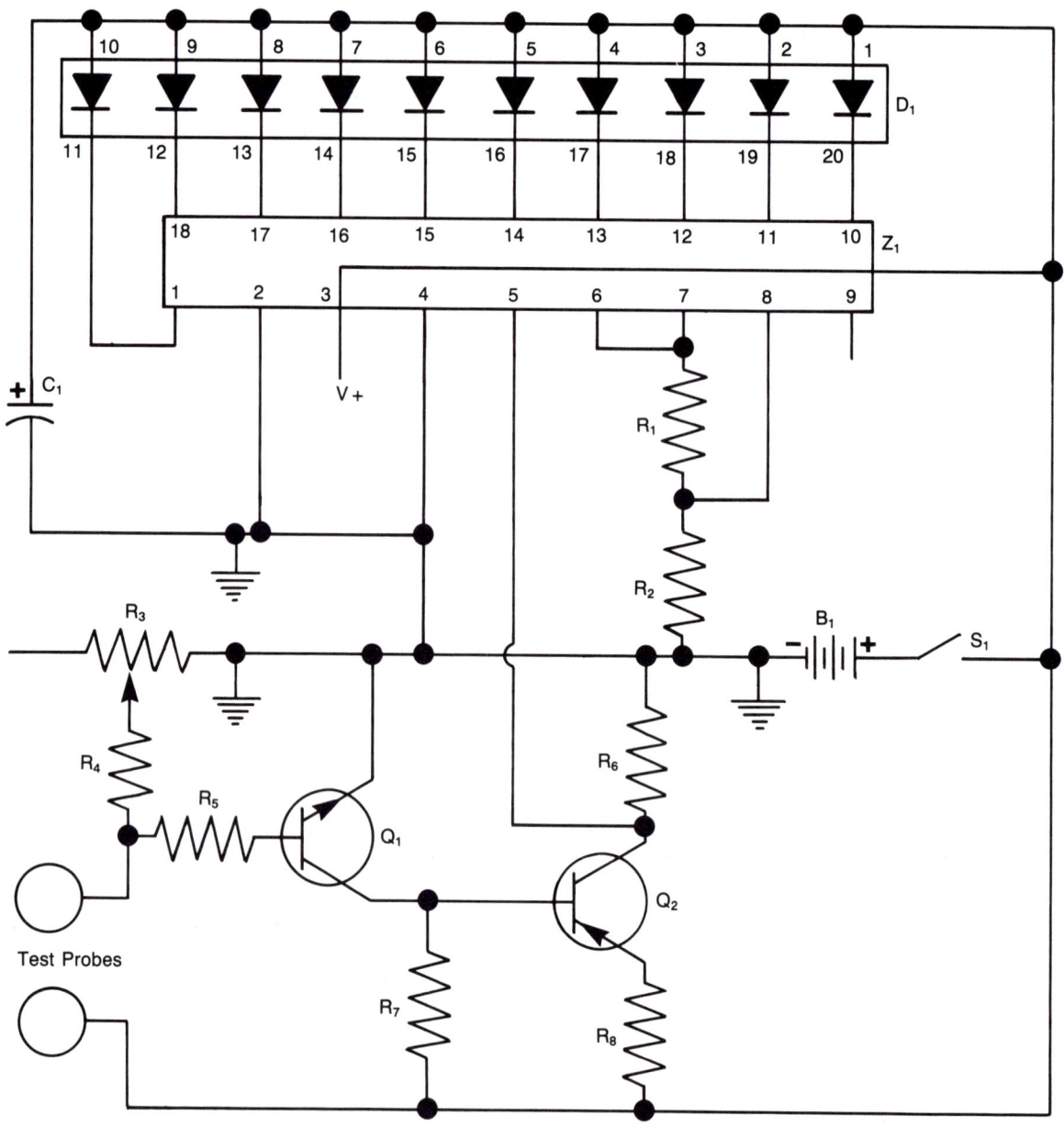

Fig. 35-6 Your friends and you can have some fun with this Bargraph Lie Detector.

Chapter 36
Capacitive Reactance

Another Type of Reactance

Devices that oppose AC current differently than DC current have reactance. Like coils, capacitors also develop special opposition to the flow of AC current that is different than their opposition to DC current. This quality of capacitors is called **capacitive reactance**.

Figure 36-1 will help you understand capacitive reactance. After a short charging period, the capacitor blocks the DC current. In the AC circuit, however, the capacitor acts more like a special resistance (reactance) which limits current flow but does not totally block it. (See Chapter 34 to learn more about capacitance).

Factors Affecting Capacitive Reactance

Capacitive reactance depends on:
1) Amount of capacitance.
2) Frequency of the AC current.

Let's look at the amount of capacitance in the circuit first. The three factors that determine the amount of capacitance are size of plates, spacing between plates, and quality of the dielectric. If the plates are large and closely spaced with good dielectric material, then the capacitor will have a

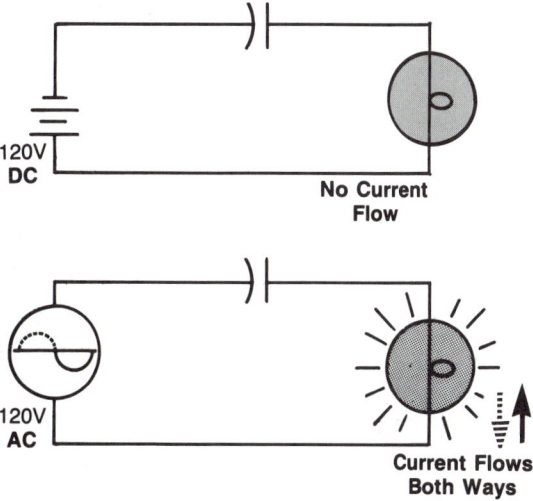

Fig. 36-1 The DC power source is unable to pump electrons through the capacitor to light the bulb, but the AC source makes electrons flow first one way and then the other to charge the capacitor plates alternately and the bulb lights.

high value. That is, it will be able to charge very quickly with many electrons being stored on the negative plate. Such a capacitor will allow many electrons to flow through the lamp in the circuit on each alternation of the AC generator. On the other hand, the lower value capacitor will not be able to store charges quickly and the amount of current flowing in the circuit will be reduced. Thus, the first factor affecting capacitive reactance is the amount of capacitance in the circuit.

You learned from your study of inductive reactance that the higher the value of inductance, the more reactance the coil will create. This is another way in which coils and capacitors are opposites. The higher value capacitor allows more current to flow in the circuit, so it must be creating less reactance. In other words, the higher the value of the capacitor, the lower the capacitive reactance.

The other factor is the frequency of the AC current (Fig. 36-2). Since high frequency currents change directions more rapidly than lower frequency ones, each alternation of the generator's cycle lasts a shorter time. This means that during each alternation there is much less time for the generator to force electrons onto the capacitor's plates. Look at the graph in Fig. 36-2. The plates will become very crowded with electrons if the charging current is applied for a long time. But early in the first time constant there is a very small number of electrons stored on the plate. There is room for plenty more. The reason the graph goes up more slowly in the later time constants is that once the plates become crowded, it is difficult to push more electrons onto them. With AC current, each alternation will only last a short time, so this graph would normally not be considered. The point is, high frequency AC currents will always be pumping electrons onto almost totally empty plates. (That is because each alternation only lasts a very short time). This means that high frequency AC currents will see capacitors as very minor sources of opposition to their flow. On the other hand, a very low frequency AC current will find more opposition because the plates of the capacitor will become more full during each alternation and the generator will have to "work" harder on each alternation to try to force those last few electrons onto the already crowded plates. Therefore you can see that the higher the frequency of the AC current, the lower the opposition to current flow. That is, higher frequency causes lower capacitive reactance. Here again, capacitive reactance differs from inductive reactance because higher frequency currents find more reactance (opposition) when coils are in the circuit but less reactance in circuits with capacitors.

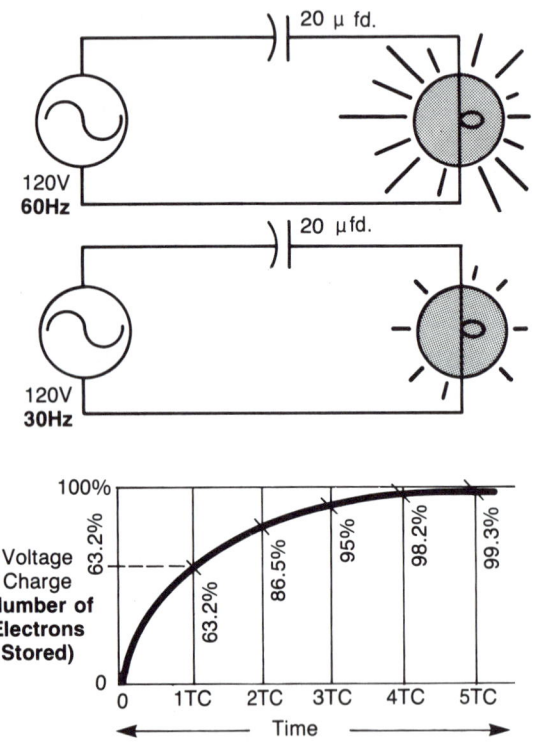

Fig. 36-2 Higher frequency currents find less opposition from capacitors because they only try to charge the plates for a very short time in each alternation.

The Formula for Capacitive Reactance

You now know that capacitive reactance is opposition to AC current produced by capacitors. You also know that it is lower for high frequency currents and for high value capacitors. Now you can calculate the actual amount of capacitive reactance in a circuit using these factors.

There is some similarity between the formulas for inductive reactance and capacitive reactance, but there are also two very important differences. Remember that both of these types of reactance depend upon similar factors, but yet they are nearly the opposites of each other. Figure 36-3 shows both formulas. The one for capacitive reactance uses C (capacitance in farads) instead of L. Also, the reciprocal is taken so that you have $\frac{1}{2\pi fc}$ instead of $2\pi fc$.

Inductive Reactance: $X_L = 2\pi fL$

Capacitive Reactance: $X_c = \frac{1}{2\pi fc}$

Fig. 36-3 The 1 over the fraction bar makes f and C affect X_c very differently than f and L affect X_L.

Anything you do to increase the total value of the **expression** below the fraction line will make the resulting value of X_C lower. (If this confuses you, remember that ¼ is smaller than ½, even though 4 is larger than 2; ⅛ is even smaller than ¼ for the same reason.) The larger the **denominator** (expression below the fraction bar), the lower the total value of the fraction. Thus, if f (frequency) is increased in both formulas, the resulting value of X_L (inductive reactance) will be higher, but the value of X_C (capacitive reactance) will be lower. This is correct according to what you know of the effect of frequency on capacitive and inductive reactances. Inductive reactance is a **direct proportion** and capacitive reactance is an **inverse proportion**. Similarly, if the value of the capacitor is increased, the resulting value of capacitive reactance (X_C) will be lower. The numerator 1 accounts for the fact that capacitive reactance is an inverse proportion.

Calculating Capacitive Reactance

Solving problems with the capacitive reactance formula is very simple. The steps are nearly the same as for inductive reactance (Fig. 36-4). First, substitute the true value for f and C. The frequency should be in hertz (Hz, or cycles per second) and the capacitance should be in farads. Second, multiply the four terms 2, π (pi ≈ 3.14), f, and C which are under the fraction bar. Divide the answer into 1 to find the value of capacitive reactance (X_C).

If you worked the problem with paper

$X_C = \frac{1}{2\pi fc}$

$X_C = \frac{1}{2 \times 3.14 \times 60 \times .00002}$ Substitute Values

$X_C = \frac{1}{.007536}$ Multiply

$X_C = 132.7\ \Omega$ Divide, & Label the answer

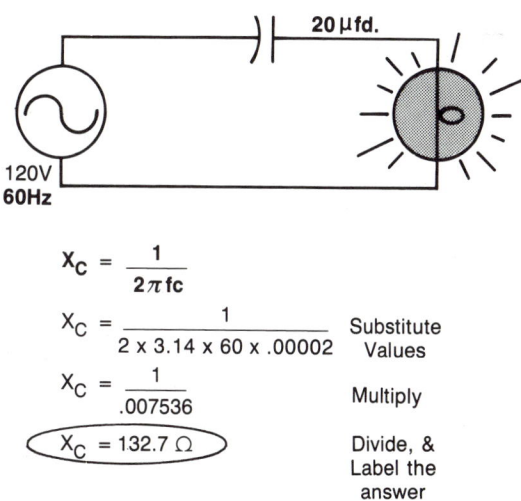

Fig. 36-4 Solving for capacitive reactance is very simple, but a calculator may help with the long decimal numbers.

and pencil, you probably noticed that it is easy to get confused by the long decimal values. Using an electronic calculator may cause other errors if you are careless or forget to check your answers. To avoid such errors, follow these steps: (Fig. 36-4 will again serve as the example)

1. Enter the values 2, × 3.14, × 60, × .00002 and then press =. This will give you the answer to the first part of the problem. Copy this value on your paper as if you were doing the problem manually.
2. Store this value in the calculator's memory and clear the display. Then enter 1 and press the ÷ key. Recall the value you stored and press =. The value of capacitive reactance should now be displayed. Write it on your paper. (**Note:** this process will work on even the simplest calculators. Your machine may have special keys or functions that will simplify it.)
3. This is the most important step. Repeat the problem and check each step with the values you have written down. Analyze the answer to each step and the final value to see that they make sense. If your answers fail either of these tests, do the problem again. Make sure you label your final answer.

Work this problem with your calculator to test this process.

Changes in Frequency and Capacitance

Find the value of capacitive reactance for the circuit in Fig. 36-5. The only difference in this and the last problem is that the capacitance is lower. Would you expect this circuit to have more or less reactance than the previous one? Test your answer by solving the problem.

Frequency is the other factor affecting capacitive reactance. Find the capacitive reactance for the circuit in Fig. 36-6. Should

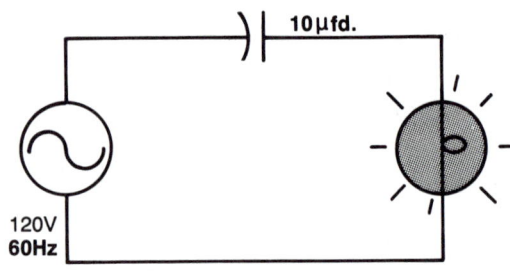

Fig. 36-5 Solve for X_c. You should find a value of X_c = 265 Ω.

the higher frequency cause greater or less capacitive reactance than you had in the first problem? Prove your answer by solving the problem. When you have worked the problem correctly using 1000 Hz, try it again using a frequency of 30 Hz. You should get a higher value of capacitive reactance.

You proved that DC current (with a frequency of 0) resulted in no inductive reactance regardless of the value of the coil

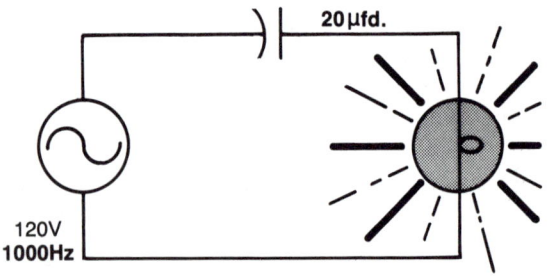

Fig. 36-6 Solve this problem to show the effect of frequency on X_c. Did you find that the capacitive reactance is X_c = 7.96 Ω? With so little opposition to current flow in the circuit, it is no wonder that the bulb glows so brightly.

Series: $X_{C_T} = X_{C_1} + X_{C_2} + \ldots X_{C_N}$

Parallel: $X_{C_T} = 1 \div \left(\dfrac{1}{X_{C_1}} + \dfrac{1}{X_{C_2}} + \ldots \dfrac{1}{X_{C_N}} \right)$

Fig. 36-7 These familiar formulas find the total reactance of capacitors in series and parallel.

in a circuit. If capacitors are the opposite of coils, you should essentially have infinite capacitive reactance in circuits with a DC power source.

You have worked problems proving that the lower the frequency goes, the greater the capacitive reactance becomes (approaching infinity as the frequency and capacitance become lower and lower). When you substitute 0 for the value of frequency, the problem is impossible to solve. Multiplying any number (or group of numbers) by 0 gives an answer of 0. Mathematically, nothing can be properly divided by 0 because there is no number by which you could multiply the 0 to get the value needed. Some math teachers emphasize this point by saying that the eleventh commandment is "thou shalt not divide by 0." Anyway, this proves that capacitive reactance is not a factor in DC circuits. The capacitor totally blocks the flow of the DC current.

Capacitive Reactances in Series and Parallel

Capacitive reactances may be combined in series and parallel networks. When this is done, two formulas (one for series, the other for parallel circuits) are used. The formulas are just like the ones for resistances and inductive reactances in series and parallel. Figure 36-7 shows both of these formulas. You should already be familiar with them by now.

Summary

Capacitive reactance is a special form of opposition to the flow of AC current developed by capacitors. It depends upon the value of capacitance and the frequency of the AC current in the circuit. A simple formula is used to find the value of capacitive reactance for a given circuit. The next chapter shows how capacitors and inductors may be used together in some very important ways.

Important Terms

capacitive reactance
expression
denominator
direct proportion
inverse proportion

Review Questions

1. Explain what causes capacitive reactance.

2. Why do we say that capacitive reactance is the opposite of inductive reactance? Demonstrate this mathematically.

3. Why is there no capacitive reactance in DC circuits? What do capacitors do to DC currents?

Chapter 37

Resonance

Two Reactances in the Same Circuit

Figure 37-1 shows a circuit that has a resistance with a capacitor and a coil. This circuit has three types of opposition to current flow: resistance, capacitive reactance, and inductive reactance. You may be wondering about the combined effects the two reactances and the resistance will have, or whether the current or the voltage will lag in the circuit. The answer is that it depends on the frequency of the AC current from the power source.

Due to the **phase shift** (current and voltage lags) caused by the coil and the capacitor, the peaks of the AC voltage and current are not appearing at the same time. If the circuit only had simple resistance (with no reactances), then the current and voltage would always be in phase as shown in Fig. 37-2. In purely inductive or capaci-

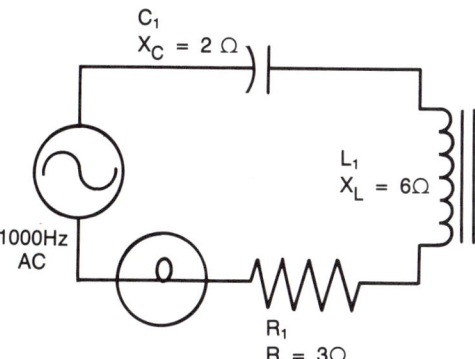

Fig. 37-1 A circuit with three types of opposition to current flow. For the sake of simplicity, assume that the resistance of the lamp is included in the 3 ohm rating of R_1.

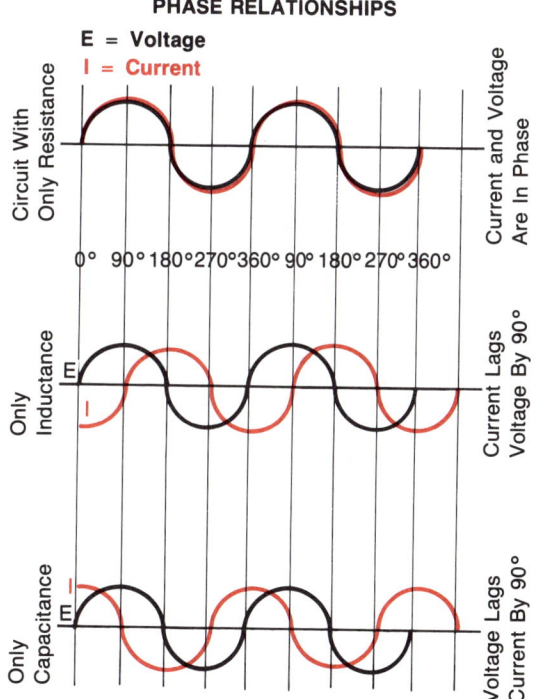

Fig. 37-2 Current and voltage are perfectly in phase (together) in purely resistive circuits, but they are out of phase in circuits which have reactances.

tive circuits, however, there is a 90 degree phase shift. When resistance and either reactance are present in the same circuit, there will still be a lag (phase shift), but it will be less than 90 degrees. That is, the peaks of the current and voltage will not occur together exactly, but they will not be as far apart as in Fig. 37-2.

Total Opposition to Current Flow

It is possible to draw one graph which shows the total opposition to current flow caused by resistance (R), capacitive reactance (X_C), and inductive reactance (X_L). Such a graph is shown in Fig. 37-3. This type of graph is called a **vector diagram**. The three lines with arrows on their ends are called **vectors**. They each point in a meaningful direction and their lengths tell you the values they represent. Notice that the vector labled X_L is 6 units long and points upward. This vector represents the value of the inductive reactance. Likewise, the capacitive reactance vector is 2 units long and points downward. The opposite direction is used because the effects of capacitance and inductance are opposites. The R vector represents resistance.

The vectors do not all point in the same direction because their directions represent the phase relationships of the qualities they depict. The resistance vector points straight ahead because there is no phase shift caused by resistance. The X_L vector is turned counterclockwise exactly 90 degrees to represent the +90° phase shift caused by the coil. In the same way, the X_C vector is turned −90° to show the phase shift caused by the capacitor.

If there were no phase shifts, and if resistance and the two reactances all affected the circuit in exactly the same way, you would be able to find the total opposition to current flow by adding the three values. For this circuit, you would get 6 + 3 + 2 = 11 if you just added, but 11 ohms is not the true total opposition to current flow in this circuit. When you combine the three sources of opposition in the circuit, you must account for the phase shift and the fact that the coil and the capacitor are opposites.

Figure 37-4 shows another vector diagram for the same circuit. It is a little different than the one you saw before. Since capacitance and inductance are opposites, if both are present in the same circuit, they work against each other. In this case, there

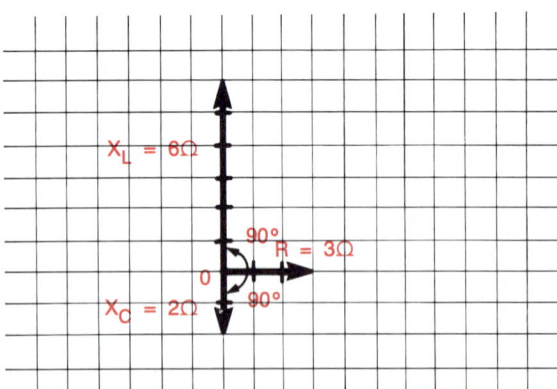

Fig. 37-3 Vector diagram which shows the relationship of X_L, X_C, and R.

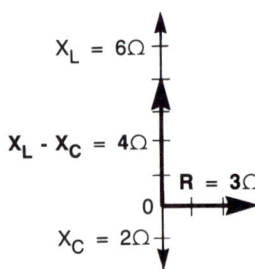

Fig. 37-4 A new vector ($X_L - X_C$) is developed by combining the effects of capacitive and inductive reactance by simple subtraction.

230

was a lot more inductive reactance than capacitive reactance, so the inductor is having a much greater effect in the circuit. This means the current will lag the voltage (because the circuit is mostly inductive), but it will lag by much less than 90°.

It is possible to find the combined effect of inductive and capacitive reactances by subtracting the X_C from the X_L value. This is done in Fig. 37-4 and labeled the new vector X_L-X_C. This new vector is 4 units long (6 − 2 = 4), and it points upward because the inductive reactance was the greater value. If X_C had been the larger value, the vector would have pointed downward and the voltage would have lagged the current.

Still one more step must be taken to find the actual total opposition to current flow. Figure 37-5 shows how to complete this step. Draw straight lines parallel to the graph lines from the ends of the two vectors until they **intersect** (cross) each other. The point at which they intersect is very important because it is used to draw a new vector. The new vector is labeled Z and has two meanings:

1) If you measure its exact length, it will tell you the exact value of the total opposition to current flow (from all sources) in the circuit. In this case, the length of the Z vector is 5 units. Therefore the total opposition to current is 5 ohms — not 11 as you got when you simply added the values.

2) The angle at the bottom of the Z vector tells exactly how much phase shift there will be between current and voltage. In this particular circuit, if you measure this angle with a protractor, you find that the phase shift is about 53°. Since this circuit is affected more by inductance than capacitance, you would say that the current is lagging the voltage by 53°.

Impedance

The quality represented by the new Z vector is called **impedance**. Impedance is the total opposition to current flow in an AC circuit. It takes into account all sources of opposition. Since it is the total opposition, impedance is measured in ohms, just as resistance and reactances are. By using the vector diagram, you can now measure the impedance (Z) vector and see that the circuit in Fig. 37-1 has 5 ohms of impedance and that the current lags the voltage by about 53°.

Using the Vector Diagram

To find the total impedance of another circuit by the vector diagram method, you will need either a piece of graph paper or two pieces of regular lined notebook paper. If you are using notebook paper, turn one sheet on its side and place it on top of the second sheet so that the lines make a grid for drawing a graph. Tape the two pieces together, carefully keeping them at 90 degrees to each other.

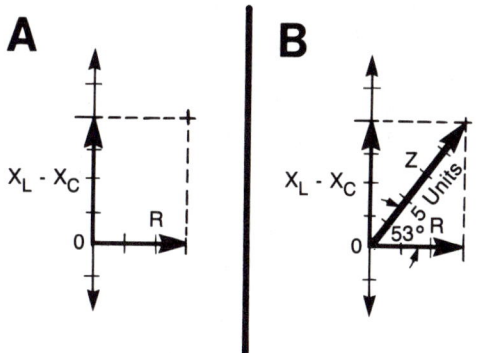

Fig. 37-5 Two more steps complete the vector diagram: A) Draw extender lines parallel to the two vectors until they cross each other (intersect). B) Draw the new Z vector from the zero point to the intersection located above.

Draw a vector diagram for the circuit in Fig. 37-6.

1) Draw the basic diagram with all three vectors in full length and proper direction.
2) Subtract X_C from X_L and draw a new, shorter vector that shows the combined effects of $X_L - X_C$. The new vector must point in the correct direction. Is this circuit affected most by inductance or capacitance? The sign (+ or −) of your answer after subtracting will tell you if you are thinking logically.

3) Draw the extender lines and then the new impedance (Z) vector. Tear a scrap off the corner of your paper and use its lines as a ruler to measure your Z vector. How much impedance does this circuit have? If you have a protractor, measure the phase angle (between the R and Z vectors). Which lags — current or voltage? How much does it lag? (If you do not have a protractor, make a guess by comparing this with the earlier problem. It must be less than 90).

If you have properly drawn and interpreted the vector diagram, your answer will be Z = 5 ohms with a phase angle of about −37° with the voltage lagging the current (because this is mostly a capacitive circuit). The vector for $X_L - X_C$ should point in the downward direction. Compare your graph to the one in Fig. 37-7.

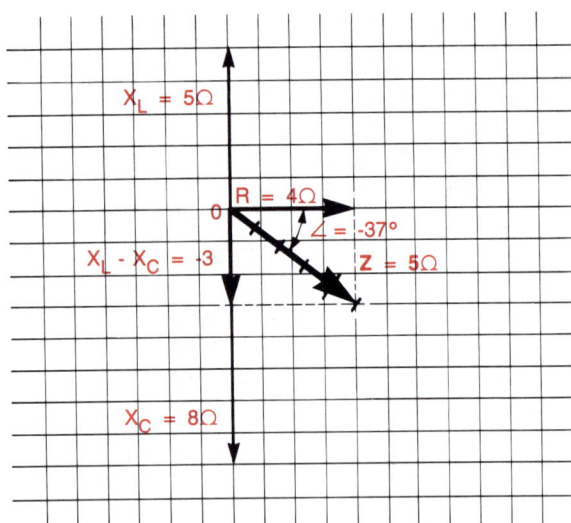

Fig. 37-7 Vector diagram solution for the problem given in Fig. 37-6. Does yours look like this? Measure all vectors carefully.

Frequency Makes a Difference

Look again at the circuit in Fig. 37-1. Notice that the frequency of the AC generator is 1000 Hz. Remember, the whole relationship of the three oppositions to current flow depends on the frequency of the AC current. So if the frequency of the original circuit were to be cut from 1000 Hz to only 60 Hz, what would happen to X_L and X_C? If you said that X_L would go down but X_C would go up, you are right. In fact, if you were to use the new values for X_L and X_C (based on a frequency of 60 Hz), you might get a vector diagram with a much different look. The graph shown in Fig. 37-8 is what you would have for this circuit with a frequency of 60 Hz.

Now you can see how important the frequency of an AC circuit can be. In the last example, by just changing the frequency, you made the coil almost negligible (forgettable) and the capacitor became a very large source of opposition. Notice also,

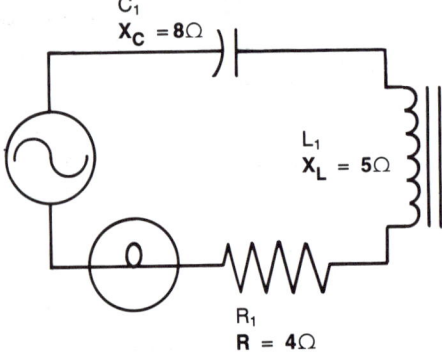

Fig. 37-6 Draw a vector diagram for this circuit. Can you find the impedance by using your graph?

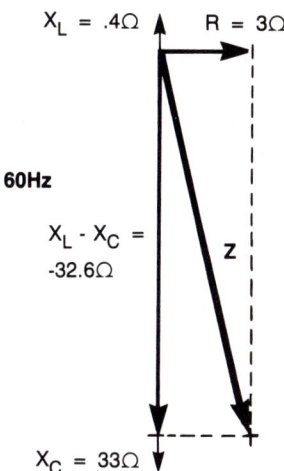

Fig. 37-8 With a much lower frequency, the capacitive reactance of our circuit is far greater than the inductive reactance. Nothing but the frequency of the AC current has been changed in the circuit, but look at the difference *that* has made in the impedance!

that the Z vector (impedance) is much longer (proportionally) for this circuit with a frequency of 60 Hz than at 1000 Hz. Figure 37-9 shows vector diagrams for the same circuit with different frequencies of AC current.

Resonance

As you examine the different graphs in Fig. 37-9, you will notice that the Z (impedance) vectors are much longer for the ones in which there is either a long X_L or a long X_C vector than the graphs in which the X_L and X_C vectors are both about the same length. When the X_L and X_C vectors are nearly the same length, they tell you that the inductive reactance and the capacitive reactance are both having about equal effect in the circuit. In fact, if they are equal in length, then X_L will cancel out the effect of X_C. (Remember that they are

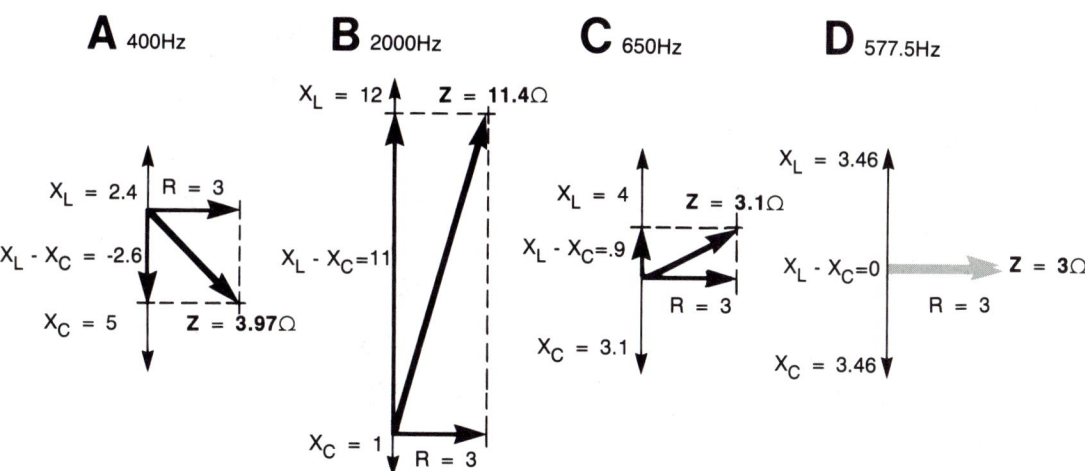

Fig. 37-9 These are all vector diagrams of the same circuit operating at different frequencies. Graph D shows the resonant frequency of this circuit. Note that the Z vector is only as long as the R vector and there is no phase shift so the Z vector points straight ahead with the R vector.

Chapter 37 *Resonance*

opposites.) When this occurs, the circuit is **resonant**. Actually, any circuit which has both inductance and capacitance is a **resonant circuit**, but each resonant circuit will resonate only at its one particular resonant frequency.

The unique frequency, for each circuit, which causes X_L to exactly equal and cancel out X_C is the **resonant frequency** for that circuit. If the frequency of the AC generator is changed, both X_L and X_C will be affected. In turn, they will change the length of the Z vector. The Z vector for a circuit will be almost exactly the length of the R vector (resistance alone) when the circuit is operating at its resonant frequency.

The circuit in Fig. 37-1 resonates at 577.5 Hz. When the AC source is supplying current at any frequency other than the resonant frequency, the Z vector — and the total impedance it represents — will be larger. When the circuit is resonating, there is almost no opposition to current flow except the simple resistance. This is true even though both capacitance and inductance are present in the circuit with the resistance.

A Use of Resonance

Resonant circuits make it possible to select one frequency from all others. For instance, there are probably several radio stations that broadcast signals strong enough to be received at your home. When you turn on your radio, all of those signals will be trying to make your radio receive them. You use a resonant circuit to choose only one station you want to listen to. Each radio station transmits its signals at a different frequency. Let's assume that there are only three stations in your area:
- 670 KHz... Easy listening station
- 1260 KHz... Country music station
- 1310 KHz... Rock music station.

The circuit you use to select your favorite station has both a capacitor and a coil. It is a resonant circuit. The capacitor is a variable type (you can change its value).

With the capacitor set so that you can hear the rock station clearly, the resonant circuit would be resonant to the frequency of 1310 KHz. The other two frequencies would try to make your radio play their broadcasts, too, but the resonant circuit would only select the one frequency you want to hear. Vector diagrams in Fig. 37-10 show the impedance that the three different frequencies will find if they try to produce current flow in the coil and capacitor of the resonant circuit. Notice how much shorter the Z vector is for the chosen frequency of 1310 KHz. The other two frequencies are having a lot more trouble in exciting the circuit.

If you wanted to hear the country station, you would turn the tuning knob which would change the value of the capacitor. This would then make the circuit resonant to a frequency of 1260 KHz. If you made a graph of this, you would notice the Z vector

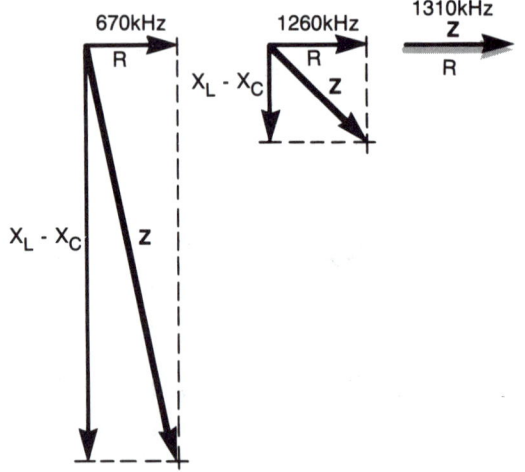

Fig. 37-10 These three vector diagrams are all for the same circuit when it is tuned to the 1310 KHz station. Notice how much impedance the other frequencies meet (as shown by their longer Z vectors).

is now shortest for 1260 KHz and the other frequencies find much more opposition.

The tuning circuits in radios, televisions, and other devices are resonant circuits. They depend on the interacting effects of X_L, X_C, and R to select certain frequencies and ignore others. Resonance is a very important concept. Be sure you understand this chapter well before you continue to the next one.

Summary

Each circuit that has both a coil and a capacitor is a resonant circuit. It will resonant at one frequency, unless the value of the coil or capacitor can be changed. Vector diagrams may be drawn to show the effects of resistance, inductive reactance, and capacitive reactance in a resonant circuit. The impedance, which is the total opposition to current flow from all sources in an AC circuit, is represented by the Z vector. Even the phase angle and whether or not current or voltage lags may be determined from the vector diagram. When the circuit is operating at its resonant frequency, the Z vector is the shortest that it can be — the same length as the R vector. When the circuit is operating at some frequency other than its reasonant frequency, either the X_L or X_C vector will be much longer than its opposing counterpart and this will cause the Z vector to point upward or downward at a sharper angle and be longer than for the resonant frequency. You will explore resonant circuits more fully — and you will learn how to actually find the resonant frequency of a circuit — in the next couple chapters.

Important Terms

phase shift
vector diagram
vector
intersect
impedance
resonant circuit
resonant frequency

Review Questions

1. What is impedance? Why is it measured in ohms?

2. Why does the X_L vector point up while the X_C vector points down?

3. Why is the Z vector's length affected by the frequency of the AC current? It is shortest when the circuit is operating at its resonant frequency — why is this so?

Careers: Radar Technician

Technicians knowledgeable in electronics and communications are often employed to care for and operate radar systems. Radar is a detection system which uses radio waves. The waves are produced by a transmitter which sends them out. The waves first strike the object, bounce back, and are detected by a receiver.

Radar can be used for both simple detection and for tracking the speed and motion of an object. Airports, military bases, planes, ships, space vehicles, and even police cars equipped to detect speeders all use radar. The radar systems could not work without resonant circuits.

Training for careers dealing with radar should include mathematics, science, and electronics. The military service is a good place to learn about radar and its uses.

Chapter 38

Series Resonant Circuits

Finding Impedance by the Formula Method

Figure 38-1 shows the circuit and vector diagram studied in the last chapter. Though the graph method of solution works well for simple problems such as this, it would be difficult to draw accurate graphs representative of real-life electronics circuits. The measurements would need to be very precise, and the values would usually be large numerals or contain several decimal places. There is a faster method to find impedance. Fig. 38-2 applies a simple mathematical formula to find the impedance of the circuit. Notice that the letters are the same ones used for some time now.

- Z = total circuit impedance in ohms
- R = resistance in ohms
- X_L = inductive reactance in ohms
- X_C = capacitive reactance in ohms

Let's solve the problem in Fig. 38-2. First, substitute the real values into the formula. Second, perform the operations enclosed by parentheses. Third, square the two values. If you are using a calculator with a X^2 key, enter the values (one at a time) and press that key. If your calculator does not do squares automatically, multiply each number by itself to square it. Record the squared values. Fourth, add the two values you have found. Fifth, to find the true impedance, use your calculator to find the square root. This may be done by pressing the key marked $\sqrt{}$, $\sqrt{x}$, or SQ RT on most calculators immediately after the addition is completed. If your calculator does not have the ability to find square roots, your teacher can show you how to find approximate values. The last step is to record the answer and label it properly.

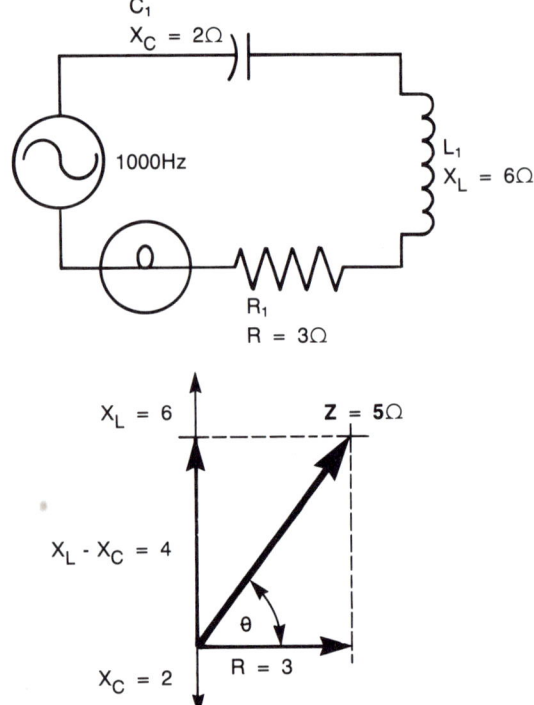

Fig. 38-1 A circuit and the vector diagram showing its impedance and phase angle. The circuit is not operating at its resonant frequency.

$Z = \sqrt{R^2 + (X_L - X_C)^2}$	Copy formula.
1) $Z = \sqrt{3^2 + (6-2)^2}$	Substitute values.
2) $Z = \sqrt{3^2 + (4)^2}$	Subtract the values inside the parentheses.
3) $Z = \sqrt{9 + 16}$	Square values.
4) $Z = \sqrt{25}$	Add.
5) $Z = 5\ \Omega$	Find square root and label answer.

Fig. 38-2 A formula may be used to find impedance. The formula allows you to do vector addition without drawing the graphs.

Let's look at what you have just done. First, does this answer agree with the answer found by using vector diagrams? Of course it does. The formula is a mathematical way of combining all the factors (R, X_L, and X_C) in the proper way to find Z. If you are an observant math student, you will recognize that this is a real-life application of the **Pythagorean theorem**. This theory states that the square of the length of the hypotenuse (side of a right triangle opposite the right angle) equals the squares of the lengths of the other two sides.

The formula is necessary because simple addition will not work. Remember that the vectors pointed in different directions, so addition would give wrong answers. Refer to Figs. 37-6 and 37-7 and use your calculator to find the impedance of that circuit by the formula method. Carefully follow the steps just discussed. Do you get the same answer as when the problem was solved by the graph method? Did you notice that the value you got for step 2 (subtracting) was a minus value? It is because there is more capacitive reactance than inductive reactance. When you square this negative value in the next step, though, you will get a positive value ($-\times-$ is $+$).

Finding Phase Angle

Another formula permits you to find the phase angle. Figure 38-3 shows this formula and its use. The formula says "the cosine of angle θ (**theta**) is equal to the value found by dividing resistance (R) by impedance (Z)." So, to actually find the phase angle (θ), you must first find the value of $\frac{R}{Z}$ and then look it up in a Table of Trigonometric Functions (trig table). Some calculators find

Cosine $\theta = \dfrac{R}{Z}$	Copy formula.
Cosine $\theta = \dfrac{3}{5}$	Substitute values.
Cosine $\theta = .6$	Divide.
$\theta = 53°$	Look value up in trig table.

Fig. 38-3 The phase angle may also be found mathematically.

trig values. Though more accurate values can be found with a little more work, for practical purposes the value 0.6 most closely indicates an angle of 53°, so you can record 53° as the approximate phase angle. Again, notice that this is the same angle found by the graph method. Test your understanding of this process by finding the phase angle of the circuit in Fig. 37-6 from the previous chapter. Check your answer against the vector diagram in Fig. 37-7.

Frequency of Resonance

Yet another formula is needed to be able to fully understand and work with resonant circuits. This formula is used to find the frequency at which a circuit will resonate (Fig. 38-4). This formula accounts for the factors that determine the frequency of resonance: value of inductance and value of capacitance. Remember that resonance occurs only when the values of X_L and X_C are exactly equal (though opposite). Since the frequency of the AC current helps to determine X_L and X_C, as the frequency changes one of these values will go up while the other one goes down. Therefore circuits with both coils and capacitors will

$$f_r = \frac{1}{2\pi\sqrt{LC}}$$

Fig. 38-4 Formula for finding the frequency of resonance.

resonate at only one frequency. Review the vector diagrams in Figs. 37-8 and 37-9 to again see the effect frequency has on X_L and X_C. The formula for frequency of resonance can be derived from the formulas for both X_L and X_C. It is very easy to work this formula with a calculator.

Figure 38-5 shows a sample problem in which the formula is used to find the frequency of resonance. The steps are identified in the figure. Always enter values in basic units (farads and henrys). When you can follow the process on your own calculator, find the frequency of resonance for the original example circuit. The needed values of L and C are labeled on Fig. 38-6. Check to be sure that your answer agrees with Figure 37-9.

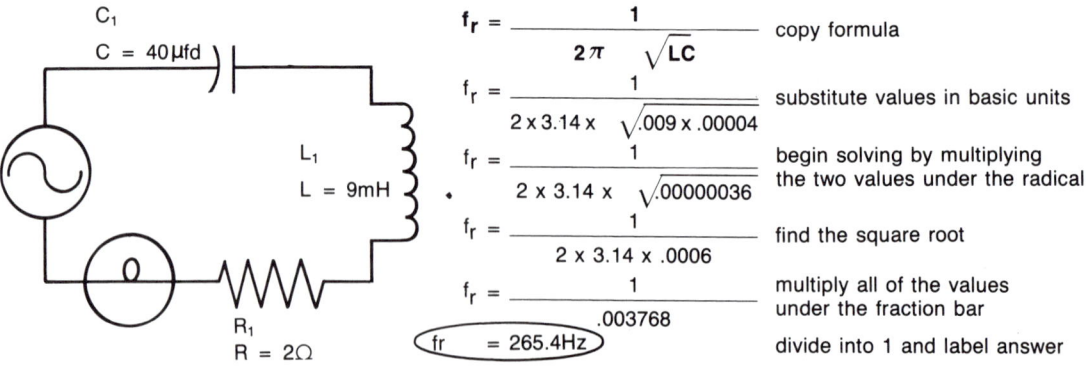

Fig. 38-5 The formula for finding the frequency of resonance is easy to work, but a calculator will be helpful for finding the square root and handling numbers with several decimal places.

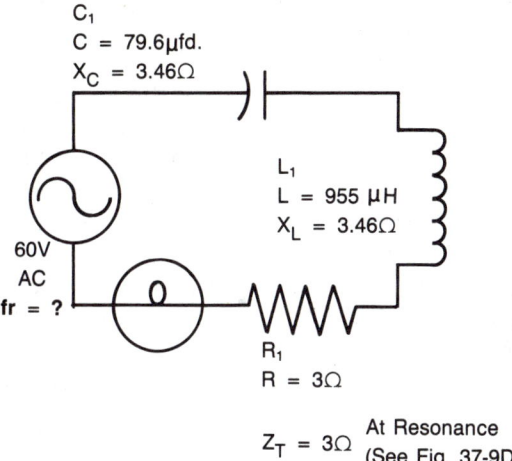

Fig. 38-6 Use the formula to find the frequency of resonance for this circuit. Notice that the impedance is only 3 ohms when the circuit operates at its resonant frequency.

Voltage Drops in a Series Resonant Circuit

It is often necessary to find the individual voltage drops of each component in a circuit. This is done by finding the current of the circuit and then using the value of resistance or reactance for each component in a simple Ohm's Law calculation. Find the total circuit current by dividing the total voltage by the total opposition to current flow (that is the *impedance*). See Fig. 38-7. For the example circuit of Figure 38-6 you will find a current of 20 A. Now use this value to find the voltage dropped by each component. This is done in Fig. 38-8. The formulas used are all natural outgrowths of the Ohm's Law formula: $E = I \times R$

The three answers in Fig. 38-8 may confuse you at first because the source voltage is only 60 V, and you have found that some of the parts are dropping more than that by themselves! The sum of the three drops is nearly 200 V! You cannot simply add the values of voltage for the same reason that you could not just add up the reactances to find impedance. The values that you just found are the maximum values of voltage that will occur in each component, but they will not remain at their maximum values continuously. In

$$I_T = \frac{E_T}{Z_T}$$
$$I_T = \frac{60}{3}$$
$$I_T = 20 \text{ A}$$

Fig. 38-7 Finding the total circuit current flow. You must consider the total opposition to current flow from all sources (impedance).

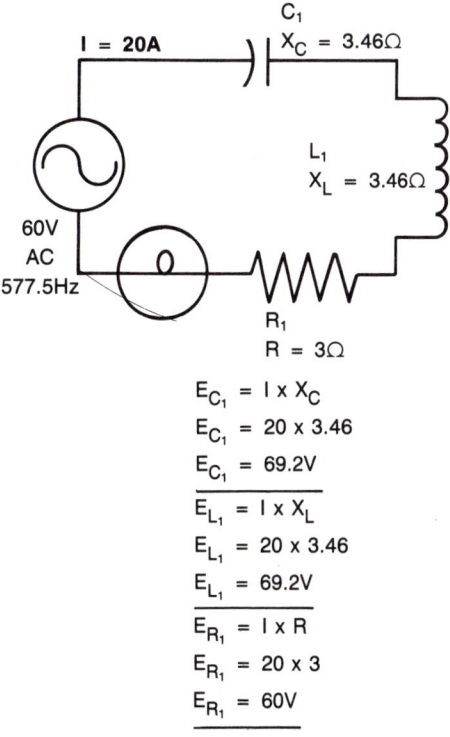

$E_{C_1} = I \times X_C$
$E_{C_1} = 20 \times 3.46$
$E_{C_1} = 69.2V$

$E_{L_1} = I \times X_L$
$E_{L_1} = 20 \times 3.46$
$E_{L_1} = 69.2V$

$E_{R_1} = I \times R$
$E_{R_1} = 20 \times 3$
$E_{R_1} = 60V$

Fig. 38-8 Use the value of total circuit current (found in Fig. 38-7) to find the voltage dropped by each component.

Chapter 38 Series Resonance Circuits

fact, due to phase shift, the capacitor will have its maximum voltage at the time when the inductor has almost no voltage drop. See Fig. 38-9.

Likewise, since the voltage leads the current in the inductor while the voltage lags the current in the capacitor, the inductor will have maximum voltage when the capacitor has nearly none. The resistor will have its maximum voltage at the same time it has maximum current in each alternation of the AC current. It will be 90° out of phase with both the coil and the capacitor.

Even though it seems that you have more voltage than you really do, there is nothing wrong with the circuit. If you were to measure the voltages of the components with a good electronic volt meter, you would find the same values found mathematically, but the source would still only measure 60 V. You cannot simply add these three voltage drops, but vector addition would result in an answer that equals the source voltage. Vector addition of these values is done by the formula method in Fig. 38-10.

I —— Current in all parts of the circuit
E_{R_1} ···· Voltage of the resistor, in phase with current
E_{L_1} ···· Voltage of the coil, voltage leads current by 90°
E_{C_1} —— Voltage of the capacitor, voltage lags current by 90°

Fig. 38-9 The maximum voltages in R_1, L_1 and C_1 all occur at different times due to the phase shifts of the coil and the capacitor.

Relationship of All Factors

What accounts for these high voltage values? Actually, the fact that the circuit is resonating is the reason for the high values

Esource $= \sqrt{(E_R)^2 + (E_{X_L} - E_{X_C})^2}$ Copy formula.

Esource $= \sqrt{(60)^2 + (69.2 - 69.2)^2}$ Substitute values.

Esource $= \sqrt{(60)^2 + (0)^2}$ Subtract.

Esource $= \sqrt{3600 + 0}$ Square values and remove parentheses.

Esource $= \sqrt{3600}$ Add.

Esource $= 60$ V   Find square root and label answer.

Fig. 38-10 Though the simple sum of the voltage drops around a resonant circuit far exceeds the source voltage (and is incorrect), vector addition will work. Notice the similarity between this formula and the one in Fig. 38-2

of all three components. All the factors depend on each other in the following ways:
- The values of X_L and X_C depend on the frequency and the values of L and C.
- The value of Z depends on X_L and X_C.
- The frequency of resonance depends on the values of C and L.
- The current depends on the source voltage value and the total impedance (Z).
- The voltage drop of each component depends on its resistance or reactance and the circuit current.

If you were to change the frequency of the generator, you would upset the balance between X_C and X_L. This would take the circuit out of resonance and make either X_L or X_C much larger than its counterpart. The component which then had the greater reactance value would also drop much more voltage than its counterpart and the whole circuit would be affected by its phase shift. (Remember that while the circuit was resonating, there was no total phase shift because the Z vector fell directly on top of the R vector.) Finally, the total circuit current flow would go down because there would be more impedance.

Likewise, changing the value of either C or L would cause you to have new values for X_L and X_C and a new resonant frequency too! Resonance is a very delicate balance of several factors.

Summary Statements for Series Resonant Circuits

When a series circuit is operating at its resonant frequency, the following things will happen:
1. The total impedance (Z) of the circuit will be very low and will equal the value of the resistance alone.
2. The current flow from the source is very high because it is limited only by the resistor.
3. The voltage drop across both the coil and the capacitor will be very high and may even be higher than the source voltage.
4. The total phase angle (as seen by the generating source) will be 0°, or in phase.

Summary

Simple formulas may be used to find impedance, phase angle, and the frequency of resonance. When series circuits resonate, high current flows, each component may drop high voltages, and total impedance is very low. Carefully compare the findings of the problems worked in this chapter by the formula methods with those solved by vector diagrams in the previous chapter to ensure that you understand these important concepts.

Important Terms

Pythagorean theorem   θ (theta)

Review Questions

1. Why do we usually use the formula method instead of the graph method to find impedance?
2. What happens when a series circuit resonates? (Mention: impedance, current, voltage drops, and phase angle)
3. What factors does the resonant frequency of a circuit depend upon?

Project: AM Transmitting Siren

Parts List

1	33 K-ohm resistor	R_1
1	100 K-ohm potentiometer	R_2
1	56 K-ohm resistor	R_3
1	330 K-ohm resistor	R_4
1	0.02 microfarad capacitor	C_1
1	100 microfarad capacitor	C_2
1	SPST momentary switch	S_1
1	SPST switch	S_2
1	9 V battery with connector	B_1
1	ECG 289A NPN transistor	Q_1
1	hand-wound coil (see text)	L_1

suitable case, PC board, wires, and hardware as required

Notes

The coil (L_1) is part of a tuned resonant circuit which makes this siren transmit to nearby AM radio receivers. The circuit must be very close to the AM receiver. The coil must be wound very carefully to ensure that the circuit is on the proper frequency.

Begin making the coil by carefully cutting a 16 penny nail to a length of 2½ inches. Then, neatly wrap one layer of electrical tape around the nail — keeping it smooth. Wind 24 gauge magnet wire evenly on the nail. Count carefully and always wind in the same direction. You must wind 275 turns of wire; bring out a loop which will be the center tap; then continue winding another 275 turns in the same direction to complete the coil. Clean the enamel insulation from the ends of the three leads as shown in the drawing. Cover the coil with electrical tape if you wish.

The potentiometer (R_2) may be used to adjust the pitch of the siren. Other components, especially C_2, may be changed slightly to alter other operating characteristics such as sustain time. The AM radio should be tuned to a blank spot on the dial and the siren circuit should be held close to the receiver.

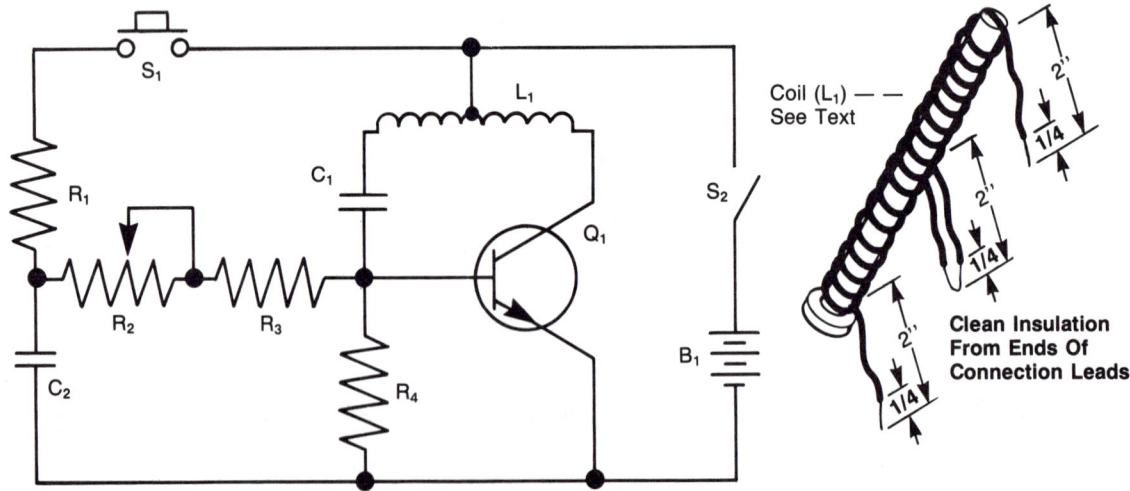

Fig. 38-11 Startle your friends by turning a common AM radio into a piercing siren with this simple circuit — no connection needed!

Chapter 39
Parallel Resonant Circuits

What Happens in Parallel Resonant Circuits?

There is very little impedance in series resonant circuits when the circuit operates at the resonant frequency. High current is permitted to flow through the circuit. This is not true of **parallel resonant circuits** (Figure 39-1). The part of this circuit between points A and B is called a **tank** because the resonant frequency will be captured and held there while all other frequencies are allowed to flow through it. So if the generator is producing AC current at the resonant frequency, that current is blocked by the tank. The current is not permitted to travel from A to B through the tank. But when the generator is producing AC current at any other frequency, the current can flow from A to B with little opposition. This is the exact opposite of what happened in the series resonant circuits you studied in the last chapter. They offered almost no impedance at their resonant frequencies and high impedance to all other nonresonant frequencies.

How Does a Tank Circuit Work?

Figure 39-2 illustrates what happens when you apply one pulse of current to a tank circuit. A battery and a push-button switch are supplying just one quick pulse of current so that you can see what each pulse will do. When the pulse is first supplied, it does two things at the same time: charges the capacitor and builds a field around the coil (frame A). Frame B shows what happens as soon as the push button is released (removing the power pulse from the circuit). Now, no current is made to flow from A to B by the battery, and there is no electrical pressure (voltage) to keep the capacitor charged. Therefore, the capacitor discharges, and its extra electrons from the top plate travel through the coil to get to the other plate. While these electrons from the discharging capacitor travel through the coil, they act just as the original current pulse did to keep the field built up around the coil.

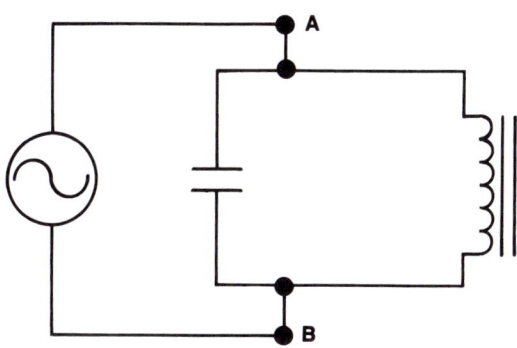

Fig. 39-1 A parallel resonant tank circuit.

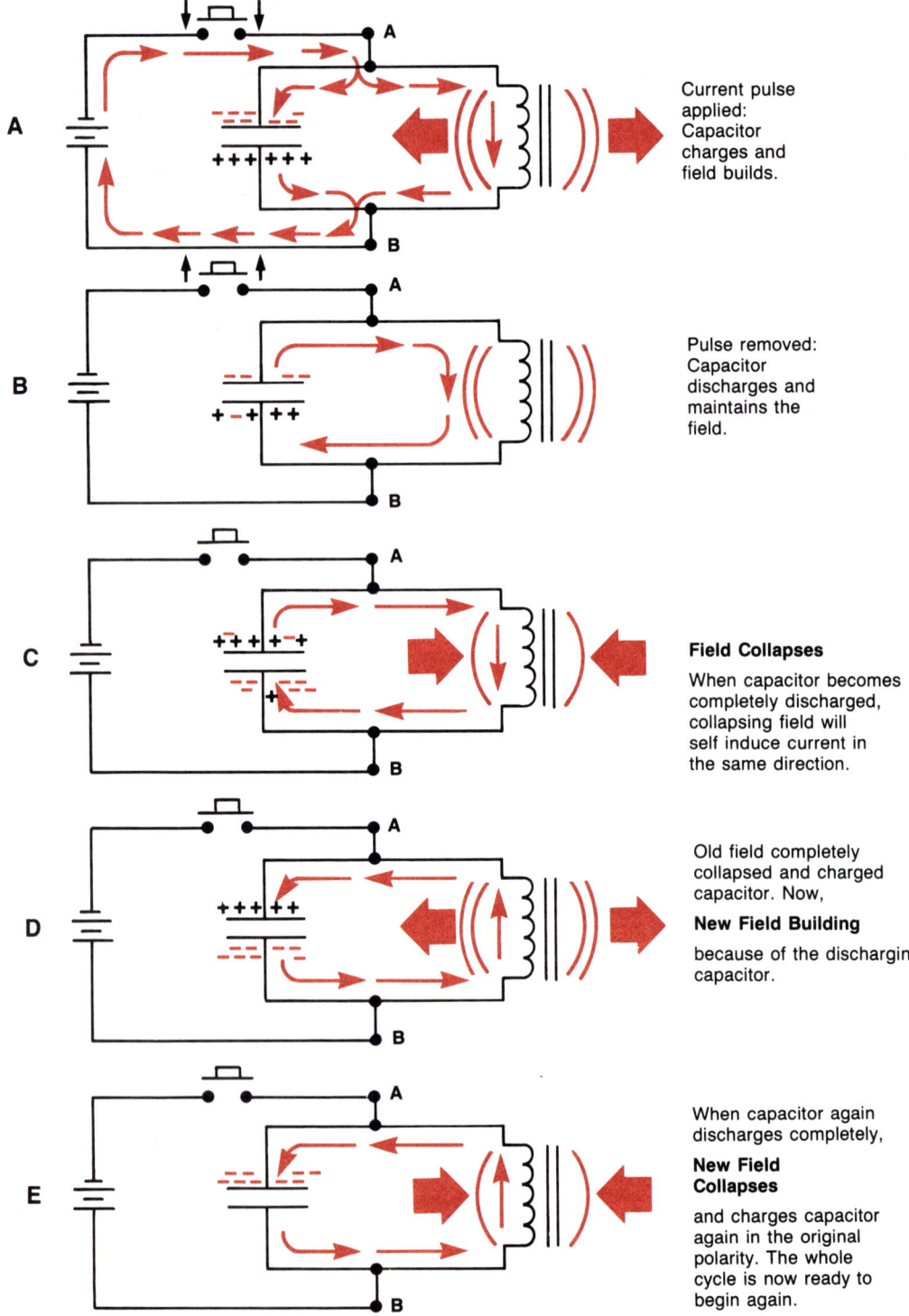

Fig. 39-2 Working action of a resonant tank circuit after one pulse of current.

When the capacitor fully discharges, there will be no current flowing which can keep the field built up. What will happen to it? If you said it will collapse, you are right. When the field collapses (frame C), it will self-induce a new current going in the same direction as the original one. The self-induced (counter-counter EMF) will make enough current flow to charge the capacitor in the opposite direction! When the field completely collapses, however, there will be no pressure to keep the capacitor charged. The capacitor will then discharge itself back through the coil in the opposite direction (frame D). Again, the current from the discharging capacitor will build another new field around the coil.

When the capacitor becomes completely discharged, the field will be at its greatest strength and will collapse because there is no current from the discharged capacitor to maintain it. Frame E shows the new field collapsing and self-inducing a current to re-charge the capacitor in the original direction as it had been first charged by the pulse from the battery. When the field completely collapses, the capacitor will again be fully charged and the cycle will begin again. However, there will be no need for a new pulse from the battery because the extra electrons on the top plate of the capacitor will be sufficient to build a new field around the coil.

This series of events forms one complete cycle of the tank circuit operation. Once begun, this same cycle will repeat many times. If everything could be perfect, and you did not lose some energy due to simple resistance, the cycle might repeat itself forever. When you pluck a guitar string, it vibrates at a certain frequency and produces a sound. If you pluck the G string, its sound will always be the pitch (frequency) of the musical note G, but it will be loud at first and slowly get quieter. If you watch the string closely, you can see that it vibrates back and forth over a great distance at first and then, as the sound weakens, the distance the string travels becomes smaller and smaller with each vibration. Since the string keeps the same frequency (384 Hz for G) it is **oscillating** at its resonant frequency. Another string, the A string, resonates at 220 Hz and produces a sound which is lower in pitch. Figure 39-3 shows graphically what is happening to the guitar string. The continuous oscillations in the top line are uniform in length (time period) and height (strength or volume). The guitar string acts more like the lower line which shows **damped oscillations**. Notice these are also perfectly uniform in length (period or frequency), but they become weaker (as shown by decreasing height) with each cycle.

The tank circuit does the same thing. One quick pulse from the battery begins the cycle and it will repeat over and over again, producing damped oscillations until all of the energy is wasted by circuit resistance and other losses. This is called the **flywheel effect**. It is similar to what happens when a pendulum is started into motion. It tries to continue swinging at the same frequency until all of the starting energy is used up. The unique thing is that the oscillations will maintain a special frequency, even as they become smaller in height! The special fre-

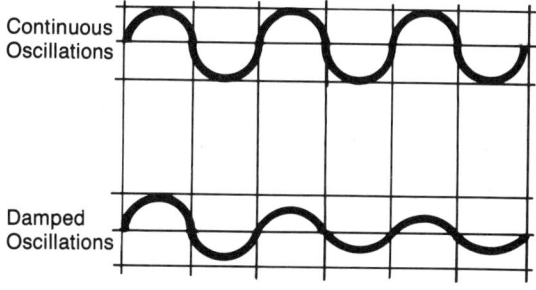

Fig. 39-3 Tank circuits and guitar strings both produce damped oscillations.

quency the tank circuit creates is its resonant frequency. Just as with series resonant circuits, the resonant frequency of the tank depends only upon the values of the coil and the capacitor. Likewise, the resonant frequency may be found by the formula:

$$f_r = \frac{1}{2\pi \sqrt{LC}}$$

All of these damped oscillations resulted from just one pluck of the guitar string, or one pulse of current from the battery. What would happen if you continued to pluck the guitar string repeatedly, every time that it vibrated, perfectly in time with its natural frequency of vibration (oscillation)? The guitar string should produce oscillations at its resonant frequency (the musical tone of G, 384 Hz) continuously without any damping. The same is true of the tank circuit. If you remove the battery and push-button switch and replace them both with an AC generator which produces a current at the circuit's resonant frequency, then the oscillations can be maintained indefinitely.

Resonant and Nonresonant Frequencies

Figure 39-4 shows the tank circuit with an AC generator. Recall the operation of the tank circuit caused by the single pulse from the battery in Figure 39-2. The battery wanted to push electrons from point A to point B, and it did so only while the button was pushed. All during the flywheel cycles of damped oscillations, which were caused by that one little pulse, no more current flowed between points A and B.

The generator in Figure 39-4 will be trying to pump current between points A and B continuously at its own operating frequency. If the generator is operating at the resonant frequency of the tank circuit, every time it produces a new peak of current at point A it will meet a peak of current produced by the tank circuit itself. Thus, the tank circuit will fight against the efforts of the generator to send currents between points A and B. When the generator operates at any frequency other than the resonant frequency, each pulse of current from it will not be met at point A by a pulse from the tank. Instead, it may even be helped through the tank to point B by the effects of either the capacitor or the coil.

Low frequency currents will be able to pass easily through the coil and higher frequencies will find little opposition from the capacitor. Only the resonant frequency will be blocked from flowing between points A and B. The tank will oscillate at the resonant frequency and block that frequency from passing while letting other frequencies work their way through easily.

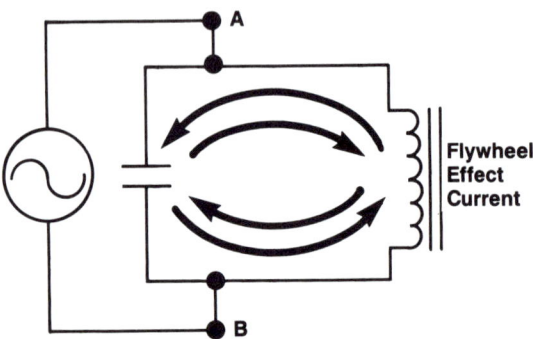

Fig. 39-4 Tank circuit with an AC generator as its power source. If the generator produces current at the resonant frequency, two things will happen: 1) Almost no current will flow between A & B. 2) Continuous oscillations will occur inside the tank circuit.

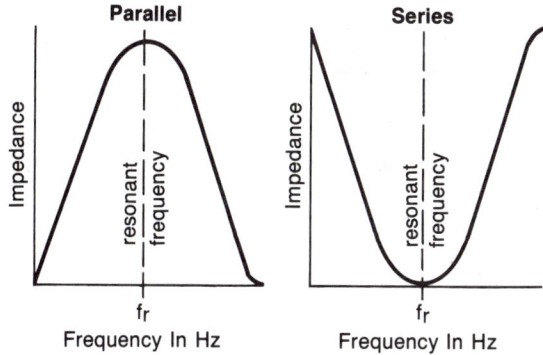

Fig. 39-5 Impedance is greatest in parallel circuits at the resonant frequency, but it is lowest in series circuits when they resonate.

Comparing Series and Parallel Resonant Circuits

The graphs in Fig. 39-5 show the impedance of series and parallel circuits. The following comparisons can be made between series and parallel resonant circuits:

1. Impedance (Z) — series circuits have very low impedance to their resonant frequency. Parallel circuits have extremely high impedance (opposition to current flow) when their resonant frequency is applied.

2. Current Flow — series circuits allow much current to flow at their resonant frequency while opposing the flow of other frequencies. Parallel resonant circuits, however, will allow any frequency except their resonant frequency to pass through the tank easily. The tank blocks the passage of the resonant frequency. There are high currents inside the tank at the resonant frequency, but that frequency cannot pass through the tank. Therefore at the resonant frequency, little current will be drawn from the source.

3. Phase Angle — the current and voltage of both series and parallel circuits will be in phase as seen by the source, but there are opposite 90° phase shifts in the capacitor and the coil of both circuits.

Uses of Tank Circuits

Chapter 37 discussed one tank circuit — the one used to tune radios. Figure 39-6 shows this circuit. All of the frequencies which are produced by nearby radio stations enter the antenna and are passed to ground through the primary of transformer L_1. They will all try to cause current flow in the tank circuit, but only the resonant frequency will be successful in creating the flywheel effect in the tank. This frequency and the information (music or news) it carries will be sent to the other radio circuits while the nonresonant frequencies are practically ignored.

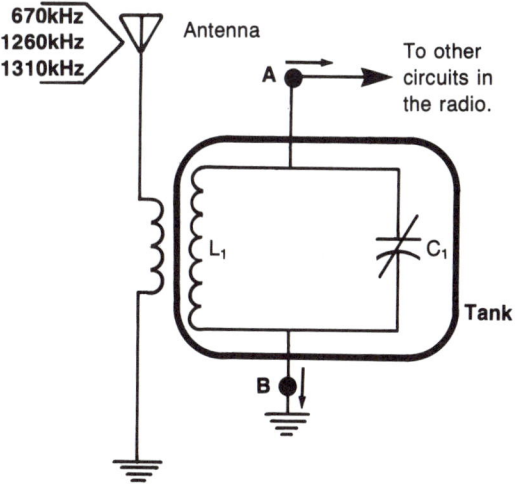

Fig. 39-6 The tank in this radio tuning circuit passes any frequency except the desired resonant one from point A to B (ground). The resonant frequency is blocked from ground by the tank and therefore must go on to the other radio circuits. The circuit may be made to resonate to different frequencies by changing the setting of the variable capacitor to select different stations.

Another use of resonant circuits is in filtering. Figure 39-7 shows some possible filtering applications of series and parallel resonant circuits.

Summary

Parallel resonant circuits differ greatly from series resonant circuits. Series circuits pass their resonant frequency while parallel (tank) circuits block the flow of their resonant frequency. The frequency of resonance for both circuits depends on the values of their coils and capacitors. Parallel resonant circuits find many applications in tuning and filtering circuits. You have now studied all of the basic principles that will be covered. The rest of this text will deal with semiconductor components and circuits. The next chapter begins with diodes.

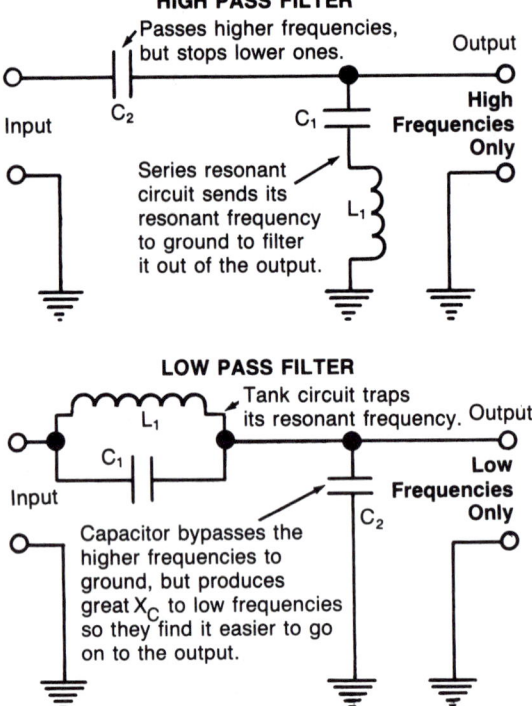

Fig. 39-7 Both of these circuits filter out certain AC frequencies. In each one, the resonant section selects one particular frequency to block or pass and the other capacitor (C_2) directs other frequencies which are higher than the resonant one to either ground or the output as desired.

Important Terms

parallel resonant circuit oscillating flywheel effect
tank damped oscillation

Review Questions

1. Explain what happens in a tank circuit when one pulse of current is sent to it. (You may use the illustrations in Fig. 39-2 as an aid in answering this question.)

2. How is the flywheel effect similar to what happens when we pluck a guitar string? What are damped oscillations?

3. Why does the tank block its resonant frequency while series circuits pass their resonant frequencies?

Pep Talk: History

All the subjects you study in school have a purpose. History and other social science courses help you to understand American culture and its development. Ours is a very technological society. Much of this technological growth is due to the development of modern electronics. Having knowledge of the successes and failures of previous generations is essential for us to wisely use modern technology.

Our democratic government allows you to advance yourself and to control your own fate, but you can only make use of and preserve this freedom if you are informed and interested. Some of the modern technological miracles (such as television, high-speed computerized typesetting equipment that makes it possible for newspapers to bring you really timely news, and other electronic communications equipment) would lose their value if you did not know about their heritage. No matter what you plan to be, all of your school subjects are important to you — even history!

Section 7
Control Devices

Chapter 40
Diodes

Semiconductor Materials

Early in this text two types of materials were discussed: conductors, which allow free electrons to travel easily, and insulators, which have no electrons that are free to move around. Both of these materials are useful for controlling the flow of electrons. Conductors allow current flow. Insulators do not. Remember, too, the other class of materials that exists — the semiconductors. Semiconductors are not very good for use as either a conductor or an insulator. The most important semiconductors are silicon, germanium, and carbon. These three materials are made up of atoms with four electrons in their outermost shell. Fig. 40-1 shows a single silicon atom and a sample of material with nine silicon atoms. Notice how the atoms are bound together so that they share the electrons in their outermost shells. The electrons in the outermost shells of atoms are called **valence** electrons. When many atoms share electrons as in Fig. 40-1, those atoms are in **covalent bonds**. Semiconductor materials are crystals made up of atoms with four valence electrons held together in covalent bonds. They share their four outermost electrons.

Semiconductor materials alone do not have enough free electrons to allow easy current flow, but they do have enough to

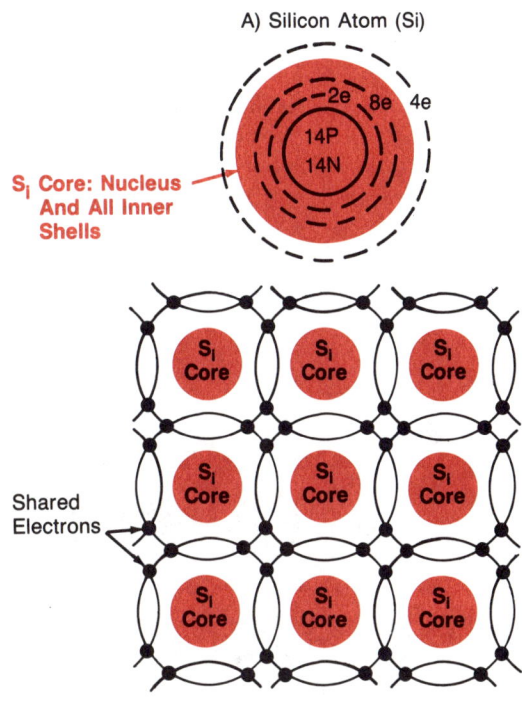

Fig. 40-1 A) The core of the silicon atom includes the nucleus and both inner shells. B) Atoms which are squeezed together tightly tend to share electrons in their outermost shells. Note that these atoms are simplified by using the cores. The shared electrons are called valence electrons, and they create covalent bonds.

allow current flow when the voltage (pressure) becomes great. Therefore they are not good for conductors or insulators. In fact, semiconductor materials by themselves are relatively useless in electronics. However, semiconductors can be very useful after being altered.

Doping Semiconductor Crystals

By heating a very pure sample of semiconductor material (silicon) until it is just about to melt and then pumping boron gas into the furnace, some of the boron atoms will diffuse into the silicon. This process is called **doping**. Diffusion is one way to dope semiconductor crystals. Others include the grown junction, alloy, and expitaxial methods. These methods are simply different ways of doping semiconductor crystals so that they are no longer pure. They will have a few atoms with either less than or more than four valence electrons. After doping, the material will be like one of the two samples shown in Fig. 40-2. If you used boron gas to dope the material, as in the example, the silicon would have too few electrons to share evenly (frame A). This is called **P-type** material because it has too few electrons. Likewise, you could diffuse phosphorus into the pure silicon and get a crystal with a few extra electrons called **N-type** material (Fig. 40-2 B). Other doping materials are listed in Fig. 40-3.

The PN Junction

An electronic device called a **PN junction** is formed when P-type material and N-type material are joined together. A PN junction is commonly called a **diode**. You know

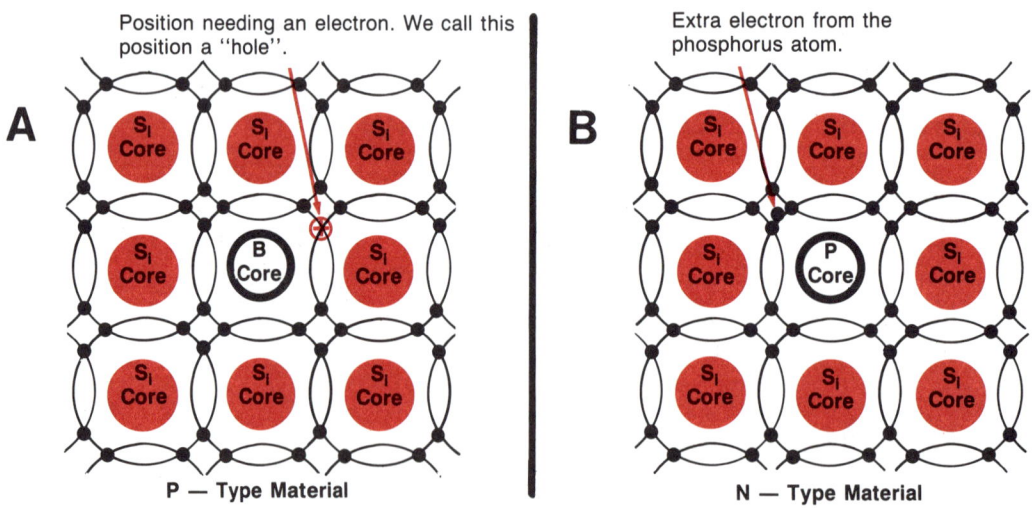

Fig. 40-2 Pure silicon may be doped with boron to have holes (too few electrons) or with phosphorus to have excess electrons. When this is done we make P-type and N-type semiconductor materials.

Materials Used in Doping Semiconductors

Used to make P-type material (trivalents)		
Material	**Symbol**	**Valence Electrons**
Aluminum	Al	3
Boron	B	3
Gallium	Ga	3
Indium	In	3

Used to make N-type material (pentavalents)		
Antimony	Sb	5
Arsenic	As	5
Bismuth	Bi	5
Phosphorus	P	5

Fig. 40-3 These materials are all used for doping silicon and germanium to make semiconductor components.

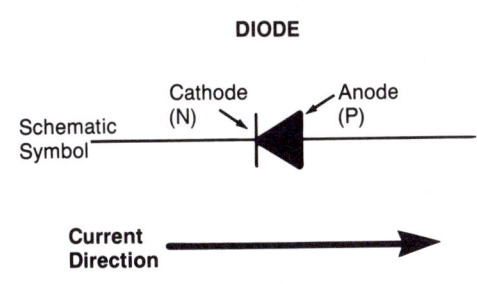

Fig. 40-4 When AC like this is applied to a diode, the output of the diode will be pulsating DC like this because the diode permits current to flow in only one direction.

diodes allow current to flow in only one direction and can be used to change AC to pulsating DC. Fig. 40-4 shows this action and the schematic symbol for a diode. Until now, the question of how diodes do their amazing job has been unanswered. The other components you have studied can pass current in either direction equally well.

A typical PN junction (diode) is shown in Fig. 40-5. The N and P materials actually join each other at the junction. The ⊕ symbols in the P material are called **holes**. Holes are places that need an electron. Holes are caused by the doping process. P material has holes and N material has extra electrons. Frame B shows what happens at the junction. A few of the extra electrons from the N material that are closest to the junction jump across the junction to fill the nearest holes on the opposite side. The electrons were negative charges and the holes were positive ones, so there was an attraction to pull the electrons across the junction. When the electrons filled the closest holes, those holes became neutral because now those atoms have just the right number of electrons to share evenly, four each. The slight + charge of the holes to the left of the junction is not great enough to pull the "newly moved" electrons out of their neutral homes, so the electrons remain. Likewise, the atoms left behind had too many electrons, but now they also have just the right number. These atoms are considered neutral. They do not attract electrons from the N material that lies further to the right of the junction. The gray area around the junction is almost all neutral, except for a slight positive charge on the right edge and a slight negative charge on the left edge. These charges, called barrier potential, and the neutral area between them form the **depletion zone**. The depletion zone separates the N (−) material from the P (+) material by such a great distance that no electrons can jump across from the N to the P unless voltage is applied.

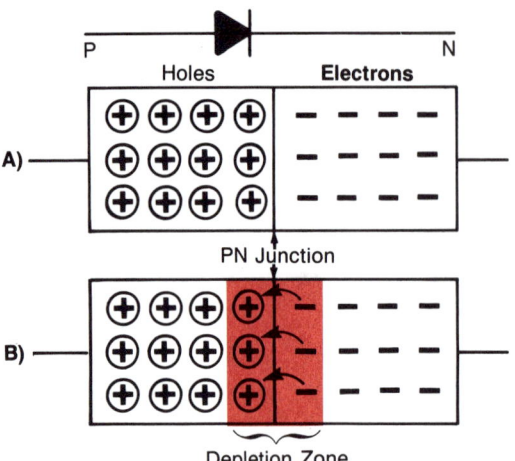

Fig. 40-5 In *A*, a PN junction (diode) is shown. *B* shows that the holes and extra electrons nearest the junction combine to form a neutral area called the depletion zone.

Action of a Diode When Current Is Applied

What happens in the diode when voltage tries to make current flow through it? That all depends on the direction the current flows. Fig. 40-6 shows what will happen if the battery is placed into the circuit so that its negative terminal is closest to the N section (the cathode) of the diode. The voltage from the battery is easily able to overcome the insulating quality of the thin depletion zone.

With this extra push from the negative battery terminal (and extra pull from the positive terminal), electrons are made to flow from the N region, through the depletion zone and into the P region. From there, they are drawn to the positive battery terminal. Thus, current flows easily when the battery is in this position. This condition is called **forward bias**, meaning that there is voltage applied to make current go forward through the diode.

Hole Flow Current

Notice that the schematic symbol "appears" to be drawn backwards. But it is not! The arrow-shaped anode (P region) points in the way positive charges (holes) travel. This may seem confusing, but when an electron leaves one atom to go to another, the remaining atom "wants" another electron. The atom the electron went to, formerly a hole, loses its positive charge. In other words, when electrons move, they leave spaces which are new holes. When studying semiconductor materials, think of current flow in terms of what the holes are doing rather than the electrons. Even though the actual atoms do not move around a lot, negative and positive charges do move. You can speak of "hole current" flow as the flow of positive

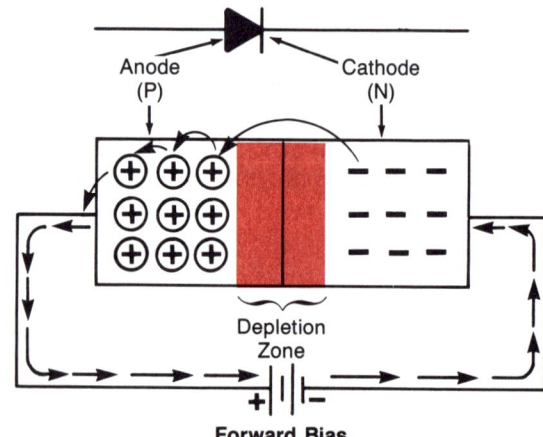

Fig. 40-6 In this simple circuit a battery is used to FORWARD BIAS the diode and cause current to flow.

charges from + to − (backward to the electron current flow which you have used throughout this book). The flow of holes is called **conventional current**. The arrow shape of the diode's schematic symbol points in the direction of conventional current flow — so, even though it seems backward to you, it does indicate the direction of travel of holes through the PN junction.

Reverse Bias

When the battery is connected so that its negative terminal is closest to the diode's P region, the diode is **reverse biased**. Fig. 40-7 illustrates what occurs when you reverse bias a PN junction. A few electrons move from the cathode (N region) toward the battery's positive terminal. Other free electrons move nearer to the end of the diode (away from the depletion zone). At the same time, some holes move to the negative battery terminal, and others move away from the depletion zone toward the anode end of the diode. This only involves the movement of a few holes and electrons. No real current flow is established.

When the extra electrons and holes move away from the depletion zone, they leave neutral atoms behind. What you have done by reverse biasing the diode is to make its depletion zone much wider. Current has a much more difficult time traveling through this extra-wide depletion zone. No current flows in this reverse direction unless great amounts of voltage are applied — enough voltage to overcome the widened depletion zone and the effects of the P and N regions. So, current does flow when the diode is forward biased, but it does not normally flow when we reverse bias the diode. Thus diodes pass current in only one direction.

Special Diodes

Fig. 40-8 shows the graph of current when a diode has reverse and then forward bias. Most diodes conduct current very easily in the forward direction when forward bias is applied. The right side of the graph shows the forward bias condition and the forward current it causes to flow. The current curve rises sharply when only a small amount of forward bias voltage is applied, just as we would expect.

The left side of the graph shows what happens when reverse biasing voltage is applied. When only a little reverse bias voltage is applied (just to the left of the center line), no reverse current is permitted to flow. This is why the current curve stays so close to the horizontal axis (zero current line).

As the reverse bias voltage increases, still no reverse current is allowed to flow. Notice the current curve remains close to the horizontal axis for quite a long way — it is almost straight. Then, suddenly, the current curve takes a sharp dive.

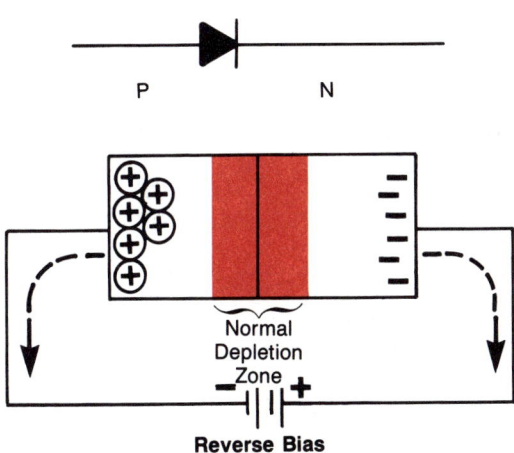

Fig. 40-7 Reverse bias widens the depletion zone, as holes and electrons gravitate toward opposite ends of the diode. The diode does not permit current to flow when it is reverse biased.

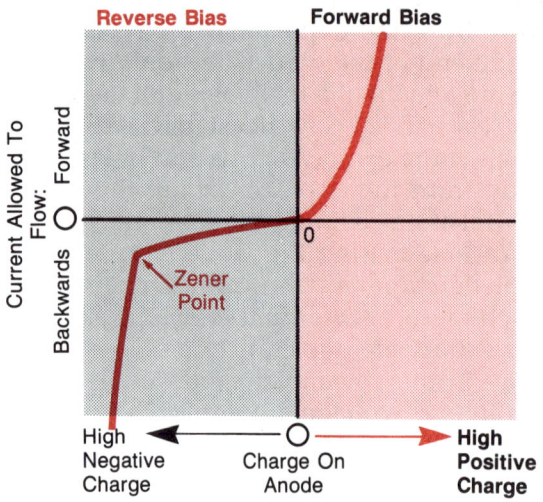

Fig. 40-8 Graph of the current curve of a diode when reverse and forward biased.

reverse biased (the electron flow from − to + cannot travel from anode to cathode).

If the voltage becomes greater than the zener point, say 13 V, the extra voltage will form current flow through the zener diode. This current will flow through both the diode and the series resistor (R_1), but not through the load (R_L). Since more than the normal amount of current is flowing through R_1, that resistor will drop the extra 1 volt and leave only the desired 12 V to power the load. In reality, the power

The downward curve indicates that much reverse current is being permitted to flow. The point at which reverse current is first permitted to flow is called the **zener point**. It indicates how much reverse voltage is needed to cause backward current to flow through the diode. Another name for the zener point is "breakdown" potential.

All diodes have a zener point, but it is usually positioned far beyond the normal operating reverse voltage of the diode. A few diodes, however, are designed and doped just the correct amount to position their zener point in a meaningful position. Such diodes are called **zener diodes**. They are used in voltage controlling circuits (Fig. 40-9). If you wanted a 12 V output from the power supply, but the output varied a little over 12 V, you could use the zener diode to regulate the voltage. Notice that the schematic symbol for a zener diode has little flags on the ends of the cathode line. The voltage of the zener point is often also labeled. As long as the voltage stays at 12 V or below, no current will be allowed to flow through the zener diode because it is

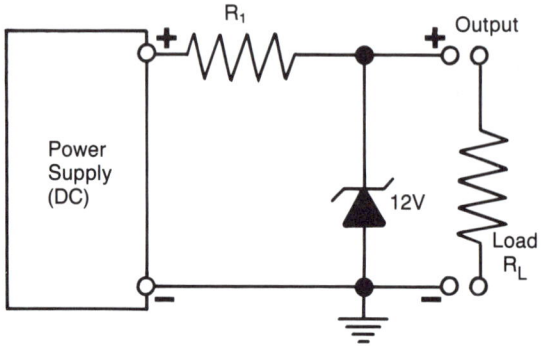

Fig. 40-9 The zener diode helps regulate the voltage output to 12 V by conducting current in reverse when the power supply produces over 12 V.

supply would always be operated a little above the 12 V level and R_1 would always be dropping some extra voltage.

Another special type of diode is the **tunnel diode**. Fig. 40-10 shows the symbol and the current graph of a tunnel diode. Tunnel diodes are very useful in high speed switching operations because of their rapid current rise and sharp peak.

Still another special diode is the **varactor**. The varactor is like a variable capacitor that changes its own capacitance value when the voltage changes. The symbol for a

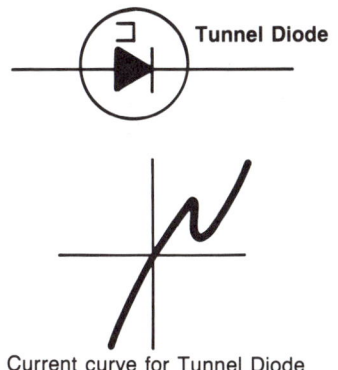

Fig. 40-10 A tunnel diode and its unusual current curve.

varactor is shown in Fig. 40-11. Varactors are used in many new push-button television channel selectors.

Remember from the previous chapter that resonant circuits are used to tune-in radio and television stations by frequency selection. One way to do this is to use varactors. The digital circuits operated by the calculator-type keyboard send a certain level of voltage to the varactors for each station selected. The varactors then change their own capacitance values to make the tuner circuits resonate to the desired frequency. This system is much better than the old tuners which depended upon many switch-like contacts to select the proper coils and capacitors for each station. The varactor tuners do not have moving parts to wear out or contacts that get dirty and loose.

The **LED**, or light emitting diode, is used extensively in digital electronic products. They glow whenever current flows through them. You have seen them as indicators for many devices, and LEDs are also often arranged to make lighted numerals for calculators, meters, and watches. The schematic symbol for a single LED and a drawing of a seven-segment LED readout appear in Fig. 40-12.

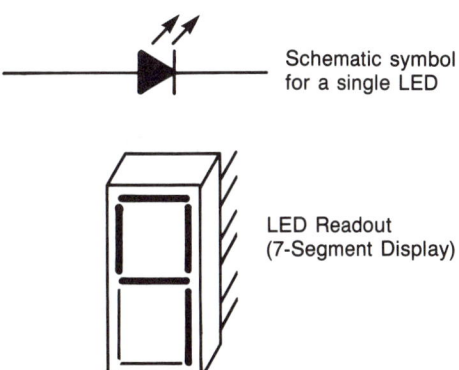

Fig. 40-12 By selecting various segments to light, the LED Readout can display numerals 0–9 and some letters.

Fig. 40-11 Varactor: A voltage variable capacitor used in modern television tuner circuits.

Summary

The simplest semiconductor devices are diodes. A diode is a PN junction. The P material has been doped to have fewer than the normal number of electrons and the N region has too many electrons. A depletion zone (neutral area) is created between the P and N regions. When forward biased, current flows through the junction easily, but when you turn the battery backward (reverse bias) no current is allowed to flow until the voltage is increased to the zener point (breakdown potential). The arrow in the diode's schematic symbol points in the direction of conventional (hole flow) current. Special diodes include the zener diode, tunnel diode, varactor, and LED. The next chapter covers how diodes and other components are used in power supply circuits.

Important Terms

valence
covalent bonds
doping
P-type materials
N-type materials
PN junction

diodes
holes
depletion zone
forward bias
conventional current
reverse bias

zener point
zener diode
tunnel diode
varactor
LED

Review Questions

1. How are diodes made? What is special about a PN junction?

2. What happens when we forward and reverse bias a PN junction? What does the depletion zone do?

3. Explain what is special about a zener diode. Name a use for zener diodes.

Chapter 41
Power Supplies

Putting Diodes to Work

One important job of diodes is changing AC to DC. The current that the power company supplies to your home is AC current. This AC current is fine for electrical devices such as heaters, lights, and vacuum cleaners, but it is not possible to use AC in the working circuits of radios, televisions, computers, or other electronic equipment. For this reason, the first circuit in electronic items is usually a **power supply** circuit which changes the 120 V AC current to a DC current of a certain required voltage.

Power supplies perform two main jobs: they change the AC to DC, and they change the voltage level to provide the voltage needed. Power supplies do not "make" power out of "thin air," they just convert it from one form and level to another. Most power supplies use diodes to change the AC to DC. Transformers are often used to change the voltage level, but resistors and zener diodes are also used in many power supplies. Fig. 41-1 shows the simplest DC power supply. It produces 120 V pulsating DC. During each alternation, the AC generator will try to cause current flow through the circuit. When the generator is trying to push electrons through the circuit counterclockwise (upward through the load and against the direction of the diode's arrow-shaped anode), the diode acts like a low value of resistance and allows those pulses of current to pass. So, the alternations (pulses) on the top of the sine wave are allowed to flow in the circuit's output.

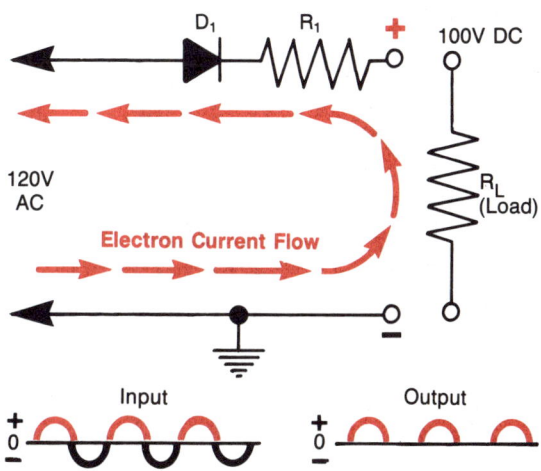

Fig. 41-1 In this simple DC power supply, the AC is changed to pulsating DC by a single diode and a dropping resistor (R_1) lowers the output voltage to the desired 100 V level. The "load" is whatever you intend to power. It could even be the rest of the circuits in a TV set. You can represent the entire load with one resistance symbol.

When the alternations on the bottom of the sine wave are sent to the power supply, however, they will try to make current go clockwise through the circuit. This means that they will try to go through the diode in the reverse direction (the same direction as the symbol's arrow). These pulses will reverse bias the diode and increase the width of its depletion zone. The diode will not let them pass through. So, the output will have the pulses on top of the center line but nothing under the center line. This is called pulsating DC current. Resistor R_1 will drop the voltage to 100 V. Resistor R_L (load) represents whatever circuits we are powering with our power supply.

Full-Wave Rectification

The process of changing AC to DC is called **rectification**. The diodes used in this process are often called "rectifiers." The circuit just discussed is more accurately called a **half-wave rectifier** because only half of the AC waveform was rectified (the pulses on top). The pulses below the center line were lost. This means that you lost (wasted) about half of the power available and the output was not smooth enough for many circuits.

Fig. 41-2 is a simple **full-wave rectifier** power supply. The colored halves of the AC sinewave input show that you can see how the circuit operates on each alternation. The first colored pulse begins at generator terminal A. During this pulse, electron current flows through diode D_1, but it is not allowed to flow through D_2 to the top of the load or D_3 back to the generator. Instead, it is forced to travel on through the ground path conductor to enter the load from the bottom. The current flows upward through the load and then back to the diodes. Both diodes D_2 and D_4 are aimed correctly for the electrons to flow through them from right to left, but this particular colored pulse originally came from terminal A of the generator. This pulse wants to get to terminal B to complete its circuit; so it ignores D_2 and travels through D_4 back to the B terminal of the generator. The graph of the circuit output shows pulse number 1 above the 0 line to represent the current going through the load in an upward direction (+).

The next input pulse from the generator is the (black-colored) second alternation below the 0 line on the input graph. This pulse will flow from the generator in the opposite direction, so it leaves the generator from terminal B. When it reaches diodes D_3 and D_4, it is forced to go through D_3 to the ground path. It follows the ground conductor to the bottom of the load (R_L) and then travels upward through the load. From there it goes to diodes D_2 and D_4 and passes through D_2 to the generator's A terminal.

Graphing pulse number two on the output graph, you will get a positive pulse because the current during this alternation went upward through the load just as pulse 1 did. Pulses 3 and 5 will act just as alternation 1 did and pulses 4 and 6 will do what number 2 did. The output graph shows that you did not lose any of the original pulses. You rectified the "full" wave. Thus, this circuit is a full-wave rectifier. Resistor R_1, serves to set the

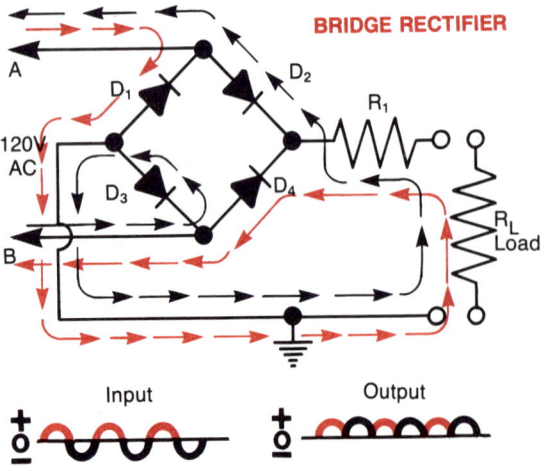

Fig. 41-2 The bridge rectifier circuit provides full-wave rectification.

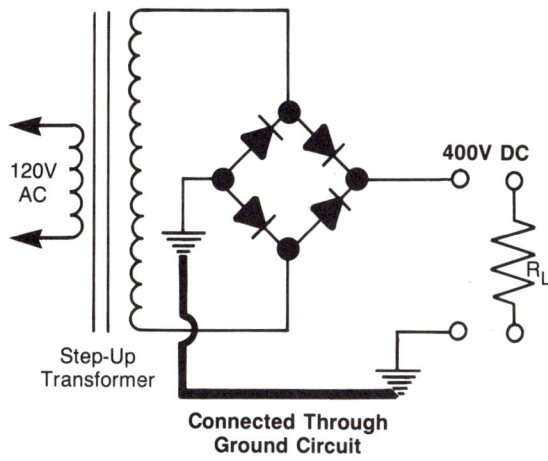

Fig. 41-3 A transformer may be used to step the AC (input) voltage up or down before it is rectified by the bridge. All of the "ground" symbols (⏚) in a given circuit are connected together by heavy wires or the metal case of the equipment.

voltage output to the desired level, just as in the half-wave rectifier.

This particular arrangement of diodes is called a **bridge circuit**. The full-wave bridge circuit has been used in many television sets because it gives full-wave rectification and an output that is easily smoothed to pure DC (straight line graph) by filtering circuits. Filtering the half-wave pulsating DC of Fig. 41-1 would be more difficult.

Transformers in Power Supplies

Resistors and zener diodes can reduce the output voltage of DC power supplies, but they are unable to increase the voltage. One common way to produce a power supply with an output voltage of more than 120 V is to use a step-up transformer before the current is changed from AC to DC. The transformer will only work with AC current. (If you do not remember why trans-

formers require AC, review Chapter 30.) Fig. 41-3 shows a simple power supply for producing 400 V DC. If you wanted a low voltage power supply, you could use a step-down transformer in the same circuit arrangement.

Full-Wave Rectifier with Only Two Diodes

Though diodes now are very inexpensive and prewired diode bridges in a single enclosure are readily available, this was not always so. The circuit shown in Fig. 41-4 is an alternate way to produce full-wave rectified DC. The circuit was originally developed when vacuum tubes were in use and was still very popular in the 1960s and 70s, before semiconductors became so inexpensive. You will certainly encounter it if you do repair work.

The central important element (besides the diodes) is the center tapped transformer.

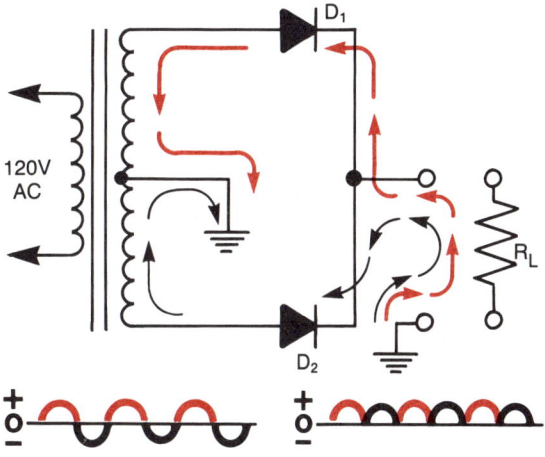

Fig. 41-4 If a center tapped transformer is used, you can get full-wave rectification by using only two diodes.

Chapter 41 Power Supplies

As the current in the primary of the transformer alternates, it will try to make the current in both halves of the secondary flow in a downward then upward direction. The current will not be able to do this though, because the colored pulses, when they try to go downward in the lower half of the secondary, will be reverse biasing D_2. This diode (D_2) will not let the colored pulses travel in the lower half of the secondary coil.

The colored pulses will, instead, travel through the ground circuit (often the metal chassis of the equipment) to the bottom end of the load (R_L). Current will enter the load and flow through it in an upward direction. From here it will return to the secondary through diode D_1, which will let it pass because it is forward biased. The gray pulses will be prevented from flowing in the upper half of the transformer's secondary coil by diode D_1. They will follow a path through the ground circuit, upward through the load, and back to the secondary through diode D_2. The load received upward pulses during both alternations, so you have produced full-wave rectified DC.

The vacuum tube equivalent of this circuit was very popular, and nearly all early television sets used it. Now, since center tapped transformers are quite expensive (and diode bridges are fairly inexpensive) it is used less often.

Multiple Output Power Supply

A power supply with several AC and DC outputs is shown in Fig. 41-5. There could be many different outputs, depending on the number of secondary windings of the transformer. All of the DC outputs share the same ground connection. More resistors could be used to make additional outputs of lower voltage values. Filtering networks, made up of capacitors and/or coils, are usually added to smooth the ripples from the rectified DC outputs.

Voltage Doublers

There is a way to get both rectification and voltage output greater than the input voltage without the use of transformers. Circuits that can do this are called **voltage doublers.** A full-wave voltage doubler circuit appears in Fig. 41-6. The positive alternations (colored) leave the generator's A terminal to travel through diode D_2 and charge capacitor C_2 before they return to the generator's B terminal. The negative alternations (black) come from the lower

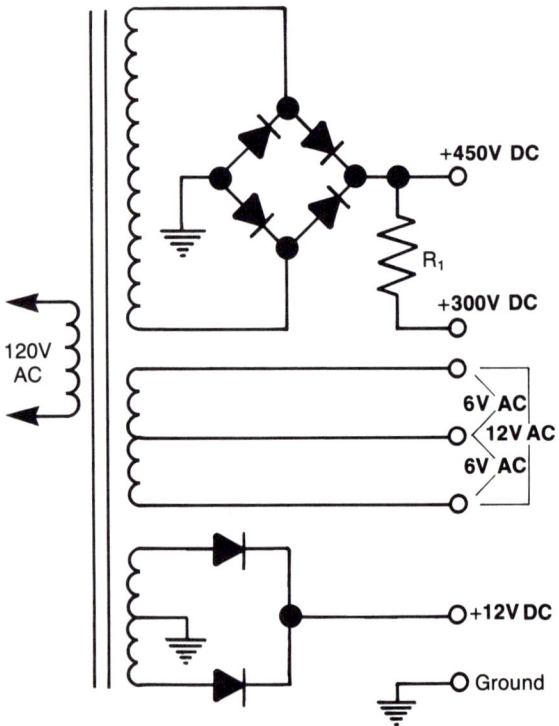

Fig. 41-5 A power supply with several outputs. The ground connection is the − terminal for all of the DC outputs.

generator terminal (B), and are permitted to go through C_1 by diode D_1. The two alternations are thereby separated from each other by the diodes and are forced to charge only one capacitor each. Both capacitors will charge to about 120 V with pulsating DC, but their capacitive action will smooth the pulses to look more like the two output curves shown for capacitors C_1 and C_2. Since the final output of the whole circuit is measured from points C to E, it will add both of the capacitors' waveforms together. The total output is roughly twice the input voltage, or a fairly smooth 230 V DC. The actual voltage can be considerably higher, depending partly upon the load and the values of the capacitors. Because they allow the output of high DC voltages without expensive (and heavy) transformers, voltage doubler circuits are very popular in modern television sets. There are also half-wave doubler circuits which use diodes and capacitors. Figure 41-7 shows a half-wave doubler.

Voltage Regulators

One other type of DC power supply circuit is the **voltage regulator**. Voltage regulators have the capability to adjust their output to compensate for changes in their input voltage and the amount of load placed on their output. In this age of computers and digital circuits, it is often very important that the DC supply voltage be stable. Early voltage regulator circuits, made up of several transistors and other

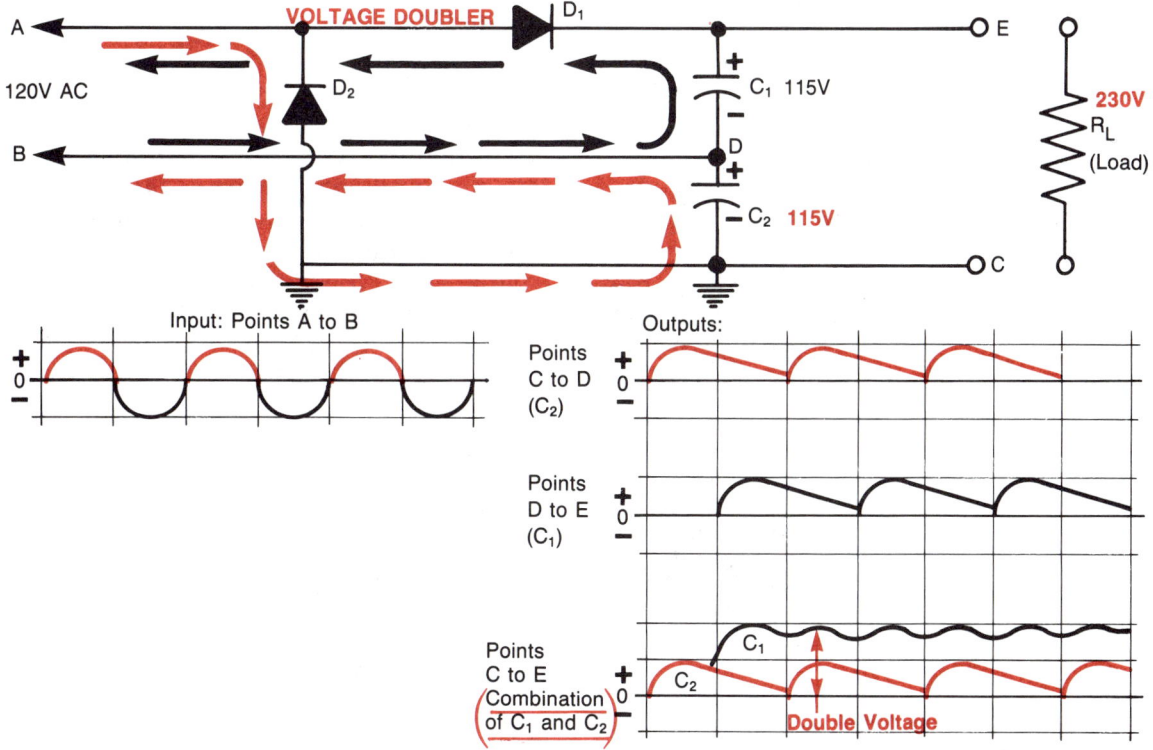

Fig. 41-6 The voltage doubler circuit gives a DC output which is about twice the AC input voltage. The separate pulsating DC outputs across C_1 and C_2 are added in the final output (from points C to E).

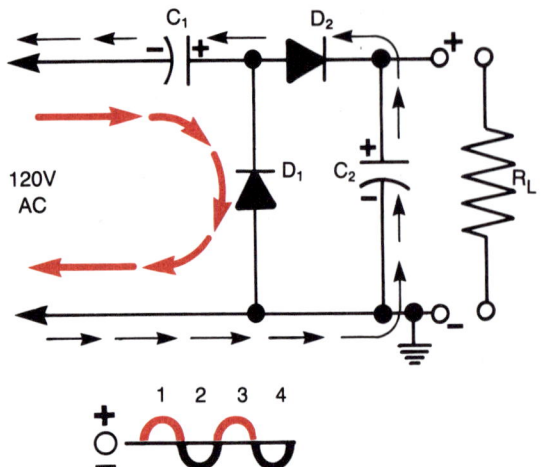

Fig. 41-7 Alternation number 1 is allowed to flow only through D_1, so it charges C_1. Pulse 2 (black) charges C_2 and passes through D_2 back to the generator. While this (black) pulse is present, capacitor C_1 is still holding its charge from pulse 1, so you have a charged capacitor in series with another applied voltage. The output will be a pulsating DC which is about double the input voltage.

A frequently used numbering system labels these regulator ICs 78××. The ×× would be replaced with the desired voltage. For instance, a 7812 will produce a very smooth and stable 12 V DC output, a 7805 gives 5 V output. The simple circuit in Fig. 41-8 uses a 7815 to give a regulated output of 15 V DC. Some special regulator chips allow the addition of a potentiometer to the circuit to make an adjustable regulated power supply — a very helpful device to have for experimentation and repair work in electronics.

Summary

Most electronic devices need DC current for their internal circuits. Power supplies are the circuits which change AC to DC and set the proper voltage level to meet the needs of electronic circuits. The DC power supply is usually one of the first circuits in electronic devices. Half-wave rectifiers only change half of the AC power cycle into DC, but they can be produced with one simple diode. Full-wave rectifiers require either a four-diode bridge or a center tapped transformer and two diodes. Resistors and zener diodes, transformers, voltage doubler circuits, and voltage regulator ICs are all methods which may be used to change the voltage output of a DC power supply.

components, were expensive. Now, however, there are a variety of inexpensive voltage regulator ICs (integrated circuits). These are complete voltage regulator circuits contained in one small package that looks like a medium-sized power transistor. They are sold by many companies.

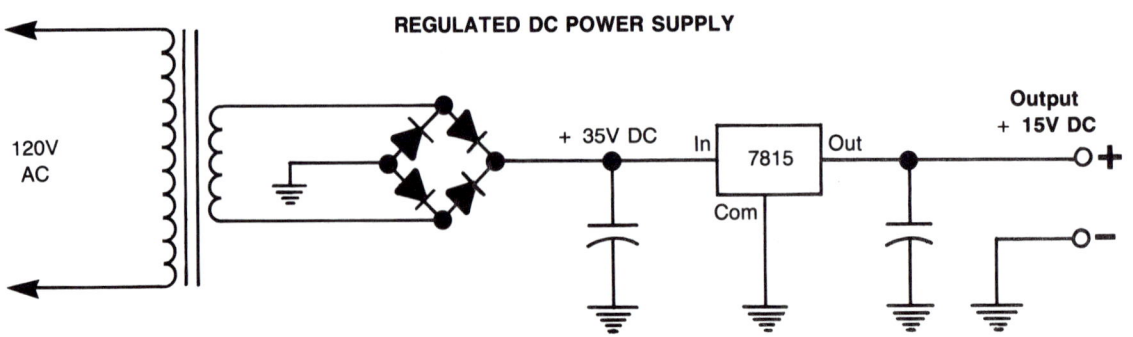

Fig. 41-8 The regulator IC chip makes regulated power supplies easy to build and economical.

Important Terms

power supply
rectification
half-wave rectification
full-wave rectifier
bridge rectifier
voltage doubler
voltage regulator

Review Questions

1. What are power supplies and what do they do?
2. Tell the difference between half-wave and full-wave rectification. Refer to Fig. 41-2 and explain how a bridge rectifier circuit works.
3. Why are voltage doublers, bridge rectifier circuits, and regulator ICs so popular now?

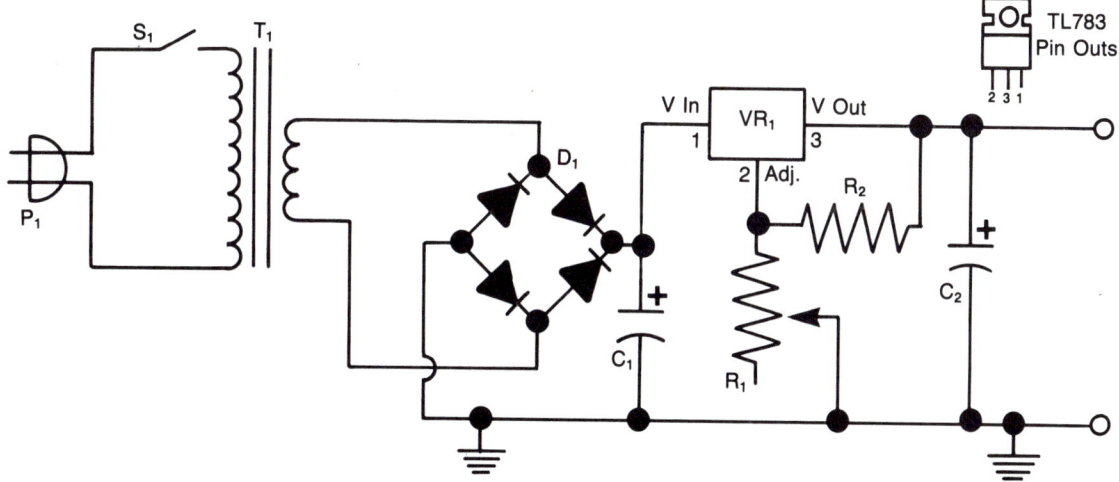

Fig. 41-9 A 2–30 Volt Variable Regulated Power Supply makes a nice project.

Project: Variable Regulated Power Supply

Parts list

1	120 VAC line cord and plug	P_1
1	SPST switch	S_1
1	2.5 V, 2A power transformer	T_1
1	2000 microfarad, 50 VDC capacitors	C_1
1	100 microfarad, 35 VDC capacitor	Cx_2
1	50V full-wave bridge rectifier, 2A+	D_1
1	10 K-ohm potentiometer	R_1
1	330 ohm resistor	R_2
1	LM317 or TL783 (Radio Shack 276-1778 or 276-1763) Adjustable Voltage Regulator	VR_1

Suitable enclosure, PC board, wire, and hardware is required. The output terminal should be selected to meet your needs.

Chapter 41 *Power Supplies*

Notes

This highly dependable, variable regulated power supply consists of a simple voltage doubler circuit which feeds a voltage regulator IC. Be very careful to use the proper leads of the IC for input, output, and adjust — check the package to be sure. The potentiometer allows accurate selection of any regulated DC voltage from 2 V to 30 V. Circuit changes can be made to expand the range; consult the IC's technical data sheet for more information.

Chapter 42
Vacuum Tubes

Vacuum Tubes

Today's electronic marvels use diodes and transistors. These important devices owe their development to vacuum tubes. The Fleming Valve was the first vacuum tube. It was a vacuum tube diode invented in 1904 by Sir J. A. Fleming in England. The first amplification tube was made by Dr. Lee De Forest in 1906. Vacuum tubes, like the ones in Fig. 42-1, were the original building blocks for radios, televisions, and even computers, but they are not used as often today because much smaller semiconductors have been developed that can work more efficiently.

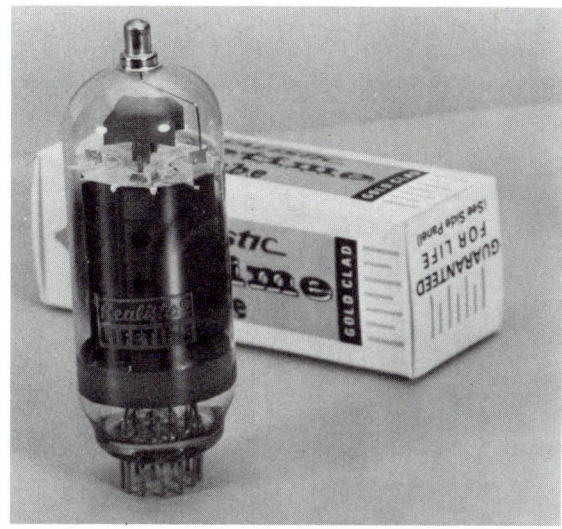

Fig. 42-1 Electron tubes, generally called vacuum tubes, come in many types and sizes.

Vacuum Tube Diodes

Before the semiconductor diodes were invented, there were diodes which performed the job quite differently. Figure 42-2 shows symbols for two types of **vacuum tube diodes.** The circle in each symbol represents the glass or metal tube around the other parts. This case may either be evacuated (have the air removed to form a vacuum) or filled with a special gas to prevent the internal parts from burning up during operation.

Tube A has just two parts: a filament and a plate. The filament acts like the cathode in a semiconductor diode. It does this by becoming so hot that electrons will try to jump away from it. This is called **thermionic emission**. If the filament has a strong negative charge and the plate has a high positive charge, then the electrons will be able to jump all the way from the filament (through the vacuum space) to the plate. The plate acts as an anode. So, current can pass from the cathode (filament) to the anode, just as in a semi-

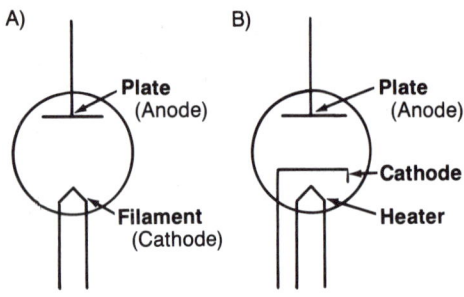

Fig. 42-2 Diodes have only two "working" parts: A cathode and an anode. The cathode is sometimes combined with the heater (filament). The anode is called the plate.

electrons which try to jump across the vacuum area backward. Thus, the diode will not pass current in reverse. It passes current in only one direction, just like a semiconductor does.

The tube in part B is also a diode. It appears that this tube has three parts, but only two of them are electronically important parts. The cathode and plate (anode) control the direction of current flow. The heater merely warms the cathode up so that the electrons will easily jump from it. When electrons jump from the surface of a cathode, they create a **space charge**. The space charge is a group of unattached electrons hovering around the cathode which can be drawn to other parts of a vacuum tube. If the plate has a high enough positive charge, the space charge electrons will be attracted to it.

Vacuum Tube Power Supplies

Power supplies just like the ones in Chapter 41 may be produced using vacuum tube diodes — and they were for years. The only difference is that they must have extra windings in their transformers (or separate batteries) to provide current for the heaters of the tubes. This is a waste of power and the tubes are much larger than today's semiconductors. Figure 42-3 shows a vacuum tube power supply circuit the exact equivalent of the one studied in Fig. 41-4. The tube used in this circuit is a 5U4 dual-diode. The 5 tells you that the filament (heater) voltage is 5 V. In this tube there are really two diodes in one glass case. Both diodes share the same filament. This is a full-wave rectifier. It was one of the most popular power supply circuits in use until semiconductors made bridges more practical.

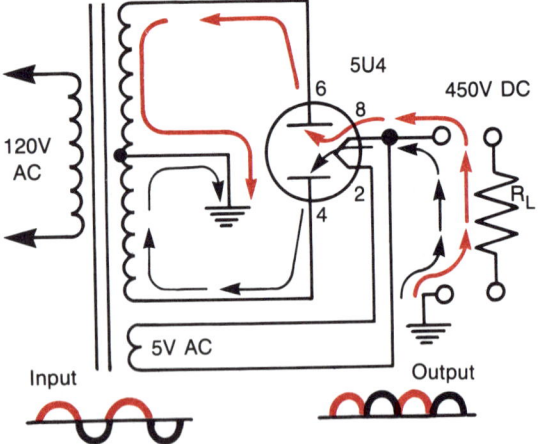

Fig. 42-3 A vacuum tube full-wave rectifier which operates like the circuit in Fig. 41-4. Compare the two circuits.

conductor diode. But, if the battery polarity is reversed so that the plate is negative, then the heat of the filament will repel any

Triode Vaccum Tubes

Another job of tubes and semiconductors is **amplification**. Amplification is the process of controlling a large amount of power with a smaller one. When you play a record on a stereo system, a very small amount of electrical energy is produced by the cartridge in the turntable's tone arm. If this signal were sent directly to the speaker, however, you would hear no sound because it is far too weak. The signal is sent first to the "amplifier" (which is actually several amplifier circuits connected together).

An amplifier uses a small (weak) signal to control a larger (stronger) amount of power. So, the output of the amplifier is really a signal which resembles the input one, but is larger. The greater power of the output signal is not created by the amplifier circuits, it comes from the power supply circuit. The amplifier just controls this greater power. The amplified signal is then used to drive the speakers and produce the desired (louder) sound.

The simplest tube that can amplify (that is, control large values of power with smaller ones) is the **triode**. Figure 42-4 shows a very basic triode amplifier circuit. Note that the triode has three working parts: cathode, grid, and plate. The heater (H) has its own little A battery (usually really a separate winding in the power supply's transformer). This heater current may be either AC or DC and, in general, it will not be of any importance in the operation of the circuit, except to heat the cathode and make it give off a space charge. The cathode, labeled K, gives off electrons when heated. The plate (P) functions the same in a triode as it did in the diode. It is a large metal plate with a high positive charge that attracts electrons from the cathode. The B battery is a high voltage source (usually the 300 to 450 volt DC output from the power supply in real-life circuits). The battery makes the plate have a very high positive charge so that electrons will be drawn to it. This battery is called the B+ source. It creates the major current flow in the circuit which goes through the load R_L. If the cathode and plate were the only two parts of the tube, we would have a simple diode, and current would flow continuously without change.

The important new part in the triode is the **control grid** (G). In an actual tube, the grid looks like the rack in an oven except that it is very tiny and it is turned up on its edge. By carefully adjusting the potentiometer, R_1, you can make a very small current flow from the little C battery, through the cathode to the grid, and then through the potentiometer and back to the C battery. The grid is very close to the cathode. The plate, with its large positive charge, is always trying to make many electrons flow from the cathode to the plate.

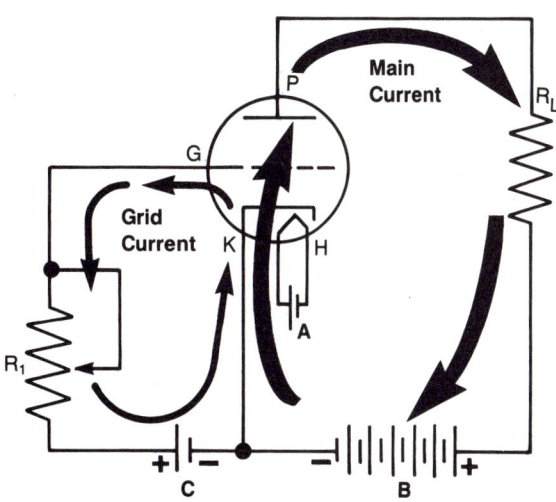

Fig. 42-4 The small (C battery) grid current controls the much larger main (B battery) current. This is what amplification really is.

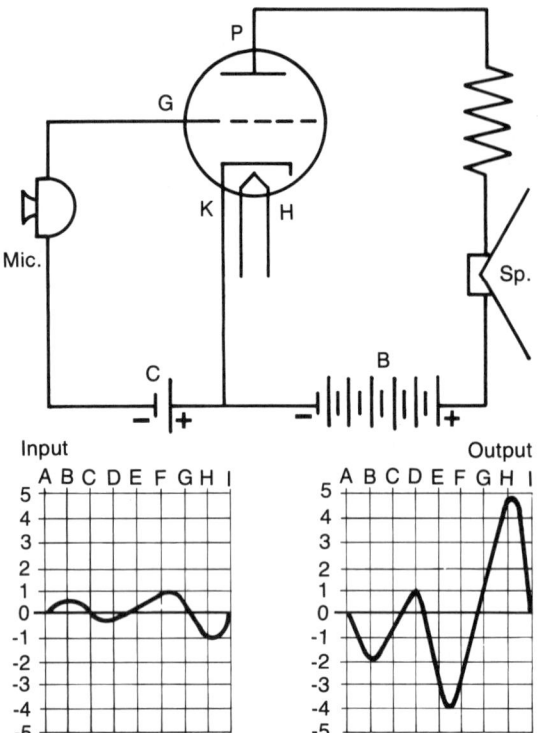

Fig. 42-5 When the triode amplifies, there is a 180° phase shift between the input and the output currents. Actual circuits would not generally use actual C and B batteries like these examples. Instead, the C and B voltages would be obtained by using dropping resistors from the power supply.

If the grid becomes positively charged, it will have a great effect on the cathode's space charge of electrons, because it is so close to them. When the C battery puts a small + charge on the grid, many more than the normal amount of electrons are drawn away from the cathode. This horde of electrons goes racing toward the grid with great momentum. A few of the electrons actually hit the tiny wires in the grid and become the grid (C battery) current. All of the others, however, will go racing right past the thin grid wires and on toward the plate. These extra electrons which you drew off the cathode add to the normal plate (B battery circuit) current. So, when you put a very small charge on the grid, you control a large current through the load (R_L).

For every one electron drawn through the grid circuit, 100 more are sent on through the plate circuit. So, if the potentiometer (R_1) were adjusted to allow two electrons through the grid circuit, then 200 would go through the plate circuit and the load (R_L). Adjusting further to pull five electrons through the grid would drive 500 more through the plate. You would be controlling a large current (100 to 500 electrons) with a very small one (one to five electrons). In other words, you would be amplifying!

Another amplification circuit is shown in Fig. 42-5. In this circuit, the triode is amplifying the signal from a microphone to control a loudspeaker. Notice there is no heater battery. It is common practice to omit the heater battery (A) from schematics because people knowledgeable in electronics know that it exists, and the drawings are simpler without it. Notice also that the C battery, which is used to bias the tube, is reversed. In actual working circuits, if the bias on the grid becomes positive (as in the last example) the plate would become saturated with electrons, overloading most circuits and causing distorted output. When the grid goes slightly negative (as in this circuit) some of the electrons normally attracted to the positive plate will be repelled back to the cathode by the grid. So, the tube works much like a water faucet. Water (current) is trying to flow all of the time, but you can limit the flow by using a few pounds of pressure on the faucet handle (a few electrons in the grid circuit).

When you speak into a microphone, it creates a small AC current which changes the charge on the grid, as the potentiometer did in the other circuit. This AC signal makes the charge on the grid vary from a very slight negative charge to a greater negative charge. The input graph shows a simple signal you could input to the circuit by making sounds into the microphone. As the charge on the grid changes in response to the input signal, the amount of current allowed to flow in the plate circuit will vary in inverse proportion to the tiny input signal. The output current, though, will be tremendously higher than the input current. The signal may be inverted back by sending it to another amplifier stage before using it to drive the speaker. Again, a small (grid) current was used to control a much larger (plate) current; we amplified!

Other Types of Tubes

Vacuum tubes may have greater amplification by adding more grids to them. The **tetrode** and **pentode** are both common tube types. See Fig. 42-6. The tetrode's fourth part, the **screen grid**, usually has a high positive charge to attract more electrons from the cathode and increase current flow. Internal capacitance between the parts is also reduced in the tetrode. The **suppressor grid** in the pentode has a negative charge. When the screen grid accelerates the flow of electrons to the plate, they sometimes hit the plate so hard that some bounce off the plate and go to the screen grid. This upsets the desired amplification. The negative charge on the suppressor grid repels these rebounding electrons back to the plate. The beam power tube (Fig. 42-7) may be either a tetrode or a pentode with a special shape and arrangement of the parts to increase power handling ability.

Double Duty Tubes

Two or more separate tubes are often placed in one glass envelope. The tubes all share one heater. This arrangement saves space and reduces the amount of power wasted in heating tubes. Figure 42-8 shows the schematic symbol of one such tube — a 6BH11. What is the heater voltage of this tube? You are right if you said 6 volts. This tube has three separate tubes in one glass case. The little numbers around the circle indicate the pin numbers for each part. Which two parts are connected to pin 11?

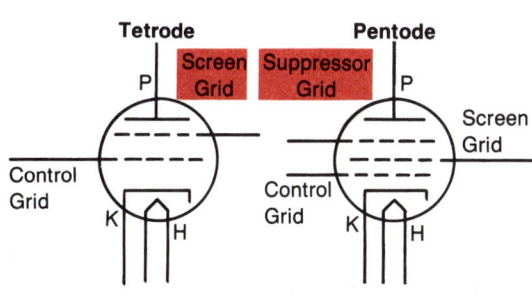

Fig. 42-6 The tetrode and pentode have 4 and 5 working parts each. The control grid is the signal input grid, just as in a triode.

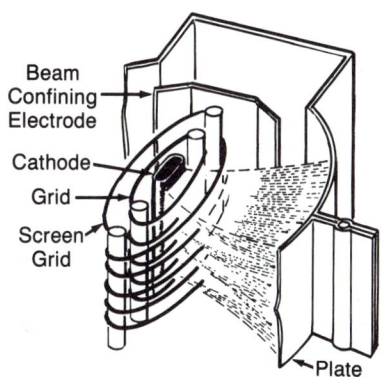

Fig. 42-7 Internal parts of a beam power tube. The heater is inside of the hollow cathode — this is true in most tubes.

What are the three tube types inside this case? Would this tube be used for rectifying or amplifying?

Special Tubes

The screen on which the picture is displayed in a television set is a very large vacuum tube called a **cathode ray tube**. The information in this chapter will help you understand how the television and its cathode ray tube operate. You will study them in Chapter 57.

Some other special tubes are used as indicators. A common type in early model radios was the "electric eye" tube which indicated when the station was tuned in for best reception. LEDs and small meters usually do this job now.

Summary

Vacuum tubes were the forerunners of modern transistors and diodes. They may be used to rectify and to amplify, just as semiconductors may be. Amplification is the process of controlling a large amount of power with a smaller one — so a tube that does this is really a control device. Common tube types include diodes, tetrodes, and pentodes as identified by the number of "working" parts each has except for the heater. The next chapter discusses transistors — the semiconductor equivalent of the triode — and their ability to amplify.

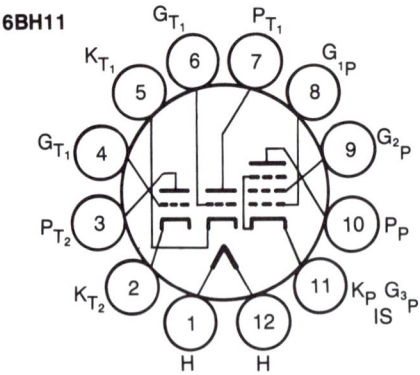

Fig. 42-8 This tube was used in many television sets in the 1970s. You may find pin diagrams like this one, and other important information about most general purpose tubes from a *Receiving Tube Manual* published by tube manufacturers.

Important Terms

vacuum tube diodes
thermionic emission
space charge
amplification

triode
control grid
tetrode
pentode

screen grid
suppressor grid
cathode ray tube

Review Questions

1. Draw the schematic symbol of a triode vacuum tube and label its parts. What two other parts would be added to make it a pentode?
2. Explain what is wrong with this statement: "An amplifier takes a small signal and makes it bigger." Make a true statement that tells what an amplifier does.
3. Refer to Fig. 42-5 and explain how a triode tube can be used as an amplifier.

History: Fleming's Valve

Sir John A. Fleming, son of a Congregational minister, worked with Marconi and was a consultant to Edison's London office. He put to work something which Edison had really noticed first — thermionic emission (the giving off of electrons by a hot filament). This process is also called the Edison Effect. Fleming's first practical vacuum tube was the diode rectifier (1904). He was knighted in England for his discovery of the valve. He lived to the age of 97, and saw his discovery improved upon by Lee De Forest.

De Forest's father was also a minister. Being from the Deep South less than 25 years after the Civil War, De Forest and his family were not well thought of because his father ran a school for Black children. His father wanted him to become a minister, but young Lee was interested in science. He earned his doctor's of philosophy in 1899. Like Fleming, he had a long life — dying at the age of 88 in 1961. He saw his important invention flourish and then be almost totally replaced by the transistor.

Chapter 43

Transistors

Two Types of Transistors

You have learned the basic principle of amplification and how triode tubes can be used to amplify. This chapter covers the semiconductor equivalent of the triode vacuum tube — the **transistor**. Just before 1950, three scientists who worked in Bell Telephone laboratories demonstrated the functional transistor. These men, John Bardeen, W. H. Brattain, and William Shockley were awarded the Nobel Prize for their discovery which changed the world of electronics.

Put P and N material together to form a PN junction, and you have a diode. Carry this one step further and you can make a transistor. There are two basic types of transistors: the **PNP** and the **NPN** types. Either one may be used to amplify. In fact, some very simple one-transistor circuits will work with either type if the battery polarity is reversed.

Schematic symbols and pictorial representations of both types are shown in Fig. 43-1. Both types have the same three parts: the **emitter** (which is like the cathode of a triode vacuum tube), the **base** (which does the job of a tube's grid), and the **collector** (like the plate in a tube). Each of these parts has a separate lead (wire) for connecting it in the circuit.

You can tell whether the transistor is NPN or PNP by the direction the arrow on the emitter symbol is pointing. A memory aid, "NPN means Never Points iN," reminds you that the emitter arrow in the NPN type points outward from the base rather than inward to it. As was true with diodes, the arrow in the symbol points in the direction of conventional (hole) current flow. That is, it points opposite to the electron flow current which you usually consider. Figure 43-2 shows several styles of transistors. Note that all of these case styles have three connections (leads).

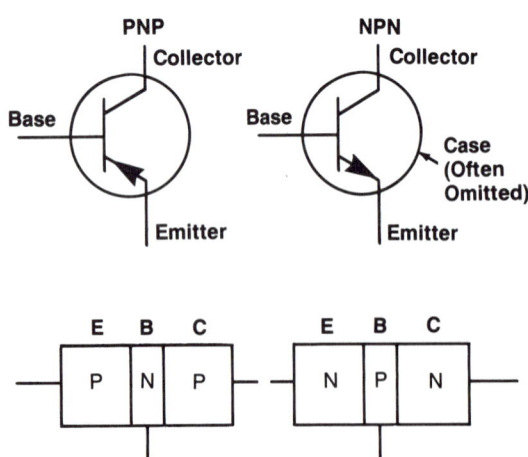

Fig. 43-1 The two types of transistor both have three parts: emitter, base, and collector. The direction in which the little arrow on the emitter points identifies the device as either NPN or PNP.

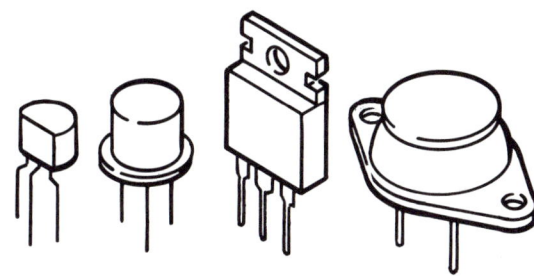

Fig. 43-2 Different case styles of transistors. The larger ones can handle more power. Sometimes the case itself is one of the 3 connections.

How Transistors Amplify

Look at the simplified NPN amplifier circuit shown in Fig. 43-3. The actual schematic of the example circuit and a pictorial view are shown so you understand what happens inside of the transistor. Separate C and B batteries are used to simplify the discussion, but remember that these different voltage levels are obtained by using dropping resistors from the power supply or main battery in a real circuit. Before reading on, turn back to Chapter 42 and compare this circuit to the one in Fig. 42-4 to see how the two circuits are similar.

The goal of this circuit is to amplify — to use the tiny C current (emitter to base current) to control the much larger B current (emitter to collector). Before the batteries are connected into the circuit, the condition of the transistor will be as shown in Fig. 43-3. There will be many extra free electrons in both the emitter and the collector (N regions) and only a few extra holes in the base (P region). This is because the base is very thin and is very lightly doped compared with the emitter and collector regions.

When you first connect the B battery, a strange thing will happen. As Fig. 43-4 shows, the electrical pressure (voltage) of the B battery forces electrons to move from the emitter, through the base, and to the collector. The B battery is trying to do this

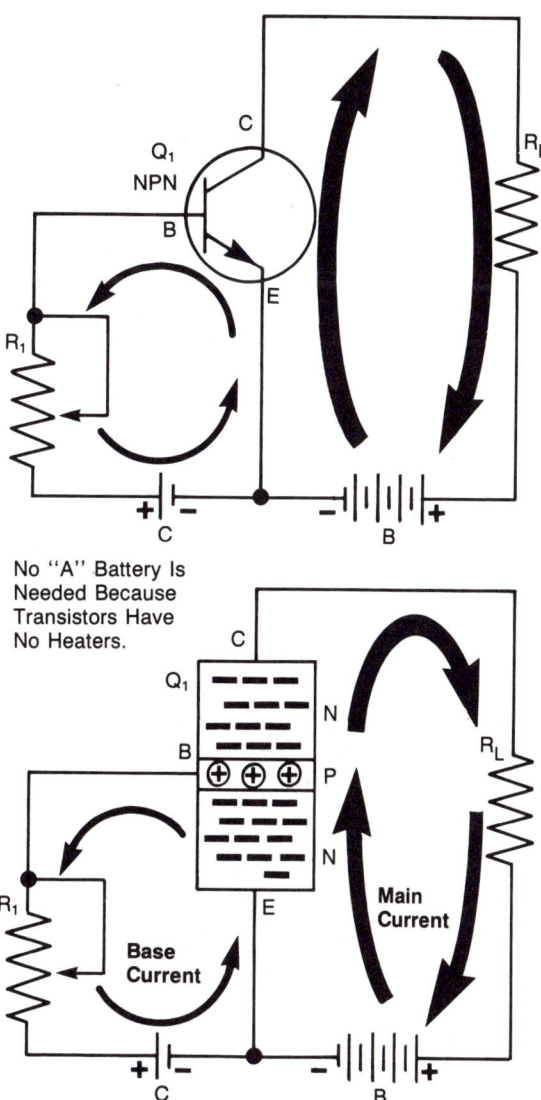

Fig. 43-3 Transistors use a small base current (C) to control a larger main current (B). This is the way they amplify.

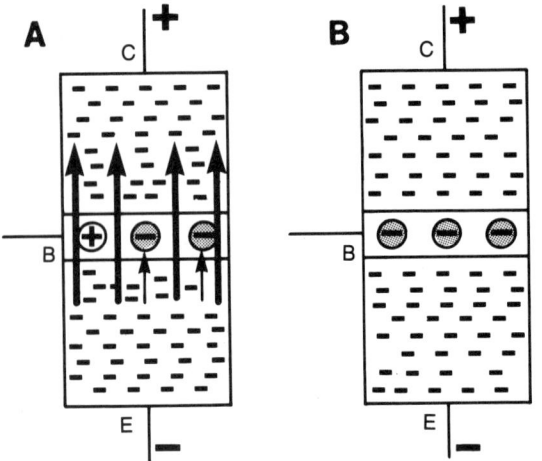

Fig. 43-4 A) Electrons stampede toward open holes, most miss the holes and continue to the collector, but a few are caught. B) All the holes are now filled, so current flow stops until more holes are available.

all of the time, but it is only successful when there are some extra holes in the base. Thousands of electrons will move from the first N region (emitter) to the other N region (collector) as long as there are a few holes in the base. But when the electrons start moving through the P material of the base, some of them will be captured by the holes in the base. The electrons have no natural desire to go to the other N region (collector) from the emitter, but they may be attracted into motion by the presence of positively charged holes in the base. Once they begin to move, they move very rapidly and seem to stampede into the base region. The base is so thin (and so lightly doped) that many of the electrons which come hurtling into it just go right out the other side where they become attracted by the positive terminal of the powerful B battery. Once this happens, they form a current flow through the load.

All of this happens in an instant. When the B battery is first connected, an avalanche of electrons races through the transistor. For every one electron that gets caught by a hole in the base region, maybe 50 more go on through to become B battery current in the load (R_L). The holes do catch a few electrons, however, and when all of the holes have made a capture, they are all neutralized! When all of the holes are filled, there is no longer any positive attraction near enough to the electrons in the emitter (N) to cause them to begin stampeding. In fact, as soon as all of the holes are filled, no more current flows (frame B of Fig. 43-4). Even the pressure of the B battery is not great enough to force electrons through the base when there are no empty holes.

What could you do to make about 50 more electrons flow from the emitter to the collector? If you said put one more extra hole into the base, you are right. Here is where the C battery comes in. The C battery is turned so that its positive end is closest to the base of the transistor. This means that when the potentiometer (R_1) is adjusted correctly, the C battery can draw electrons out of the base region. When an electron is removed from the base, it leaves a hole behind. Each new hole in the base will entice about 50 more electrons to stampede through to the collector and become main current flow.

A tiny current from emitter to base (through the C battery) can control the much larger main current (B battery current) from emitter to collector. By adjusting the potentiometer (R_1), you can draw a few, many, or no electrons from the base. Whatever you do to the number of electrons leaving the base, there will be about 50 times that much change in the number of electrons flowing through the load (R_L). This is **amplification**. You have controlled a large current with a smaller one.

How Does a PNP Transistor Work?

You may wonder how a PNP transistor works. The answer is that it works the same, but sort of backward. Figure 43-5 shows a circuit with a PNP transistor. The main (B battery) current is always trying to flow, from the collector to the emitter. This main current is controlled by the smaller current in the base circuit. By adjusting the potentiometer (R_1), you can vary the small base current. The PNP transistor will use this input signal (the adjustment of R_1) to control the main current.

Hole Current

Many books and reference materials that you may read discuss how PNP transistors work in terms of conventional (hole flow) current. This may confuse you at first, but consider that each electron is met by a hole when it flows forward and it leaves a new hole behind it when it moves. You could imagine it as the holes rushing across the base from the emitter to meet the electrons at the collector. In the NPN transistor, the emitter emitted electrons, but in a PNP it emits holes!

The important principle is that a change of a few more or less electrons in the base will control a much larger flow of electrons (and holes) between the two P regions. The other important thing to notice is that the C and B batteries are both turned in exactly the opposite directions as compared to the ones in the NPN transistor amplifier. Compare Figs. 43-3 and 43-5. Notice the battery polarities and the direction the current flows in each circuit.

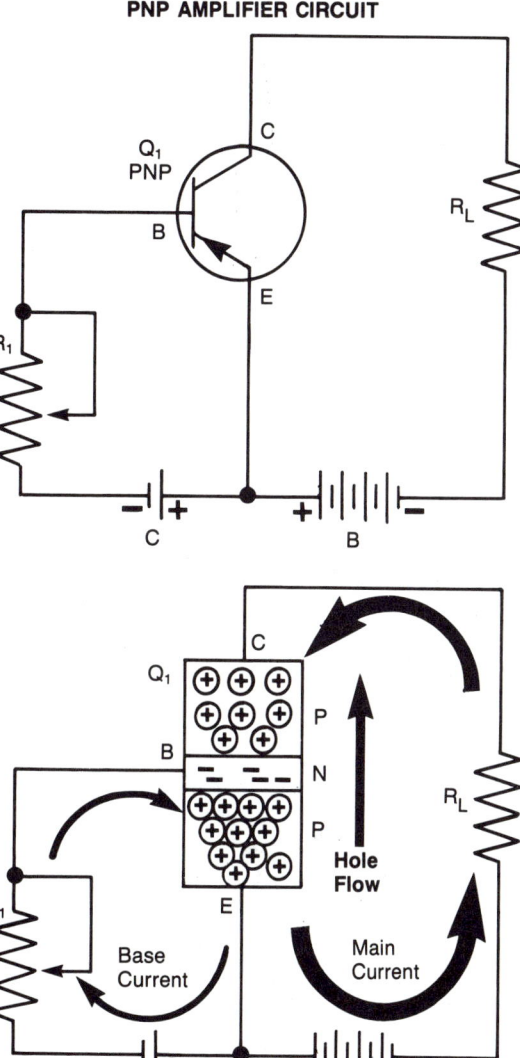

Fig. 43-5 A PNP transistor can amplify too. Inside the transistor, holes will be moving in the opposite direction to the electrons.

An Actual Transistor Amplifier Circuit

Figure 43-6 shows two stages of audio amplification from a television set. The input signal comes from the audio detector circuit on the left. Capacitor C_1 and potentiometer R_1 form the tone control. When R_1 is adjusted to a high resistance, C_1 has little effect. If R_1 is changed to a lower resistance, then higher frequency sound signals (which the capacitor can easily pass) are allowed to go to ground through C_1. This gives the output less treble and allows the bass to sound "boomy." Capacitors C_2 and C_3 are used to pass the signal current (AC in-

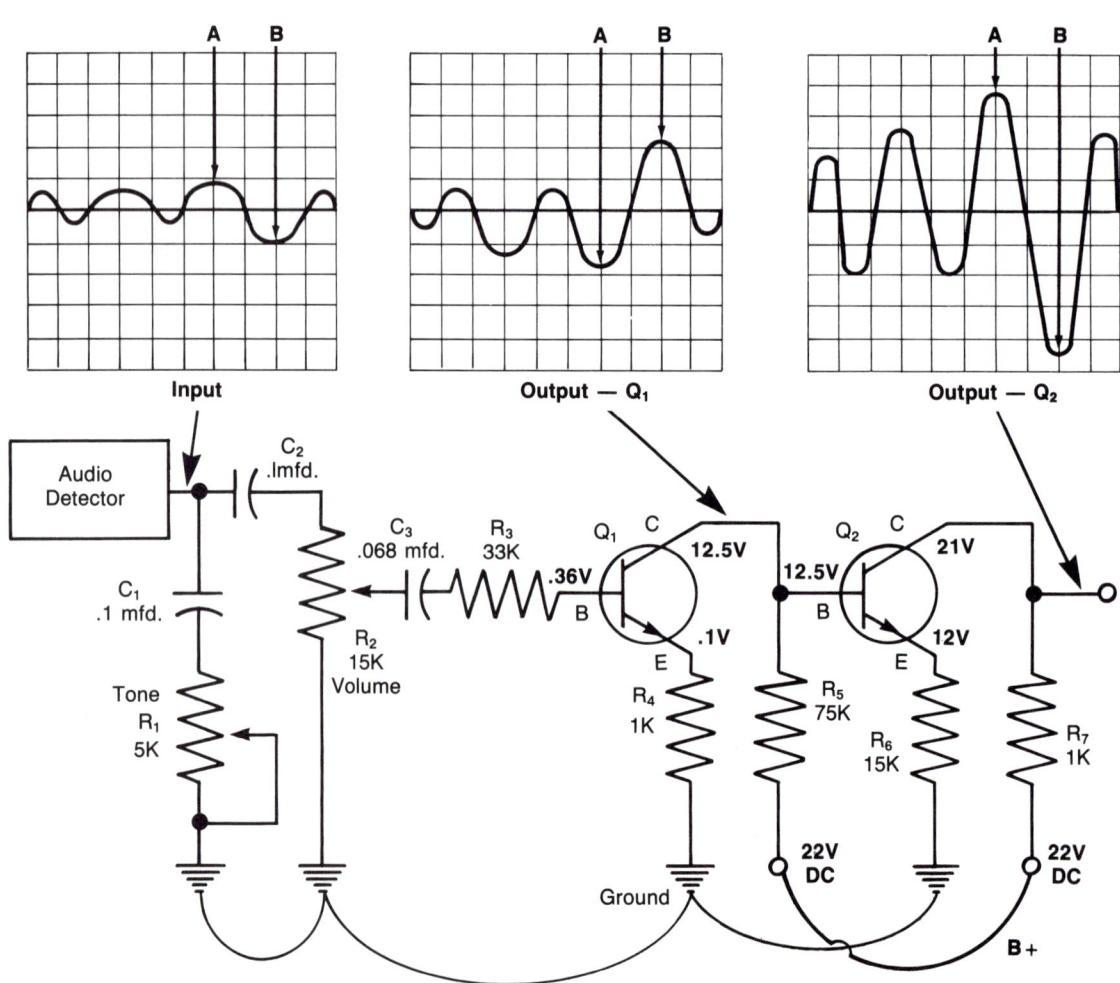

Fig. 43-6 In a real circuit, resistors drop the B+ (source) voltage from the power supply down to the proper levels needed to bias each transistor. Notice the effects of peaks A and B of the input signal on the outputs of Q_1 and Q_2. Each transistor amplifies the signal and inverts it.

formation), but block the DC operating currents from going between the sections (stages) of the circuit. The potentiometer labeled R_2 adjusts how much input signal is fed to the transistors. It is a volume control. The signal enters the base of transistor Q_1 through R_3. The "main current" of Q_1 flows from ground, through R_4 to the emitter, through the transistor and then back to the B battery (power supply — +22 volt DC source) through R_5. Thus, R_5 is the "load" resistor for this stage of the circuit. The tiny input signal current controls this larger current through R_5, exactly as was explained for Fig. 43-3. The amplified signal leaving Q_1 from the top of R_5 is fed to the base of Q_2. Here it serves as an input to Q_2. Notice that R_7 is much smaller than R_5, but they are both connected to the 22 volt source (B+). The voltage at the collector of Q_1 is much lower than at the collector of Q_2 because of the different values of these two resistors.

In order for an NPN transistor to work, it must be properly biased with a higher collector voltage than its emitter and base voltages. In this type of circuit, the base voltage should be very close to, but a little higher than the emitter voltage. The collector voltage should be much higher than both E and B. The resistors on the emitters and collectors work together to make the voltages correct. The emitter resistors are often called bias resistors. The main current for Q_2 goes from ground to R_6, through the transistor by R_7, and then back to the B+ (22 V DC) source. The signal from Q_1 controls the main current of Q_2. Thus it is amplified again. This means that the output of Q_2 is much, much larger than the input signal was to Q_1. It has also been inverted twice — once by Q_1 and again by Q_2 — so that it is now back in phase. The output is taken from the top of R_7 and sent on to other amplification stages before going to the speaker. A few other resistors which help in the operation of the remaining stages were omitted to simplify this part of the amplifier circuit.

Advantages of Transistors

Vacuum tubes work well and can do about the same things that transistors can do, but they are large, and their filaments (heaters) waste a lot of energy just to heat them. Heat is usually a form of energy waste in electronic circuits.

Transistors have many advantages. They are cheaper to produce, smaller in size, operate at lower voltages, and do not need to be heated. The small, lightweight electronic circuits known today would be impossible without transistors, because the power supply alone in most tube versions of the same circuit would weigh several pounds and use far too much current to be operated by small batteries.

Summary

Transistors come in two main types: NPN and PNP. Both types amplify well, but the circuits where they are used differ. Transistor amplification circuits use a small (C) current to control a higher (B) current — much like the triode tubes. Transistors are smaller and lighter than tubes, and they require much less power. The next two chapters discuss light sensitive and special semiconductor devices.

Important Terms

transistor
PNP
NPN
emitter
base
collector
amplification

Review Questions

1. How can you tell which way the current will flow when looking at a transistor's schematic symbol? Which type is the NPN?

2. Which parts of an NPN transistor are most similar (in function) to the cathode, grid, and plate of a triode vacuum tube?

3. Refer to Fig. 43-3 and explain how an NPN transistor amplifies. Tell the purposes of R_4 and R_5 in Fig. 43-6.

History: Bardeen, Brattain, and Shockley

The three men who invented the transistor worked for Bell Laboratories. John Bardeen was born in 1908 in Wisconsin. He was the first person to win two Nobel prizes in physics. He shared one for the transistor (1956) and another in 1972 for work with super-conductivity.

Walter H. Brattain was born in China of American parents in 1902. He spent his youth on a cattle ranch. During World War II, he worked on the magnetic detection of submarines for Bell Laboratories.

William B. Shockley was born in London, England, in 1910. He studied in America and earned his doctorate at the Massachusetts Institute of Technology (MIT) in 1932. He helped in the discovery of semiconductor diodes and transistors. His main contribution had to do with doping semiconductors. He was criticized for his personal belief that intelligence is greatly affected by heredity.

These three men changed the shape and direction of the electronics industry. Their combined contribution cannot be overestimated.

Chapter 44
Light Sensitive Components

Photodiodes

You already know about one type of diode which has something to do with light: the LED, or light emitting diode. The LED gives off light when current passes through it. Another diode, the **photodiode**, can do the opposite. When light strikes it, it will allow current to flow. The operation of a photodiode is shown in Fig. 44-1. The battery is always trying to force current through the diode, but the diode is reverse biased and allows no current flow. Whenever **photons**, particles of light, strike atoms in any semiconductor material, they knock bound electrons loose. All **optoelectronic** (light operated) **devices** work on this principle. So, when light photons hit bound electrons in the photodiode's depletion zone, they separate pairs of electrons and holes. The holes move to the negative battery terminal and the electrons go toward the positive terminal — passing through the motor on the way. If enough photons strike the photodiode, the motor will receive enough current to turn because each photon will produce a moving electron. So, if it is dark, the diode will stop the battery from running the motor because it is reverse biased and will not let the current flow in this backward direction. However, when light strikes the diode, it will allow current to flow and run the motor. There will be one electron to form current flow for each photon which enters the diode.

The strange thing about this example is that you could take the battery out of the circuit, and a small amount of current would still be produced by the light striking the diode. The electrical pressure (voltage) would be quite low, so current would only be allowed to flow when the circuit resistance was very low. However, some current would be produced. This is how the photovoltaic cell (solar cell) which you learned about in Chapter 6 works. It is really a special arrangement of photodiodes.

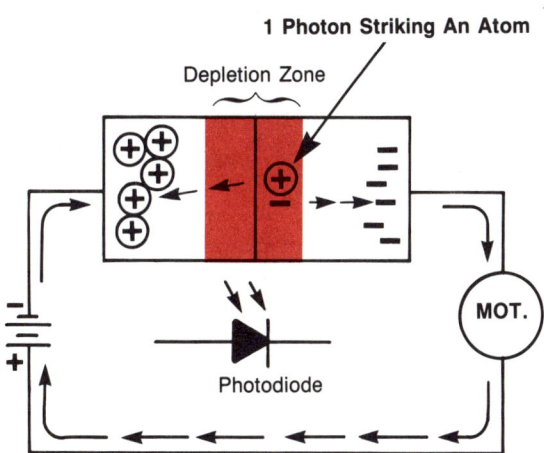

Fig. 44-1 When a photon of light enters the photodiode, an electron is knocked loose to create current flow. Note the schematic symbol of the photodiode and the direction of the "light" arrows.

The special name for photodiodes used to control the flow of current rather than to produce new current is the **photoconductive diode**. Photoconductive diodes and photovoltaic cells are both important devices in electronics.

Phototransistor

The same circuit just described is shown again in Fig. 44-2, except that the photodiode has been replaced with an NPN **phototransistor**. When transistors and their ability to amplify were discussed, it was said that each new hole in the base of an NPN transistor will allow many electrons to flow through the transistor until an electron is captured to fill the new hole. New holes are made by pumping a few electrons out of the base with a C battery.

Look closely at the circuit in Fig. 44-2 and notice that there is nothing connected to the base of the transistor. There isn't even a wire for it! This is because the base is controlled with light. A tiny lens or window in the case of the transistor will let light shine into the base region. The photons of light will each free one electron and one hole, just as they did in the photodiode. What is different, however, is that instead of those single electrons merely becoming a current flow themselves, each one which is freed to flow will leave behind a hole that will encourage 50 or so more electrons to stampede across the transistor trying to catch the hole. These 50 electrons will add to the current flow. So, if you were to shoot five photons into the circuit with the photodiode, five electrons would have flowed through the motor, but with the phototransistor, those same five photons will result in a current flow of about 250 electrons! The phototransistor, then, will permit current flow when light strikes it, but it will also amplify the effect of the light at the same time. The circuit now uses a few photons of light to control a large current flow.

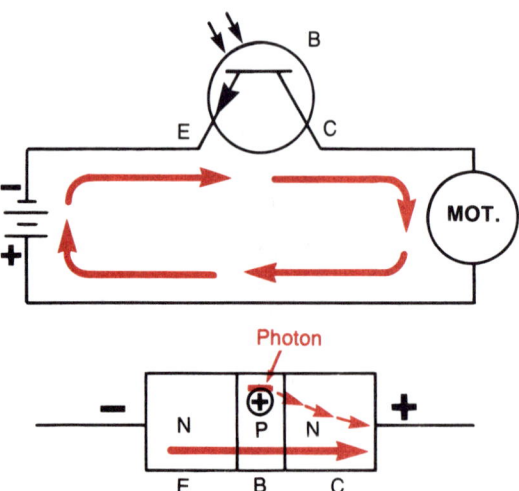

Fig. 44-2 When a photon strikes a bound electron, it separates the electron and leaves a hole. The hole causes current to flow which amplifies the effect of the photon.

Photoresistors and Phototubes

Still another light-activated control device is the **photoresistor** or CdS cell. This is a resistance made with cadmium sulfide which varies in resistive value when struck by light. A typical cell will have a range of about 5 megohms in darkness to 100 ohms in bright light. Figure 44-3 shows a photoresistor and its schematic symbol.

Before semiconductors became available, special vacuum tube diodes called **phototubes** were used in applications that required light sensing.

Uses of Optoelectronic Devices

Some applications of optoelectronic devices are obvious. Jobs like turning on street lights, making the picture on a television brighter when the room is brightly lit, and counting items on an assembly line are easily understood by people with no training in electronics. Another application includes detecting the presence of too much smoke in an industrial exhaust stack by having a beam of light pass through the stack on its way to a phototransistor. There are, however, more subtle uses for these devices that are very important in modern electronics.

Isolation

Often, in complex electronic equipment, such as computers, it is important to be able to pass information (small AC signal currents) from one section of the circuit to another with no danger of passing other currents (such as DC operating currents). This type of isolation is sometimes necessary between computers and the devices they monitor and control. A special electronic device, the **opto-coupler** (short for optically coupled isolator, OCI) meets this need very well. The opto-coupler is actually a LED and a light-sensing device inside the same case. See Fig. 44-4. The LED is aimed so that it will shine on the sensor any time pulses of current in the diode cause it to glow. Thus, the light will carry the important "information" without carrying any other currents or "noise" which could cause interference. Since the "information" in computer circuits consists of a series of very fast pulses of on and off current, the opto-coupler can be made with either photodiodes or phototransistors.

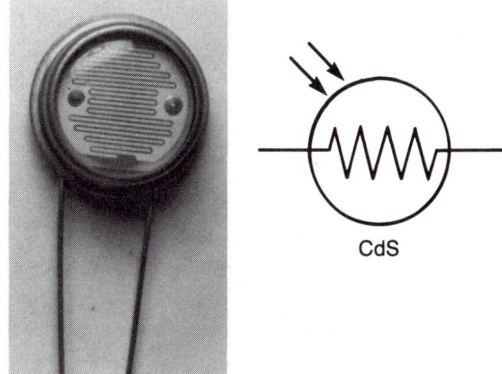

Fig. 44-3 A photoresistive (CdS) cell and its symbol.

A special opto-coupler has the LED and the sensor separated by an open slot. This device can be used to "read" a chopper wheel, as in Fig. 44-5, to detect the motion of a turning shaft. The number of pulses detected per second are used by computer circuits to determine the speed of the shaft's rotation and to keep track of how many rotations are completed. Such mechanisms are used where very accurate control of the turning shaft is important.

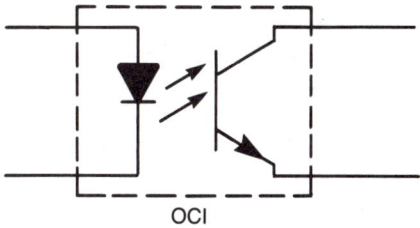

Fig. 44-4 Schematic symbol for an opto-coupler.

Chapter 44 Light Sensitive Elements

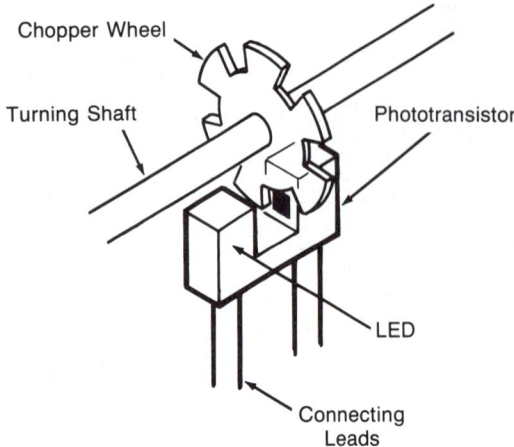

Fig. 44-5 Monitoring movement of a chopper wheel with an open opto-coupler.

Fiber Optics

Another similar use of these principles is the transmission of information over long distances by fiber optic cables. **Fiber optic cables** are made up of many single transparent fibers. If light is shone into one end of a fiber, it will travel through the fiber and cause the other end to glow. The sides of the cable will not be affected very much by the light traveling through the length of the fiber. Several fibers bundled together to form a cable will not interfere with each other. If varying or pulsating light sources are placed on one end of the cable and sensors are placed on the other end, information can be sent through the cable by means of the light beam. The telephone company has made much use of this principle.

Lasers

The **laser** is a highly confined beam of intense light which travels in a straight line. Lasers can be used to do work, like burning cancer tissue in knifeless surgical operations, or they can be used to carry information. The laser carries information like a fiber optic cable, except that no cable is needed when an open "line of sight" exists between the light emitter and the sensor. When the laser must be bent to go around a curve, the beam may also be sent through fiber optic cables. These cables are smaller than the wire cables carrying the same number of lines would need to be. Also, they can carry information signals greater distances than wire cables before the signals need to be strengthened by amplification.

Summary

Though semiconductors and vacuum tubes can detect light, semiconductors, have become far more popular. Both photodiodes and phototransistors conduct current when light strikes them, but phototransistors also amplify the effect of the light photons that strike them. Optoelectronic devices may be used to monitor processes and light levels for computer control circuits as well as to isolate circuits from each other and still pass information. Lasers and fiber optic cables make it possible to carry information over great distances on light beams.

Important Terms

photodiode
photon
optoelectronic devices
photoconductive diode
phototransistor
photoresistor
phototube
opto-coupler
fiber optic cable
laser

Review Questions

1. What is the difference between a photodiode and a phototransistor?
2. What do opto-couplers do? Where are they found?

Project: Light Meter with Bargraph Display

Parts List

1	10-Element LED Bargraph display	D_1
1	LM3914 LED Bar Display Driver IC	Z_1
1	2.2 microfarad capacitor	C_1
1	1.2 K-ohm resistor	R_1
1	3.3 K-ohm resistor	R_2
1	10 K-ohm potentiometer	R_3
1	15 K-ohm resistor	R_4
1	47 K-ohm resistor	R_5
1	10 K-ohm resistor	R_6
1	1 K-ohm resistor	R_7
1	470 ohm resistor	R_8
1	ECG 289A NPN transistor	Q_1
1	ECG 290A PNP transistor	Q_2
1	9 V battery with connector	B_1
1	SPST switch	S_1
1	CdS photo resistive cell	CdS_1

Suitable enclosure, wires, PC board and hardware as required.

Notes

This circuit is very similar to the one in the Lie Detector circuit given in Chapter 35. The main difference is that the test probes have been replaced with a photoresistive cell to measure light intensity. Adjust the potentiometer carefully and mark the setting that gives the best meter range. Then calibrate the meter by marking, beside the bargraph, dots according to the amount of light each one represents. The CdS cell could even be mounted in a "probe" with long leads so that the meter could tell the amount of light in a remote location — such as another room. What other types of sensors could be used with this circuit? Try some other resistive sensors to see if you can design your own special meter.

LIGHT METER WITH BARGRAPH DISPLAY

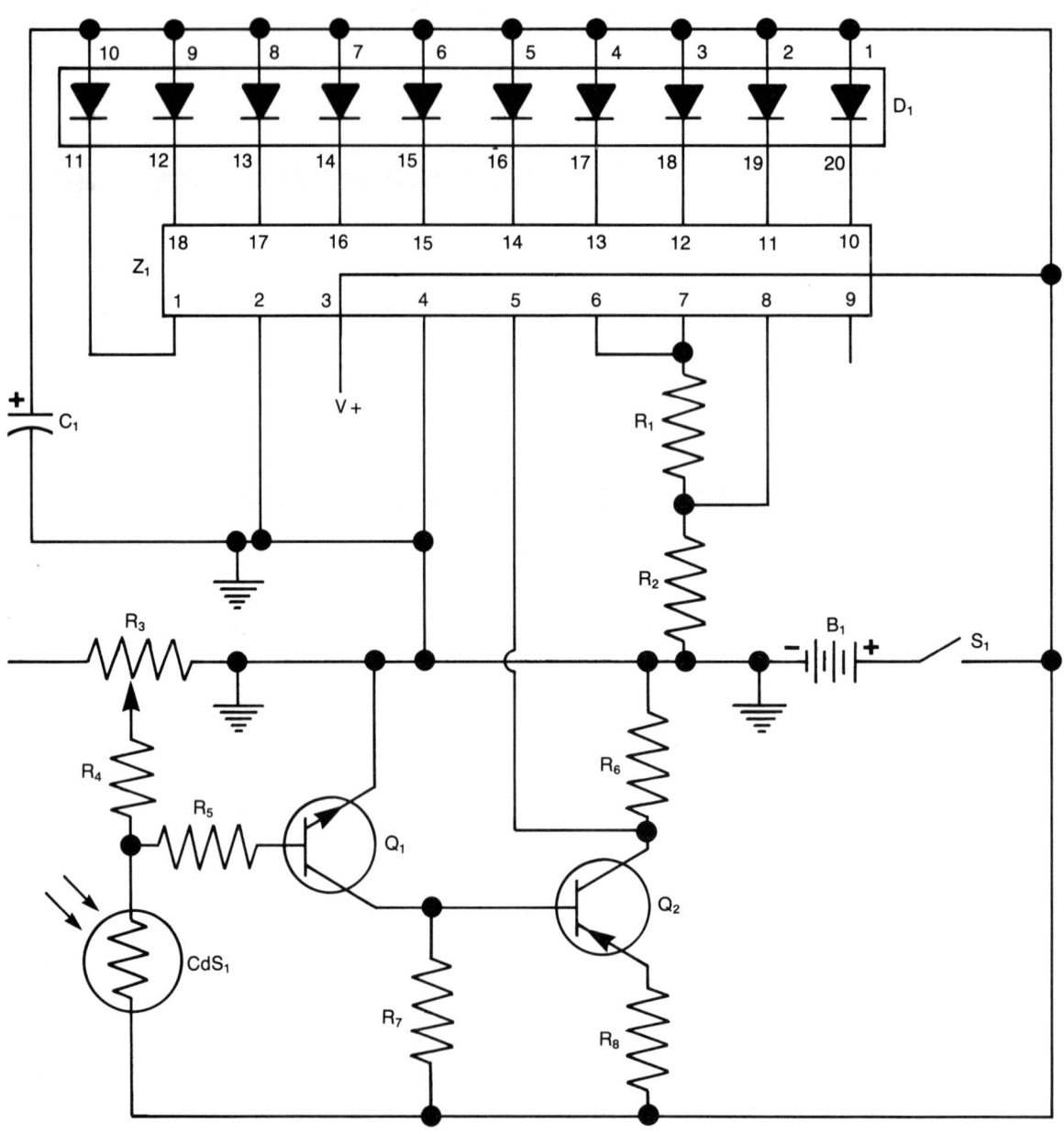

Chapter 45
Other Semiconductor Devices

Another Type of Transistor

The transistors you studied in Chapter 43 are the most common types. There are, however, special transistors that do their job in a slightly different way. Both NPN and PNP are classified as **bipolar** devices. This is because the main current must pass through both P and N regions (two poles) in order for them to do their job. The other type of transistor is the **field-effect transistor**, or FET. Another term for FETs is "unipolar" transistors. To be even more precise, there are two common types of FETs: the **JFET** (junction field-effect transistor), and the **IGFET** (insulated gate field-effect transistor). They work in a similar manner, but there are some differences.

Let's consider the JFET first. Figure 45-1 shows the schematic symbol for a typical JFET and a drawing to illustrate its operation. The current tries to go from the source to the drain through the solid bar of N-type material. It does not have to go through any junctions or jump any barriers between P and N materials. However, the collar of P material around the center of the bar can affect current flow through the bar. The path through the bar which the electrons take is called the "channel."

Figure 45-2 shows what happens if you charge the gate with a negative voltage. The negative charge will reverse bias the junction between the gate (P material band) and the bar of N material. This causes a

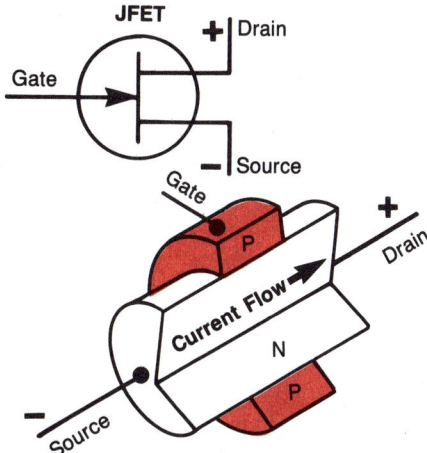

Fig. 45-1 The unipolar JFET transistor does not require its main current to cross any PN junctions.

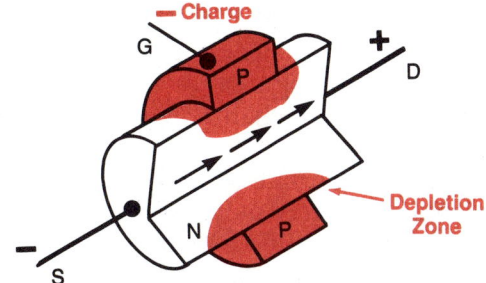

Fig. 45-2 A negative charge on the gate forms a depletion zone which pinches the main current channel. The greater the charge is, the narrower the channel will be.

Chapter 45 *Other Semiconductor Devices* 287

depletion zone (just like the one in any common diode) to be set up near the junction of the two types of material. The depletion zone has fewer free electrons than the normal N material deeper within the bar. It will not allow current to flow through itself. Since the current cannot pass through the depleted area, it will have a narrower channel to pass through.

If you increase the negative charge on the gate (collar), you will make this depletion zone extend deeper and deeper into the bar and "choke" the channel smaller and smaller. The smaller the normal N material channel is, the more difficult it will be for current to go through it from source to drain. If a strong enough negative charge is placed on the gate, you can stop current flow altogether. So, the JFET amplifies by controlling the main (channel) current from source to drain with a varying voltage charge on the gate. The presence of a voltage charge on the gate, with little or no actual current flow into the gate, is all that is needed to control the main current. In this way, the JFET is a little more like a vacuum tube than a regular bipolar transistor is, because the bipolars require some base current for control.

A simpler, and more common, production shape for JFETs is shown in Fig. 45-3. The slab on the bottom labeled "substrate" is just the foundation on which the device is built. It would usually be P material in an N-channel JFET like the one shown. The light color depletion zone is shown. The brighter color current flow arrows mark the channel and the direction of current flow.

A JFET made on a bar of P material with a gate made from N material is also possible. The symbol for this P-channel is the same as the one shown for the N-channel JFET, except that the arrow on the gate will point away from the base. The P-channel JFET works the same, too. However, to understand it, you must reverse your thinking

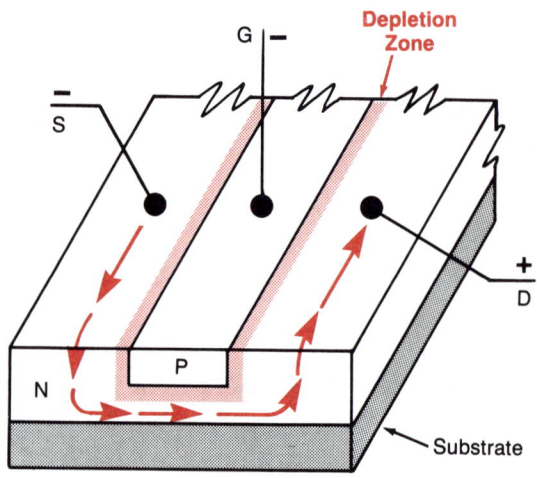

Fig. 45-3 JFETs may be produced more simply in this shape.

and consider the depletion zone choking off the reverse flow of holes instead of the forward flow of electrons.

Insulated Gate FETs

The main difference between JFETs and IGFETs is that the gate in an IGFET is totally insulated from the source and drain. Therefore there is no PN junction between the gate and the rest of the device. The insulation is made from a metal oxide. The entire family of IGFETs and integrated circuits made with this special insulation process are called **metal-oxide-semiconductors** or MOS devices. MOSFETs are the most common types of IGFETs.

The structure of a MOSFET is shown in Fig. 45-4. Notice that the gate of the MOSFET is labeled with both + and − charge signs. This is the main advantage of the MOSFET. In the JFET, if the gate ever becomes positively charged (instead of its normal − charge), the PN junction between the gate and the channel will be forward biased and electrons will be "sucked" out of

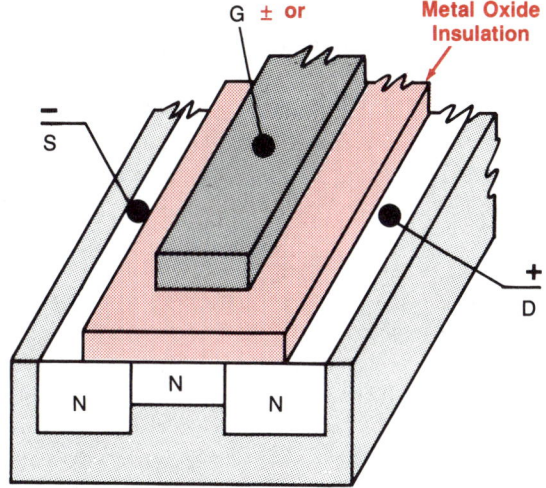

Fig. 45-4 The MOSFET allows the use of both + and − charges on the gate because of its metal oxide insulation.

the channel and through the gate circuit. This would upset normal current flow through the channel.

In the MOSFET, the metal oxide insulator prevents this from happening. It does not matter whether the gate is charged − or +, electrons cannot go through the insulation layer to get to the gate. Notice also that the channel is thinner than the source and drain areas and separated from them by junction lines. The channel is diffused into the substrate material separately and it is not doped quite as heavily as the source and drain. In operation, if the charge on the gate is negative, it repels electrons away from the gate area. This does not permit them easy entrance into the channel and restricts current flow from source to drain.

On the other hand, if the gate is charged positively, electrons are attracted into the channel area where they can become influenced by the stronger + charge of the B+ battery (connected to the drain) and then be drawn through the channel. So, + charges on the gate increase main current flow and − gate charges decrease it. You can control current flow over a wide range by changing the voltage charge on the gate.

The schematic symbols for N-channel MOSFETs appear in Fig. 45-5. The one with the broken center line (B) is called an enhancement type. The ones discussed so far are depletion types (A). The enhancement type has no actual channel of N material. Instead, the electrons must force their way through the substrate in response to the charges placed on the gate. Again, these are all N-channel devices. P-channel ones would have the little arrows pointing outward from the gate and would require the B+ battery to be connected to the source instead of the drain.

The Importance of MOSFETs

When integrated circuits are made, small size is very important. MOS devices are much smaller than bipolar ones, so many more transistors can be produced on a single IC chip by using MOS technology.

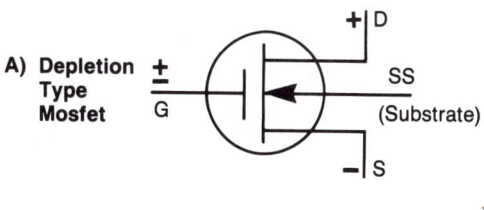

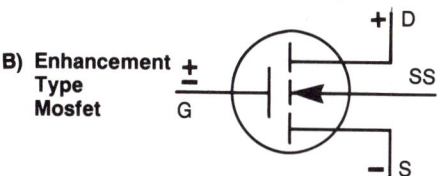

Fig. 45-5 Two N-channel MOSFETs. The SS terminal (substrate) may be internally connected or labeled "B" (for bulk). P-channel MOSFETs would have the arrow pointing outwards.

Chapter 45 Other Semiconductor Devices

There is a drawback, though. Bipolar transistors are pretty rugged electronically. MOS devices can be ruined very easily by the release of static charges. It is very important not to touch the connection pins (leads) of MOS devices with your hands to prevent accidental destruction. The chips usually come packaged with their pins surrounded by conducting foam. Ground yourself and your soldering iron. Avoid removing the foam until the last possible minute to protect these electrically fragile chips.

Silicon Controlled Rectifiers

The schematic symbol and an illustration of a **silicon controlled rectifier** (SCR) are shown in Fig. 45-6. When there is no voltage charge on the gate, no current can flow between the cathode and anode in either direction. This is because one of the three junctions will be reverse biased either way. However, if current is trying to flow from the cathode to the anode, and a quick pulse of positive current is applied to the gate, the SCR will "turn on" and conduct current. This will not happen if the main current is trying to go in reverse (from anode to cathode). One quick little pulse of gate current will turn the SCR on (if forward biased), but it will not turn off when the gate pulse is lost. This is different from a transistor which needs continuous base current to maintain conduction. As long as the main current continues to flow forward, it will be allowed to flow without interruption after just one gate pulse to turn it on. If the main current stops or reverses, it cannot be started again unless the SCR is forward biased and another gate pulse is applied. SCRs are frequently used in control circuits, such as the motor speed control circuit shown in Fig. 45-7. This circuit is good for slowing down electric drills.

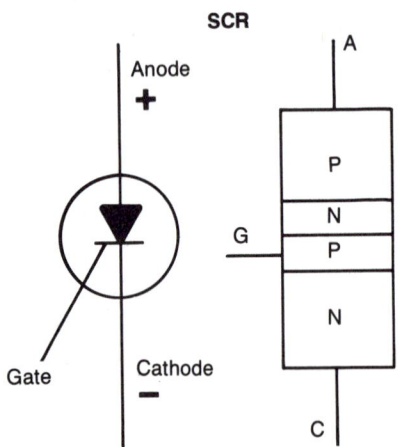

Fig. 45-6 The SCR (silicon controlled rectifier) is a very useful control device.

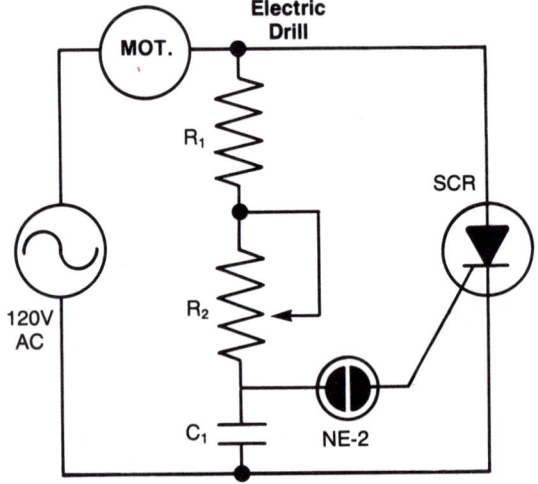

Fig. 45-7 A motor speed control circuit which uses an SCR.

The Triac

One drawback of the SCR is its inability to pass current in both directions. This is an advantage in some circuits, but a serious shortcoming in others. There is another device, a **triac**, which acts somewhat like two SCRs taking turns when AC current reverses. Since the triac allows current flow in either direction, the two ends are labeled main terminals 1 and 2 (MT_1 and MT_2) instead of anode and cathode. Figure 45-8 shows the internal structure of the triac and a schematic that illustrates approximately how the device is similar to two SCRs working together. SCR 1 conducts in one direction and SCR 2 conducts in the other direction, but they both need only one triggering pulse from their shared gate to turn them on.

Summary

In addition to regular bipolar diodes and transistors, there are unipolar devices and other special semiconductors. JFETs and MOSFETs are important because they can use voltage charges (without gate or base current flow) to control their channel (main) current. MOS technology is important because it allows smaller size and greater complexity in IC chip design, but MOS devices can be ruined by static charges. SCRs and triacs will continue to conduct after one quick gate pulse turns them on, unless their main current falls below the needed level. The SCR only passes main current in one direction, but the triac allows current to flow both ways.

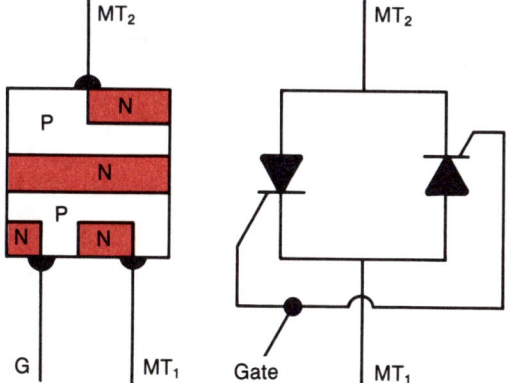

Fig. 45-8 The triac is much like 2 SCRs working together and sharing one gate.

Important Terms

bipolar
field-effect transistor (FET)
JFET
IGFET
metal-oxide-semiconductors
MOSFET
silicon controlled rectifier SCR
triac

Review Questions

1. How does a JFET work? What is the difference between this and a regular bipolar transistor?

2. What is unique about MOSFETs? Why are they used in integrated circuits?

3. Describe the operation of a typical SCR. How does the triac differ from the SCR?

Careers: Aerospace Careers

Space exploration has opened up a wide variety of careers unheard of when your parents were born. These careers include technicians, engineers, scientists, skilled machinists, model makers (for wind tunnel tests), doctors, public relations specialists, administrators, and countless others who support the famous astronauts who actually fly the space vehicles.

Training for these careers varies greatly with the postion, but competition for job openings is high because of good pay and benefits and the glamor associated with the space program. Electronics is a very important aspect of the space program — many advances in modern electronics are direct outgrowths of research in space.

If you want to become involved with the space exploration program, contact the nearest National Aeronautics and Space Administration (NASA) facility. Training usually will include science, mathematics, electronics or some other skill area. Good communication skills are also necessary.

Systems and Applications

Section 8

Chapter 46

Digital Electronics

Digital Circuits

Many modern wonders of the electronic age are made possible by special circuits called **digital circuits**. As the name implies, these circuits have something to do with numbers (digits). Digital circuits process and use numbers and make decisions. The numbers must be converted to a code of 1s and 0s (called binary) before the digital circuits can use them. Digital circuits are the basis for modern computers, calculators, electronic watches, and many control devices.

The AND Gate

The simplest of all logic circuits is the **AND gate**. Figure 46–1 shows what an AND gate does by using an example circuit made up of switches. The **truth table** for an AND gate and its schematic symbol are also shown. The truth table shows what the output of the circuit (gate) will be with all possible inputs. The A and B switches are the inputs. Point C in the circuit is the output. Any time that point C has a positive charge, the light will burn and you say our output is 1. When no positive charge is available at point C (when either or both of the A and B switches is open), the output is 0 and the lamp will not light. In the truth

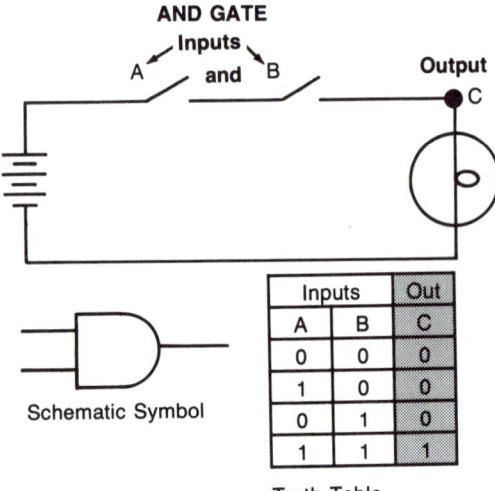

Fig. 46-1 An example circuit for explaining the AND gate. The truth table shows what will happen to the output under all possible conditions of the inputs.

table, a closed switch (positive pulse) and positive outputs at C are represented by the number 1. The number 0 represents an open switch or no output. These two conditions, 1 and 0, could also be labeled yes and no, go and no-go, on and off, + and −, or any other convenient two-choice labels. The 1 and 0 are commonly used in digital electronics. Digital gates and circuits may process numbers and other types of information if they are first converted to this **binary** (1s and 0s) **code**.

No current will flow if switches A and B are both open (0 and 0). There will be no positive output at C. This is the first line of the truth table. Likewise, if only one of the two switches is closed, no positive output will appear at C. The center two lines of the truth table show these two possible conditions. The bottom line shows what happens when both A and B are closed. So, the job of an AND gate is to give an output of 1 when all of the inputs are also 1. There could be more inputs than two (additional switches in series), but they would all have to be closed (have an input of 1) for you to get an output of 1. If any switch is left open, the output will be 0.

Figure 46-2 shows diodes and transistors used to produce AND gates. The other gates discussed in this chapter can all be made with semiconductors, but switches will be used as examples for the sake of simplicity.

A

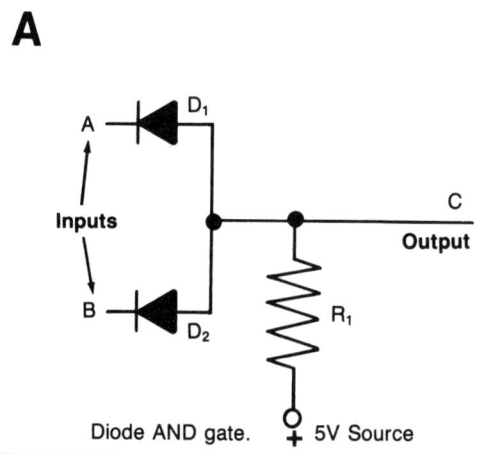

Diode AND gate.

INPUTS		OUT	COMMENTS:
A	B	C	
0	0	0	— pulses at A & B allow D_1 & D_2 to both conduct current to the +5V source, so R_1 drops the full 5V and output at C = 0
1	0	0	+ pulse on D_1 cuts it off, but — pulse on D_2 allows it to still conduct, C = 0
0	1	0	Reverse of above, D_1 is conducting and C still = 0
1	1	1	+ pulses on both D_1 AND D_2, both diodes are cut off. No current flows and the full pressure of the 5V source is available as an output at C

B

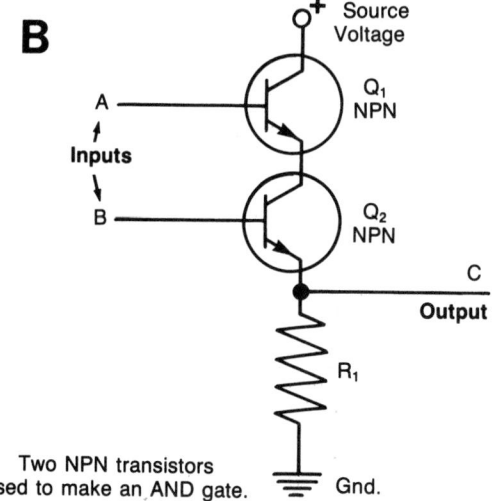

Two NPN transistors used to make an AND gate.

INPUTS		OUT	COMMENTS:
A	B	C	
0	0	0	Both transistors are cut off by — pulses on their bases
1	0	0	+ pulse at A so Q_1 does conduct, but current still cannot flow from ground to the + V source because Q_2 is still cut off
0	1	0	Reverse of above, Q_2 now conducts, but Q_1 cut off
1	1	1	+ pulses at both A AND B, both transistors conduct current from the ground to the +V source. So, the + V drop across R_1 can be detected at C as an output.

Fig. 46-2 Real AND gates are made with transistors or diodes.

The OR Gate

By putting the two switches in parallel with each other, you can make an **OR gate**. The OR gate, its truth table, and schematic symbol are shown in Fig. 46-3. The output of the OR gate is very different from that of the AND gate. If both switches are open, the output is 0. However, there will be an output of 1 if either one or both of the switches is closed.

In real circuits, transistors and diodes would be used to make the decisions represented by the switches in the example circuits. It would be very tedious to draw the number of gates needed in most actual digital circuits. These symbols simplify the schematics of digital equipment greatly.

The Inverter

An **inverter** is a circuit with an output that is the opposite of its input. The typical transistor amplifier circuits discussed in Chapter 43 are inverters because their outputs are 180° out of phase with the inputs. In digital circuits, amplifiers are often used to control one current with another one of the same size or to just invert a signal rather than to gain greater power. The schematic symbol for an inverter and its truth table are shown in Fig. 46-4. Notice that it only has one input and one output. The bar over the A for the output tells you that the output is the inverted form of the input. The bar means "not," as in the example which would be read "not A."

Inverters may be used with other gates (or designed into them) to make different outputs. The opposite of the AND gate is the **NAND gate**. You can read this as "the Not AND gate." A NAND gate can be made by placing an inverter directly behind a regular AND gate, as the schematic symbol shown in Fig. 46-5 infers, or by using a special circuit arrangement. Notice that the output columns of the truth tables for the AND and NAND gates are exactly opposites. The reason that the output is labeled C BAR ($\bar{C}$) instead of plain C is shown by the shaded example in the figure.

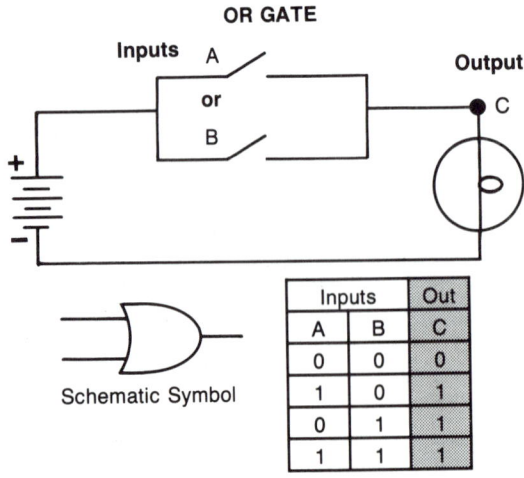

Fig. 46-3 The OR gate and its binary coded truth table.

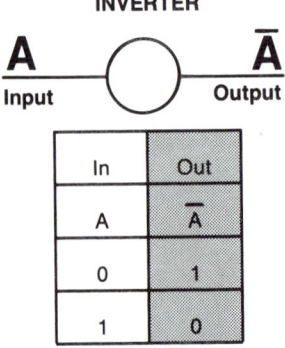

Fig. 46-4 An inverter gives an output which is the opposite of its input.

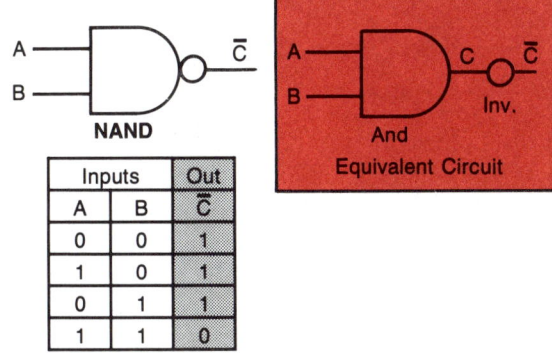

Fig. 46-5 The NAND gate is the logical opposite of the AND gate — it is an inverted AND.

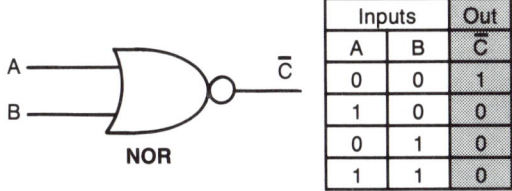

Fig. 46-6 The NOR gate is a "Not OR" gate — the logical opposite of an OR.

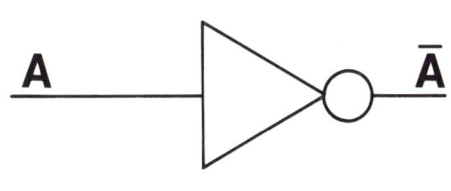

Fig. 46-7 Inverters usually are shown as inverting amplifiers when they "stand alone" on schematic diagrams.

You may wonder whether there is a **NOR gate** — since there is a NAND gate. Yes, one is shown in Fig. 46-6 with its truth table.

When an inverter is found in a schematic without directly being a part of any other gate, the symbol for an amplifier is usually placed with it. This is because the inverter is often an amplifier itself which just inverts. The symbol for an amplifier is a triangle, so the symbol shown in Fig. 46-7 is often used for an inverter.

It is also possible to use inverters on the inputs of gates to change their operation. Figure 46-8 shows that an AND gate may be forced to operate as a NOR gate. Compare the truth table for this converted NOR gate to those of normal AND and NOR gates. (The dotted lines merely show where the inputs and outputs are labeled.) Other gates also may be altered by inverting their inputs — see if you can draw the symbol and complete the truth table for a NAND gate made by inverting the inputs of an OR gate.

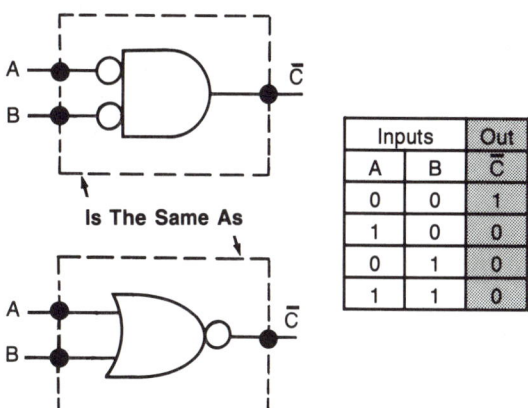

Fig. 46-8 By inverting the inputs, an AND gate can be used as a NOR gate.

The Exclusive OR and NOR Gates

An "exclusive" gate does a similar job to that of a regular OR or NOR gate, but it reverses the output when *both* A and B inputs are "on" (set to 1). The symbols for an exclusive OR gate **(XOR)** and an exclusive NOR gate **(XNOR)** are shown in Fig. 46-9 with their truth tables.

The Flip-Flop

One other basic device is needed to make digital circuits. The **flip-flop** is a simple device which flips or flops from an output of 1 or 0 to the opposite state every time it gets an input pulse. The simplest type of flip-flop would have only one input and one output, but such a device would often not be useful. The most commonly used type is the J-K flip-flop. A J-K flip-flop and its truth table are shown in Fig. 46-10. It has seven leads, and these are their functions:

• Q = The main output of the flip-flop. This output will flip from 1 to 0 to 1, etc., as instructed by the device.
• $\bar{Q}$ = An inverted output. Whatever state Q is in, $\bar{Q}$ will be the opposite.
• CLK = Clock. The **clock** is an input which will be continuously pulsing to 1, then 0, and then back at a steady rate. The clock is an important part of any digital circuit because it makes sure that all decisions affecting each other are made at the proper time. The J-K flip-flop requires that the clock be at a certain state for all decisions.
• PS = Preset. This is a special input which will preset the output (Q) to the logic state of 1.
• CLR = Clear. This is another special input which will clear the flip-flop. A pulse (1) on the CLR input will set the output (Q) to 0.

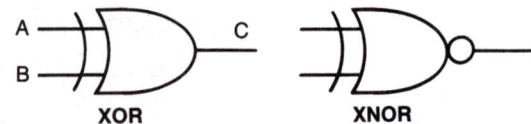

Inputs		Out
A	B	C
0	0	0
1	0	1
0	1	1
1	1	0

Inputs		Out
A	B	C
0	0	1
1	0	0
0	1	0
1	1	1

Fig. 46-9 The exclusive OR and NOR gates (XOR, and XNOR) are useful in some circuits.

• J = Main Input number one. The "information" the flip-flop is to use will be put into it on two inputs: J and K. Input pulses to the J and K inputs are ignored unless the CLK is at 1. You would say that J and K are enabled by a 1 CLK pulse. The PS and CLR inputs do not depend on clock pulses — they override the clock.
• K = Main Input number two. If the CLK is at 1 and a 1 pulse is sent at the same time into both J and K, then the Q output will either flip or flop.

The truth table shows what happens when the J and K inputs are pulsed. Note that the CLK must be in the logic 1 state for the inputs to have any effect. When the flip-flop flips or flops, it **toggles**. This word has the same meaning as a toggle switch. The flip-flop will stay in the condition to which you toggle it until it is purposely changed. The J-K flip-flop will toggle (change from whatever its present state is to the opposite state) each time that it gets inputs of logic 1 from all three inputs: J, K, and CLK. This is the main job of the flip-flop — to toggle when commanded to do so by the inputs.

298

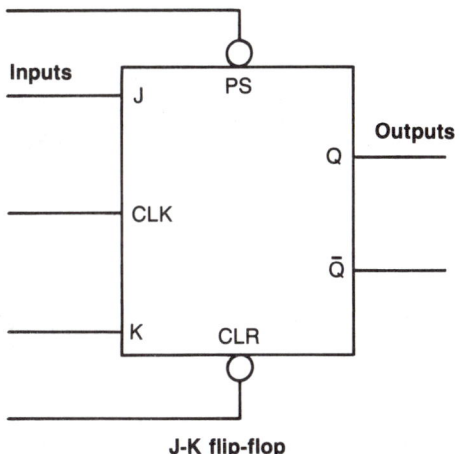

Fig. 46-10 The Q output of the J-K flip-flop toggles when inputs CLK, J, and K are all given a 1 pulse together.

Using Gates to Make Decisions

So far, only the use of individual gates to make simple "yes and no" (1 or 0) decisions has been discussed. To make more complicated decisions, as done in a computer, many gates must be linked together, and all decisions must be converted into the "1 or 0" format which the gates can handle. The "1 or 0" coded numerical information is called **binary numbers**. Binary means that there are two. Two possible states exist: 1 or 0 (yes or no). All of the numbers you normally use can be rewritten in the binary form. For instance, the number 5 may be coded into binary as 0101. The binary code for the number 2 is 0010.

Figure 46-11 is a circuit which can count from 0 to 16 by the binary method. In this circuit, the CLK input of FF-A is used as the main input for the whole circuit. In other words, it is counting the number of steady clock pulses. The four lamps labeled A, B, C, and D will be the output in the example. The lamps will display the number counted in the four place binary code. You can also call this a four **bit** code.

Let's begin the explanation of how the circuit works by applying a pulse to the CLR (clear) terminal of all four flip-flops. (The CLR, PS, and Q̄ terminals are not shown — but they do exist). This will set all of the Q outputs of the four flip-flops to 0 so that none of the lamps (A through D) will be lighted. Thus, after resetting, the output lamps will display the binary form of the number 0 (no lamps lighted). Notice that each of the CLK inputs is inverted before going into the flip-flops.

When the first positive pulse is sent to the input (IN), it is inverted before it

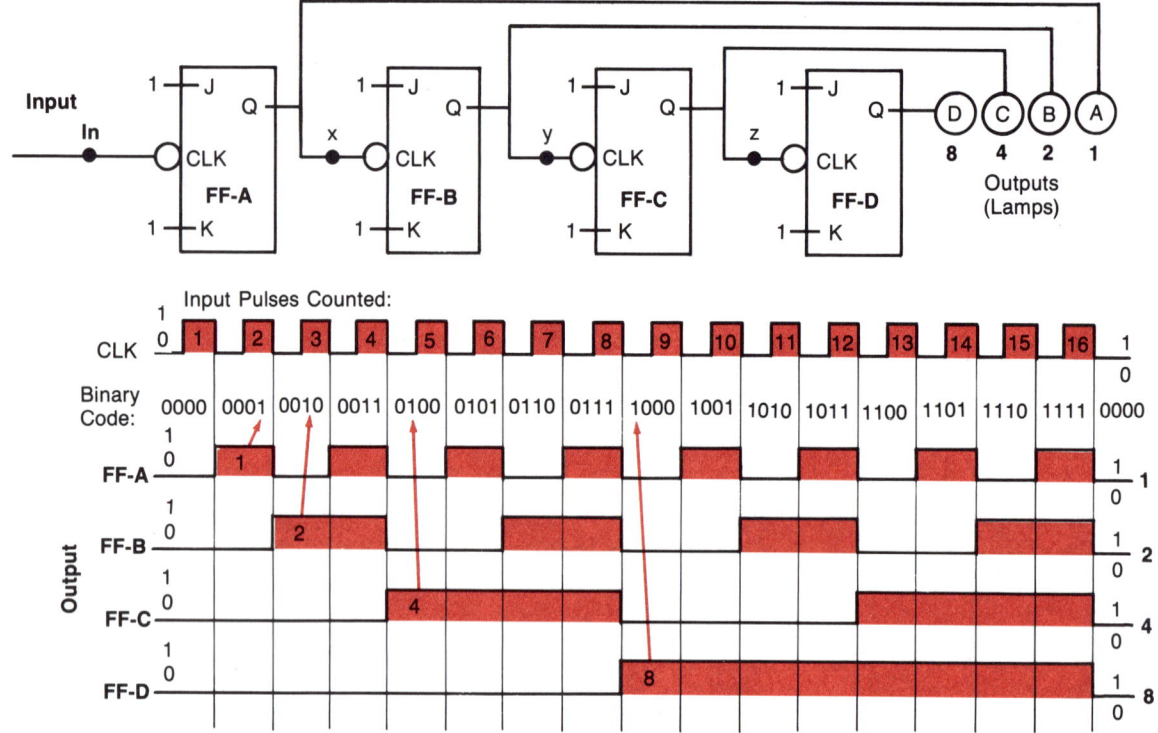

Fig. 46-11 The four flip-flops in this circuit are connected so that they can count the steady clock pulses. The number of pulses completed is displayed in binary code on the four lamps at the right. Inputs J and K are permanently tied to logic 1 to enable the CLK pulses to be counted.

reaches the CLK which interprets it as logic 0. Both the J and K inputs of all four flip-flops are permanently tied together to logic 1 in this example. The 0 pulse on the CLK input has no effect on the output. When the input pulse (at IN) goes "down" to logic 0, its inverted form (1) will enter the CLK input of FF-A and toggle its Q output from 0 to 1. This will light lamp A and supply a logic 1 (high) pulse to the X input of FF-B. The CLK input of FF-B, however, is also inverted, so a logic 1 (high) at point X becomes a logic 0 (low) before it goes into the CLK input of FF-B. Therefore, the Q output of FF-B remains unchanged at 0.

So, the first low (0) pulse you put into the whole circuit at point IN will light lamp A and all other lamps and flip-flops will be unaffected by it. The lamp code will read 0001. This is the binary form of the number 1. Only the A (number 1) lamp is lighted. The code would be read as $0 + 0 + 0 + 1 = 1$. The graph at the bottom of the figure shows the states of the Q outputs (and lamps) of all four flip-flops. You are now ready to enter the second positive (high) pulse. Only the FF-A waveform shows a high pulse at this point.

If you enter a second high pulse, there is no change until the pulse goes down. Then, with this second low pulse, the Q output of FF-A will toggle back to the 0 (low) condition. This Q output of 0 will have two effects. First, it will turn off lamp A.

Second, it will go to point X and then its inverted form (1) will trigger the CLK input of FF-B and toggle its Q output from 0 to 1. That means that lamp number two (B) will now be lighted, but lamp one (A) will be off. The binary code will be 0010 which is the number 2 in binary (0 + 0 + 2 + 0 = 2).

A third pulse will toggle FF-A and have no effect on FF-B or the others (because of the inverted inputs). Thus, the third pulse will relight lamp A and leave lamp B on as well. The code of 0011 represents 0 + 0 + 2 + 1 = 3. The fourth pulse will cause FF-A to have an output of 0. FF-B will also go down to 0. This will turn off lamps A and B and trigger FF-C to toggle and light lamp C (4). The code will read 0100 which is the number 4 in binary (0 + 4 + 0 + 0 = 4). The circuit has now counted the four pulses which you have put into its input (IN) and shown their number in the binary code. Additional pulses will be counted in the same manner. The limit of this circuit's counting capability is 16 because the binary code 1111 (all four lamps lighted) is as high as the circuit can go. That code (1111) would be interpreted as 8 + 4 + 2 + 1 = 15. One more pulse applied after reaching 1111 (15) will reset the whole circuit back to 0000 which can either be interpreted as 0 or as 16.

Mathematical Operations in Binary

The binary system can be used for more than simple counting. Figure 46-12 shows the conversion of numbers between the base 10 system and the binary system (base 2). The figure uses a five bit code so you can handle numbers up to 32 now instead of just 16 as in the earlier examples.

Figure 46-13A shows how to add two binary numbers. When two 1s are added, their value is 0 plus a 1 to carry to the next

Decimal Value of Each Bit

Decimal Numbers	16	8	4	2	1
1	0	0	0	0	1
2	0	0	0	1	0
3	0	0	0	1	1
4	0	0	1	0	0
5	0	0	1	0	1
6	0	0	1	1	0
7	0	0	1	1	1
8	0	1	0	0	0
9	0	1	0	0	1
10	0	1	0	1	0
11	0	1	0	1	1
12	0	1	1	0	0
15	0	1	1	1	1
16	1	0	0	0	0
17	1	0	0	0	1
24	1	1	0	0	0
30	1	1	1	1	0
31	1	1	1	1	1
32	0	0	0	0	0

One way to convert from base 10 to binary: To change the number 5 to binary code, use division by 2 —

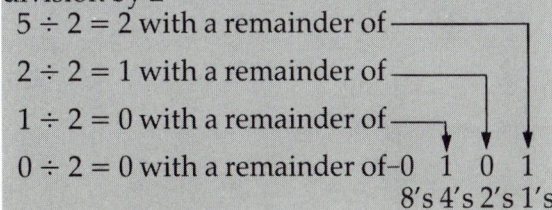

5 ÷ 2 = 2 with a remainder of
2 ÷ 2 = 1 with a remainder of
1 ÷ 2 = 0 with a remainder of
0 ÷ 2 = 0 with a remainder of–0 1 0 1
                                  8's 4's 2's 1's

Fig. 46-12 Selected examples of base ten numbers are shown converted to binary in this table. The number 5 is converted to binary at the bottom as an example of one way to convert from decimal to binary form.

bit. A 1 plus a 0 equals a 1 with nothing to carry, and 0 + 0 equals 0 with no carrying. In the example, in the 2 column you add three 1s. This gives 1 to record as the answer and another 1 to carry. The answer is found by interpreting the coded result of the binary addition. Try adding the numbers 5 and 19 on your paper. Did you get 24 when you interpreted your answer? If not, did you use the correct input codes: 5 = 00101 and 19 = 10011. The coded answer should be 11000.

Numbers may also be subtracted in binary. The example in Fig. 46-13B is divided into five columns (identified with letters A through E) so that you can explain what happens in each column. Column A is easy: 1 minus 1 equals 0, so you record 0 as the answer in that column. Likewise, column B will have an answer of 0. Column C is a little tricky. If these were the regular numbers that you are used to working with, and you had to subtract 1 from 0, you would borrow from the D column so you could complete the operation. You do about the same thing in binary. You find the answer for column C by borrowing from column D. When you do, column C has a value of 2 and column D is left with a value of 0. Then, 2 minus 1 equals 1 for the answer in column C. Column D will now be 0 minus 0 which equals 0. Column E is very easy. Try one on your own for practice.

Even multiplication and division may be done in the binary system. Examples of binary multiplication and division are shown in Fig. 46-14, but this is not quite how digital circuits perform these operations. All decisions made by digital circuits must be reduced to a series of simple "yes or no" (1 or 0) choices. The powerful capability of such circuits comes from the tremendous speed at which thousands of these simple choices may be made. Even very complex mathematical problems, with large numbers, may be solved by digital circuits which use the binary number system. An electronic calculator first converts your base 10 inputs to binary, solves your problem, and then converts the answer back to base 10 so you can read it easily. The circuits operate so quickly that all of this is done almost instantly.

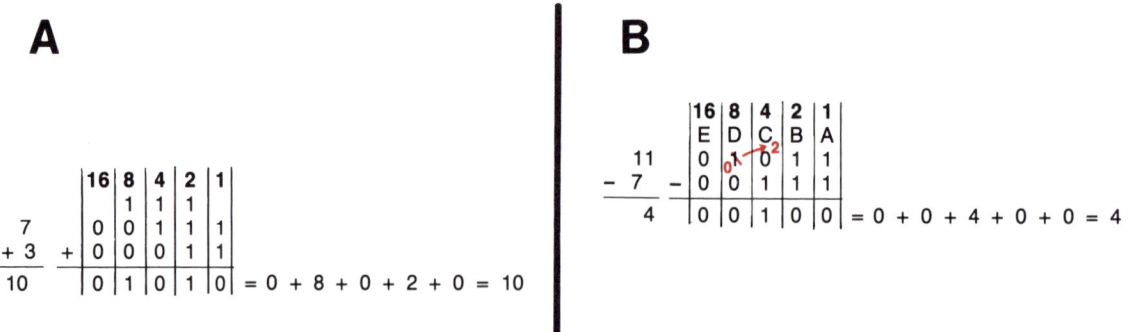

Fig. 46-13 A) Adding two numbers in binary. Notice that you can "carry" in binary when any column has a total of 2 or more. B) Subtraction in binary is simple also. You have to borrow from column D to complete the subtraction in column C.

Multiplication:

```
           16  8  4  2  1
      5         1  0  1
    x 3    x    0  1  1
     ---       ---------
     15        1  0  1
              1  0  1
          + 0  0  0
          ---------------
            0  1  1  1  1  = 0 + 8 + 4 + 2 + 1 = 15
```

Division:

```
                        1  0  1    = 4 + 0 + 1 = 5
   7 ) 35   111 ) 1  0  0  0  1  1
              -   1  1  1 ↓  ↓
                  ---------
                     1  1  1
                  -  1  1  1
                     ---------
                            0
```

Fig. 46-14 Multiplication and division are both also possible in binary.

Summary

Logic gates may be used to make binary decisions. The most common gates are AND, OR, NAND, NOR, XOR, and XNOR. The gates are usually made from transistors and diodes, but they may be represented by simple switches for the purpose of explanation. Inverters and flip-flops are also very important devices in digital circuits. The J-K flip-flop is a very common type. Clock pulses are often used to ensure that decisions are made in a logical order. The next chapter explains how the many transistors, diodes, gates, and flip-flops needed to produce digital circuits may be made onto a single semi-conductor chip called an integrated circuit.

Important Terms

digital circuit
AND gate
truth table
binary code
OR gate
inverter
NAND gate
NOR gate
XOR
XNOR
flip-flop
clock
diodes
binary numbers
bit

Review Questions

1. Draw the symbols and truth tables for the following gates: AND, OR, NAND, NOR, XOR. Try to do this without looking in the book first, then check your logic.
2. Name and explain the functions of the seven leads on a J-K flip-flop. What is the purpose of flip-flops in digital circuits?
3. Convert the following numbers and problems to binary, solve the problems, and then interpret the binary coded answers back to base ten numbers: 3+5 = ? 14−4=? 19+11=? 54+2=? 3×4=?

Chapter 47

Integrated Circuits

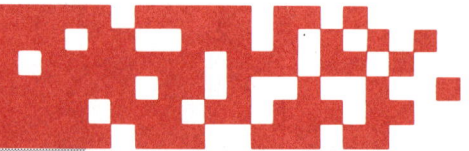

A Whole Circuit on a Chip

Much of the progress in electronics since 1950 has been directly related to three factors: smaller components, faster operating circuits, and lower cost. Many devices you take for granted now (like an electronic, pocket-sized calculator) would have been impossible in 1950 because technology had not learned how to utilize small semiconductors. Even if a calculator could have been built, it would have been very large, would have cost thousands of dollars, and would have operated much more slowly than the present ones do. This chapter discusses integrated circuits — the key to miniaturization in electronics.

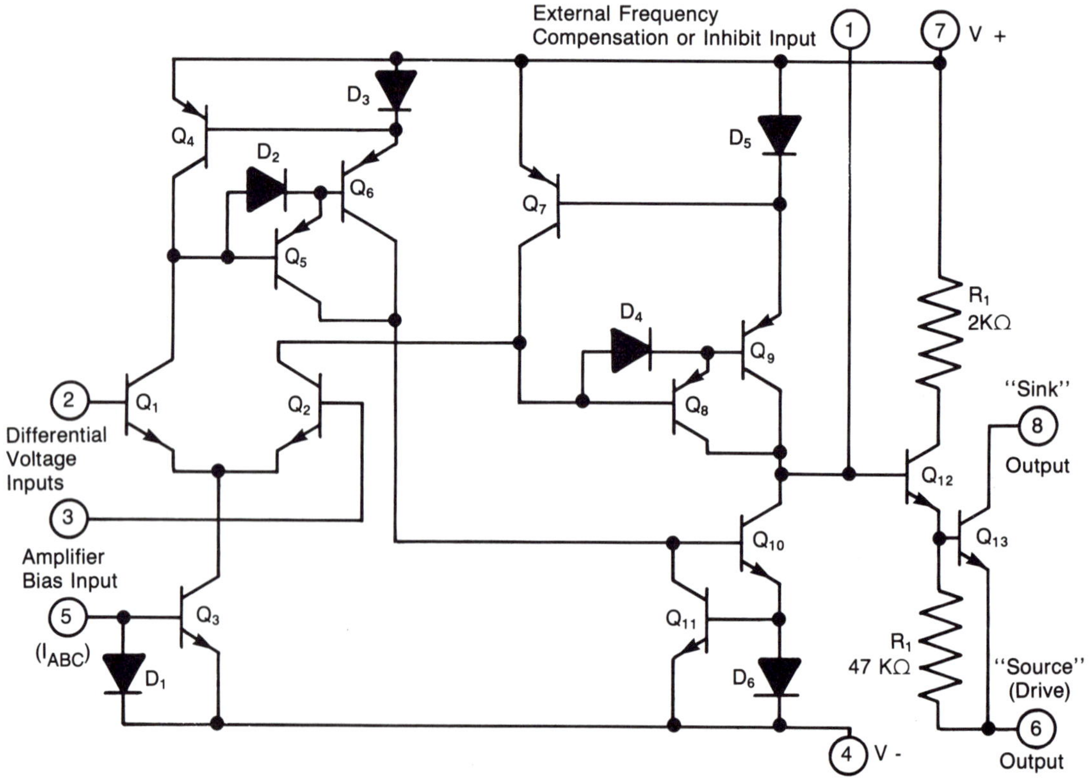

Fig. 47-1 The whole circuit in this schematic is contained on a tiny IC chip.

An **integrated circuit** (IC) is actually a combination of many components on a single semiconductor "chip." See Fig. 47–1. Diodes, transistors, and even resistors may be built into ICs. However, the resistances sometimes take up more space than they are worth and do not have as much precision as regular (discrete) resistors. Capacitors, coils, and transformers require very large amounts of space and cannot be easily contained on ICs. But, transistors and diodes — and the gates, amplifiers, flip-flops, and inverters which you can make from them — are easily produced on IC chips.

In fact, when a single transistor is produced at the factory, it is actually produced on a slice that contains many transistors just like itself. The transistor chips are separated and sold individually. The manufacture of an IC containing several transistors and diodes is about the same, except the transistors produced on the chip may be different types, and they are not separated from each other. Of course, the next step in making smaller and smaller ICs allows you to make many ICs on one slice and then separate them. Figure 47–2 shows a slice that contains more than a thousand ICs before separation and finishing. From here the chips will have tiny leads attached and will be enclosed in a case of plastic or other suitable material.

The actual chips (or slices) are produced by the same methods discussed in chapter 40, except that the masks used to select where various N and P regions are to be deposited have holes in them for many different devices. The N and P regions are formed by diffusion doping or a similar process and switching devices. A typical digital IC might contain several of the same device. Other digital ICs may contain a more complex circuit, such as the 74193 which is a binary counting circuit, all on one chip. Power supplies and circuits or devices which permit the chip to communicate with people or other circuits must be added to make the chip function.

Linear ICs might include amplifiers, oscillators, timers, and other similar circuits. Some popular examples include the 741 OP Amp (discussed later in this chapter), the 555 timer, and the LM386 which is a complete 2-watt stereo amplifier on a chip. Most linear ICs have some analog functions.

Complex ICs like the CPU chip (Central Processing Unit) of a microcomputer (the CPU chip is the main "brain chip" which

Types of ICs and Their Uses

There are two broad families of ICs: **digital** and **linear**. The digital ICs contain mostly gates, flip-flops, and other digital

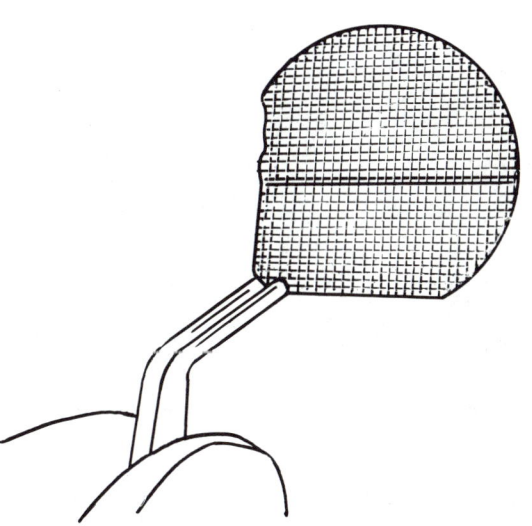

Fig. 47-2 This "slice" contains over a thousand IC chips which will be separated.

Chapter 47 Integrated Circuits

makes the decisions) may be hard to classify as purely digital or linear because they can have some of both types of circuits. These chips are the result of **LSI** (Large Scale Integration) technology. An LSI chip may have hundreds of gates and transistors on a single chip. They usually do have more digital features and functions than they have linear ones. For example, the 8048 microprocessor chip has all of the main circuits to serve as the CPU of a microcomputer together with the timer (clock), some memory, and some other features — all on one chip! A typical CPU chip is shown in Fig. 47-3. The chips around it are mostly memory and interfacing devices. The MOS chips used to make digital clocks and music synthesizers are additional examples of LSI technology.

Operational Amplifiers

An important linear building block often used today is the **operational amplifier** (OP Amp). OP Amps are easily made on IC chips. They are complete amplifier circuits which have high gain and use direct coupling (this means that fewer extra resistors, capacitors, and coils will be needed in the circuits in which OP Amps are used). The schematic symbol for an OP Amp is shown in Fig. 47-4. Notice that there are two inputs to the OP Amp. The input labeled + is a normal input. The input labeled — is an inverting input which is useful in many circuits. Frequently, many OP Amps are used in the same circuit. This is why some IC chips will have more than one OP Amp on one chip.

Other Common IC Chips

In Chapter 41 two common types of ICs were discussed — bridge rectifiers and voltage regulators. Also, opto-couplers were discussed in Chapter 44. Other IC chips are produced for use as computer memory, input/output (I/O), and drive units. Still other chips are used as buffers and circuits for converting data from analog to binary (digital) form or vice versa. Buffers are

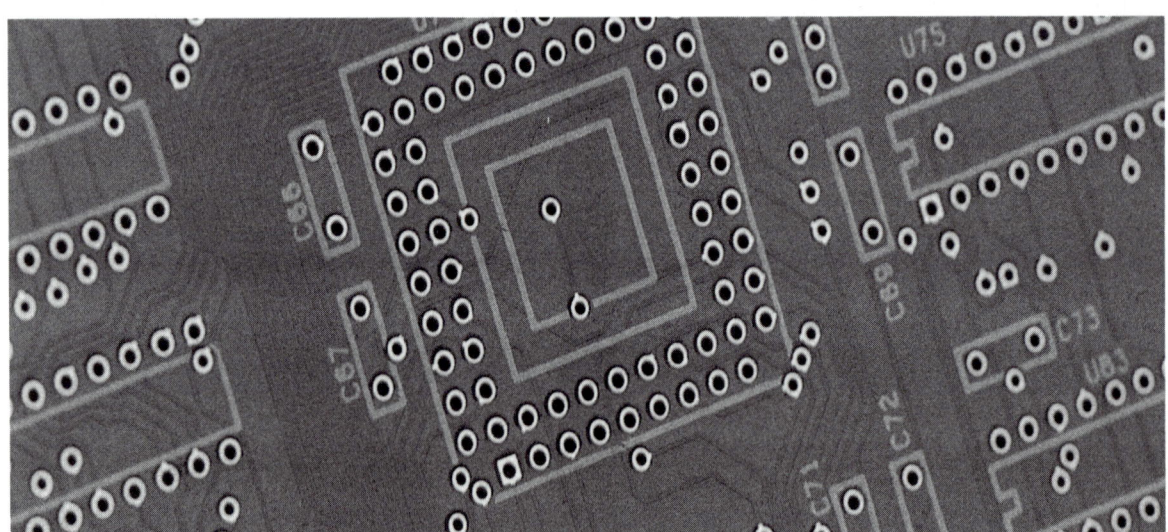

Fig. 47-3 The main circuit board (motherboard) is shown here.

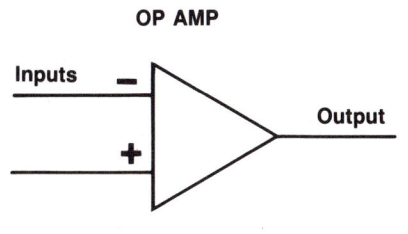

Fig. 47-4 The OP Amp is a very important building block in many types of circuits.

placed between two circuits which normally would not be able to communicate with each other easily. Sometimes they also perform temporary storage or isolation functions. Converters may be digital to analog (D/A) or analog to digital (A/D) depending on their intended use. Some chips also are used to communicate with input and output devices such as keyboards, CRTs, disk drives (for storage), and other I/O devices. Most of the chips which aid in communication, act as buffers, or serve as A/D and D/A converters are known as **interface devices**. Interface means to help in communication by converting from one form to another.

Some special memory chips called EPROMs may be programmed to remember certain instructions permanently unless the user wishes to erase the instructions. The EPROM is a custom memory chip which can be very versatile. They are used when the same instructions are needed frequently by a circuit. Special input circuits called EPROM programmers ("EPROM burners") are used to enter the instructions into the chip.

The instructions will be remembered by the chip and released for use by other circuits until the EPROM is erased. Ultraviolet light can also erase EPROMs, so be careful to keep them covered when used outdoors or in brightly lit areas.

Summary

Integrated circuits have made many important advances in electronics possible. As new technology allows developers to pack more and more devices on a single chip, ICs promise to bring even more modern miracles our way. At the same time, production costs of ICs keep going down to make these aids available to more and more people.

Integrated circuits may be broadly grouped into two types: digital and linear. Digital types include the basic gates, flip-flops, and other devices which operate in binary. Linear ICs, such as the popular OP Amp, can handle analog data. With LSI technology, a whole computer may be built on one chip. Other chips are used as buffers, D/A or A/D converters, and interface devices.

Important Terms

integrated circuit
digital IC
linear IC
LSI chip
operational amplifier
interface devices

Review Questions

1. What are the advantages of using ICs instead of individual transistors and diodes in the production of microcomputers calculators, and video games?

2. Explain the basic difference between digital and linear ICs. Give a few examples of each and explain their functions.

3. Why is it difficult to produce resistances, capacitors, and coils on IC chips? What do the initals LSI mean?

Project: The Simplest Logic Probe

Parts List

1	330 ohm resistor	R_1
1	7404 inverter IC	Z_1
1	LTL-52RG Bi-color LED (Radio Shack # 276-014)	D_1
1	test probe (may be homemade)	P_1
2	microgator clips (or similar)	
	hardware, wires, and enclosure as desired	

Notes

This is probably the simplest logic probe which gives three clearly identifiable indications. Connect the + clip to the high (+5 V) side of the circuit under test and the — clip to ground. Touch the desired test point with the test probe and observe the LED. The indications will be:

- Red = high (+5 V)
- Green = low (0 V)
- Flashing red, then green = pulsing at a rate slower than 15 or 20 Hz
- Yellow = pulsing at a higher frequency rate

The unit is so small and simple to build that you probably can construct it without a PC board, but you may want one to protect the circuit. The finished circuit could be enclosed in a small plastic tube, left open, encased in shrink wrap, or encapsulated in a manner similar to the "Fire and Ice Voltage Tester" (project number 1, in Chapter 3). The test probe can be made in this same way, too.

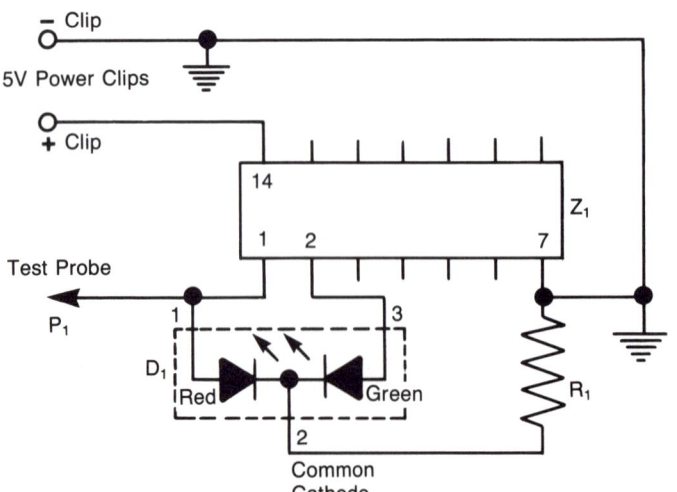
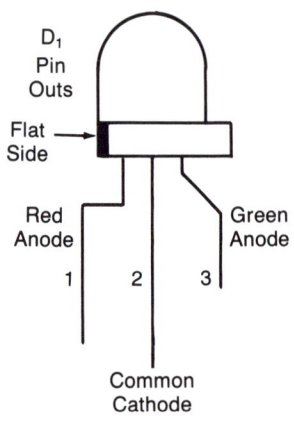

Fig. 47-5 The Simplest Logic Probe is a very handy little test device.

Chapter 48
Computer Hardware and Software

Types of Computers

In the last two chapters, you learned about digital logic gates and integrated circuits. The electronic computer which these components and circuits make possible has had a great effect on people's ability to store and use information and to solve problems. A wide range of computers is available today for industry and the home. Many modern automobiles have computer circuits. Even the popular video game is really a computer circuit which is used for a special purpose. This chapter discusses computers and how we teach the machine to do what we want it to do.

The first thing to consider when selecting a computer for a home or business is the "size" of the computer. Size means far more than mere physical bulkiness. In fact, some very powerful computers are relatively small. Size includes the following factors:

- Amount of computing power.
- Computing speed.
- Memory size.
- Word size handled.
- Memory storage system.
- Cost.
- Special capabilities.
- Physical size.

The computer circuits considered here range from handheld calculators to the large **mainframe** computers used by industry and government. The popular **microcomputer** (Fig. 48–1) commonly found in small businesses and private homes falls between these two extremes. Let's examine the factors listed above by comparing microcomputers to other types of machines.

Computing Power

The microcomputer has much more computing power than a calculator, but considerably less than the large mainframe

Fig. 48-1 This IBM PC is a popular microcomputer for homes and offices.

computers used in industry. There are also minicomputers and midicomputers which are more "powerful" than a microcomputer.

Computing Speed

A microcomputer can do many of the jobs that can be done by larger computers, but it takes much longer for the smaller machine to do the job. Often, the size or the complexity of a job makes it impractical to wait for a microcomputer to perform the task. Faster computers are usually more powerful, and more costly.

Memory Size

A microcomputer may have from 16 K to 512 K bytes of RAM. This statement has two important new terms: **RAM** means Random Access Memory. That is the memory which may be freely written into or read from during the use of the machine. Another type of memory is **ROM** (Read Only Memory) which can only be read from but cannot be written into. The EPROM you studied in Chapter 47 is a special Erasable Programmable ROM chip. **Bytes** are like "chunks" of data. Remember, a single 1 or 0 (binary) decision is called a **bit**. A group of bits, say eight of them organized together, is called a byte. If your microcomputer has 64 K of RAM, it means that it has a random access memory capable of storing about 64,000 bytes of information. Many popular microcomputers have 64 K or more of RAM. More expensive computers usually have more RAM than small ones.

Word Size

The **word size** tells how many bits can be included in each word (chunk of data) handled by the computer. The average microcomputer has a word size of 8 or 16 bits. Large mainframe computers are able to use words as long as 64 bits, while many small calculators only use 4 bit words.

Memory Storage System

Programs and data used by the computer frequently need permanent or semi-permanent storage. The most commonly available form of storage for microcomputers is the **floppy disk** (Fig. 48-2). Floppy disks can store between 80 K and 960 K bytes per disk, depending on the size of the disk and several factors having to do with the format in which the disk stores information. When one disk becomes full, the user may remove it and store additional data on another disk. The disks must be handled very carefully to prevent damage that might result in loss of stored data.

The disk stores data just as audio information is stored on a tape in a stereo system. It is stored in the form of magnetic impressions. In fact, a less expensive system of storage used by many home computer enthusiasts is to store data and programs on cassette tapes. Tapes are also used by mainframes for mass storage, but the tapes and the equipment to "play" them are much larger and more expensive than for home use.

Fig. 48-2 Floppy disks can store a lot of data. The user of this Apple microcomputer is changing the disk.

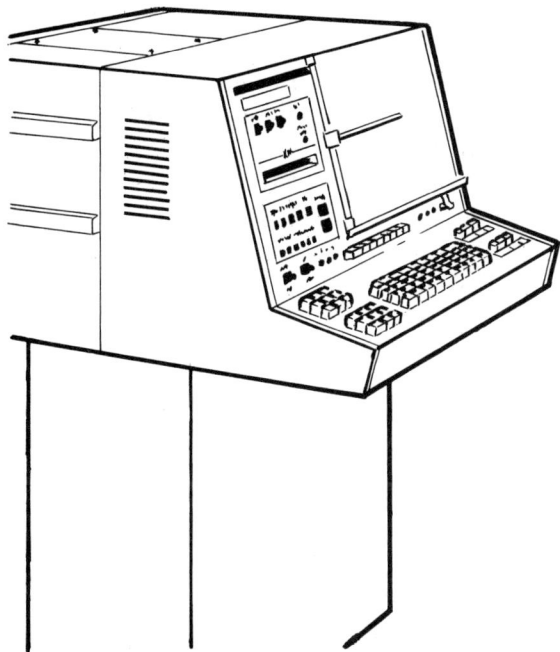

Fig. 48-3 This computerized photo typesetting machine for the printing industry cartridges instruct the computer circuits how to space letters of different type styles.

Information also may be stored on cartridges which contain pre-programmed ROM chips that are "plugged" into special sockets when they are needed. Many video games and some "dedicated" industrial computers such as the computerized typesetting machine in Fig. 48-3 use this type of memory storage.

Another important storage system is the **hard disk**. The hard disk stays inside the machine. It can store about 10 or more megabytes of data. The user does not change the hard disk.

Many older computers used punched paper tapes and cards for mass storage. These systems are still in use, but they do not operate as fast as magnetic storage systems do.

Cost

The speed and power of a machine, the amount of memory, and the size of the words it uses are all factors in determining its cost. Likewise, a hard disk memory storage system is much more expensive than a cassette tape player. A good microcomputer with 64 K or more RAM and a dual disk drive storage system may be purchased inexpensively. Adding special features, additional memory, and pre-packaged software (programs) increases the cost. The cost of all machines has fallen over the past 25 years, and the machines available have become more powerful and more capable.

Special Capabilities

Some computers are better at certain things than others. Some computers are easily portable. Some have built-in communications capabilities which allow them to "talk" to other computers over telephone lines. Any special features such as these will increase the cost of the machine.

Physical Size

Generally, computers are made as small as they can be made. Often the size of the screen and the keyboard are the factors that determine the minimum size of a microcomputer. Minicomputers and mainframes are much larger than microcomputers, but they operate so quickly they are capable of serving several users at the same time. Each user will operate a terminal that is connected to the main machine through cables or telephone lines.

Software

Computers need to be instructed as to what to do and how to do it. This is done with **software**. Developing the instructions for computers is called **programming** and each set of instructions is a program. There are different types of programs. The first type is the **language** (which is a "system program"). The most popular language for use in microcomputers in the home is BASIC. Industrial computers also use other languages such as FORTRAN, COBOL, PL-1, PASCAL, and others. Each language has special capabilities and features which make it effective for certain uses. BASIC works very well for most home and business microcomputer uses.

Another type of program teaches the computer how to interact with its own input and output devices. A very common program of this type is CP/M which is an **operating system** that instructs the computer as to how and when it should read and write information on the disks, keyboard, screen, or other input/output devices. The operating system is frequently stored either in ROM inside the computer or on a disk which must be placed into the machine and "read" before the computer can do anything. CP/M may be stored on the same disk with other programs so that the user does not always have to use one disk to "boot up" the system and then exchange it for another disk to actually complete computing work. Some other popular operating systems are designated with the initials "DOS," for Disk Operating System.

The computer user may write his or her own programs and store them in any of the mass storage systems available. These programs then become software for the machine just as prepackaged or purchased programs and languages. A short program which could be used to teach people to add appears in Fig. 48-4. This program was written in the language BASIC. Large programs may have thousands of instructions.

Still one other type of software which is available is software packages. This text was written on a very popular word processing software package called *Wordstar*. When the *Wordstar* disk is placed into one of the computer's disk drives, the computer loses some of its capabilities and gains other ones because it is instructed by the software package to do certain things. Another disk is placed in the second drive for storage of the data. (In this case, it was the text information typed into the machine). The information is printed onto paper by a separate machine, a printer. The copy was checked for mistakes by a person who then corrected any errors by calling the needed pages back to the screen from the disk and typing only the necessary changes. When the "file" is reprinted, the corrected version will be printed. The final copy is error free without anyone having to retype large sections of it. Other software packages include special programs for drawing graphs, dealing with money and accounting, computer games, instructional packages, and communications software.

```
10  REM — PROGRAM ADDER — addition practice
    program
20  PRINT "ADDITION PRACTICE PROGRAM"
30  PRINT
40  LET C=0
50  PRINT "Directions: Type in two numbers separated
    by a comma"
60  PRINT "            and then hit 'RETURN' "
70  INPUT A, B
80  PRINT "What is the sum of "; A;" plus ";B;" ?"
90  INPUT X
100 IF X=A+B THEN 180
110 LET C = C + 1
120 IF C=3 THEN 250
130 IF C=4 THEN 270
140 IF C=2 THEN 230
150 PRINT "Sorry! That is not the correct answer."
160 PRINT "Let's try again"
170 GOTO 80
180 PRINT "Very Good!! How about another one? Type 1
    for yes or 0 for no."
190 INPUT Y
200 IF Y=1 THEN 30
210 PRINT "O.K. — So Long"
220 END
230 PRINT "Wrong Again! Give it another shot."
240 GOTO 80
250 PRINT "You're Hopeless!! Try one more time"
260 GOTO 80
270 PRINT "Go read the chapter on addition in the math
    book."
280 GOTO 220
```

Fig. 48-4 This simple program in BASIC teaches people how to add. If you do not already know how to program in BASIC, find a friend who does and go through the program together to see how logic and step-by-step procedures are used to instruct the machine.

A Complete Microcomputer

Figure 48-5 shows the block diagram of a complete microcomputer system. The main "brain" of the computer is, of course, the CPU IC in the center of the diagram. This chip makes the decisions and instructs the other parts how and when to act. This system has four I/O devices: the monitor (CRT screen), keyboard, modem (which allows you to use telephone lines for I/O with other computers), and printer. The ROM and RAM are used to deliver needed information to the CPU and the RAM also can store information put in by the keyboard or disks. Mass storage is on the two floppy disks. The various sections of the computer system (blocks) are connected by cables and printed circuit board conductor tracks which carry the information signals to needed locations.

Summary

Computers are very valuable tools in our high tech society. They range in size from small handheld machines to huge, powerful mainframes that cost millions of dollars. Microcomputers have become so inexpensive that many small businesses and private homes have relatively powerful machines and prepackaged software. Software includes languages, operating systems, packaged programs and user-written programs. Most computers have both RAM and ROM memory as well as some mass memory storage system. Floppy disks have become very popular for use in microcomputers.

Important Terms

mainframe	byte	software
microcomputer	word size	programming
RAM	floppy disk	language
ROM	hard disk	operating system

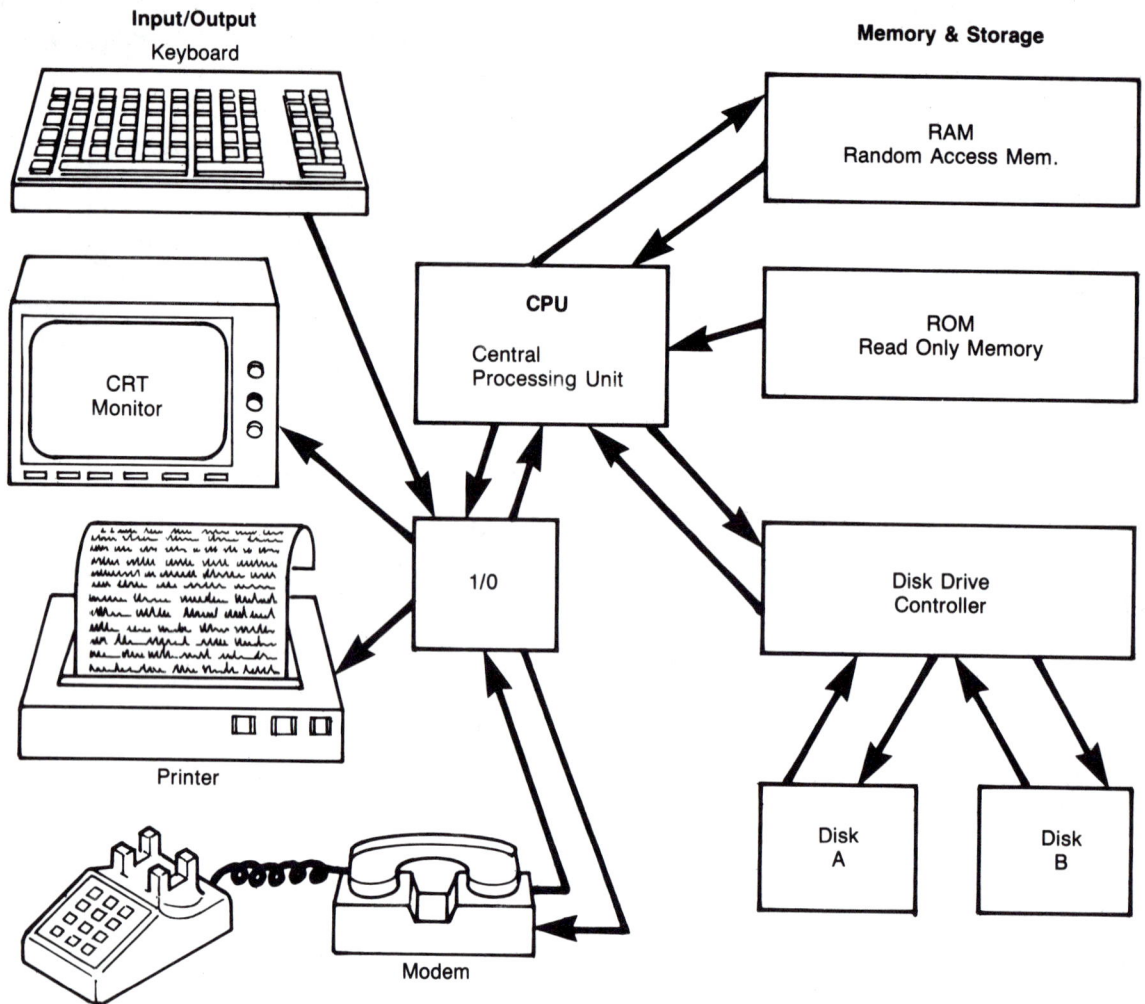

Fig. 48-5 A simple block diagram of a complete microcomputer system. The power supplies are omitted.

Review Questions

1. Name any microcomputer you are familiar with and describe its mass storage system, I/O devices, amount of RAM, and special features.

2. Which machine would be more capable — one with 16 K or one with 128 K? What do these figures mean?

3. What would happen if a floppy disk were placed near a large magnet or a large motor which builds up powerful moving magnetic fields? List two other types of mass storage systems.

Chapter 49
Amplifiers

The Importance of Transistor Bias

The circuit you studied in Chapter 43 is redrawn in Fig. 49-1 This circuit is called the **common emitter** amplification circuit. To quickly review, recall that changing the setting of R_1 will cause more or less emitter-to-base current (C battery current) to flow. Each electron pulled out of the base will leave a new hole. The new hole will cause a stampede of electrons to race from the B battery's negative terminal toward the positively charged hole. One of these electrons will be captured by the hole, but all of the rest will race past it. Those electrons will go out through the collector to return to the B battery's positive terminal through the load resistance (R_L). The main current is the collector current. It is larger than the base current which controls it.

If you were to graph the base current and the collector current together, you would find a curved line similar to the colored one in Fig. 49-2. By selecting different values of

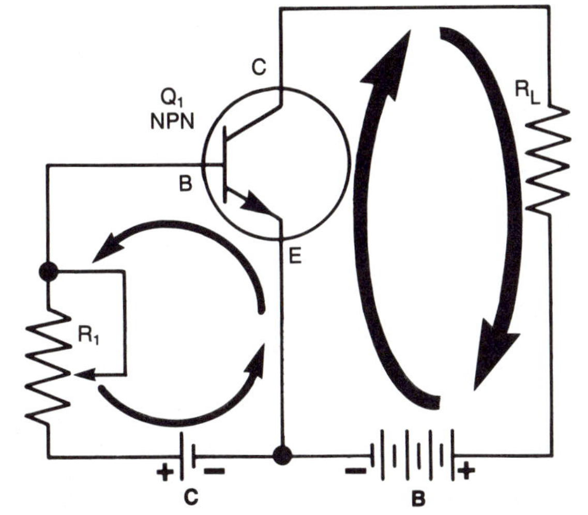

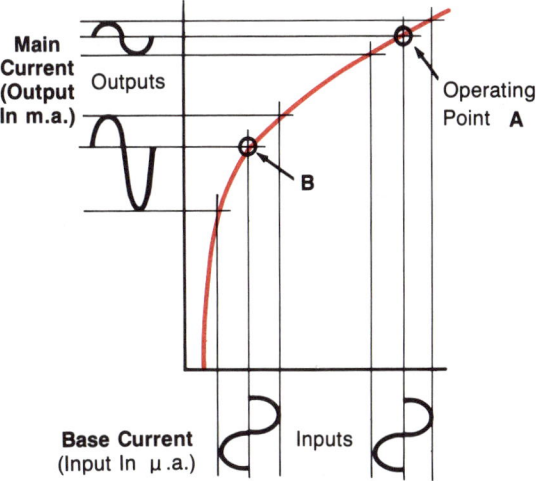

Fig. 49-2 When operating point A is used, the output has the exact same shape as the input — it is just larger. (Note that the base current is in μ.a., but the main current is in m.a. So, even though the output *appears* smaller, it is really much larger than the input.) But, when operating point B is used, one part of the sine wave is stretched out much longer than it should be.

Fig. 49-1 A common emitter amplification circuit with an NPN transistor.

C battery (or biasing resistors in a real circuit), you could make the central operating point of your amplifier circuit fall anywhere along this curve. It is important the operating point be near the center of a long, straight section of the curve. The figure shows why. If a curved section is used, the output will not accurately reflect input. Instead, the output will be distorted.

Distortion also occurs when the operating point is selected so close to the limits of possible operation that the transistor begins to **clip** the output. This is because the input signal either reverse biases the base (causing it to cut off) or saturates it (causing maximum possible current to flow) as shown in Fig. 49-3. So, selection of biasing resistors is very important for getting linear amplification without distortion.

The common emitter circuit has the following characteristics:

- High power gain because there is both current gain and voltage gain.
- Medium input and output impedances.
- 180 degree phase inversion.
- Usefulness in many applications.

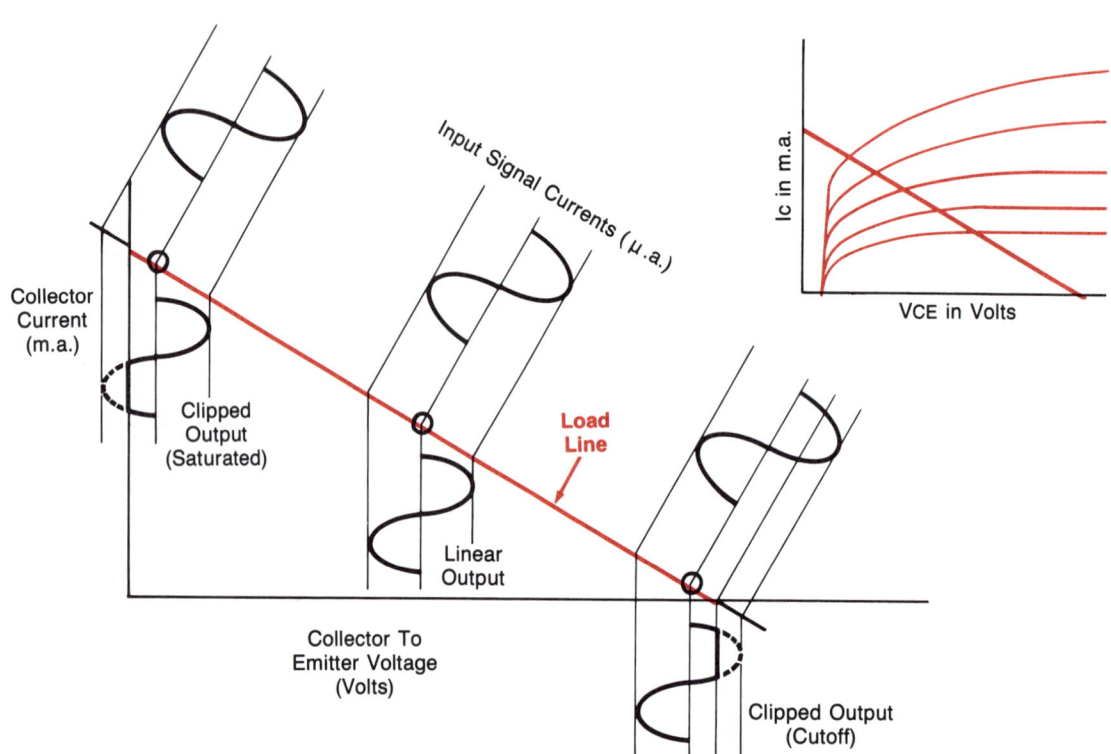

Fig. 49-3 The load line extends from the maximum possible collector voltage for the circuit (the B+ voltage) on the horizontal axis to the value of collector current which would flow if the transistor created zero resistance to current flow. An operating point chosen near the center of the load line avoids clipping of the output. The small graph in the corner is a family of curves for the circuit. Each curve shows what will happen for a different level of base current (I_B). Load lines are usually superimposed on (drawn on top of) these graphs — we left the curve family out of our drawing to make it less confusing.

Classes of Amplifiers

The amplifier circuit discussed so far is a **Class A** amplifier. In a Class A amplifier, the operating point is selected so that it is near the center of the load line. The full input signal then is reflected in the output — none of it is clipped. Class A is a very common usage, but it is not the only one.

Sometimes there are advantages to placing the operating point right on the cutoff line and allowing the amplifier to only amplify and output half of the input signal. In this case, as shown in Fig. 49-4, the amplifier both amplifies and acts as a half-wave rectifier for the input signal at the same time. This is called a **Class B** amplifier. Class B amplifiers give an output during 180° of each 360° input signal. One common use of Class B amplifiers is to put two of them together in an arrangement called **push-pull**. In this arrangement, one of the transistors amplifies only the top half of the input signal and the other transistor amplifies only the bottom half. The total output combines the two amplified halves to reproduce the whole input signal in a much larger form. See Fig. 49-5. These circuits are very popular in stereo systems and other audio output circuits.

Two other classes are possible. **Class C** biases the transistor a little below the cutoff

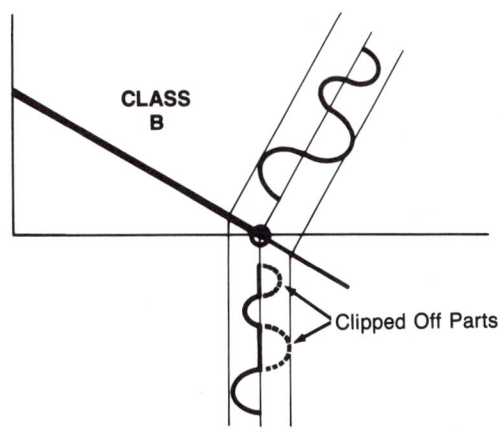

Fig. 49-4 The Class B amplifier only conducts half of the input signal.

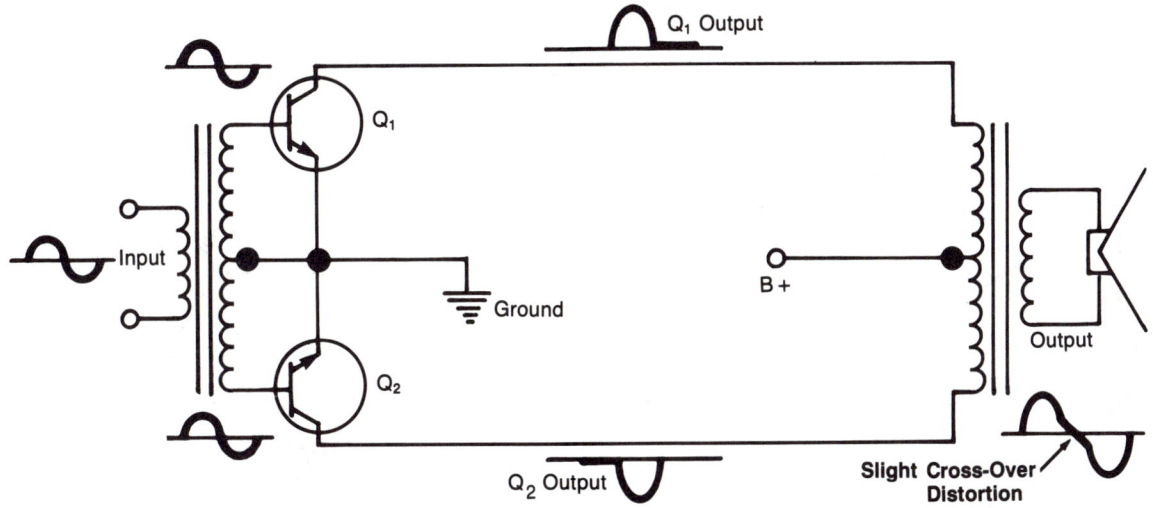

Fig. 49-5 The push-pull amplifier combines two Class B amplifiers to give a higher power output than possible with only one transistor.

point so that it conducts for less than half of the input signal. Figure 49-6 shows this condition. Class C is not used a great deal, although it has found some use in tuned RF (radio frequency) amplifiers. The figure also shows the biasing of **Class AB**. Class AB allows part of the signal half which is being cut off to be conducted. Class AB is sort of a compromise between Classes A and B. Class AB may be used in push-pull circuits, but the output power will be lower than when Class B is used.

Common Base Circuit

The amplifiers discussed so far are all common emitter circuits. The name common emitter indicates that both the input and output signals are found in the emitter. The input current goes from emitter to base, and the output goes from emitter to collector. So, both input and output signals travel through the lead (wire) into the emitter. The common emitter circuit is used often because it gives the highest power gain of the three possible circuits. There is both voltage gain and current gain in the common emitter amplifier.

The **common base** circuit, shown in Fig. 49-7, does not yield any current gain, but it does give some voltage gain. Its advantage is that it has very low input impedance and very high output impedance. Remember that impedance is the total opposition to AC current flow. The very low input impedance means that even very weak signals can cause an output to be produced. This makes the common base circuit excellent for uses such as amplifying the weak RF signals from the antenna circuit of a radio. The impedance of the antenna circuit itself is also low — about 50 ohms. The common base circuit is often a good one for matching the impedance of one section (stage) of a complex circuit to another one. When impedance is not closely matched, there is a power loss.

Common base RF amplifiers will frequently have tuned (tank) circuits on both their input and output sides. Another advantage of the common base circuit is that the output signal is in phase with the input signal. There is no phase inversion as found in the common emitter circuit. Notice that the base lead is the only wire in this circuit which will carry both the input signal current and the main (output) current; this is why it is called a common base circuit.

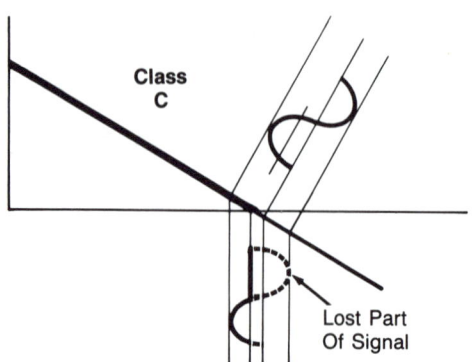
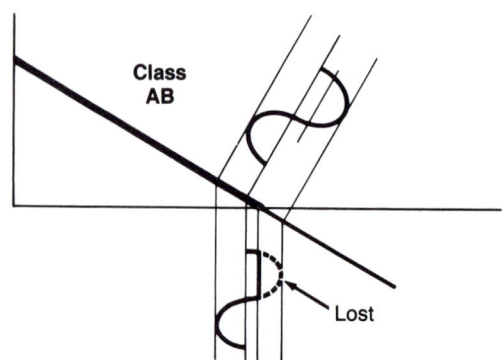

Fig. 49-6 Classes C and AB are formed by moving the operating point. This is done by changing the bias on the transistor.

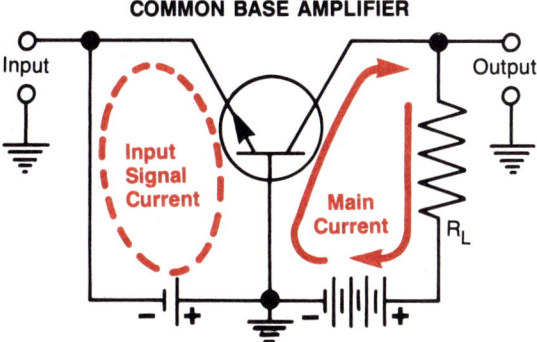

Fig. 49-7 The common base amplifier has the following characteristics: low input impedance, high output impedance, no phase inversion, no current gain, and large voltage gain. It is good for use in impedance matching and RF amps.

The Common Collector Amplifier

The third type of amplifier is the **common collector** circuit. A common collector amplifier is shown in Fig. 49-8. Notice that the only wire that carries both input and output currents is the collector lead. The tiny input signal current cannot go through the emitter because the load resistor (R_L) stops it. The input signal current can go through capacitor C_1, but the DC power supply current of the battery (B_2) is blocked by the capacitor to prevent a short circuit through the ground path. The output is taken from the top end of the emitter resistor (R_L). Common collector circuits are opposite of common base circuits. Common collector amplifiers have:

- Very high input impedance.
- Low output impedance.
- No voltage gain.
- High current gain.
- No phase inversion.

These characteristics make the common collector good for impedance matching. It is also good for isolating the input (source) circuit from the loading effects of other parts of the circuit. This is because the high input impedance will only allow a very small input signal current to flow.

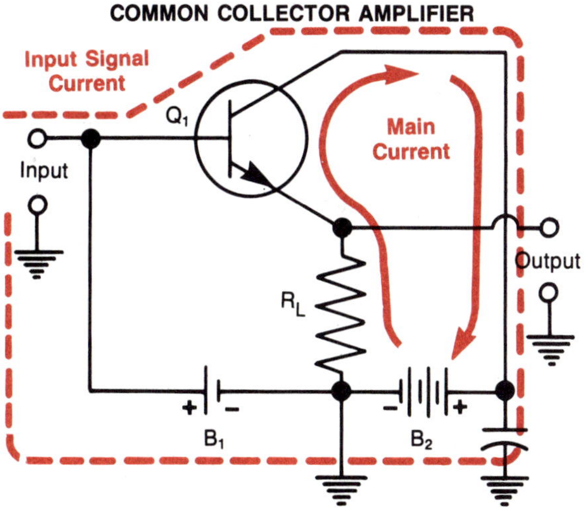

PNP Transistors in Amplifiers

We used NPN transistors in all of the examples. It is possible to use PNP transistors to make all three types of amplifiers. In most cases, reversing battery polarities will be the major changes needed in the circuits. Likewise, vacuum tubes can be used in equivalent circuits:

- The grounded cathode is like the common emitter (grounded emitter) circuit.
- The grounded grid is like the common base (grounded base) circuit.
- The grounded plate is like the common collector (grounded collector) circuit.

These circuits are drawn in Fig. 49-9. Vacuum tubes may also be used to produce amplifiers of each class (A, B, C, and AB), and in a push-pull mode.

Fig. 49-8 The common collector circuit is also good for impedance matching, but its impedance is highest on the input side.

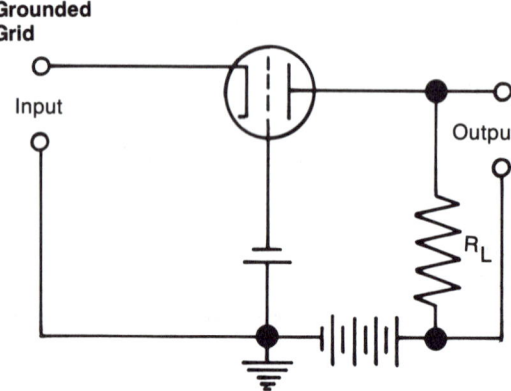

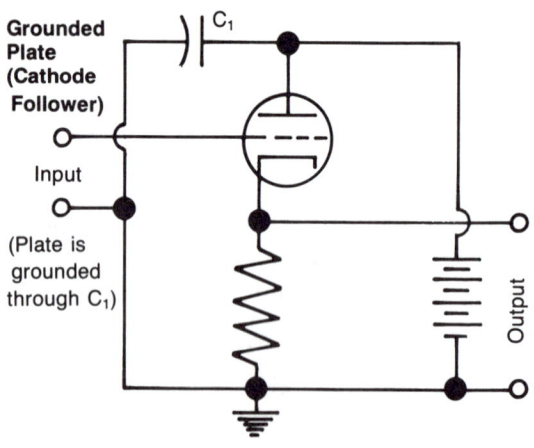

Summary

There are special amplifier circuits for special uses. The most common transistor amplifier is the common emitter circuit operating in the Class A condition. By changing the biasing of the transistor, you can make Class B, C, or AB amplifiers. Two Class B amplifiers may work together to form a push-pull circuit. Biasing must be done carefully to avoid unwanted clipping and distortion. Common base and common collector circuits are good for impedance matching and special applications.

Important Terms

common emitter
distortion
Class A amplifier
Class B amplifier
push-pull
Class C amplifier
Class AB amplifier
common base
common collector

Review Questions

1. If you wanted to make a low power audio amplifier with only one transistor, which class and type of amplifier would you produce? Why did you make these selections?

2. Draw a chart with the following headings across the top: Common Emitter, Common Base, and Common Collector. The headings down the side should be: Current Gain, Voltage Gain, Phase Inversion, Input Impedance, Output Impedance, and Typical Use. Now, fill the chart in with words like yes, no, high, low, medium, and with the uses of the three types of amplifiers.

3. What is distortion? What can cause it?

Fig. 49-9 These are vacuum tube circuits which do the same jobs as the common emitter, common base, and common collector circuits which you just studied. Compare the vacuum tube and transistor circuits to see the similarities.

Chapter 50
Putting Amplifier Stages Together

This chapter adds to your knowledge of amplification and discusses more practical circuits.

Why Use More Than One Stage?

The first answer to this question is to make an amplifier with greater output power than would be possible with only one transistor. This is probably the most important reason for building **multiple stage amplifiers**.

Another reason for using more than one stage is to overcome the problem of phase inversion when using the common emitter circuit. By putting two common emitter stages together, the problem is solved. The first stage inverts the signal to reverse phase, and the second stage inverts it again so that it is back to its original phase, as in Fig. 50-1.

Still another reason for using more than one stage is for impedance matching. If you had an input device (or circuit) with low impedance, and you wanted to amplify its signal and send it to a high impedance output device, you might first guess that a common base circuit would do the job. However, remember that this circuit has practically no current gain. You might choose to use a common base stage with one or more common emitter stages. Then, you would accomplish both goals — matching the impedance and developing the desired gain.

Cascading

Linking several amplifier stages together is called **cascading**. Most audio devices which produce high power have several cascaded stages. When stages are cascaded, the goal is to pass the signal from one stage to the next without disrupting the DC bias levels needed for proper operation of each stage. The simplest way to connect two stages together is shown in Fig. 50-2. The method illustrated in this circuit is called **direct coupling**. Notice that the signal does not have to go through any parts (except wire) to get from the collector of transistor 1 to the base of transistor 2 — it has a direct path.

Resistor R_2 serves two purposes:
1. It serves as the load resistor for Q_1.
2. It biases the base of Q_2.

The collector of Q_2 must be a higher voltage than the collector of Q_1 (which is directly tied to the base of Q_2), so resistor R_4 must have a lower value than that of R_2.

The biggest advantage of direct coupling is that it is inexpensive. Very few parts are needed to build a directly coupled amplifier. The disadvantage of direct coupling is that there is a limit to how many stages may be cascaded in this manner. The circuit sometimes may become unstable — especially when temperatures of operation vary a great deal. Direct coupling is very popular in industrial circuits and measurement circuits which often need to pass both the AC signal and the DC voltage level from stage to stage.

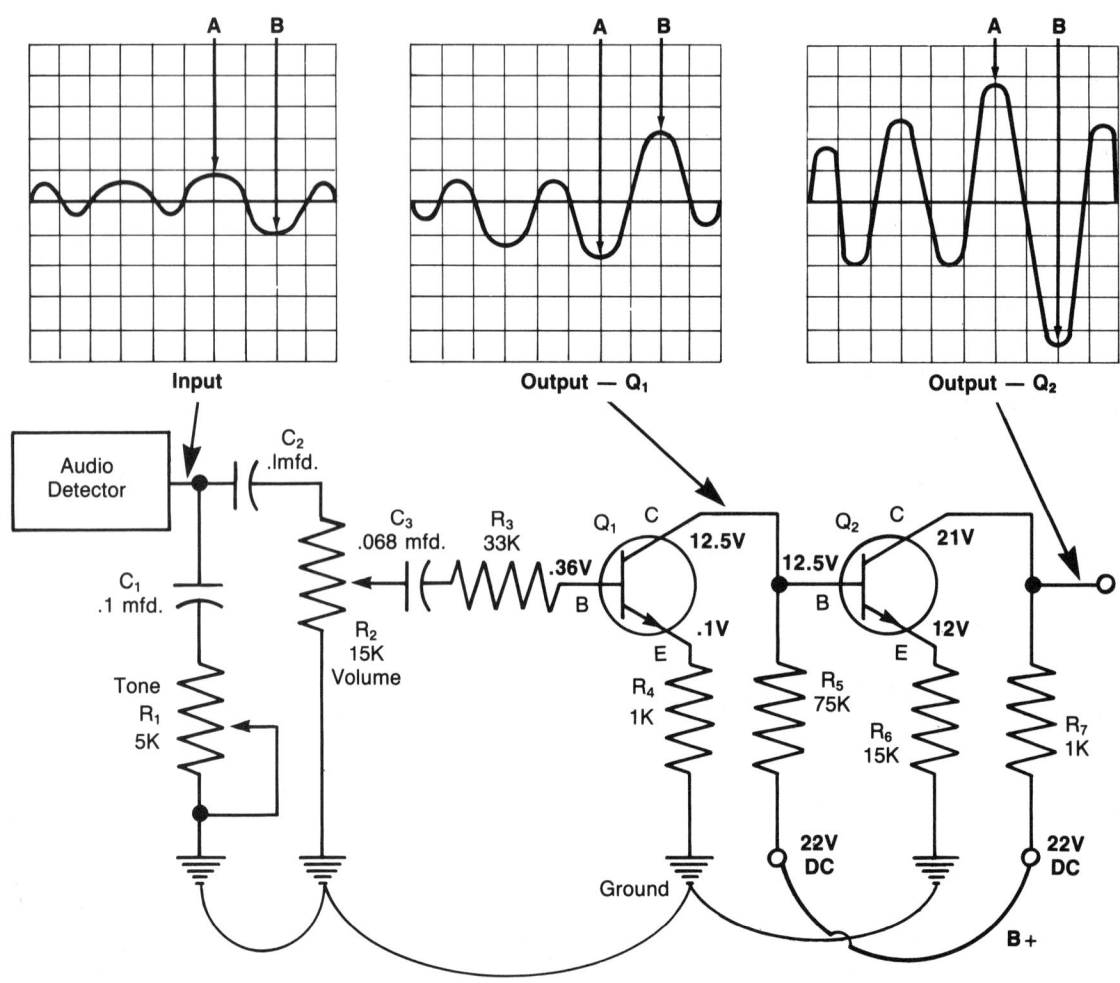

Fig. 50-1 Transistor Q_2 amplifies, and it also changes the signal back to its original phase.

Transformer Coupling

Another method of coupling, called **transformer coupling**, is shown in Fig. 50-3. Transformer coupling gets its name from the fact that there is actually a transformer between each stage. The output of one transistor is fed to the primary of the transformer. The transformer transfers the signal to its secondary by magnetic induction — with no actual electrical connection. The output of the secondary is sent to the base of the next transistor. Transformer coupling is also often called "inductive coupling."

The big advantage of transformer coupling is that the transformers can be used for impedance matching of inputs and outputs. Since transistors require changes in base current for their input signals, even a step down transformer (which steps voltage down, but current up) may be found between two stages. Many audio amplifiers use a transformer to couple the last stage to

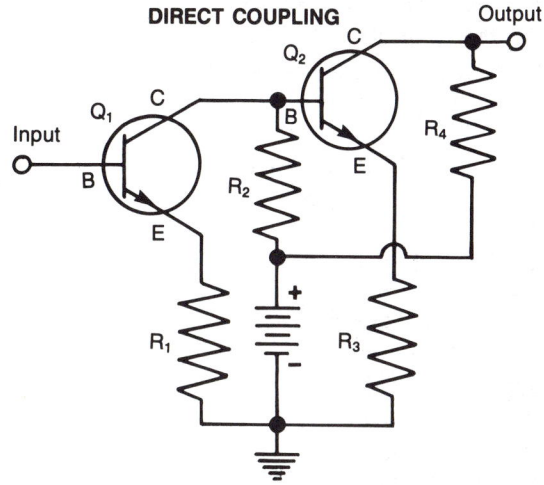

Fig. 50-2 Direct coupling is popular in industrial and measurement circuits.

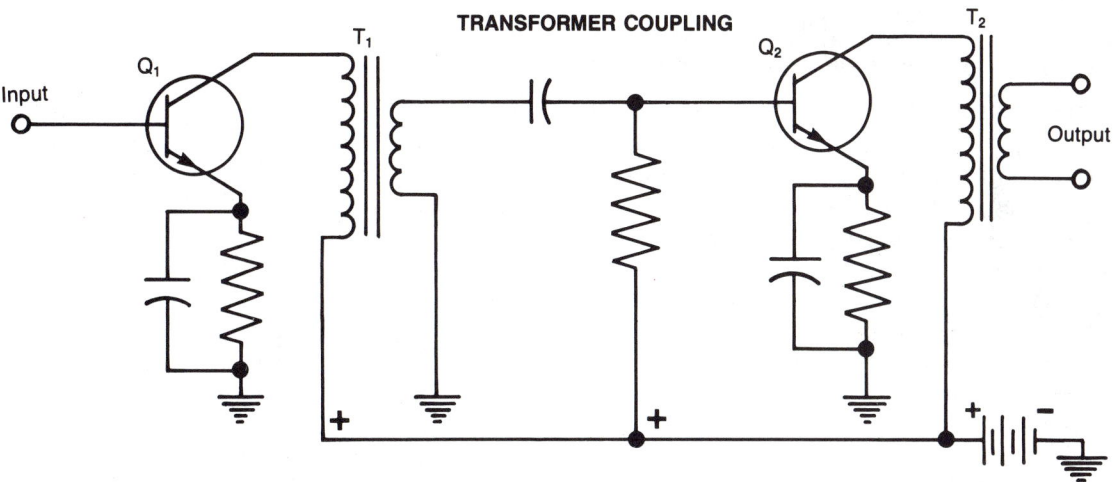

Fig. 50-3 Transformer coupling is very useful for impedance matching.

the loudspeaker, even if they use other coupling methods between the various stages within their circuits. This is because the loudspeaker needs high current but little voltage to operate.

One additional advantage of transformer coupling becomes obvious when an amplifier is designed to amplify one particular frequency while limiting other frequencies. These amplifiers generally use tuned "tank" circuits on their inputs or outputs to select the desired frequency. Since the tank circuit must have a coil, it is convenient to use a transformer for the coil of the tank circuit and then use its other coil as an inductive coupler to the next amplifier stage. Fig. 50-4, a simplified drawing of two stages from a television video IF amplifier, shows this technique. The IF circuits all operate at frequencies between 41 and 46 MHz (megahertz). The tank circuits ensure that only the desired frequencies are passed and amplified by the IF amplifier stages.

The disadvantages of transformer coupling are higher cost and poor response at higher frequencies (due to the transformer losses discussed in Chapter 30).

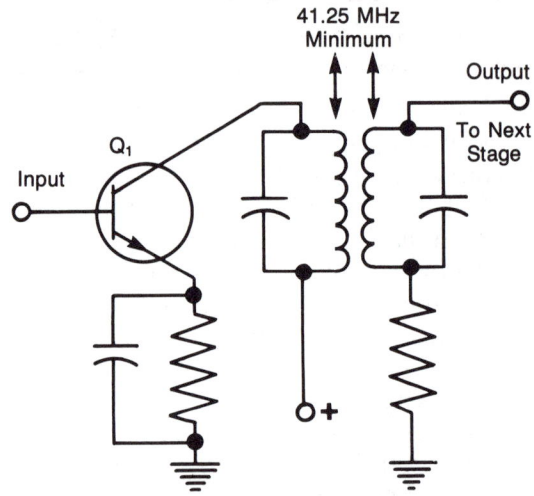

Fig. 50-4 The double ended arrows above the transformer are a movable core which allows us to tune the tank circuits to the desired frequencies — only the wanted signals will be passed along to Q_2.

Resistance-Capacitance Coupling

Resistance-capacitance coupling, or **RC coupling**, is often used with audio frequency amplifiers. Figure 50-5 will be used to explain how this method works. The most important parts are R_4 and C_3. The resistor (R_4) is the load resistor for transistor Q_1. If you were using direct coupling (as in Fig. 50-2), this resistor would also be the bias resistor for the base of Q_2. But, that would mean the collector voltage of Q_2 would have to be much higher for that transistor to amplify. By using the capacitor (C_3), the AC signal current is allowed to pass from the collector of Q_1 to the base of Q_2, but block the DC voltages. The voltage used to bias Q_2 can now be different than the collector voltage of Q_1. The DC current cannot go through the capacitor — only the AC signal current will be passed to the second stage.

RC coupling is generally less expensive than transformer coupling. It is not as good for impedance matching as transformer coupling, and it cannot pass the DC voltage level from one stage to the next. These limitations make RC coupling undesirable for some special circuits, but it is the best choice for many circuits in the audio frequency range.

Summary

When more amplification is needed than can be obtained by using one transistor, generally several transistor amplification stages will be cascaded together. The stages may be linked together by three main methods: direct coupling, transformer coupling, or resistance-capacitance coupling. Each coupling method has unique advantages and disadvantages. The push-pull circuit (Chapter 49) is one other way to use two transistors together. There are also vacuum tube versions of all these coupling methods which are similar. Refer back to Fig. 50-1 and see if you can tell which coupling method is used in that amplifier circuit. The next chapter deals with oscillators — a very important circuit that can create its own AC signal.

Important Terms

multiple stage amplifiers
cascading
direct coupling
transformer coupling
RC coupling

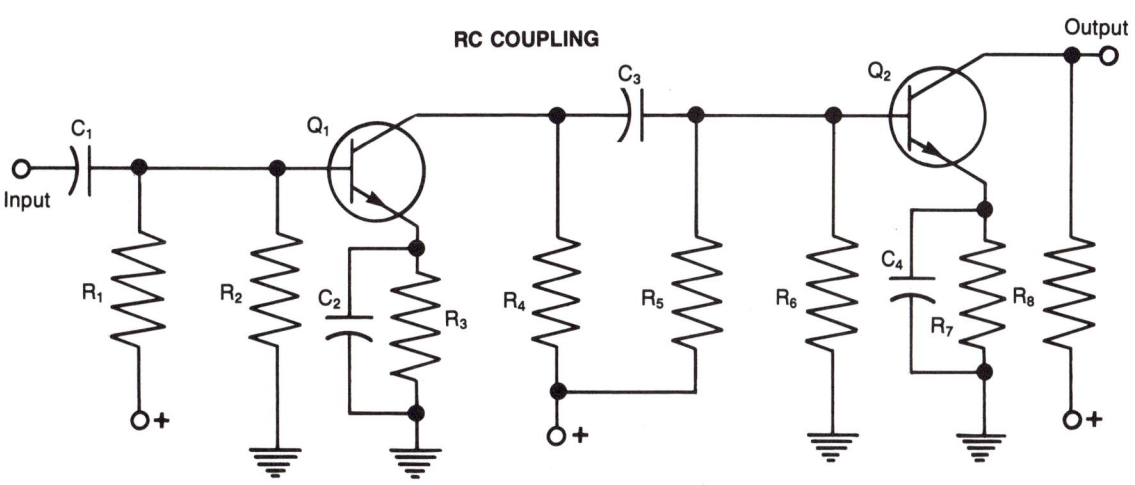

Fig. 50-5 The RC (resistance-capacitance) coupled circuit is very useful in audio amplifiers.

Review Questions

1. If you were building an electronic meter or oscilloscope that had two stages of transistor amplifiers to make it possible to measure weak signals and voltages, which type of coupling would you choose? (Hint: Remember, you must be able to pass both AC and DC voltages and signals to make the instrument useful.)

2. Which type of coupling is best for impedance matching between stages?

3. Which method of coupling allows the AC signal to pass from one stage to the next, but blocks the DC operating voltages?

Careers: Sound and Acoustics Careers

Many electronic circuits deal with sound. Televisions, radios, stereo systems, and PA systems are common examples. But, even an ultrasound machine used to make important medical tests is just another use of electronics and sound together. The serious electronics student should learn all that he or she can about sound and acoustics.

Careers in this area include operating sound equipment (like the sound technician for a performing musical group) or designing the equipment or the auditoriums in which it is used. Many people also produce, sell, and repair sound equipment. Some subjects which could help prepare you for careers dealing with sound are communications, technology education courses, mathematics, and science. Even courses in drama, music, and the performing arts might be helpful in leading to sound-related careers in the entertainment business.

Project: Light Activated Control Circuit

Parts List

1	Cds photoresistor	R_1
2	100 K-ohm resistors	R_2 & R_3
1	10 K-ohm potentiometer (a 15-turn pot is best)	R_4
1	1.2 K-ohm resistor	R_5
1	4.7 K-ohm resistor	R_6
1	SPST switch	S_1
1	9 V battery with connector	B_1
1	741 Op Amp IC	Z_1
1	ECG 289A NPN transistor	Q_1
1	5 V relay with contacts rated for intended load	L_1

PC board, wires, enclosure, and hardware as required.

Notes

The relay should operate with very low coil current. Experiment with the circuit on a breadboard before constructing it into a project — see what effect there is when you change values of the resistors. This circuit could be used for detecting intrusion, controlling devices by a light beam, or turning on your radio at the "crack of dawn." Amaze your friends by turning your stereo on or off simply by aiming a flashlight at it! If the circuit is to be used in a continuous operation situation, you may want to power it with a simple DC power supply instead of the battery; a circuit like the one in chapter 41 will work nicely. Reversing the connections to pins 2 and 3 of the 741 IC chip will make the relay close when the photoresistor is darkened instead of lit.

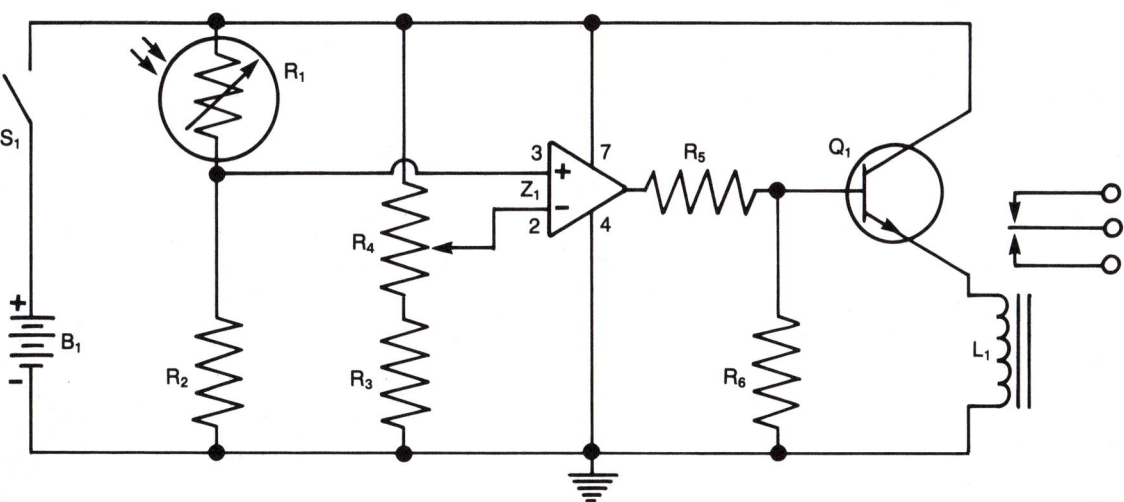

Fig. 50-6 The Light Activated Control Circuit is useful and a fun project to build. Can you tell what the purpose of R_6 is in this circuit?

Chapter 51
Oscillators

Oscillator

You know two main uses for transistors: amplifying analog signals and forming digital gates. These circuits do something to or with an input signal — either to control a large output signal or to make decisions. But where did that input signal come from in the first place? Sometimes the input signal comes from microphones, phono cartridges, switches, industrial sensors, or other sources. In other circuits the input signal may be produced by a special amplification circuit called an **oscillator**. This chapter tells how these interesting circuits do their important job.

Feedback: Friend or Foe?

Have you ever heard someone speaking through a public address (PA) system, and there was a sudden ear-splitting, squealing noise that did not stop until the amplifier was turned down very low? Fig. 51-1 diagrams what causes this loud squeal. The input signal from the person talking into the microphone is amplified in a normal manner. Then, the amplified sound (now much greater than the original input sound) travels back from the speaker to the microphone and acts as a new, and louder, input.

This new input is amplified to become a still louder output which will also travel back to the microphone for even more amplification. The process continues until it is interrupted. The loudness of the squeal

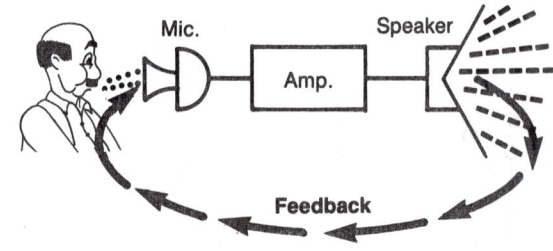

Fig. 51-1 Feedback can be an annoying problem in PA systems.

grows very quickly and then levels off to a final (but very loud) volume. How loud it will get depends upon the maximum gain of the amplifier circuits. The frequency (pitch or tone) of the squeal depends upon the frequency of resonance of the internal circuits.

This process is called **feedback**. The output signal has been "fed-back" to the input for further amplification. Audio feedback can be caused if the gain (volume control) of the amplifier is set too high for the conditions. Poor placement of components can also cause feedback. This can be if the microphone is in front of the speakers or near a wall that reflects the sound. Since the frequency and volume of the squeal become constant very quickly, the output (electrical) signal to the speaker would really be an AC waveform. The output is, therefore, actually a series of oscillations.

Feedback and oscillations are constantly at odds in most amplifier circuits. They act as an extreme form of distortion to the output. Sometimes internal parts of a circuit may break down perhaps due to high temperatures, improper input levels, or poor circuit design. These break downs cause oscillations to occur within an amplification stage. Circuit designers plan carefully to avoid these problems. There are, however, special amplification circuits that are designed specifically to cause oscillations. They have a built-in feedback path to encourage oscillations. These circuits are oscillators. The circuit's job is to produce new AC signals. So, although feedback is usually "foe" to most amplification circuits, to oscillators it is an important "friend."

The Armstrong Oscillator

The **Armstrong oscillator** is very easy to understand. An example circuit appears in Fig. 51-2. The input for the amplifier is created by the first set of damped oscillations which are set up in the tank circuit when the switch is first closed. If you do not remember how this works, review Chapter 39. This AC input is amplified by the transistor and creates an output in the "tickler coil" which will have the same frequency. The tickler coil is physically close enough to the coil in the tank circuit to inductively couple with it (as if they were a transformer). So, the output is fed-back to the input through this tickler coil.

The oscillations are not allowed to "damp" out, because there is a new output signal for each input signal. The frequency of oscillation is determined by the resonant frequency of the tank circuit. If you wanted to change the frequency, you would use a variable capacitor for C_1. If the feedback signal is in phase with the output one (to cause increased oscillations) it causes "regeneration." If it is out of phase, it causes "degeneration" — that is, it fights against continued oscillations.

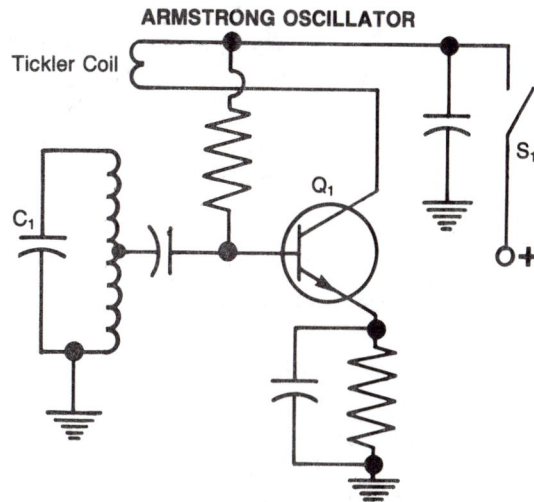

Fig. 51-2 The tickler coil gives us the needed feedback in the Armstrong oscillator.

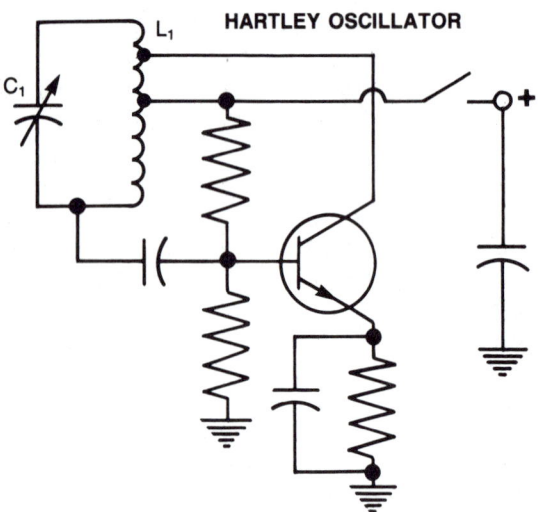

Fig. 51-3 The Hartley is very similar to the Armstrong oscillator, but it uses a tapped transformer (L_1) instead of a tickler coil.

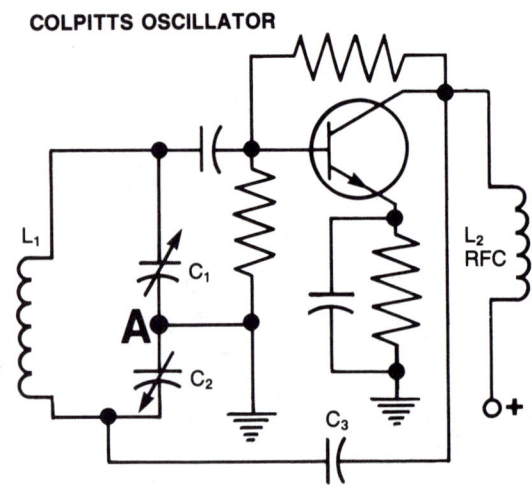

Fig. 51-4 The feedback path in a Colpitts oscillator is through capacitors C_1, C_2, and C_3.

Hartley Oscillator

A variation from the Armstrong oscillator is the **Hartley oscillator**. The main difference (as shown in Fig. 51-3) is that the Hartley uses a tapped transformer for its feedback path instead of the tickler coil used in the Armstrong. The figure shows a variable Hartley, because the capacitor in the tank circuit (C_1) is variable to change the oscillating frequency.

Colpitts Oscillator

In function, a **Colpitts oscillator** is very similar to the Hartley oscillator, except that the feedback path is connected on the capacitor side of the tank circuit. Fig. 51-4 shows a Colpitts oscillator. Capacitors C_1 and C_2 are connected in series with each other. When you connect to point A (between the two capacitors) you center tap the tank circuit just as you did by connecting into the tapped coil of the Hartley circuit. Since both capacitors are variable (and may even be "ganged" so that both are adjusted by turning one knob), this is a variable frequency oscillator. Coil L_2 is an RFC (radio frequency choke) which prevents RF signals that the oscillator creates from going to the DC power supply. RF chokes are used in many high frequency oscillators for this purpose.

Crystal Controlled Oscillator

When you studied sources of EMF, you learned how a crystal will create small pulses of current due to the piezoelectric effect when it vibrates (oscillates). The crystal can be vibrated by giving it an electrical "shock," and then you can use the resulting pulses of EMF as a time base for

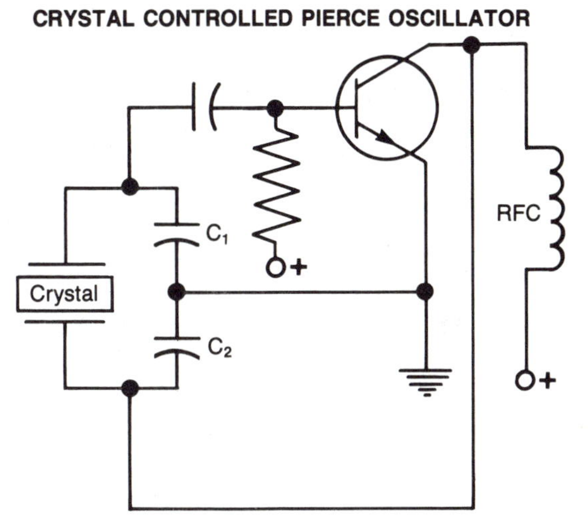

CRYSTAL CONTROLLED PIERCE OSCILLATOR

Fig. 51-5 The crystal in this Pierce oscillator helps keep it perfectly on frequency.

setting the frequency of an oscillator circuit. This is done in the **crystal controlled oscillator.** Figure 51-5 shows a Pierce Crystal Controlled oscillator.

Crystals may be used to help keep other types of oscillators on their proper frequency, too. The crystal sets the frequency of oscillation — capacitors C_1 and C_2 only determine how much feedback is applied. The capacitors are not used to change the frequency of this oscillator, even if they are variable types. Since the crystal may change slightly in frequency if its temperature changes, a "crystal oven" is sometimes used to keep the temperature constant.

Multivibrators

Two transistors may be used in a special oscillating circuit which produces a square-wave output. The circuit, shown in Fig. 51-6, is a **multivibrator.** Both transistors are selected and biased to be "switchers." Switchers either conduct at "full force" (are saturated) or are cut off. They are not good as linear (analog) amplifiers. You use switchers in making digital gates. What happens is that one of the transistors will conduct and give an output at its collector which is fed to the base of the other transistor.

Remember that there is a phase shift of 180° in a grounded emitter amplifier. When a positive pulse is fed into the base of either transistor, its output will be a negative pulse, and vice-versa. Thus, if Q_1 has a positive output (saturation) and sends it to the base of Q_2, then Q_2 will develop a negative output (cut off) to send back to the base of Q_1. This will make Q_1 develop another positive output to send to Q_2, and so on. The length of time it takes for each output pulse to be developed depends on the duration of the RC time constant of the coupling capacitors and resistors (Chapter 35). Transistor Q_2 will be held cut off as long as capacitor C_1 can hold its charge from the positive pulse it received from Q_1. The output may be taken from either transistor's collector, and it will be a square-wave.

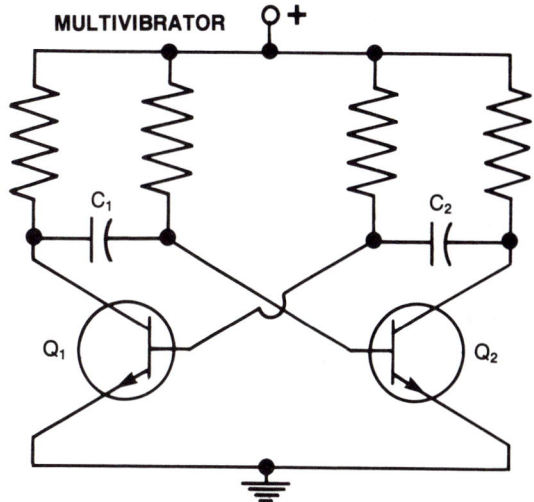

Fig. 51-6 The multivibrator produces a square-wave output.

Chapter 51 *Oscillators*

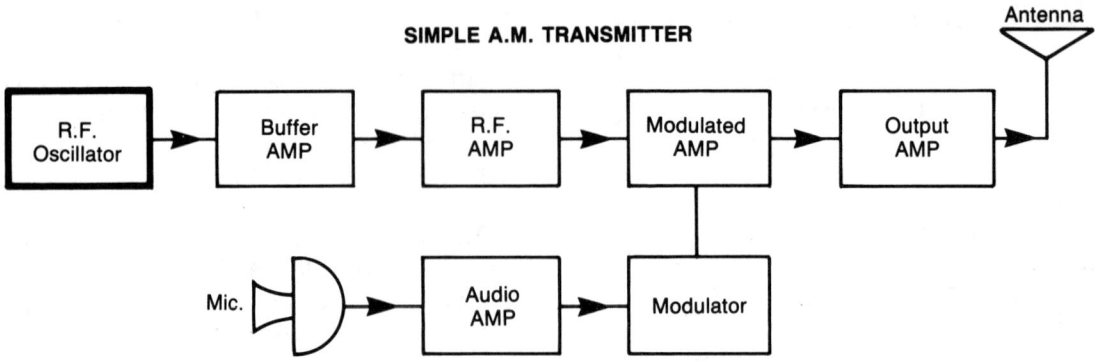

Fig. 51-7 The RF oscillator produces the radio frequency AC waveform that will carry the signal in a radio transmitter.

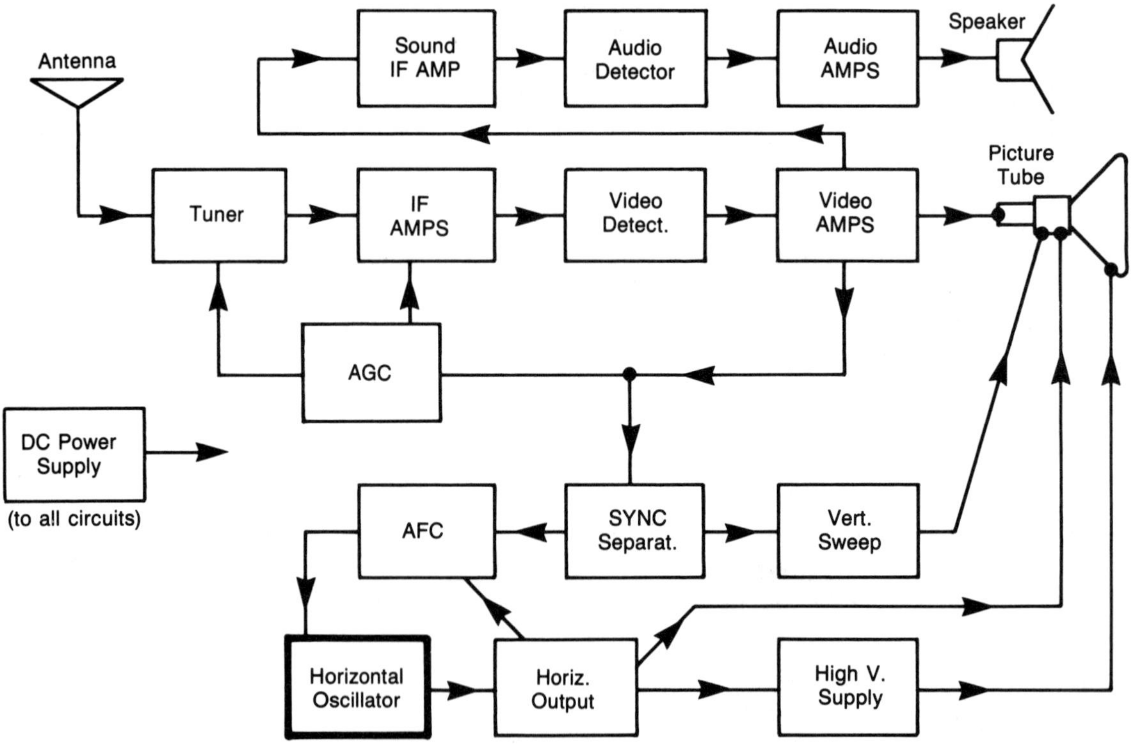

Fig. 51-8 The horizontal oscillator produces a sawtooth waveform at 15,750 Hz to guide the sweep path of the electron beam in the picture tube. Signals from the "sync separator" and AFC (automatic frequency control) help keep the horizontal oscillator in step with the TV station.

Uses of Oscillators

Oscillators are used as generators of AC signals and waveforms. Two examples are given in Figs. 51-7 and 51-8. Figure 51-7 shows the block diagram of a simplified AM radio transmitter. The oscillator in this circuit produces the AC waves which will be amplified to carry the signal to your receiver. The oscillator also sets the frequency of those waves so that your favorite station will always broadcast on its particular frequency.

Figure 51-8 is a block diagram for a black and white television set. The horizontal oscillator produces a sawtooth wave output signal which stays set at exactly 15,750 Hz. This sawtooth signal tells the beam of electrons that sweeps back and forth across the screen ("painting" the picture in 525 horizontal lines) when it should make each "stroke" across the screen. If this oscillator gets off its proper frequency, the picture will become unstable or tear apart. This happens when the electron beam makes its strokes at the wrong times.

Still another use of oscillators and multivibrators is in timing, such as the "clock" circuit needed in digital computer circuits. These oscillating circuits must keep very accurate "time," so they are carefully designed not to "drift" to the wrong frequency.

Summary

Oscillators are circuits that use feedback to produce oscillations. Their output is an AC waveform. You use them to generate AC frequencies. Unwanted feedback is a problem in regular amplifying circuits. The main types of oscillators all perform about the same job. Their most obvious differences are in the ways that feedback is coupled back to the transistor. Multivibrators use two transistors to produce a square-wave.

Important Terms

oscillator
feedback
Armstrong oscillator
Hartley oscillator
Colpitts oscillator
crystal control oscillator
multivibrator

Review Questions

1. Explain what feedback is in an audio amplification (PA) system and how to avoid it.

2. How can we use feedback for something good? Choose three oscillator circuits and tell how the feedback is obtained in each one.

3. Explain how a multivibrator works.

4. Name one use for an oscillating circuit.

Section 9
Power and Communication

Chapter 52

Power for Heating

Power

Electrical power and electrical-electronic circuits are used for many things. This section of the textbook deals with many electrical and electronics applications in power and communication. You will learn about the production and control of heat in this chapter. Because electric heating devices use a lot of power, this is one of the highest costing applications of the electric power that comes into your home.

Heat: Sometimes You Want It, Sometimes You Don't

Almost every time that heat has been mentioned in this text, it has been in reference to its bad side — that it is a form of waste in electronic circuits. The main forms of heat loss are resistive heat loss, hysteresis, and eddy currents. Strangely enough, these transformer losses (explained in Chapter 30) are caused by the very qualities of electricity that make it possible for you to use it to produce wanted heat! In fact, many electrical circuits try to make heat by the same methods as undesired (waste) heat is produced by other circuits.

The television set is a good example. If you are familiar with an old television set built before 1970, you know that it had many vacuum tubes in its circuits. If you held your hand near the top of the back cabinet cover (over the vents), you felt quite a lot of heat coming from the set.

Compare this to the amount of heat felt from a modern, fully transistorized television set the same size. Both sets do the same job in a very similar way and with about the same results. There are two things, however that are quite different about the sets:

1) The older set is wasting a lot of energy by the heat it produces.

2) The older set is wasting energy by the amount of current it draws from the 120 V AC power line. The older set may consume more than 200 watts of power, while the newer set operates on less than half that amount.

The sources of the heat in the older set are primarily found in two types of components. The most obvious heat wasters are the vacuum tubes. The tubes need heat to operate — so each one has a filament (heater) which does nothing but make heat. The second heat waster is the power supply transformer. These big old transformers had several secondaries. They gave off heat created by each of the three transformer losses mentioned.

Modern televisions eliminate the need for the tubes by using transistors. They also use semiconductor devices (diode bridges, voltage regulator ICs, and zener diodes) and resistances in voltage doubler circuits instead of the wasteful transformers found in the power supplies of older models.

Heat from Resistance

Electric heaters, irons, and cooking ranges all use the same method to produce their heat. These, and many other household items, produce heat by sending large amouts of current through resistance wires. This is **resistance heating**. Two very good resistance wires are Nichrome and Constantan. You may have seen these resistance wires glowing red or orange in a toaster or other electric heating device.

Another example of heat from resistance to the flow of current is the incandescent light bulb. When the current goes through the bulb's filament (made of tungsten), so much heat is given off that the filament glows almost white. Normal light bulbs are used in applications requiring low levels of heat. Special "infrared" (heat) lamps are used where more heat is required. For example, many fast foods restaurants use infrared lamps to keep foods warm.

The quartz heater is still one other example of heating produced by current flowing through a resistance. Fig. 52-1 shows one type of quartz heater. These devices work just like other resistance wire heaters, but the wire has a sheath (covering) of quartz crystal. The quartz acts with the heat much like a magnifying glass does with light. It redirects (focuses) heat so that it is radiated out with higher intensity.

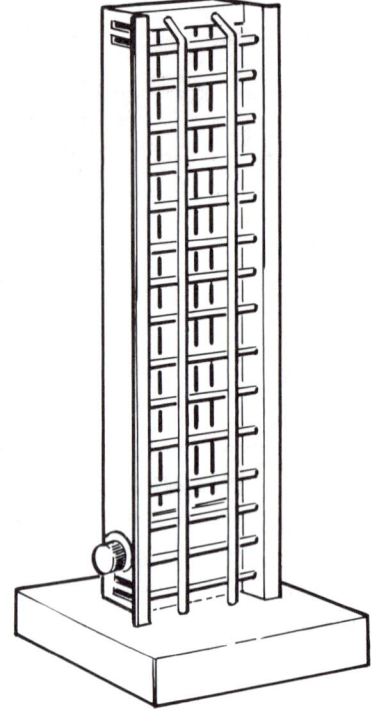

Fig. 52-1 Quartz heaters radiate more useful heat than ordinary resistance heaters.

Control and Transfer of Heat

There are three methods of heat transfer:
1) **Conduction** — Heat is transferred in solid materials through conduction. For conduction to be effective, the molecules in the material must be so closely packed together that the heat can be passed from one molecule to the next. If you hold a metal rod in your hand and heat the other end of it, the end you hold will soon become too hot to handle. This is an example of heat transferred by conduction.

You can control this type of heat transfer by selecting materials which have either densely packed or more spaciously arranged molecules for various jobs. The heat sinks used to cool power transistors are made of metal in order to conduct heat away, while the plastic handle on a soldering iron protects your hand because it is a poor conductor of heat. Many materials which conduct electric currents well will also conduct heat. See Fig. 52-2.

2) **Convection** — Convection involves the molecules traveling from place to place to take on and give off heat. In solid materials, the molecules remain relatively still, so convection will not work in them. In liquids (water) and gases (air), however, the molecules are free to do a lot of traveling. You know that "hot air rises." If you watch water boiling in a pan, you will see that the water heated by the burner at the bottom rises to the top and a circular motion is set up as cooler water flows down to the bottom. This circular flow is an example of convection currents. Fig. 52-3 shows a use of heat transfer by convection.

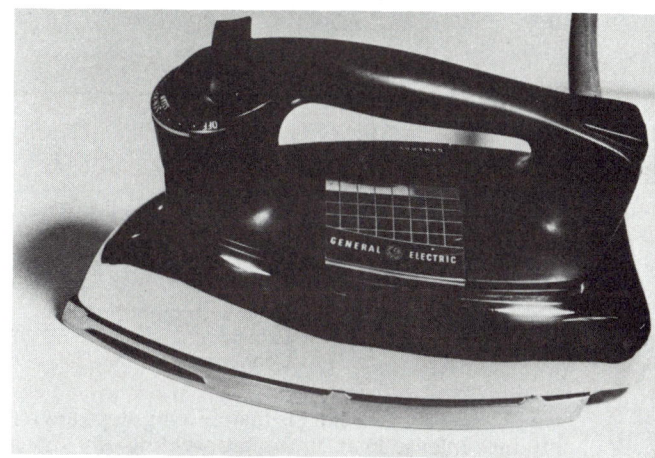

Fig. 52-2 The electric iron illustrates resistance heating, heat transfer by conduction, and heat control by insulation. Can you explain how this is so and identify the parts involved in all three?

Fig. 52-3 The air vent on this stereo receiver is placed above the warmest area in the circuit (the power transistors) to encourage heat transfer by convection. If a record or book is left on top of the vent, the convection (air) flow will be stopped and the power transistors will burn up. Repair shops make much money due to the public's lack of understanding of this simple principle.

Chapter 52   *Power for Heating*

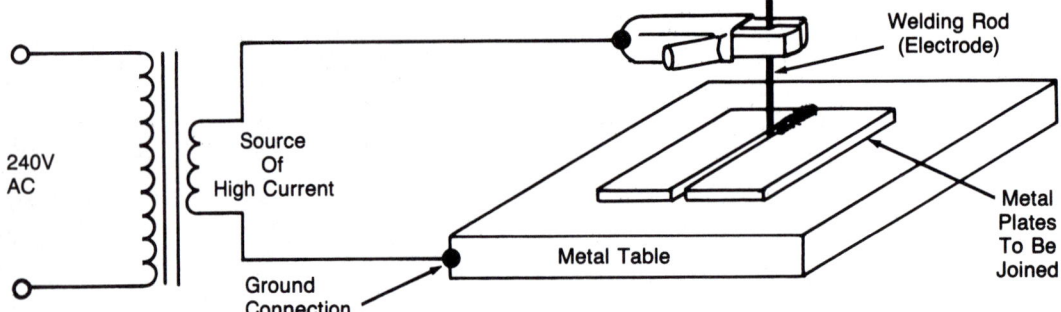

Fig. 52-4 An AC welder is mainly a big step-down transformer. The high current arcs across a small air gap to produce intense heat during arc welding.

3) **Radiation** — Radiation can take place even when there is not material, as in outer space. Heat travels in straight lines — like the rays from the sun — through both space and the air. Infrared heat lamps and quartz heaters both transfer most of their heat by radiation. To prevent radiant heat transfer, place an opaque screening material between the heat source and the place where the heat is unwanted. This may be even more effective if the screen is a light color on the heated side so that it will reflect the radiated heat away. To absorb more radiant heat, an item should be a dark color. This is why solar water heating panels are painted black.

Electric Arc Heating

Fig. 52-4 illustrates how an arc welder works. The welding machine is nothing more than a source of high AC or DC electric current. The welder touches the welding rod to the metal being welded, and a very high current starts to flow because there is almost no resistance in the circuit. As soon as the current starts to flow, the welder pulls the tip of the rod away from the metal just a little bit. The power is so great that the electrons jump across the little air gap in great numbers to form an **arc**. The arc burns very hot — so hot that the metal in both the rod and the metal parts being welded are actually melted and their metals flow together to bond.

The arc burns so intensely that it gives off great amounts of ultraviolet light. The ultraviolet light is very harmful to people's eyes. Welders wear special protective goggles for protection. Never watch welding work without this same protection on your eyes. Arc heating has also been used for melting metal in the production of steel and other industrial processes which require intense heat.

Induction Heating

Transformer losses may be used to produce heat in another process called **induction heating**. The process is shown in Fig. 52-5. The machine is really like a large transformer without a secondary coil. The item to be heated must be a metal which

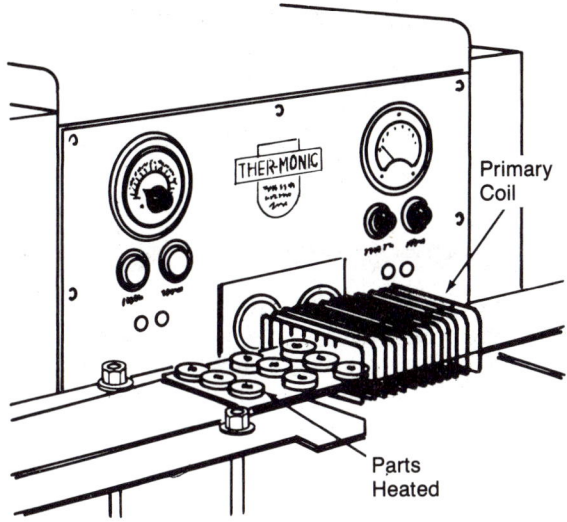

Fig. 52-5 Induction heating of a metal work piece uses the three transformer losses.

can conduct current. When the AC current flows in the primary coil, it tries to induce current in the secondary (the metal work piece). The secondary, however, is not a good transformer coil. It will have many little eddy currents, it may have resistance (copper losses), and it is likely to have enough reluctance to cause large amounts of hysteresis. All of these transformer losses will result in heat in the work piece.

Also, because it is a short circuit, very high current will flow in the work piece. It is important to remove jewelry when working around large coils (as in large transformers and motors) because the moving magnetic fields given off by these devices can heat the ring by induction heating and burn your finger off almost instantly.

Microwaves

Recently, **microwave ovens** have become a popular heat source in the home. The oven actually produces no real heat. It produces radio frequency (RF) waves in the microwave range (about 2,450 MHz). The waves shake the molecules in the food — especially water molecules — and the friction of these molecules racing around causes great amounts of heat inside the food. The air inside the oven is not heated, except for a small amount of heat which is picked up from the food by convection. Often there is a small fan in the oven. Its purpose is to make sure that the microwaves are reflected to all areas so that there will be no "cold spots" in the oven. It is really a reflecting device, not a fan. Metal pans should not be used because they could damage the oven.

The oven must keep all of its microwaves inside its cavity. Radiation can be very harmful to people and animals. Therefore make sure that the door closes correctly and that all seals and metal trim pieces are in proper place and not damaged. Additionally, there are safety features such as switches which will not allow the oven to operate when the door is not tightly shut. If any damage is detected or if the safety switches fail to operate, the oven should be inspected by a professional service facility.

Summary

Some electric circuits are especially designed to produce heat. In other circuits, heat is a form of wasted energy. Heat is generally produced by resistance, transformer losses, or microwave radiation. Heat may be transferred by radiation, conduction, and convection. Understanding these three methods of heat transfer and the ways to produce heat electrically enables you to control heat for useful purposes. The next chapter covers uses of electricity in lighting.

Important Terms

resistance heating
conduction
convection
radiation
arc
induction heating
microwave ovens

Review Questions

1. List three devices in your home which use resistance heating. How is heat transferred in each of these devices? (**Hint**: Some devices use more than one method.)

2. Can you join two pieces of plastic by arc welding? Could you heat them by induction heating? Why or why not?

3. What should you do if you find that your microwave oven still operates even after the door is damaged? Why?

Chapter 53

Power for Lighting

The Incandescent Light

The production of light was one of the very first widespread applications of electricity in homes and businesses. Today's electric lighting systems offer a broad range of lighting types to suit different applications.

The simplest and most common use of electricity for lighting is the **incandescent lamp** (Fig. 53-1). A filament of tungsten is enclosed in a glass bulb. The air inside the bulb is replaced with an inert gas which will not burn and which does not contain oxygen. This is done to make the filament last longer. It would quickly burn itself up if it had oxygen.

Advantages and disadvantages of incandescent lighting are:
- The fixtures (lamps) and their bulbs are inexpensive and easily installed.
- The bulbs are available in a wide variety of colors, shapes, sizes, and wattage ratings.
- Incandescent lights will work on either AC or DC current, but they do use a lot of power compared to some other systems.
- You can get bright light which does not badly distort the colors of the items illuminated, but glare can be a problem with this system.
- The heat given off by the bulbs can be a safety hazard, can make a room warmer than desired, or can be used for a heat source — depending on the situation.
- Bulbs may shatter if they get wet or are dropped.
- Bulbs do not last as long as some other lighting components do.

Despite the disadvantages, incandescent lighting probably is still the most common form of lighting in use today. The system is standardized, inexpensive, and easy for the homeowner to troubleshoot.

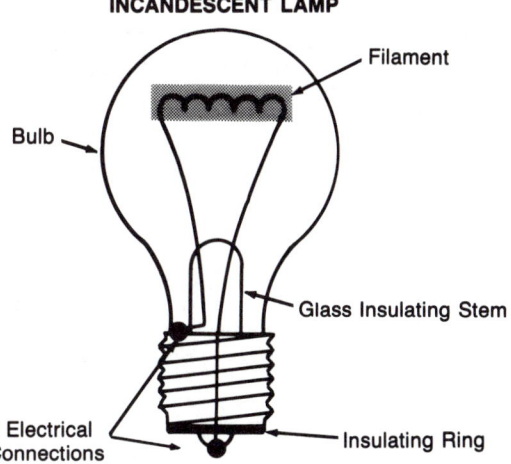

Fig. 53-1 Current enters one of the connections in the base, flows through the filament, and returns to the power source through the other connection.

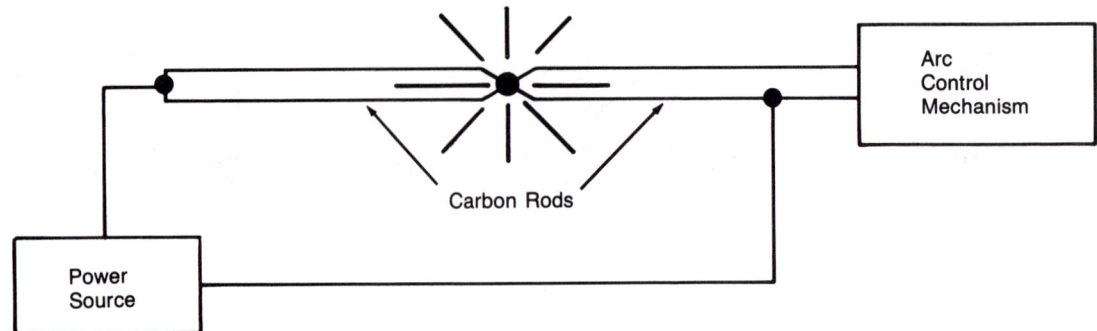

Fig. 53-2 The electric arc lamp can produce intense illumination with great amounts of ultraviolet light.

Electric Arc Lighting

Fig. 53-2 diagrams an electric arc lighting system. The principle of operation is almost the same as that of arc welding, except that a mechanism, instead of a human worker, maintains the proper arc length. These lights have been used in many applications requiring very intense light such as movie projectors, platemakers in the printing industry, and other industrial uses. They are quickly being replaced in many applications by the xenon (flash) bulb which may be pulsed quickly to produce intense light.

Fluorescent Lighting

Many schools, businesses, and public buildings use **fluorescent lighting**. The familiar long tubes are the most common shape of fluorescent lamps. Fig. 53-3 shows how a simple fluorescent light circuit works. The glass tube is coated on the inside with a light-emitting material (fluorescent or phosphorescent salts). The tube is filled with mercury vapor. The starter, an automatic switch, will remain closed to pass current until it warms up, and then it will open.

When the power switch is first turned on, current flows from the generator, through the filament on one end to the tube, through the closed starter switch, and then back to the generator through the other filament. Therefore, both filaments get hot.

The amount of actual light provided by the filaments is not very important. Their only real job is to heat the mercury vapor

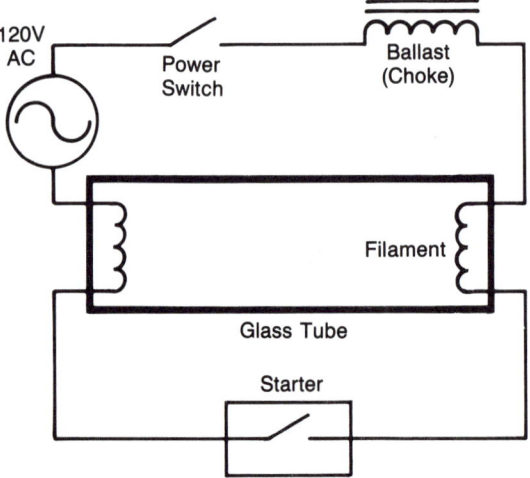

Fig. 53-3 The filaments are only heated during the starting time in this fluorescent light circuit.

inside the tube and to give off electrons by thermionic emission. As soon as conditions are right inside the tube, the starter switch opens and breaks the path of current from one filament to the other. At this time, the electrons emitted from the hot filaments race across the space inside the tube like a small lightning storm. As they move, they bump into mercury vapor atoms and make them give off ultraviolet light. But, you cannot see much of the ultraviolet light with your human eyes, so it is of little value by itself. This is where the fluorescent coating on the inside of the glass tube becomes important. The ultraviolet light rays strike the coating and cause it to glow — giving off useful, visible light.

The **ballast** is a choke (coil) which opposes rapid changes in current. Since the current is AC (always changing), the ballast helps to limit current flow and also helps supply the surge needed to start the arc of electrons through the long tube. This is only one of the possible circuits which will operate fluorescent lights. Many industrial and commercial circuits eliminate the starter and have a more complicated ballast which provides a very high voltage surge to start the arc.

A fluorescent lighting system has some advantages and disadvantages compared to incandescent lighting:
- The fixtures and the tubes are more expensive and more complicated to install and troubleshoot.
- The tubes last much longer than incandescent lamps and consume less power, but they will not last long if they are turned on and off frequently.
- The tubes are available in many shapes and colors. They produce light that is well diffused and free of glare. Also, the tubes do not get hot as incandescent bulbs do.
- Fluorescent lights do not work well in extremely high or low temperatures. They are not efficient when operated on DC. Some types are slow starting.

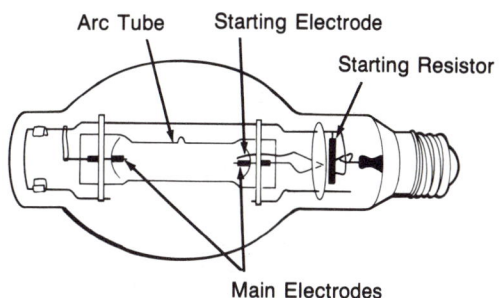

Fig. 53-4 A typical mercury-vapor lamp.

Mercury-Vapor Lights

Mercury-vapor lamps are used in outdoor applications (highways, bridges, parking lots) and in large enclosed areas (arenas, factories). They operate very efficiently. One disadvantage is that they produce a light which is very high in the blue and green range with little yellow and red. This makes things appear to be the wrong colors. Another problem is that many types require a long start-up period. Some types cannot be restarted after turning them off until they cool down. For these reasons, when used indoors, mercury-vapor lighting is usually used with some other system. A typical mercury-vapor tube is shown in Fig. 53-4. These tubes and the fixtures in which they operate are expensive, but they have very long life.

Neon Lights

Neon lights are also gas-filled tubes and come in a variety of shapes and colors. Their most common use has been in the advertising industry. The brightly lighted signs made of neon tubes once made urban areas "glitter" with color. The circuit to operate neon tubes mainly consists of a very high voltage transformer (more than 10,000 volts). Changes in style and the development of attractive, colorful signs which use acrylic plastic panels have reduced the popularity of neon signs.

Neon lamps have found another use, however. Small neon bulbs, called **neon glow lamps** are used as indicator lamps and sometimes as important parts of triggering circuits. The bulb has two electrodes. The neon gas must be ionized before the electrodes will glow. Many of these bulbs ionize at about 55 volts. If DC current is used, only one of the two electrodes will glow, but both will glow on AC.

Sodium-Vapor Lights

Another lighting system used in large, open places is **sodium-vapor**. Sodium-vapor lights are about the same as the mercury-vapor type, but they give off an orange-yellow light. If everything in the parking lot looks yellowed or almost black (and high in contrast), it is because of sodium-vapor lighting.

Pulsed-Xenon Lights

Photographers often use an electronic flash attachment for their cameras. The lamp in these units is a **xenon** (flash) type. You also have seen xenon lamps in strobe lights. When the photographer fires the flash unit, a capacitor releases a very high voltage pulse which causes current to arc through the xenon tube. The light given off is very bright and has almost all of the color range of natural, bright sunlight. These lamps are also used a great deal in the printing and graphic arts industry. They can replace carbon arc lights in platemakers for greater simplicity.

When used in "continuous" operation, however, the tube must be pulsed. The circuit which operates the tube supplies many rapid high voltage pulses — so the lamp actually flashes on and off very rapidly. The flashes are so rapid that your eyes are tricked into "seeing" light which you believe to be "continuous." Pulsed xenon lights are often cooled by fans because they do produce high amounts of heat. Even in small, battery-powered flash units for cameras, the voltage of the capacitor's discharge is hundreds of volts. Be very careful when handling these units. They can give you a real shock!

Summary

The two most common types of lighting for homes and offices are incandescent and fluorescent. Each of these systems has advantages and disadvantages which make them better suited for specific applications. Many of the disadvantages may be overcome by using the two systems together. Other lighting systems include mercury-vapor and sodium-vapor for outdoor or large area lighting. Xenon and neon are two special purpose types of lighting which would not be commonly found lighting homes and businesses.

Important Terms

incandescent lamp
fluorescent lighting
ballast

neon
neon glow lamps
mercury-vapor lamps

sodium-vapor lamps
xenon

Review Questions

1. Why are incandescent lights used so often in lighting homes? What are the advantages and disadvantages of this type of lighting?

2. Tell why schools usually use fluorescent lighting and mention its main advantages and disadvantages.

3. Tell some of the uses of each of these types of lighting: mercury-vapor, sodium-vapor, and xenon.

History: Thomas A. Edison

Edison is probably the most famous of American inventors. He has been featured in books and movies. Some of his well-known inventions include the phonograph, improvements to the telephone, a motion picture projector, and in 1879, the electric light bulb. He discovered the Edison effect which paved the way toward the development of vacuum tubes.

Edison is a true American hero in many ways. He came from a poor background, had little education, and made his own way by the "sweat of his brow." He patented nearly 1300 inventions — certainly the term "genius" applies to him. By his own definition, "genius is 1 percent inspiration and 99 percent perspiration." Many of the time-saving devices we take for granted today are direct results of Edison's dedication and efforts. His success — through effort — inspires us to maximize our own talents. His method was to keep trying in the face of failures. He once tried nearly 8000 designs for a new storage battery without success. Such undaunted determination could help us all.

Project: Flashing Numerals

Parts List

1	SPST switch	S_1
1	4.5 to 6 V battery with holder	B_1
1	100 K-ohm potentiometer	R_1
2	1 K-ohm resistors	R_2, & R_3
1	270 ohm resistor	R_4
2	220 ohm resistors	R_5, & R_6
1	555 timer IC	Z_1
1	100 microfarad capacitor	C_1
1	ECG 289A NPN transistor	Q_1
2	7-segment LED displays, common Anode	D1, & D2
2	16 pin SPST DIP switches	S_2 & S_3

Suitable enclosure, wires, hardware, and PC board as required.

Notes

This project has many novelty uses. It could be concealed in a cake to flash someone's new age on his or her birthday or the year of graduation from school. There are many options available. You could use long ribbon cables between the LED displays and the rest of the circuit to make remote operation possible. Even many letters can be formed with the two displays. If flashing is not desired, the whole flashing circuit may be left out. To operate the circuit, select the desired segments to light with the DIP switches and set the flashing rate with the potentiometer.

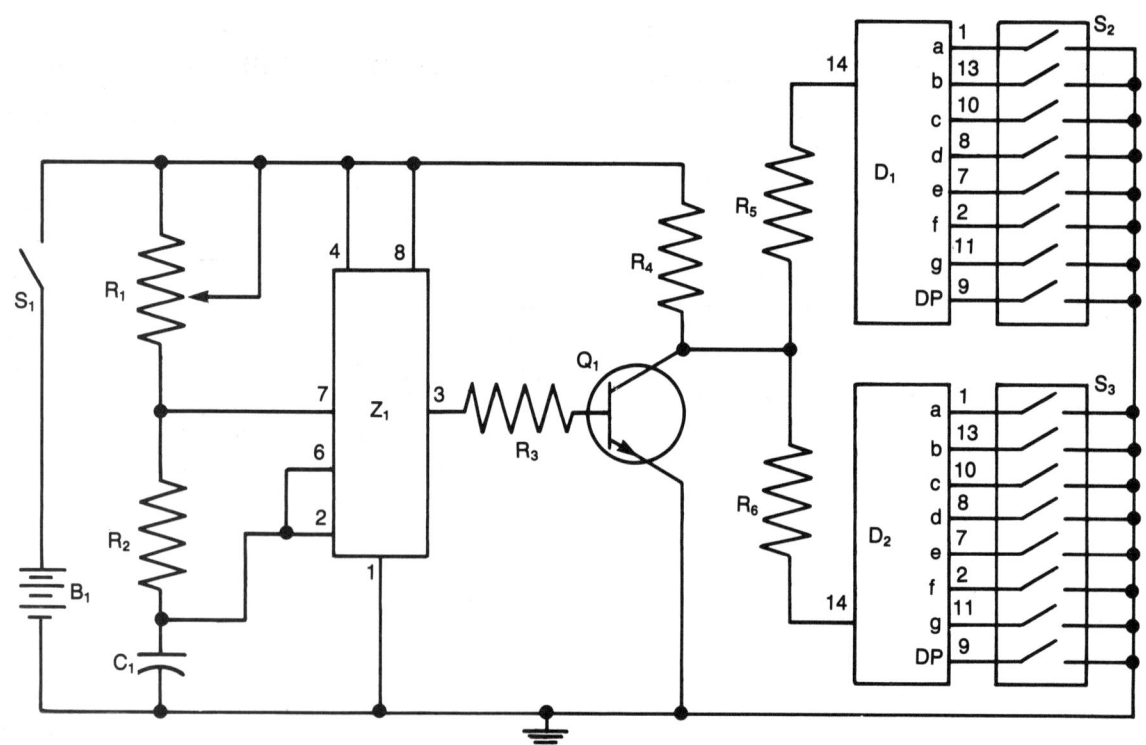

Fig. 53-5 The Flashing Numerals Project is easy to build and allows you to try some of your own design alterations.

Chapter 54
Power for Motors

Electric Motors

Lighting and heating are important uses of electrical power. The other major use of electrical power is to produce motion. Machines that convert electrical energy into rotary motion are called **electric motors**. This chapter deals with motors and how they work.

How Motors Work

All electric motors work on one principle. They use two magnetic fields (Fig. 54-1). One of the fields is developed by coils or permanent magnets that remain stationary. The other field comes from coils or magnets on a rotating shaft through the center of the motor. The shaft assembly may be called an **armature** or a **rotor**, depending on the type of motor. The coils or magnets are frequently called the **field** or the **stator**.

As Fig. 54-1 shows, the turning part is trying to line itself up with the magnetic field produced by the stationary part. When the north and south poles are as close together as possible, it will stop and remain in this position. If you could somehow make the poles of either the stationary or the turning field reverse (just as the existing poles were beginning to line up), then the rotating part would have to keep turning further to try to "catch" the repositioned field. Just as the poles begin to line up in the new position, you could reverse the poles again and force the rotation to continue for another cycle.

This is what actually happens in all motors. It happens over and over again so rapidly that the armature or rotor turns smoothly and with enough power to drive other devices attached to its shaft. Which field is reversed (the one on the shaft or the stationary one) and how it is made to reverse is what determines the type of motor.

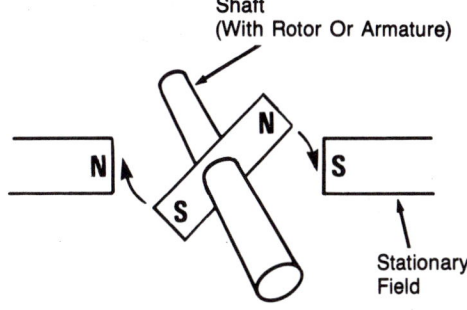

Fig. 54-1 All motors depend on the efforts of one magnetic field to align itself with another one to develop rotory motion.

Simple DC Motor

A simple **DC motor** is diagrammed in Fig. 54-2. The DC motor has a **commutator** and **brushes** which act as a reversing switch when the shaft turns. Current goes from the battery through the two field coils. These stationary coils produce magnetic fields that remain the same polarity all the time. Current also is sent from the battery through the brushes and commutator to the armature coils. Coil A will produce a north pole when the current goes through it from one direction and a south pole when the current reverses. Just as the armature begins to line up with the field coils, the commutator will have turned enough to contact the other set of brushes. Thus, the current will now be going through both armature coils in the reverse direction. The pole that had been north will now be south and will turn all the way around to the other side!

Fig. 54-2 is called a **shunt wound** DC motor because the field coils are in parallel with the armature. Its schematic appears in the bottom of the figure. It is also possible to wind the motor in series. The schematic for a **series wound** DC motor is shown in Fig. 54-3. The series DC motor is sometimes called a "universal" motor because it will also operate on AC. One drawback of the series motor is that it tends to speed up and

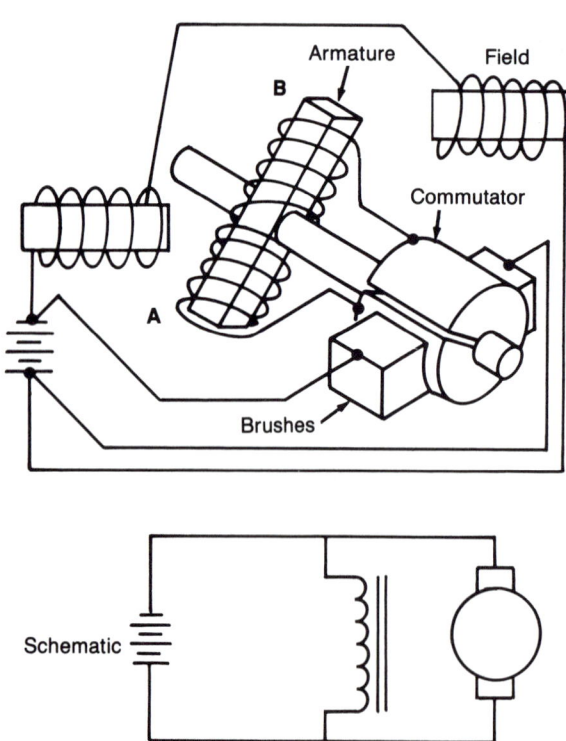

Fig. 54-2 Simple shunt-wound DC motor. The brushes carry current to the armature. The commutator makes this current reverse in polarity as the shaft rotates.

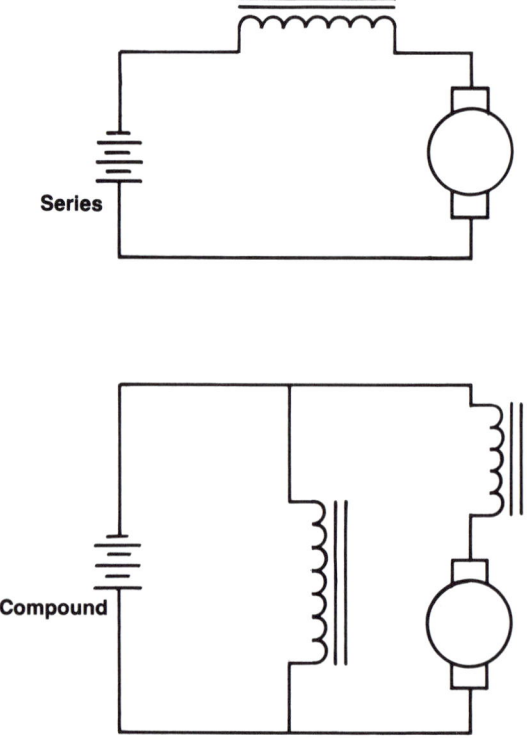

Fig. 54-3 Series and compound DC motors operate on the same principle as shunt wound motors, but their speed and torque characteristics are different.

slow down when the load becomes lighter or heavier. In fact, if operated under total no-load conditions with high voltage, the series motor could turn so fast that it would damage itself. Shunt motors are much more constant in their speed with varying loads.

The figure also shows the diagram of a **compound** DC motor. The compound motor has two field coils: one in series with the armature and the other one across it (in parallel with it). Compound motors combine some of the advantages and disadvantages of both series and shunt motors to provide the needed torque and speed characteristics for specific applications.

AC Induction Motors

Most AC motors depend on the normal direction changes of the AC current for their field reversing action. But, reversing the current was only one of the two jobs of the brushes and commutator in the DC motor. The other job was to get the current into the moving armature. The method AC induction motors use to do this is truly ingenious. Fig. 54-4 shows the turning part of an AC induction motor. This turning part is called the rotor in an AC induction motor. Specifically, this is a "squirrel-cage rotor." The rotor is made up of laminated pieces. Grooves are cut lengthwise along the rotor's surface. Copper bars are placed in the grooves. The copper bars are all joined on the ends by two rings of copper. If the laminated core were removed, the copper parts would look like a cage. This is why it is called a squirrel-cage rotor. The rest of the motor is just an arrangement of stator (stationary field) coils and metal poles to concentrate their effects.

There is no electrical connection to the rotor whatsoever. The rotor is not a permanent magnet, either. How do you make the rotor have the changing magnetic fields needed to force its motion? As the

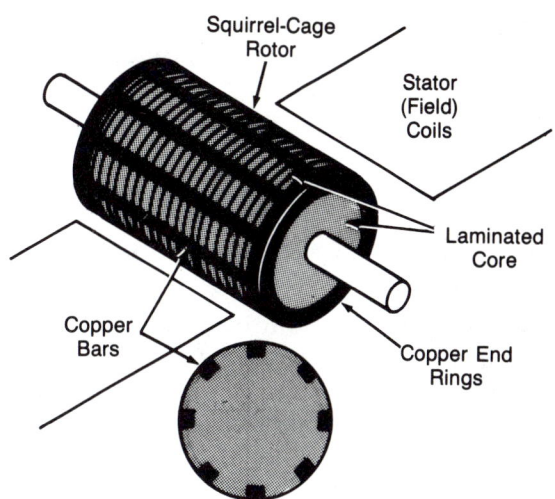

Fig. 54-4 The squirrel-cage rotor is not electrically connected to the stator in an AC induction motor.

stator (stationary) field coils carry AC current, they develop rapidly changing magnetic fields around the rotor. These changing magnetic fields cut through the copper bars in the rotor. Since the copper rings on the ends of the rotor connect all of the copper bars together, the copper bars actually form many one-loop coils. New electric currents are set up in these one-loop coils by magnetic induction (just as currents are developed in the secondary coils of transformers). Thus, the changing magnetic fields created by the stator coils induce current into the copper bars of the rotor.

These new, induced currents in the one-loop coils (copper bars) meet very little resistance. They can be very high currents. When they flow, they also set up magnetic fields around their coils (the copper bars). These new magnetic fields are concentrated by the laminations of the rotor which serve as an iron core (just like the ones you studied in transformers). So, you get both current and changing magnetic fields into the rotor of an AC induction motor by the process of electromagnetic induction without any electrical connection between the rotor and the power source.

Starting AC Motors

Once they are running, AC induction motors can do a lot of work very effectively. They have a problem starting, though. If you just placed a squirrel-cage rotor in a stator field and applied current, it would likely sit there, hum, and create a lot of heat. If you gave it one little twist with your hand, it would begin to power itself and run smoothly. This is because the fields in the rotor must be moving to be a little "behind" the fields in the stator. This is called "slip." If the fields are right together, there is nothing to cause the rotor to turn.

Therefore all AC induction motors have some method of starting. One method involves a **centrifugal switch** which supplies current to a special starting coil when the motor is still or running slowly. When the rotor gets up to proper speed, the switch automatically opens and removes current from the starting coils. Other starting methods include the use of capacitors, three phase current (which develops slip automatically and needs no help to start the motor), and shaded poles.

Shading causes the field that the pole produces to move across the surface of the pole. A shaded pole is created by cutting a slot in the core material of the pole and installing a one-loop coil around that part of the pole (Fig. 54-5). When the current flows in the main coil, it induces current into the shading loop. This current will be out of phase with the current produced by the main coil. So the magnetic field which it produces in the shaded section of the pole (core) also will be out of phase with the field in the rest of the pole.

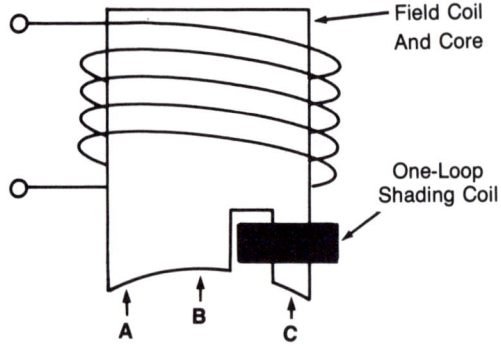

Fig. 54-5 The magnetic field in a shaded pole will first be strongest at A, then at B, and finally at C.

The field will seem to move from point A to B and then to C as each of these regions becomes stronger. This moving field will be enough to start the rotor turning so that normal operation can begin.

Induction Motor Characteristics

Once started, most induction motors run at a relatively constant speed. They develop extra torque to overcome increases in loads (until they are overloaded). The different types of AC induction motors vary greatly in their starting characteristics. Capacitor start and three phase types start with high torque and can be started with heavy loads already applied. The shaded pole and permanent split phase types have very little starting torque and are only suited for light duty use such as in fans. When selecting a motor, you must consider all of the conditions under which it will operate.

Special Types of Motors

Synchronous motors (Fig. 54-6) are used in applications not requiring high torque. They do demand very constant speed, though. They are specially designed to synchronize ("lock in") with the AC generator's cycles and to keep perfect pace with the 60 Hz AC cycles.

Servo motors are designed to keep in perfect step with a controlling source. These have been used in radar systems to ensure that the screen shows what the antenna is aimed at in the proper location. Stepper motors also can be used to turn a selected amount and then stop at a chosen point. The main difference is that their input must be a series of pulses. Both types are used in industrial robots and specialized equipment that demands great accuracy.

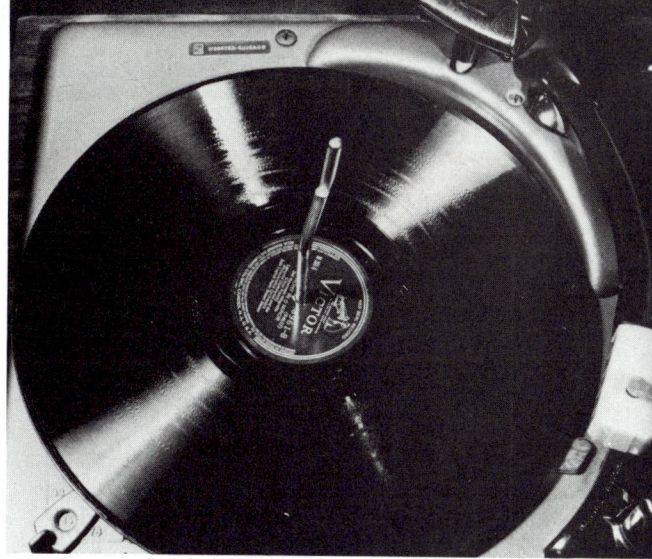

Fig. 54-6 The small AC synchronous motor keeps the turntable turning at the correct speed so that the records will sound natural.

Summary

Electric motors are a very important part of daily life. Almost everything that operates from electricity and involves motion uses some sort of electric motor. All motors work on the same principle. Like poles repel and unlike poles attract. The main differences between the types of motors has to do with how the poles are set up and changed. DC motors use a commutator and brushes to get current into the armature and reverse it as needed. The universal DC motor also may run on AC. AC induction motors have no electrical connection between their stator coils and their rotor. The current is transferred to the rotor by magnetic induction, just as it is in a transformer. AC induction motors require a starting system because they do not self start. The type of motor selected for a given job should depend upon the characteristics of the motor and the conditions under which the work is to be done.

Important Terms

electric motors	stator	shunt wound
armature	DC motor	series wound
rotor	commutator	compound
field	brushes	centrifugal switch

Review Questions

1. What parts must a motor have in order to operate on both AC and DC current? Which type of motor can do this?
2. How are the magnetic fields in the rotor of an AC induction motor set up?
3. Which type of motor would you choose for a small AC powered clock? electric drill? window fan?

Chapter 55
Communication by Telephone

Fig. 55-1 The first telephone had little in common with the phones of today.

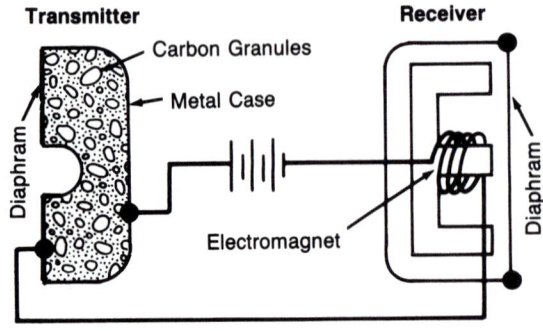

Fig. 55-2 Simplified one-way telephone circuit.

Communication Devices

This chapter begins a series of four chapters on electrical and electronic devices for communications, beginning with the telephone. Few modern conveniences are more necessary and more taken for granted than the telephone.

Early Telephones

The first telephone was invented by Alexander Graham Bell in 1875. Fig. 55-1 shows this strange looking instrument. It did not have a separate **transmitter** (microphone) and **receiver** (speaker or earphone) like today's telephones. It also did not have any method of alerting other phones that someone was trying to make contact. This prototype merely showed that it was possible to send the sound of the human voice from one place to another by means of electrical currents in wires. **Ringers** (bells) and **dials** were added as the phones became more complicated.

How is Sound Carried through Wires?

A diagram which shows how sound can be carried electrically appears in Fig. 55-2. In this circuit, some steady current would be flowing all the time. It would be very low, though, because the carbon granules in the transmitter form a high resistance. As you speak into the transmitter, the sound waves from your voice compress (squeeze) the granules closer together. The closer together the granules are, the better they will conduct current. That is, when the grains are squeezed tightly, they have less resistance and allow greater current flow. When the current flow becomes greater, the coil in the receiver develops a stronger magnetic attraction. The magnetic strength of the coil will change as you speak into the transmitter. It will change at an audio frequency rate. The coil will alternately attract and release the diaphragm in the receiver. So, the diaphragm in the receiver will be moved at the same rate and with the same amount of force as the one in the transmitter. This second diaphragm will move the air around itself to create sound.

Speakers in radios and stereo systems do the same thing. They have large paper or plastic cones. The example circuit shown could only carry voice information from left to right. A similar circuit with the transmitter and receiver in the opposite positions would be required to bring signals back the other way.

Making Connections

The basic method used to carry the voice signals has been understood for 100 years. The methods used to make connections with other phones have changed dramatically. When the phone system was small, with only a few phones in each town, all connections were made by **operators** who actually plugged your line in with the line of the phone you wished to reach. The operator also sent the current to "ring" the other phone.

As the system grew, it became impractical to have all connections made in this way. The dial phone made it possible for the user to dial a number (now seven digits) to reach any other phone in a dialing area. The operator was only needed for long-distance calls (calls outside the bounds of your dialing area). Eventually, it even became possible to direct-dial long distance calls by dialing 1 and the area code before dialing the number. Now, the operator is only needed when you want to place a collect call or some other special type of call, such as person-to-person.

The actual telephone "instruments" in homes and offices changed quite a bit over the years. Early phones had a crank which turned a hand-operated generator. The generator developed the power to ring the ringer at the operator's switchboard. Later, phones had dials which were used to select the other phones you wanted to reach by number codes. Today, many phones have push buttons instead of dials.

The Telephone Exchange

Until recently, **telephone exchanges** hummed with the sounds of thousands of relays making contact. The relays did the jobs operators had done in the telephone's early days. They connected the lines of parties who were trying to make contact. Now, most of this work is handled by digital circuits and computers. The solid state circuits operate much faster. They cost less to build, repair, and replace. They do not waste power like the old relays did (making magnetic fields).

What the Dial or Keypad Does

When you dial a number, or press it into a keypad, your telephone changes the number into signals that can be used by the circuits at the exchange to make the proper connections. The signals of the two systems

Fig. 55-3 Telephone communication has progressed tremendously since the days of switchboard operators.

are different. The dial sends a series of pulses as its signal. The keypad on newer phones is called a **touch-tone** dial. As the name implies, it develops a tone for each number pressed. One problem that the phone company had to overcome in developing the new system was to make it possible for both systems to work together. If this had not been done, the old phones would had to have been replaced immediately when the system was changed. The cost and complexity of such a change would have been enormous.

System compatibility is very important. Be sure that you buy and install good quality telephone equipment in your home. The installation must be done properly to avoid tying up the lines with malfunctioning equipment.

A great variety of telephone styles are now available from many companies. There are remote phones which send radio signals from the handheld receiver to a base (which is wired to the phone lines). These phones free the user from tangled wires and allow travel further from the base. Automobile telephones are more popular now than ever before. There is a great deal of competition among several companies that sell telephone equipment or services (such as long distance networks). It is possible to telephone practically anywhere in the world where there is another phone.

Special Equipment and Services

The device that links computers together through the telephone lines is called a **modem**. Early modems were small, cradle-shaped devices in which the handset of a phone would rest while the modem was in use. The user would dial the number to contact the desired equipment and then place the handset in the cradle. Many modern modems hook directly to the phone lines. They allow the number to be dialed from the computer's keyboard automatically.

Many modern phones have memory capabilities that allow the user to preset certain numbers so that they may be dialed automatically by pressing just one button. This service is also available in many exchanges to people who do not have special phones. A higher fee is charged by the phone company which allows you to store the selected numbers in their equipment at the exchange office.

Summary

The telephone is an important part of everyday life. It is both a personal convenience and a necessary business device. Changes in telephone design over the years have resulted in a system that operates quickly and efficiently. Making connections between phones demands a complicated, computerized network. At one time, operator assistance was needed for all calls. Most calls are completed by direct dialing. The next chapter deals with communication by radio.

Important Terms

transmitter
receiver
ringer
dials
operator
telephone exchange
touch-tone
modem

Review Questions

1. What are the differences between Bell's first telephone and the phone in your home today? What is the purpose of each visible improvement?

2. Why is system compatibility important?

3. How is the human voice transferred from one location to another through phone lines?

History: Alexander Graham Bell

Alexander Graham Bell was born in Scotland in 1847. He lived much of his life in Canada and the United States. He had worked with his father in teaching deaf people to speak. Bell's interest in inventing an aid for deaf people led to the invention of the telephone. He displayed his invention at the 1876 Centennial Exposition in Philadelphia. Just as Edison improved Bell's telephone, Bell improved Edison's phonograph.

Bell lived to see the impact of his invention. In 1915 he spoke over the first transcontinental telephone line. The person on the other end of the line once again heard those unforgettable words: "Watson, please come here. I want you." The person on the other end, 3000 miles away, was, indeed, his old assistant Thomas Watson. Watson had heard those very words 40 years before when an accident caused Bell to call for help. They were the first words ever carried by wires. Bell's desire to help deaf people has made an impact on all of us.

Chapter 56
Communication by Radio

The Radio Transmitter

Telephones were limited by their need for wires to connect all of the phones together. The next step was to find a way to send voice signals through the air without wires. This is accomplished by radio transmitters and receivers. Now, many telephone conversations make at least part of their journey by means of radio signals.

You have already seen the block diagram of a basic AM radio transmitter. The very first stage of a transmitter is the oscillator (Fig. 56-1). The oscillator develops an AC waveform of the proper frequency for transmission. Each radio station operates on one assigned frequency. Tank and other resonant circuits at various places in the transmitter make sure no signals are transmitted that are not of the station's correct frequency.

The AC waveform is amplified and fed to the power amplifier. At the same time the oscillator is creating the radio frequency (RF) waveform, sound (speech and music) is being sent into the audio amplifier. This audio signal is much, much lower in frequency than the RF signal from the oscillator. The RF signal in the example AM broadcast station will be 1310 KHz. The audio signals will vary between 16 Hz and 20,000 Hz. This means that even the highest possible audio inputs will only be about 1/60th of the frequency of the RF signal. In other words, the RF signal will peak about 60 times for every one peak of the audio input frequency (or more for the lower audio frequencies).

The two signals are shown in Fig. 56-2 for comparison. The RF is a steady signal which stays the same frequency (number of peaks per inch in the graph) and amplitude (height) for its entire length. The audio signal changes in both frequency and amplitude. The lower the audio frequency, the longer each peak. Peak A is a very low frequency, peak B is medium, and peak C is a high frequency. The amplitude of the audio signal also changes. Amplitude varies with the loudness of the input. Thus, peak B was the loudest (most powerful) input of the three. The graph, then, shows three peaks of audio input:

Fig. 56-1 A simple AM transmitter.

- A — low frequency, low volume.
- B — medium frequency, high volume.
- C — high frequency, medium volume.

It is impossible to actually draw the graphs of both the RF and audio frequencies in the same scale on the page in your text. If both graphs were to the same scale, there would be about 60 peaks of the RF signal in the space occupied by the C peak!

Modulation

The **modulator** makes the power output amplifier's output vary at the rate and amplitude of the audio signal. After modulation occurs, the combined (modulated) signal looks like Fig 56-3. The frequency of the signal is the RF frequency. The amplitude, however, changes at the audio rate. This is what is transmitted through the antenna: A waveform with a frequency of 1310 KHz but which changes in amplitude (height). This is called **amplitude modulation** or AM. This is the type of signal transmitted by all AM radio stations.

RF Carrier (from oscillator)

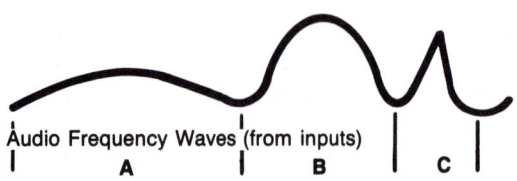
Audio Frequency Waves (from inputs)
A   B   C

Fig. 56-2 The RF and audio frequency inputs.

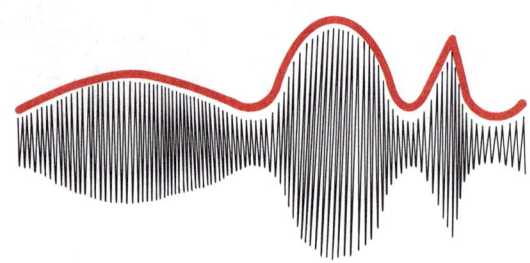

Fig. 56-3 The modulated signal. The colored line is not transmitted. It was drawn to call your attention to the fact that the shape of the signal is identical to the audio input.

Receiving the AM Signal

The device you call a radio is more correctly a radio frequency **receiver**. When you tune it to the imaginary station (1310 KHz), you will receive the signals transmitted. By changing the resonant (tank) circuit, you can receive other stations which broadcast on different frequencies. (Review Chapters 37–39 if you do not remember how this is done.) The block diagram of an AM receiver appears in Fig. 56-4. The signal is received by the antenna and amplified by the RF amplifier.

The local oscillator produces a steady RF signal which is exactly 455 KHz higher in frequency than the incoming frequency of the selected station. In the example (1310 KHz), the local oscillator signal will be 1765 KHz. If you subtract the incoming RF signal from the local oscillator frequency, you get 455 KHz: 1765 − 1310 = 455 KHz. When you change the tuning capacitor to select a different station, you tune resonant circuits in both the RF amplifier and the local oscillator. So, when you change the station to 780 KHz, the local oscillator will also

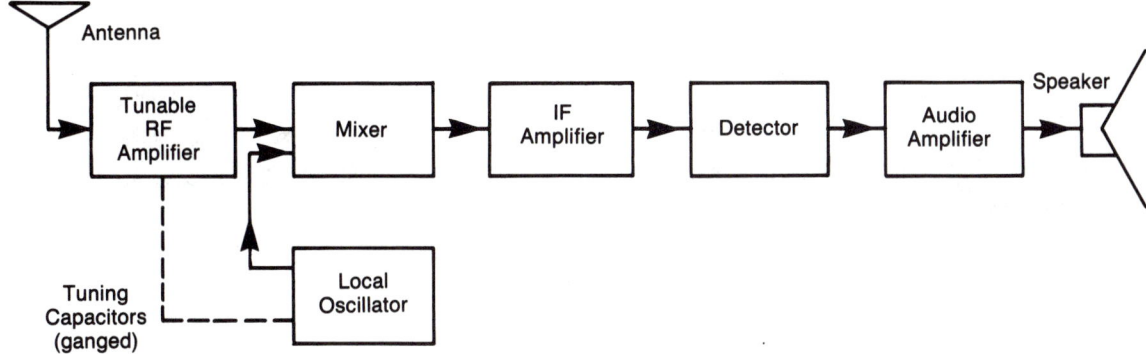

Fig. 56-4 Block diagram of an AM superheterodyne receiver — the device we commonly call an AM radio.

change to 1235 KHz to give a difference of 455 KHz. The difference must always be 455 KHz.

The two signals are sent into the mixer circuit. The mixer combines them in such a way that its output will be an amplitude modulated signal of 455 KHz. This signal will look exactly like the modulated signal in Fig. 56-3, except that the internal RF part will be 455 KHz instead of 1310 KHz. Receivers which use this tuning system are called **superheterodyne** receivers. They are the most popular type.

Why Change to 455 KHz?

The output of the mixer is the **intermediate frequency** or IF. The reason you change the incoming signal to an intermediate frequency is to make the stages which amplify this signal simpler. Some antique radios that did not use the superheterodyne system had resonant circuits in each amplification stage. The listener had to turn several tuning knobs each time the station was changed. These radios were very difficult to operate. Changing to the IF frequency allows you to make all of the remaining stages operate at 455 KHz. This permits you to tune the radio by adjusting one simple control.

The next stage is the IF amplifier. It amplifies the intermediate frequency signal. There may be more than one IF amplification stage, but they will all operate at 455 KHz.

The Detector

After the IF signal is amplified, the audio signal is separated from the IF "carrier" in a circuit called the **detector**. The whole IF (modulated) signal is fed into the detector. The most important part of the detector circuit is actually a diode. When the modulated IF signal goes through the diode, it is rectified. In other words, only the top half of this AC signal will pass through the diode. The output of the detector will be a half-wave rectified, pulsating DC. See Fig. 56-5. The remaining circuits in the radio receiver will not respond to the high frequency (455 KHz IF) part of the signal. They will respond to the audio frequency portion. That is, the imaginary line across the top of the detected signal. "Demodulation" refers to the whole process of detecting and then ignoring the IF portion. Finally, the demodulated audio frequency signal goes to the audio amplifier. Then it is converted back to sound waves by the speaker.

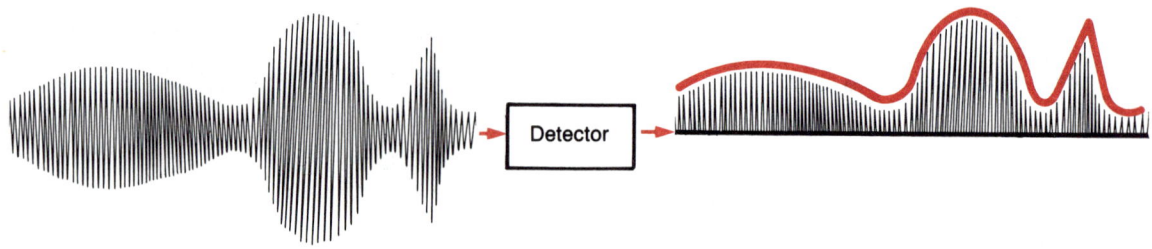

Fig. 56-5 The output of the detector is a pulsating DC which is still amplitude modulated. It is too difficult to draw this small, but the bottoms of the rectified signal are all supposed to be flat. As before, the light colored line at the top is not part of the signal.

FM Radio

The transmitter and receiver just discussed are both AM radio circuits. Another system of transmitting radio waves is called FM or **frequency modulation**. When you modulated the AM signal, you made the RF carrier wave change in amplitude as directed by the audio input signal. In FM radio, you change the RF carrier's frequency in response to the audio input signal. Fig. 56-6 shows how frequency modulation (FM) works.

In this example, the RF carrier with a steady sine wave in the audio frequency range is being modulated. Consider the signal transmitted from an FM broadcast station which is assigned a frequency of 103 MHz. Notice that the FM stations use a much higher frequency than the AM stations. So, the RF (carrier) signal produced by the transmitter's RF oscillator will have a steady frequency of 103 MHz. However, when this RF signal is modulated, its frequency is changed, not its amplitude.

The amplitude of the FM modulated waveform stays the same for its entire length. But look closely at the spacing of the peaks in the modulated signal. They are further apart when the audio input signal is low and closer together when the audio signal is high. The only times they are the same distance apart as in the original carrier

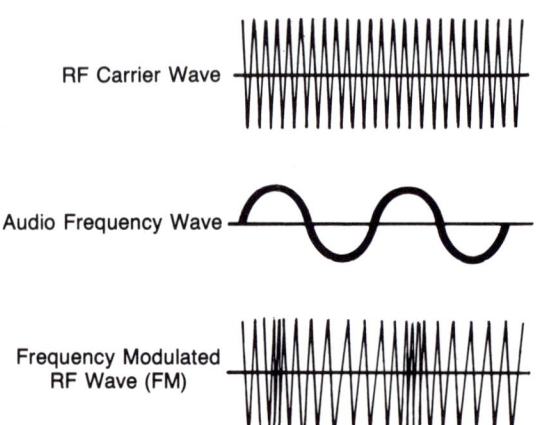

Fig. 56-6 In the FM system, the frequency of the RF carrier is modulated (changed) by the audio input signal. The amplitude (height) of the FM signal remains constant.

waveform is when the audio frequency waveform is exactly at its centerline. The FM transmitter actually sends out a signal that changes in frequency! In fact, half the time the transmitter is broadcasting on a frequency higher than the assigned one, and the other half, it is broadcasting below its assigned frequency.

If you average all of the output signals for a period of time, you will see that the average frequency is the assigned frequency of 103 MHz, but the transmitter is above this frequency or below it most of the time. For this reason, FM broadcast stations are spaced far enough apart on the FM band to allow each one an operating range ("bandwidth") of about 200 KHz.

The FM Receiver

The receiver used to listen to FM stations is very similar to the AM receiver. It uses the superheterodyne system. The IF is 10.7 MHz in most FM receivers. Two differences between AM and FM receivers, however, are that the demodulation circuit is called a **discriminator** instead of a detector, and there are stages called **limiters** between the IF amplifiers and the discriminator.

The discriminator for FM cannot be a simple diode. An example discriminator circuit appears in Fig. 56-7. The two tank circuits (A and B) are tuned to different frequencies. Tank A is tuned to a frequency higher than the incoming 10.7 MHz (frequency modulated IF) signal. Tank B is tuned to a frequency below the IF signal. When the exact 10.7 MHz frequency is input, both D_1 and D_2 will be allowed to conduct equal amounts of current by their tank circuits. When the frequency is higher than 10.7 MHz, however, D_1 will conduct more than D_2. This means that R_1 will have a greater voltage drop than R_2. Take the output from these two resistors. When the voltage of R_1 is greatest, the output will form a peak above the center line. When R_2 has the greater voltage drop, the output will be a peak below the center line. Another type of FM demodulator circuit is called the ratio detector. The ratio detector also uses two diodes. It accomplishes the same function as the discriminator in a different way.

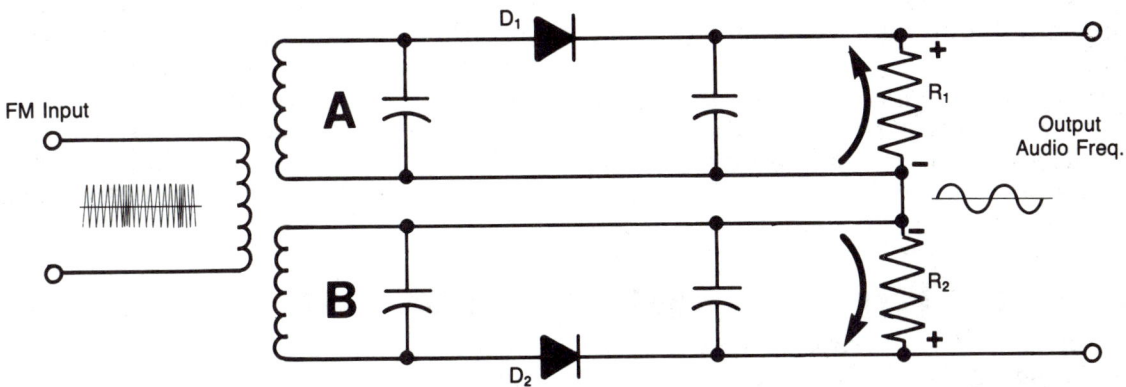

Fig. 56-7 The two tanks in an FM discriminator are tuned to frequencies above and below the carrier frequency.

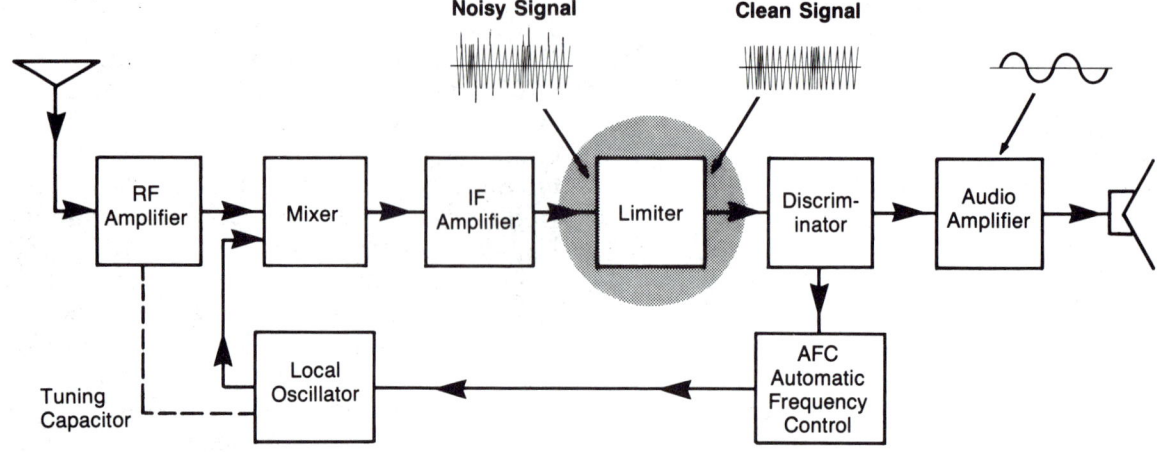

Fig. 56-8 Block diagram of FM receiver showing input and output of the limiter.

Limiting Noise

Limiters are circuits that limit the amplitude of the IF signal. Remember that all of the information in an FM signal is carried in frequency changes. The amplitude should stay the same all the time. But, if noise — in the form of amplitude peaks — enters the signal, it could be detected, amplified, and output through the speaker. Therefore limiters clip excess amplitude from all peaks (Fig. 56-8) The limiter stage is a simple amplifier circuit that is overdriven so that its output is clipped on both the positive and negative peaks.

This process cannot be used in AM radio because the important information in the signal is carried by the amplitude of the signal peaks. Limiting would erase the information! This is one advantage of FM radio. The output is "cleaner" (free of noise). The output is not affected by static caused by automotive spark plugs and other sources of interference.

Summary

Both AM and FM radio are used today. Each system requires transmitters to send signals and receivers to pick them up. Most modern radio receivers are superheterodynes. Radio receivers combine several circuits: oscillators, amplifiers, tanks, and half-wave rectifiers (detectors). In AM signals, the amplitude of the signal carries the needed information. In FM signals, the information is carried in changes of frequency above and below the assigned center frequency. Impressing information onto a carrier signal is called modulation.

Important Terms

modulator
amplitude
  modulation
receiver
superheterodyne
intermediate
  frequency
detector
frequency
  modulation
discriminator
limiter

Review Questions

1. Explain the differences in how information is carried by AM and FM radio signals.

2. What is the purpose of the superheterodyne system? How does this system work?

3. Explain the purpose and function of the AM detector, FM discriminator, and limiter.

History: Morse and Marconi

Samuel F. B. Morse is credited with the invention of the telegraph and the development of Morse Code (a form of the Morse Code is still used). Some historians believe that Joseph Henry did the electrical work. Morse refused to share fame and wealth with those who had helped him. A great achievement of his was convincing Congress to spend $30,000 to build the first telegraph line and begin the age of mass communication.

The father of radio is Guglielmo Marconi, an Italian engineer. He improved the Morse telegraph by sending the signal without wires. At first, the Italian government was not impressed, so he went to England and patented his invention. One of his early demonstrations was sending reports from the yacht races at the Kingstown Regatta. This would have been very difficult with cable-linked telegraph. Marconi also developed a "radio beam" for guiding airplanes.

Project: Fiber Optic Transmission System

Parts List

Qty	Description	Designator
2	50 K-ohm potentiometers	R_1 & R_6
2	5.6 K-ohm resistors	R_2 & R_3
2	100 K-ohm resistors	R_4 & R_9
1	1.2 K-ohm resistor	R_5
1	220 ohm resistor	R_7
2	1 K-ohm resistors	R_8 & R_{11}
1	1 Megohm resistor	R_{10}
1	10 K-ohm potentiometer	R_{12}
2	0.1 microfarad capacitors	C_1 & C_4
1	10 microfarad capacitor	C_2
1	500 microfarad capacitor	C_3
1	0.005 microfarad capacitor	C_5
1	0.05 microfarad capacitor	C_6
1	100 microfarad capacitor	C_7
2	SPST switches	S_1 & S_2
2	9 V batteries with connect.	B_1 & B_2
2	741 OP Amp IC chips	Z_1 & Z_2
1	386 amplifier IC	Z_3
1	ECG-289A NPN transistor	Q_1
1	set Fiber Optics Emitter/Detector (suggest MFOE71/MFOD72 sold as set #276-225 by Radio Shack, or similar)	D_1 & Q_2
1	Fiber Optic Cable (RS 276-228)	FOC_1
1	8 ohm speaker	SP_1

suitable enclosures, PC boards, wires, and hardware as required.

Notes

This is a challenging pair of projects. Only students who have had success in project construction should invest the money and time required to build this project. You should breadboard it before making PC boards and enclosures. You may desire to use another emitter/detector set and substitute lenses for the cable so that the signal may be sent through "free air." Another option allows experimentation with the light beam. Cut the cable about 6 inches from one end and use the two free sections to see what materials the beam can and cannot pass through. The input source may be a radio, signal generator, or other device. Even a crystal microphone will work, but use another source until you get the circuit working well. Check the packages the emitter/detector set comes in to identify the proper leads before you connect them into the circuit.

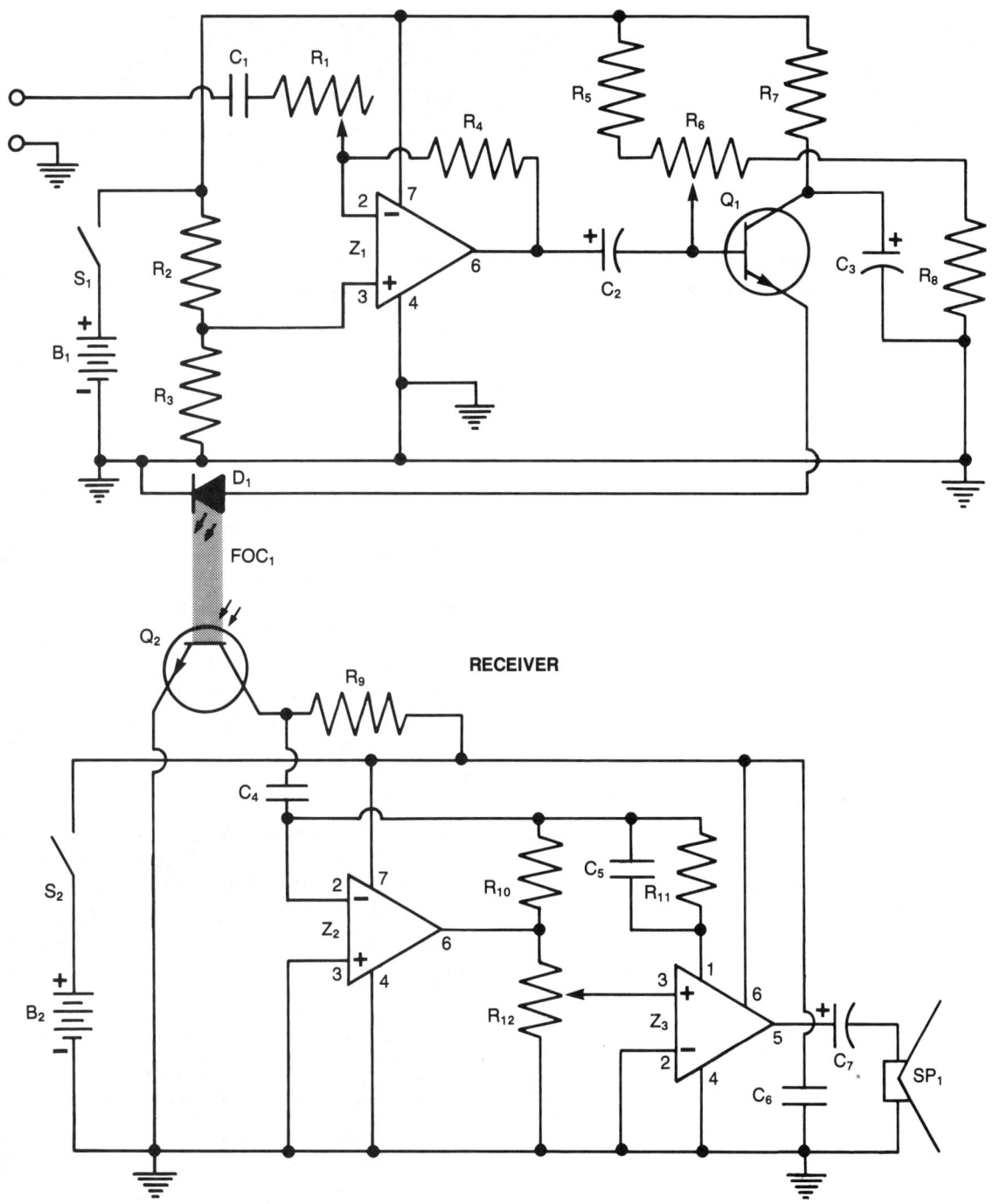

Fig. 56-9 This circuit can actually transmit audio information over a beam of infra-red light.

Chapter 57

Communication by Television

The Television Transmitter

You have learned about the transmission of audio information through wires and by means of radio frequency waves. The next logical step in the development of an advanced communication system is the ability to send visual information from one location to another. The television sends both still and moving visual images from transmitters to receivers without wires. This chapter discusses the basic television system.

The television transmitter does the same job for the TV station that the radio transmitter does for a radio station. The television transmitter modulates the carrier frequency with the information from the inputs (both video and audio). Then it sends the signal to the antenna. The video information comes from the camera. The camera changes the picture into a series of electrical impulses. These impulses represent the light and dark spots in the picture. The audio portion is carried by frequency modulation (FM). The video portion is carried by amplitude modulation (AM). The transmitter is really like two transmitters in one. Both share a common antenna. All of this information requires considerable bandwidth. For example, channel 3 has a total width from 60 to 66 MHz. The video signal's carrier is assigned 61.25 MHz. The audio information modulates a 65.75 MHz carrier.

Television Receivers

A block diagram of a black and white television receiver appears in Fig. 57-1. Let's look at it a block or two at a time to see how it operates. The signal comes in through the antenna and goes to the tuner. The tuner block actually contains an RF amplifier, local oscillator, and mixer. They are similar to those in the superheterodyne radio receiver discussed in the last chapter. From the tuner, the signal is sent to the IF amplifiers. Often there are several of these.

The IF amplifiers amplify the signal just as they do in radio circuits. From the IF amplifiers, the signal is sent to the video detector. This circuit sends two signals to the video amplifiers. The first one is the FM sound IF signal. This is just a simple frequency modulated signal like the one in the IF section of an FM radio. The other signal is called the **composite video signal**. The sound IF signal goes on to the audio circuits which are much the same as an FM radio receiver.

The Composite Video Signal

The composite video signal is a complex signal. It is shown in Fig. 57-2. Section A is the video information for one of the 525 scan lines that will "paint" the picture on the screen. The taller the peaks, the darker

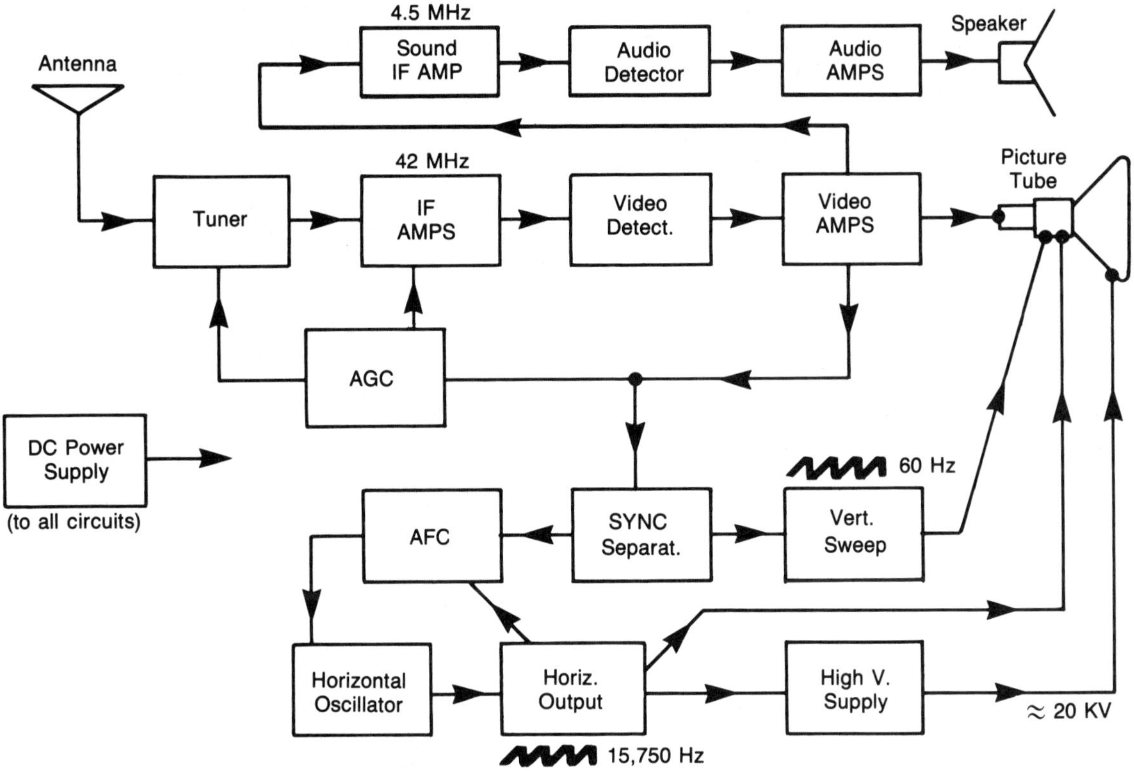

Fig. 57-1 Block diagram of a black and white television receiver. Some of the important frequencies and voltages are shown.

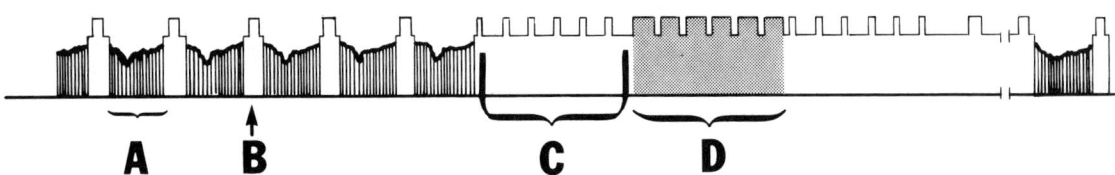

Fig. 57-2 The composite video signal has several parts: A) the video information for one of the 525 scan lines which make up the picture, B) a blanking pulse, C) equalizing pulses, D) vertical blanking pulse.

Chapter 57   *Communication by Television*   367

the screen. When this particular line is scanned, the left edge of the screen will be very dark. The screen will brighten somewhat about one-third of the way across and then get darker again. The screen is bright white only when the peaks of the signal are very near the bottom of the waveform.

After each line is scanned, the **electron beam** must travel back to the opposite edge of the screen to begin the next line. This would normally make a bright white line across the screen. The beam is cut off by the very high amplitude **blanking pulse** (section B), though this pulse must be tall enough to make the beam totally black and to trigger the synchronizing circuits so that everything happens at the correct time.

Sections C and D are equalizing pulses and the serrated vertical **synchronization pulse**. These pulses let the receiver know that the electron beam has traveled all the way to the bottom of the picture (painting one line at a time and moving down a little for each line). When the beam gets to the bottom, it needs to go back to the top. Again, you must "blank" the beam out with high amplitude pulses and signal the circuits to move the beam back. You also must allow enough time for the beam to make its return trip. All of these functions are completed by sections C and D.

Sections B, C, and D are repeated the same way no matter what the picture looks like. Only section A changes as certain areas of the screen get brighter or dimmer.

Interlaced Scanning

If the picture were painted just once (all 525 lines from top to bottom) and then started again, it would take so long that the picture would flicker. To avoid this, the odd lines are painted first (1, 3, 5, etc., to 262½). Then the beam goes back to the top to paint the even lines. See Fig. 57-3. Thus, while the odd ones are beginning to fade out, new even ones are being placed between them to keep the picture uniformly bright all of the time. This process is called **interlaced scanning**. Some of the lines are actually wasted by sections C and D of the composite video signal (about 16 lines are lost). There should be about 240 frames of single line information (each one like section A) between each set of C and D sections. After sections C and D pass, another set of 240 lines will be seen before the next set of vertical blanking and equalizing pulses.

Scanning Circuits

The electron beam in the picture tube would just sit still in the middle of the screen and make a flickering white spot, if it were not for the scanning circuits and the deflection yoke. The **yoke** is a set of coils located on the neck of the picture tube right where it begins to get bigger (Fig 57-4). Two of the coils are placed on the top and

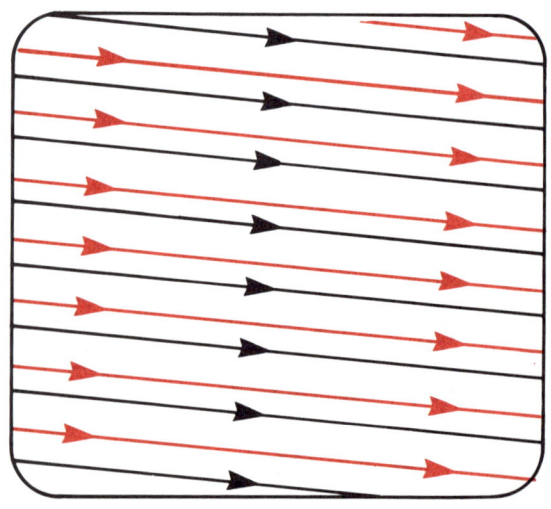

Fig. 57-3 Interlaced scanning: The black lines are painted and then the colored ones. The beam must be blanked out as it travels back for the beginning of each line and during its return trip to the top of the screen, too.

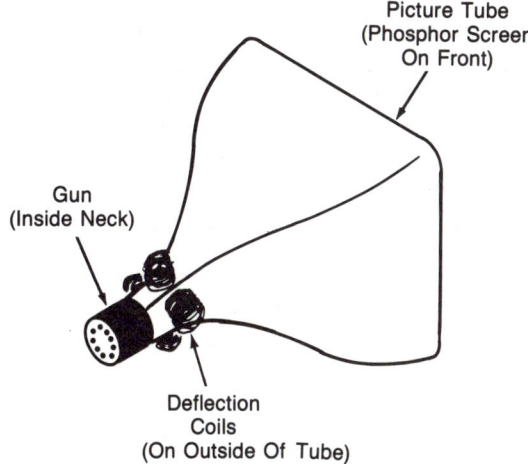

Fig. 57-4 The main parts of the television picture tube (cathode ray tube or CRT).

bottom. The other two are on the right and left sides.

If you energize one of the coils with current, it produces a strong magnetic field. This magnetic field will bend the beam so that it does not hit the center of the screen. By using the four coils together, you can place the beam at any chosen spot on the screen. Two of the coils are called **horizontal deflection coils**. The other two are the **vertical deflection coils**. The horizontal coils move the beam back and forth to paint each line. The vertical coils move the beam up and down as needed to cover the whole screen and interlace the lines correctly.

The electric currents that energize the deflection coils come from the vertical sweep and horizontal output circuits. You have read about the importance of the horizontal oscillator in an earlier chapter. Recall that its job is to produce a steady sawtooth-shaped waveform to move the beam from side to side. Timing is critical here. The sync separator senses when the sweeps should be made by reading this information from the tall synchronization pulses in the composite video signal. It sends this information to the vertical and horizontal circuits to make sure that everything is done at the correct times.

High Voltage

The only thing not included so far to make the television complete is something to make the electron beam travel to the screen. The beam is produced by a "gun," really just a special cathode. The intensity of the beam is determined and changed rapidly by the amplitude of the composite video signal. Its position on the screen is determined by the deflection coils, in response to the vertical and horizontal circuits. But the only thing that can make the electron beam jump through the vacuum between the gun and the front of the tube is a very high positive voltage charge. This charge is thousands of volts. A typical color television will have a high voltage of about 25,000 volts (depending partly on screen size).

The high voltage is developed in a unique way. The output of the horizontal output stage goes to two places: the deflection coils to move the beam from side to side and the high voltage power supply. One type of high voltage supply uses a step up transformer with a very long secondary coil. It is called the **flyback transformer**. When the 15,750 Hz horizontal sawtooth wave goes through the primary, it produces a similar output in the secondary which is very, very high in voltage. This high voltage AC wave is rectified to produce a high voltage DC. The high voltage DC is sent to the front of the picture tube where it attracts the electron beam.

Phosphors on the inside of the screen change the electron beam to visible light wherever it hits. Be very careful when working on televisions. The high voltage is very dangerous. The picture tubes have two dangers, too. They can store a high voltage charge that is dangerous even after the set is turned off. They can also **implode** if they are hit. Implode means to collapse so quickly that the pieces are forced out the other side like an explosion. Implosion is caused by the great force of the vacuum inside the tube.

Color Television

A color television has all the same circuits plus a few more to handle the color information. Usually the picture tube has three guns instead of one. One gun fires the green information, one the red, and the other the blue. If you want to make a blue dot on the screen, the blue gun fires electrons that are carefully aimed to hit only blue-colored phosphors. If the blue gun hit red phosphors, it would make red, so the aim is important. To get purple, both the red and blue guns brighten red and blue dots which are close beside each other. The dots are so small that your eyes are tricked into believing they see purple. A bright white spot is made by illuminating three dots (one red, one blue, and one green) in a close triad. If you wished to include the color circuits on the block diagram (Fig. 57-1), they would be located between the video amplifier and the picture tube.

Summary

The television is so commonplace today it is taken for granted. It is a good circuit to study because it includes many stages and has different signals to observe. The major functions accomplished in the television are: tuning in the signal, amplification, detection, scanning, producing sound, synchronization, and high voltage. These topics have been discussed only briefly. Read a book devoted entirely to television for further information. The last chapter in this section will deal with other methods of communication.

Important Terms

composite video signal
electron beam
blanking pulse
synchronization pulse
interlaced scanning
yoke
horizontal deflection coils
vertical deflection coils
flyback transformer
implode

Review Questions

1. Examine Fig 57-2 and tell what each section of the composite video signal does.
2. How many lines are scanned in each frame of the picture? (Don't forget about interlacing). What does blanking do?
3. Where does the high voltage for the picture tube come from? What are some safety precautions concerning work on televisions?

Chapter 58
Other Forms of Electronic Communication

Coded Messages

One of the earliest methods of communication which used electricity was the telegraph. By the Civil War the telegraph (Fig. 58-1) was a very important tool for business, industry, and the military. The circuit was simple by today's standards, but the impact was great. Messages were sent by a **code** of long and short signals. The international code (slightly different than the original code developed by Morse) is shown in Fig. 58-2.

Today, most people assume the use of code is old-fashioned and forgotten. The code is still very important. It is often possible to transmit code clearly enough to be understood, even though voice communication is impossible. Code may be sent by radio waves or wires. The equipment for code can be simpler in design than that needed for voice. Code is also still used by hobbyists. Even though advances are being made daily in the ability to transmit and receive voice communication, code will be around for a long time.

The Printed Word

We do not know for sure what the earliest forms of communication were. Perhaps the prehistoric people made grunts and primitive words to each other with their voices. Maybe they used signs, body language, and facial expressions.

We do know how they stored communications for others to see later. Cave pictures and hieroglyphics led eventually to

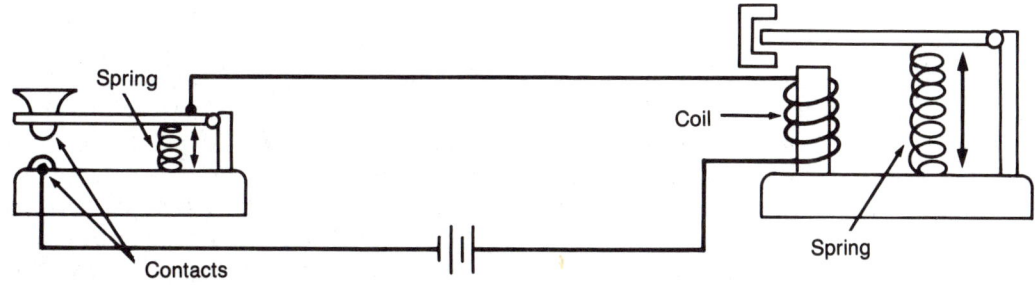

Fig. 58-1 The telegraph is a simple, but important circuit.

International Code

A .—	N —.	1 .————	
B —...	O ———	2 ..———	
C —.—.	P .——.	3 ...——	
D —..	Q ——.—	4 ....—	
E .	R .—.	5	
F ..—.	S ...	6 —....	
G ——.	T —	7 ——...	
H	U ..—	8 ———..	
I ..	V ...—	9 ————.	
J .———	W .——	0 —————	
K —.—	X —..—	Period .—.—.—	
L .—..	Y —.——	Comma ——..——	
M ——	Z ——..	Question Mark ..——..	

Fig. 58-2 The international code is still used. How would you call for help if your ship was sinking?

Fig. 58-3 The popular copying machine is a necessity for the modern office.

written words made up of characters and letters, or **printing**. Eventually, the printing press was developed by Johann Gutenberg (about 1450), and the age of mass communication began.

Electricity probably first entered the process in the form of the first motorized printing press. From such a humble beginning, it is surprising to see how much today's printing industry depends upon electricity and electronics. From the electric typewriter, common in homes and offices, to the huge newspaper press or the high-tech computerized photo typesetter, newspapers and books could not be printed as quickly or as well as they are today without electronics.

Pictures and Copies

An **electrostatic copier** is another communication tool. We often call these Xerox machines because that is the name of the company which first made the process popular. The copier causes powdery "ink" called toner to be attracted to paper by high electrostatic charges. The toner is then fused by heating so that it sticks to the paper. Some advertisements for modern copier machines are not far from right when they claim that anything eyes can see can be copied with good results.

Electronics is important in these

machines in the following ways: power for the motors, power for the exposure lamps, power for the electrostatic chargers, and power for the high temperature fuser. Electronic (usually digital) control for the whole machine ensures that everything happens at the right time and detects failures such as paper misfeeds.

It is also possible to send a picture through telephone lines by the use of a facsimile recorder. Since telephone communication can be transmitted by cables, lasers, microwaves, and even bounced from satellites, there is a wide variety in the methods of communication which these systems provide.

Radar

Radar means **Ra**dio **d**irection **a**nd **r**ange. Radar allows people to detect the presence of objects and to track their movement. It is very important for controlling air traffic and as a military tracking system. The system works on a simple principle. As shown in Fig. 58-4, timed signals are transmitted in a specific direction to the target. The radio waves (3 to 9 GHz for marine radar) hit the target ship and bounce off. They enter the antenna and are routed through the duplexer to the receiver. The duplexer lets the transmitter and receiver each take turns using the antenna. The time that the signal takes to return is measured by the internal circuits.

The indicator plots the target found on a CRT screen. The center of the screen is the location of the radar station (your ship). The distance from the center to each bright spot tells the distance a target is from the station. The direction is also indicated because the antenna is turning in perfect step with the beam on the CRT screen. This is shown by the gears and shaft connecting the two. It usually is done by servo motors coupled together electrically. One motor is on the indicator and the other one is on the

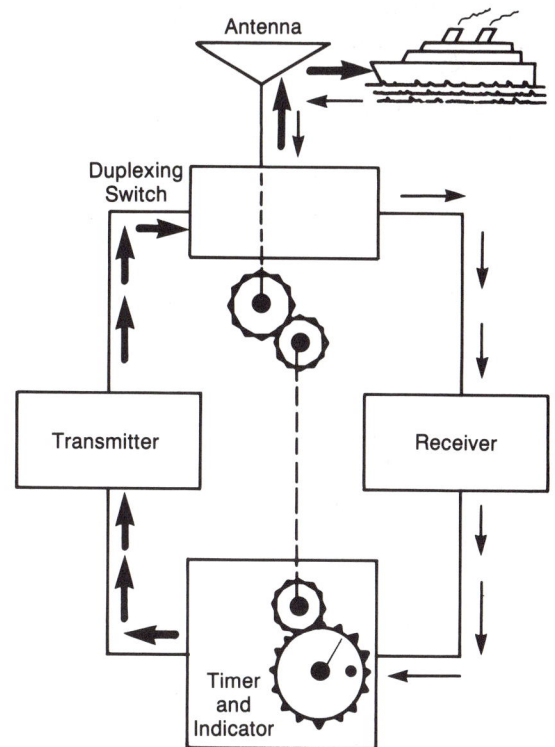

Fig. 58-4 Radar makes it possible to find the range (distance) and direction of ships at sea, even in heavy fog, by bouncing radio waves from them.

antenna. Smaller radar systems are used by the police to track speeding vehicles.

A Growth Field

Electronic communication is a broad field now, and it is growing rapidly. Here are just a few more examples of electronic communication devices:
- Hearing aids.
- Tape recorders.
- Videotape decks.
- Computerized voice synthesizers for mute people.
- Cameras
- CB radio transceivers.
- Signal lights on railroads and at traffic intersections.

- Automatic direction finders (ADF) for airplanes.
  - Electronic musical instruments.
  - Optical scanners for computers.
  - Telephone answering machines.
  - Intercom systems.
  - Laser disks.

You probably thought of others as you read this list. The communication systems of today would not be possible if it were not for the advances made in electronics. Speculate about what tomorrow will bring in the way of new electronic marvels for communication. Perhaps video telephones, a way to instantly transport people from one place to another, a way to think about something and have an implanted computer print it so you don't have to write... Where do we draw the line between science fiction and reality?

Summary

Communication includes every way we send messages to others and record data. When you pass a note to a classmate, you are communicating. When the teacher gives you a sharp look, that is communication, too. Here you are concerned with electronic communication systems. Several examples of electronic communication have been discussed. Take a few minutes to think about other forms of communication and the role which electronics plays in them. It will be difficult to name any modern communication system not dependent upon electronics in some way.

Review Questions

1. What is meant by the word communication? Give different examples.
2. Team up with a friend and send a few short, simple messages by tapping on your desk with your finger. Make sure you try SOS and your own names.
3. How is electronics used in an electrostatic copier?
4. How does a radar system work? What are some uses of radar?

Pep Talk: English — Reading and Writing

There are many technical and nontechnical careers in communication. Communication is one of the main things that separates us from the "lower" animals. Courses in English, public speaking, foreign languages, and journalism help to strengthen communication skills. Almost all careers require some skill at communication. Even in highly technical fields (such as electronics), the good communicators advance to the top.

One of our greatest freedoms in America is freedom of the press, but this freedom is wasted if our citizens cannot read well enough to benefit from the newspapers and write well enough to communicate new ideas. The next time your English teacher assigns a paper, don't view it as "just another boring assignment." Instead, look at it as an opportunity to sharpen one of your most important life skills — a chance to become more "human." Pick an interesting topic and do your best to make it a good paper.

Important Terms

code	electrostatic copier
printing	radar

Control and Other Applications

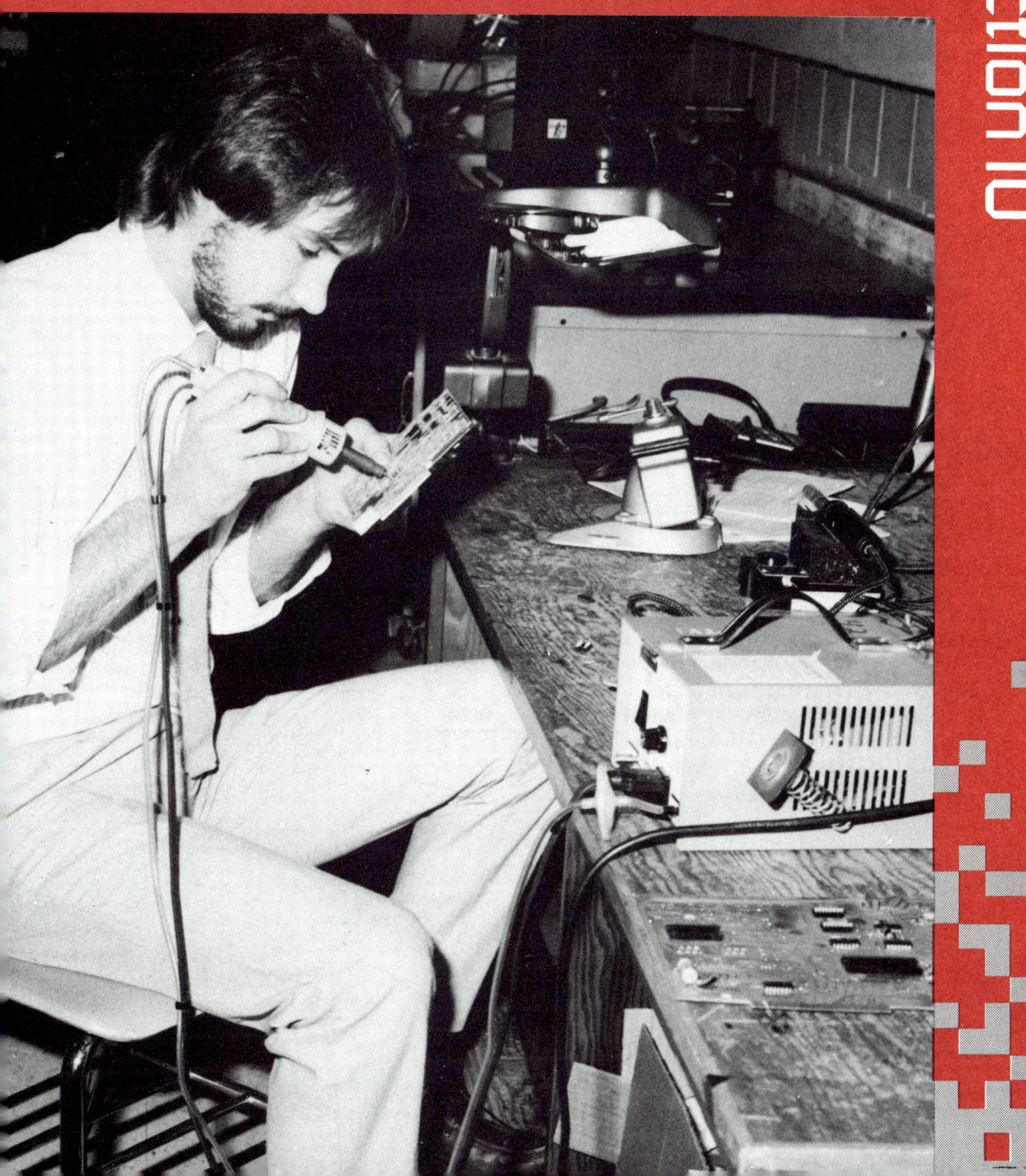

Section 10

Chapter 59
Industrial Control

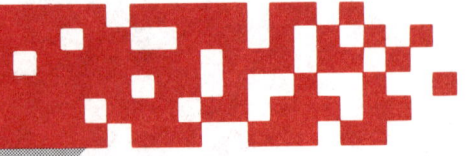

What is Industrial Control?

Uses of electricity-electronics can be classified under power, communication, control, and other applications. You have studied power (light, heat, and motors) and communication (telephone, radio, television, and other methods). This chapter is about uses of electricity in industrial control. Modern factories depend a great deal on electricity for their power and their communication systems. They also use it to control their operations and to warn of possible safety hazards.

Industrial Control includes anything from the use of a switch to turn a machine on and off to complex computer applications involving high-tech industrial robots and totally automated factories. This chapter is limited to the discussion of switches, relays, and some sensing methods used by industry.

Switches

Most of the switches used in example circuits in this text have been simple **SPST** (single-pole, single-throw) type. Switches that control more than one circuit at a time or have two or more "on" positions are also available. Fig. 59-1 shows the schematic representations of several types of switches. These are the most common types. The second switch (SPDT) is the single-pole, double-throw type. This switch has three connections (labeled 1, 2, and C). Current can be connected from the C (common) terminal to either of the other terminals (1 or 2), but not to both at the same time. The DPST (double-pole, single-throw) switch can turn two circuits on or off at the same time. The dotted line indicates that both of the **poles** (the lines with the

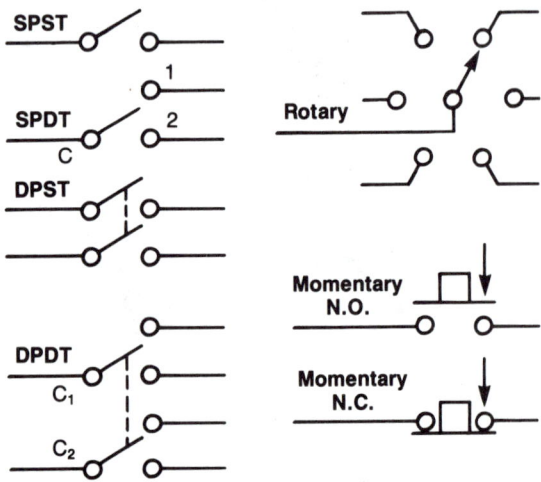

Fig. 59-1 Symbols for seven common switches.

arrows) are moved when you **throw** the switch. Since this switch can only make contact when moved (thrown) in one direction, it is a single-throw switch, but it is a double-pole switch because two contactors (poles) move together.

The **DPDT** switch is very common because of its versatility. It has two poles (contactors) which may both be thrown in two directions together. Such a switch could be used to control two circuits (Fig. 59-2). Here, the stereo receiver is located in the living room, and there are remote speakers in the bedroom. The DPDT switch makes it possible to send music to either set of speakers without mixing the separated sounds from the right and left channels. The return wires for each speaker would go back to the receiver without going through a switch. This circuit would not allow you to listen to both sets of speakers at the same time.

Can you design a switching circuit that would allow the use of speakers in either the living room alone, the bedroom alone, or both? Try to sketch one using two DPST switches. A schematic which shows such a circuit appears in Fig. 59-5 (at the end of this chapter). Compare your circuit to this one.

The rotary switch is very useful when several functions or ranges are to be selected. The example shown in Fig. 59-1 is a single-pole, six-position switch. By stacking several switch wafers together or using different numbers of poles and positions, an almost unlimited variety of switching operations can be completed with rotary switches. Examine some rotary switches in surplus equipment and meters to see how many possibilities there are.

Fig. 59-1 also shows two momentary (spring loaded, pushbutton-type) switches. The **N.O. switch** is "normally open," which means that current is allowed to pass through it only when the switch is pushed. The spring keeps it open at all other times. The **N.C. switch** (normally closed) will allow current to pass any time except when it is pushed. This switch is used when you want to interrupt a circuit which is normally on.

Relays

Now that you are familiar with the basic switch types, you should be able to understand that relays could be made to perform the functions of any of these switches.

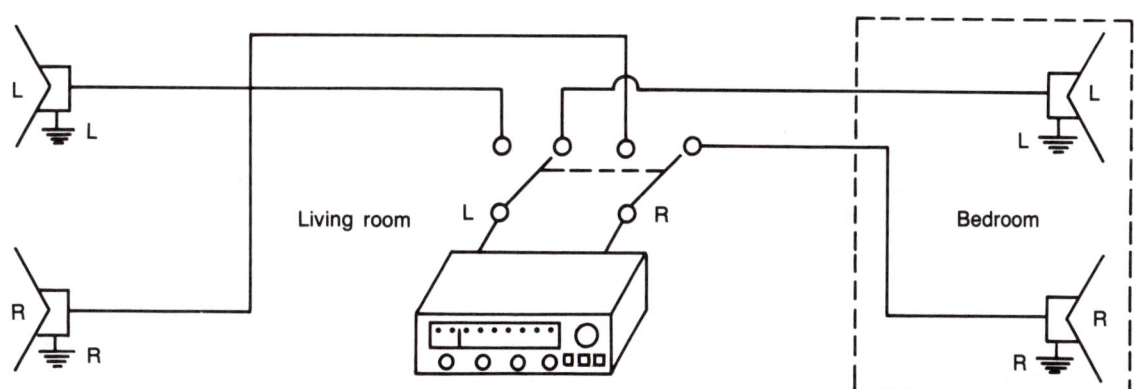

Fig. 59-2 A DPDT switch allows use of either set of speakers, but not both. Can you think of a way to overcome this problem?

Remember that relays allow you to electrically control circuits. This gives two capabilities.
1) You can control a high current or voltage circuit with a low power control circuit.
2) You can control a circuit located in one place by one located in another place.

Thus, an industrial worker could sit comfortably in a safe control room and use small (low-power circuit) switches to control the operation of high-powered equipment in a dangerous environment (such as handling hazardous chemicals). Electro-magnetic relays are still used in many applications, but solid state (semi-conductor) relays are becoming more and more common.

Telemetry

Relays and semiconductors can do many jobs for you. They need wires to carry signals and current from control devices to acting devices, however. Frequently, the needed signals are sent by radio waves.

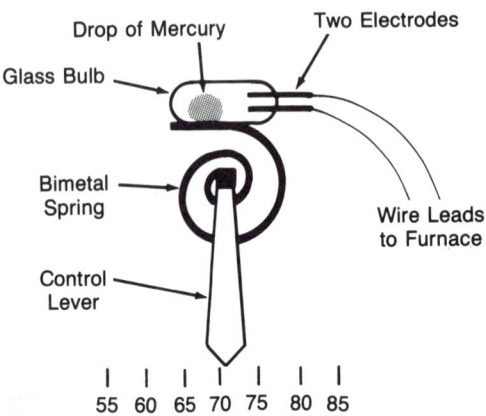

Fig. 59-3 The thermostat which controls heating or air conditioning in your home may be very similar to this one.

When radio waves are used to carry information and signals to and from industrial equipment, the process is called **telemetry**. You may have played with a radio (remote) controlled race car or other toy. Many of the same circuits and functions used to produce such a toy are also used in industrial telemetry systems.

Sensing

Sending control signals is only part of the job of industrial control systems. Detecting certain conditions is also important. Many devices used to sense important conditions or changes of conditions are called **transducers**. All of the six sources of EMF are used in various sensing devices and transducers. A few examples are:
- Photo cells are used to sense changes in light level.
- Thermocouples are used to sense the temperature of melting metals.
- Electromagnetic coils are used to detect the nearness of metals.
- Pressure transducers can be used to sense the pressure in pipe systems.
- The presence of static charges may be detected through capacitance.
- Even chemical changes can be detected electronically.

Thermostats

A very common control device found in industry and in the home is the **thermostat**. A simple **mercury switch**-type thermostat is shown in Fig. 59-3. The spring is made of two metals. It is called a bimetal spring. When the air temperature around the spring changes, the spring coils and uncoils in response to the changes. The mercury switch consists of a small glass bulb which has two metal electrodes in one end and a drop of mercury. (Mercury is a metal which is liquid at normal room temperatures.)

When the bulb is tilted down on the left, the mercury runs to that end. There is no path for current through the switch.

When you tilt the bulb the other way (down on the right), the mercury will run to the end with the electrodes. Current will then be permitted to pass from one electrode to the other through the liquid mercury. Mercury switches are used in applications other than thermostats because they are quiet. They do not make sparks which could cause explosions. The control lever sets the spring in the correct position so that the switch will close at just the desired temperature. As the air temperature changes, the spring tilts the mercury switch to turn the heating or air conditioning system on and off automatically as required.

A Complete Control System

A simplified industrial control system is shown in Fig. 59-4. This system controls a furnace for melting metal. The furnace is fueled by natural gas which enters through a pipe. The pipe has a solenoid valve which allows gas to flow only when the valve's solenoid receives power from the control circuit. Three relays (A, B, and C) control the power to the solenoid valve. Relay A is powered by a photovoltaic cell. When the photo cell receives adequate light, the relay is pulled to the closed position to allow current to flow. (Of course, amplification may be needed in a real circuit.) When the smoke in the exhaust stack becomes too dark — indicating improper burning con-

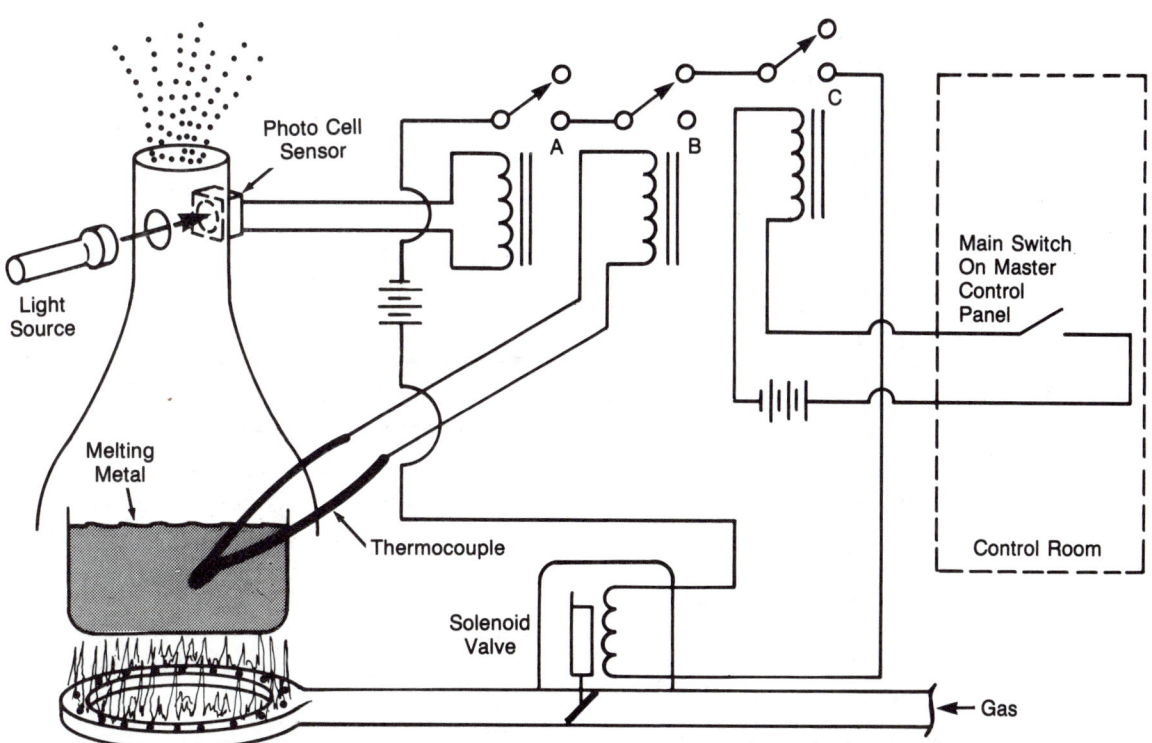

Fig. 59-4 The gas valve of this metal melting furnace is controlled by three relays.

ditions and excess pollution — the light from the light source will not reach the photo cell with enough intensity to keep the relay open. Remember, the cell does not produce as much power when it is dimly lit. As long as the exhaust is clean, the relay will be held closed by power from the photo cell, and normal operation will continue.

Relay B is controlled by the current from the thermocouple. You learned that a thermocouple produces current when it is heated. This part of the circuit is designed so that the relay passes current until the thermocouple gets so hot that it develops adequate power to pull the relay. For this reason, the relay has been wired in the N.C. (normally closed) configuration. The relay will allow current to flow to keep the solenoid valve open (and gas flowing) until the metal reaches the proper pouring temperature. Then the relay will pull and stop the flow of gas.

Finally, relay C is the master control relay. It is controlled by a switch on the master control panel. It would be left closed when the system is operating properly. It would be left opened when the system was to be shut down or in emergencies. The master control panel might even include gauges and warning lights to tell the operator which position each relay is in, the pressure of the gas, the temperature of the metal, how much pollution is being produced, and other important information. From a control panel for a major industrial system, that operator can remotely and safely monitor and control all functions of the system.

Other Control Functions

It is often necessary to control the speed of a motor or the position of a moving part on industrial machines. Likewise, control circuits are used to count items on assembly lines, to monitor flow rates, to limit various conditions, and for many other functions in industry. In addition to the convenience and safety provided by electronic control circuits, factories which use much electronic control are not plagued by mistakes made by inattentive workers or undetected equipment failures. For these reasons, most industrial plants use complex electronic control systems.

Summary

Many of the circuits and components you have already studied are combined to produce control circuits for industry. Switches, relays, thermostats, speed controls, gauges, transducers, and other sensing devices monitor and control industrial processes. Humans are protected from working in hazardous areas by using remote control and telemetry systems for industrial control. The next chapter discusses the role of computers in industry.

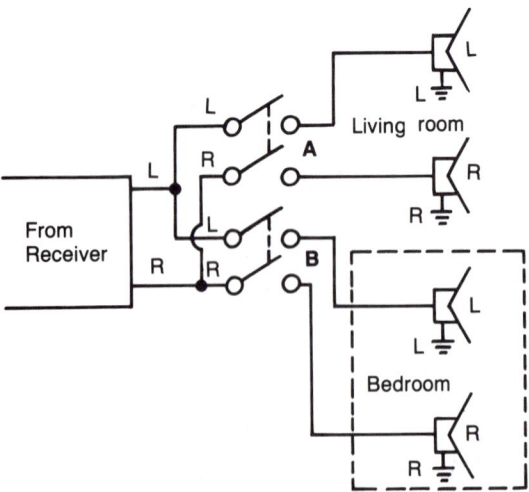

Fig. 59-5 DPST switch A controls the living room speakers while switch B controls the bedroom speakers independently.

Important Terms

industrial control
SPST
DPDT
poles
throw
N. O. Switch
N. C. Switch
telemetry
transducer
thermostat
mercury switch

Review Questions

1. Refer to the diagrams in Fig. 59-1 and tell a possible use for each type of switch shown.
2. How does a thermostat work? What does it do?
3. List the advantages of a mercury switch.

Project: Easy, Accurate Transistor Tester

Parts Lists:

1	47 K-ohm resistor	R_1
1	10 K-ohm resistor	R_2
1	33-ohm resistor	R_3
1	0.1 microfarad capacitor	C_1
1	9 V battery with connector	B_1
1	DPDT toggle switch	S_1
3	SPST momentary push switches	S_2, S_3 & S_4
1	Audio output transformer	T_1
1	Tri-color LED (RS 276-035)	D_1
1	small, 8-ohm speaker	SP_1
3	microgator clips (or similar) with colored insulators as required	

Suitable enclosure, PC board, wires, and hardware, as required.

Notes

This transistor tester is very simple to build and use. Its indications are easy to understand. Arrange the four switches by number order on the enclosure. To test an unknown transistor:

1) Set the NPN/PNP Selector (S_1) for the type of transistor you "expect" the one under test to be. If you have no way to tell which type it is, use either setting. The tester will tell you the type.

2) Connect the three test clips to the proper leads of the transistor. If you are not sure which leads are which for the transistor being tested, connect the leads in any order. The transistor tester will indicate when the proper combination of connections is found.

3) Press the "push to test" switch (S_2). Listen for a tone from the speaker. Also, watch for lighting of the tri-color LED. If either of these indications is present, then you have some type of audio transistor. If

there is neither tone nor glow, the transistor may be the other type (NPN or PNP), a special type (such as a high frequency transistor), or bad. To find out which condition exists, try pressing S_3 (High Frequency) and S_4 (Low Frequency) alternately and then together. If no indication is heard or seen, release all three push-button switches and change the NPN/PNP Selector switch to the position you have not used. Repeat the tests. Depending on the type of transistor, a good transistor will give an audio output and a visual output on the LED.

4) If the transistor never gives any indications under any of the tests, there could be three remaining causes: the test clips are connected to the improper leads (rearrange them and repeat the test), the transistor is a special type which cannot be tested, or the transistor is bad.

This tester can test most types of common transistors. Some types give different frequencies or volumes of tones, and some only give a visual indication. Most of the transistors used in the projects in this text should give a good, loud tone that changes as the frequency test buttons are pressed. The LED should also light, especially on the high frequency tests. The tester may be used for "in-circuit" tests. The indications may be a little different than normal because the circuit will have other sources of capacitance, inductance, and resistance.

There is little danger of damaging most transistors by "experimentally" finding the type and positions of the leads with the "hunt and peck" method. Feel free to let the tester find the type of unknown transistors for you. This device will be a valuable test instrument for your shop.

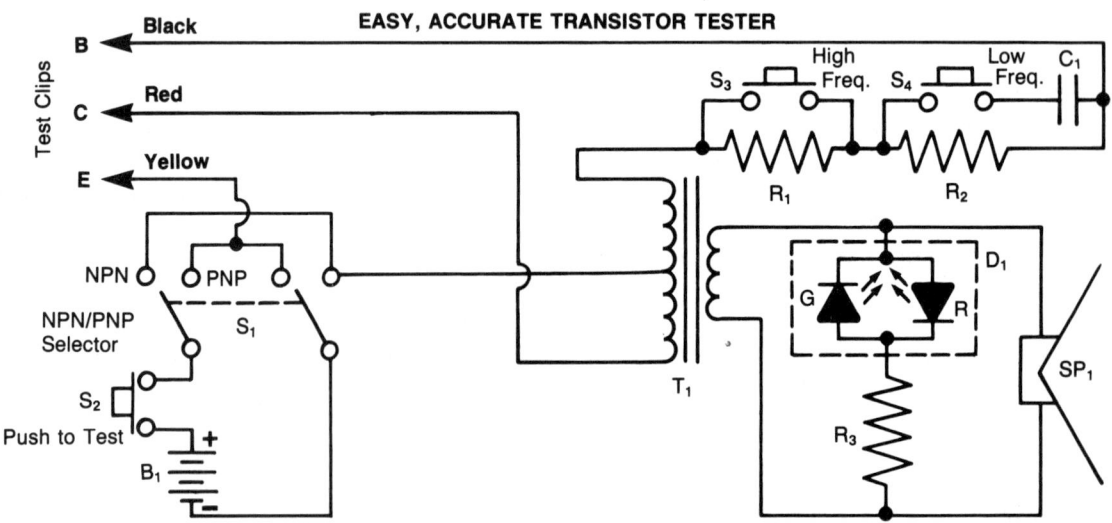

Fig. 59-6 The easy, accurate transistor tester can test most types of transistors in or out of their circuits.

Chapter 60
Data Processing

A Fast-Growing Field

Another important application of electronics in business and industry is the work done by computers. Much of this work is called **data processing**. This chapter deals with computers and their uses in industry.

Business Applications

Computers are able to store much **data** in a small amount of space (Fig. 60-1). Data is any important information needed to solve a problem. It is also information needed for **recordkeeping** purposes. Early applications of computers in industry were for recordkeeping, financial records, problem solving, and storing other information and inventories.

Data processing has grown to make it possible for even small businesses with microcomputers to have automatic billing systems and complete inventory and financial control by computer. The computer may be programmed to keep track of how much each customer owes and when payments are due. Then, the machine will generate a printed bill and automatically address it. These are simple tasks, but the machine does them so quickly that more volume of business can be handled by fewer people.

The computer can also keep track of the operation of the whole business and trends in the industry. Business executives are able to use graphs and data provided by computers in their management-level planning meetings (Fig. 60-2).

Data processing is one of the fastest growing fields in industry. There are many careers available that deal with computers and software. As you read about the applications of computers, think broadly

Fig. 60-1 The data stored in the bulky file cabinets can easily be stored on small floppy disks like the ones shown here.

about the many career opportunities which this rapidly expanding new technology includes.

Finance

The payroll of a business may also be computerized. The computer reads the information on the number of hours worked by each employee from the timeclock system. The computer then stores this information. At the end of the pay period, the computer adds up all the hours of each worker and calculates the pay. The software is programmed so that the computer figures out how much money to withhold for taxes, insurance, savings and retirement plans, and other purposes from each employee's check. Then the machine prints the check and records the information on withholdings.

At the end of the year, each employee's yearly pay and withholdings are calculated by the computer (from the records it has kept) and a statement is sent to the worker for use in figuring their own taxes.

All of this can be done with very little paper shuffling by the computer. At one time, these functions alone would have required many work hours. But efficiency is not the only benefit of the computer. Greater accuracy is just as important. You have probably heard the expression "To err is human, but to really screw up you need a computer!" In reality, humans make most of the mistakes which uninformed people blame on computers. Most of the errors, like the man who received a $30,000 bill from the telephone company, are the result of information being improperly put into the machine by human operators. All things considered, computers are crucially important for keeping the financial records of industry.

Statistics

Another data processing job of the computer is to store and sort records called **statistics**. If you were asked about a baseball player's statistics, you would quickly realize that "statistics" includes batting average and other information about his playing record. The government and many industries also keep a lot of statistics about us. For instance, if your business produced toys, you might consult the information (statistics) in your computer's memory bank about the make-up of American families before deciding whether to produce a new line of skateboards or an improved tricycle for toddlers. If the data in your computer showed that there were far more toddlers than teenagers in the population, you probably would

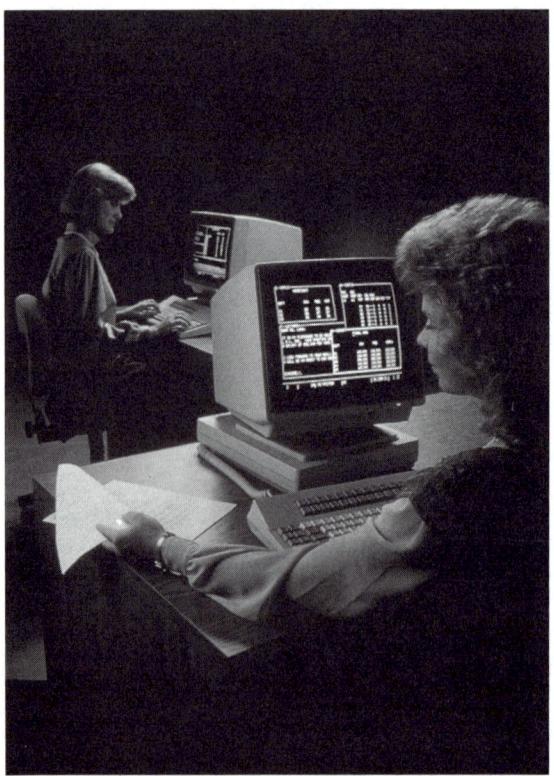

Fig. 60-2 These executives depend on computers every day to make important decisions.

develop products aimed at the younger market.

The government keeps statistics about its citizens for many reasons. These statistics help leaders predict trends, see changes in lifestyle, keep track of unemployment and government spending, trace population shifts and growth, and even predict voting patterns.

Sometimes people become angry at some of the uses of the data stored in governmental and industrial computers. They believe that it is an invasion of privacy. This issue will require even more attention as the statistical **data banks** grow more comprehensive and the machines that access them get even more powerful in coming years.

Research

The computer's role in scientific research cannot be overestimated. The computer makes it possible to conduct research and do experiments that at one time would have been impossible just because of the enormous number of work hours required to handle the data. Computers can provide answers to complex problems quickly and even control research equipment as shown in Fig. 60-3.

Education

Computers have become important in schools and colleges, too. They are used for registering students, printing grade reports, and storing academic records. School systems and colleges also use computers for their financial recordkeeping. Large mainframe computers serve as databanks and perform statistical calculations used in research projects in colleges and universities.

One unique use of computers in education, however, is **tutorial** applications.

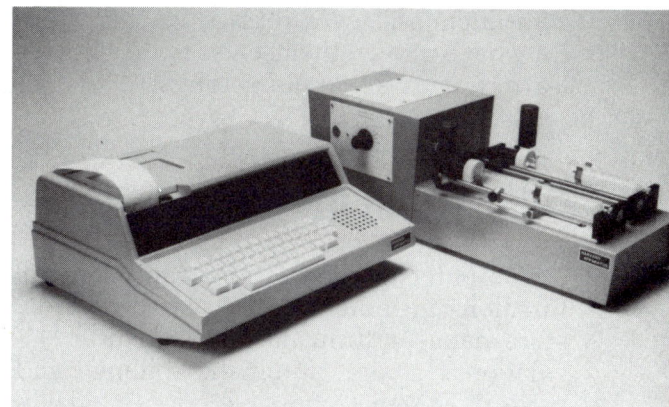

Fig. 60-3 This computerized pump, used in medical drug research, automatically changes speeds and keeps records of its activities according to input from the doctors.

Both mainframes and microcomputers are programmed with special educational software packages to teach certain subjects. In some cases, the computer is used as a special aid for students who are having difficulty learning in the traditional classroom. In other cases, all students may be taught certain skills or information by computer. See Fig. 60-4. It is unlikely that

Fig. 60-4 These students are studying by using a computer software program. Computers can be valuable aids to learning for many students.

computers will ever replace teachers. But they have proven themselves to be valuable learning aids in certain settings.

Summary

Computers perform important control functions in industry. They keep records, store data, perform needed statistical calculations, make automatic billing and payroll systems possible, and aid in research. The computers are more accurate than the human workers who operate them. There are some moral questions that need to be addressed concerning the large amounts of information about private citizens that is stored in computer databanks and the uses of such information. Even schools and colleges use computers for normal business procedures and as aids to learning. The next chapter is about industrial robots and automated factories. Computers play a very important role in these systems, too.

Important Terms

data processing
data
recordkeeping
statistics
data bank
tutorial

Review Questions

1. Imagine that you own a medium-sized shop that sells and services televisions and other electronic appliances. List several possible uses for a computer in your fast-growing business.

2. List some possible career opportunities that involve work with computers.

3. What information do you think the government and private industries should store in computer data banks? What are proper uses of this information? What dangers could there be if the information is misused?

Careers: Computer Programmer

People who write software that instructs computers about how to do their jobs are called programmers. They usually work in office environments. Because business, industry, and government are the major users of computers, these are the main sources of jobs for programmers.

Programming must be done precisely because the machine will assume that you really mean everything you say to it. Sometimes there is a lot of pressure to get programs written quickly.

To prepare for a career in programming, you should do well in mathematics classes and take courses in computer science in high school and in college. Programmers are well paid. They serve an important role in our society.

Chapter 61
Robotics

What is a Robot?

Recently much research has been directed toward the development of robots and totally automated factories. People view this trend in many ways. Some are excited about the potential for greater production with fewer "rejects" and better reliability. Others fear the loss of jobs and the social pressures that could develop from more leisuretime without activities to fill it. This chapter discusses how the robots work and what they do.

Many people get a mental picture of a science-fiction style "humanoid" when they hear the word **robot**. Indeed, science-fiction literature and media has pictured robots as machines that look like metallic humans. In actuality, many of the machines that industry refers to as robots bear very little resemblance to humans or any other animals.

Figure 61-1 shows a typical industrial robot in operation. This particular robot is just one arm that is movable in a limited

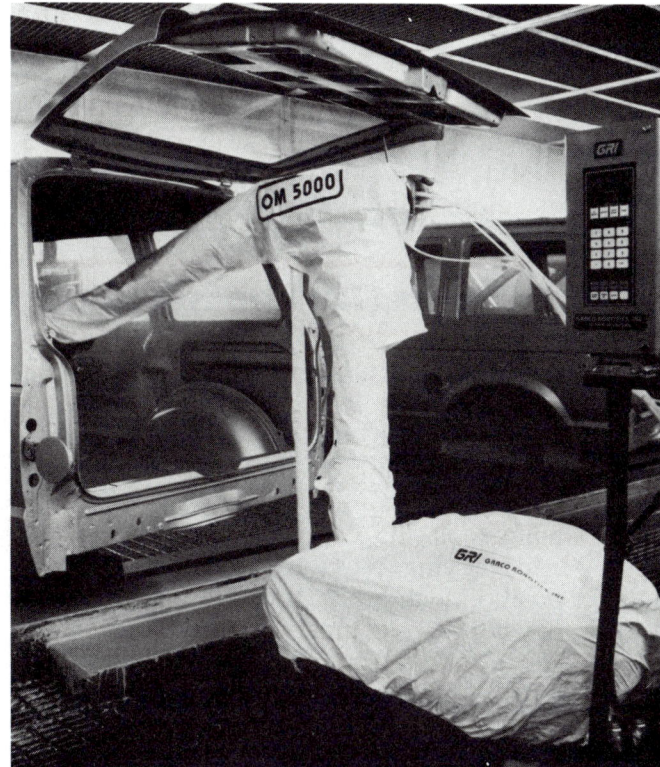

Fig. 61-1 Real industrial robots are sturdy machines with few frills — they are "strictly business."

number of directions. It is used for painting. It does not have a human-looking face, fake eyes, or a voice. It cannot answer the door or vacuum the floor. It is really very limited in what it can do. It can't walk. It can't decide what to do on its own. It gets all of its instructions from computer circuits which must be programmed by humans. But it does have some very endearing qualities in industry.

On the assembly line, it is the only worker that never complains about low pay or long hours. It never misses work because of illness or forgetting to set the alarm clock. It does its job perfectly every time, not a little too quickly one time and then with too much pressure the next. It doesn't care whether the room is too hot, or too cold, or even if the lights go out. It costs a lot to buy, but it has not asked for a raise since the day it was installed. It doesn't even care if the boss yells at it once in a while.

This may be a tough pill for you to swallow, but a real, industrial robot is not a cute little toy that runs around and makes quaint little sounds. A real robot is a machine, plain and simple.

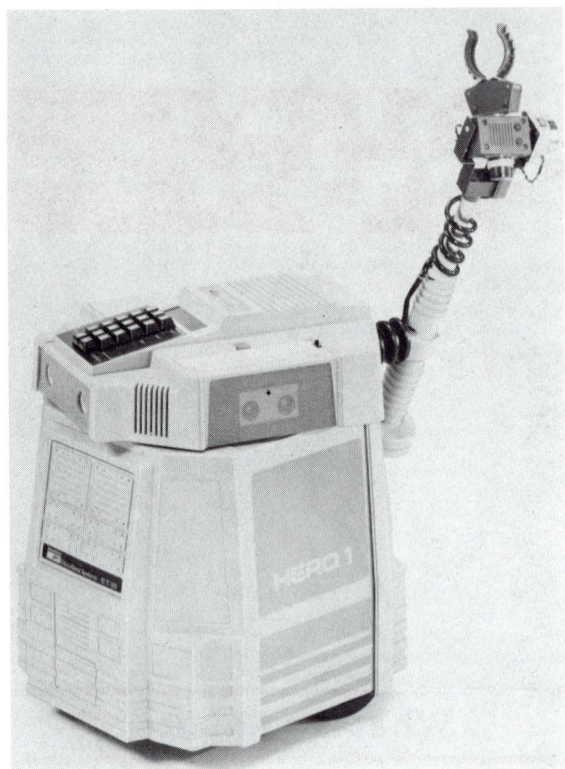

Fig. 61-2 "Hero 1" is a training robot that the user can assemble from a kit. Hero 1 has many features — even the power of speech!

What Can Robots Do?

Robots are used for a variety of industrial jobs. Many of these jobs are too boring, difficult, or hazardous to be done by humans. They have proven themselves to be of great value in welding, spraying paints, placing components into position, and moving parts from one location or position to another. It would be possible to develop big, complex robots which could run around a factory and do many different jobs. This would have many drawbacks, however.

First, the robot would be very, very expensive. Second, the more ways it has to move, the more difficult it is to make the robot strong enough to be capable of any real "work." Lastly, the more features there are in one machine, the more likely it is to break down and the more difficult it is to repair. Since dependability and efficiency are two of the main advantages of robots in the first place, making the robots complex would defeat these two purposes. Most factories that use robots have several simple ones to do one or two jobs each instead of a few complicated ones.

Robots for Learning

Figure 61-2 shows a popular robot for hobbyists and students. It is able to do many tasks: walk and talk; sense light, sound, motion, and obstructions; and maneuver items. As an aid for learning the principles of robotics it is an excellent choice. But it is not physically strong enough for many heavy industrial uses. A stronger, but simpler training robot appears in Fig. 61-3. This one is a little more similar to many common industrial robots.

Robot Motions

Figure 61-4 shows the basic motion directions of industrial robots with coordinate labels commonly used for each. Robot A is only capable of straight line motions in three directions. If it were instructed to extend and then contract to its maximum limits in all three directions, it would draw an imaginary figure in space with the end of its arm. That figure, for this robot, would be a rectangular box. The robot would be able to reach any point inside of the box, but none outside of it. Instructions would be given to this machine in cartesian coordinates just as you would use to draw a graph.

Fig. 61-3 Another popular training robot. It is not as "human" as Hero, but it simulates industrial robots very well.

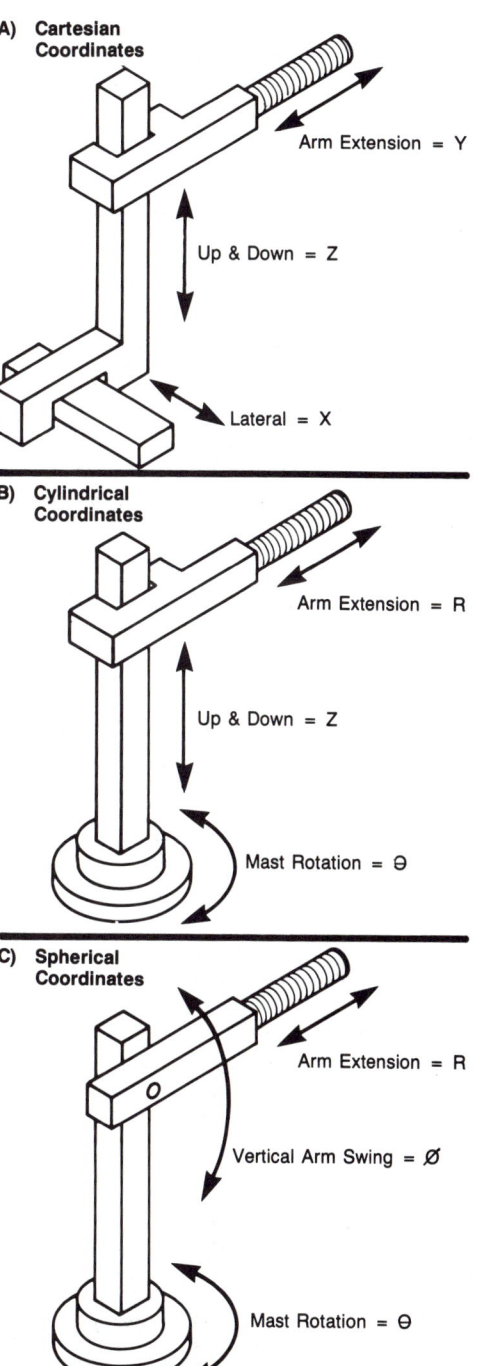

Fig. 61-4 These three diagrams illustrate some of the common ways to describe robot motions. Some industrial robots have arms with even more rotating (angular) joints than diagram C shows.

Chapter 61 *Robotics* 389

Robot B is able to make the same vertical and extension motions as robot A. It rotates on a still base rather than moving laterally on a track, however. Therefore, if you drew an imaginary figure with its full limits of motion, the figure would be a cylinder. It would not be able to reach outside of the cylinder. It could easily reach any point inside the cylinder except a hollow center (because the arm is always partially extented). You could instruct this machine using cylindrical coordinates. Since X and Y are not labeled on our drawing, you could find them mathematically. For instance, Y could be found by the formula: $Y = R \sin \theta$. If you are good at math, you probably can see how to find the X value as well.

Robot C has two rotating motions: the mast turns, and the arm swings. Use your imagination to try to figure out what sort of figure this machine would draw if its motion limits were used. Remember that it has two rotating (angular) motions. Did you determine that the figure would be a sphere? If so, good thinking!

These are only a few examples. Robots could be made with many combinations of these motions and more. Look back at the robot in Fig. 61-1. How many straight line and rotating motions can you find in this picture? The only thing to remember is that it is harder to make the robot accurate and strong if it has many motion directions. Figure 61-5 shows a robot toy styled like an actual industrial robot. Compare the motions it makes with those in Figs. 61-1 and 61-4.

Powering the Motions

Industrial robots may be powered in three basic ways: servo motors, stepper motors, and hydraulics. **Hydraulics** is the process of causing motion by using liquids (like oil) in pressurized cylinders. The fluid is pumped by a pump and then sent to operate the desired cylinders by valves.

Sometimes air is used instead of liquids. The system is then called pneumatic. Electronics can play a control function even in these robots by controlling automatic valves. Servo and stepper motors are both powered and controlled by electricity.

Motion may be controlled by either open loop or closed loop systems. In **open loop** systems, the moving part is instructed to move to a given point. The controlling computer assumes that it gets there after allowing the proper time to elapse. In the **closed loop** system, sensors give immediate feedback about the position of the device at all times during its travel. This feedback signal goes to the computer so that decisions about corrective actions can be made if the motion instructions have not been completed correctly. Obviously, the closed loop system has many advantages. It is also more costly and sometimes more difficult to program, though.

Another way to ensure accurate motion, especially in simple robots, is to set physical

Fig. 61-5 Radio Shack's Armatron is a robot toy that uses many of the same motions as some industrial robots. It even looks authentic, but it is not programmable.

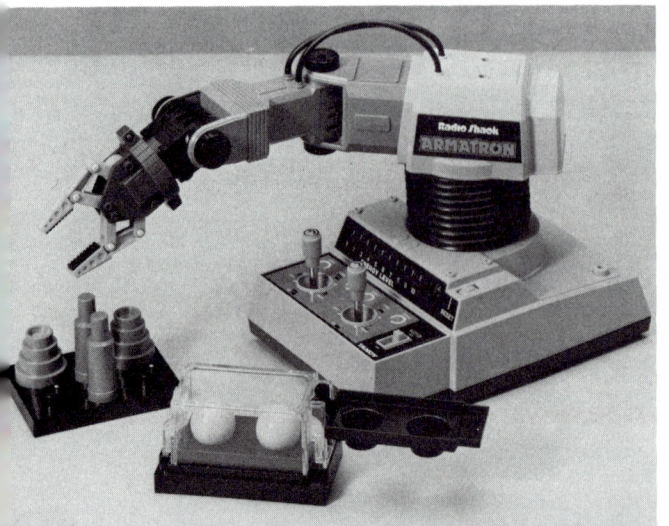

stops or limiting switches which will stop the motion of the machine when it reaches the desired position. All of these methods help to ensure **repeatability**. Repeatability is the ability of the robot to find the exact same spot repetitively. If your factory robot placed parts improperly, it would not be worth very much to you.

Teaching the Robots

The robots may be instructed how to do their jobs by three basic methods. A computer keyboard on or near the robot may be used to enter coordinates. Coordinates may also be sent directly to the robot from another computerized machine or the design department. Still another technique is to use a **teach pendant**. The teach pendant is a small control box, similar to one for a remote controlled toy, which instructs the robot (manually) how to do its job. An operator will carefully guide the machine through its entire set of operations.

When the operator is satisfied that the pattern of motions is accurate and efficient, the instructions are memorized by the computer circuits. The robot will be able to follow them repeatedly until they are again changed by the teach pendant or other inputs.

"Almost" Robots

Actually, many machines in industry have been "almost" robots for years. Even before 1970, numerical control and computerized lathes and milling machines were becoming common in industry. These machines could load their own raw stock (metal to work on), change their own bits and tools (cutters and blades), and follow complicated motion instructions which told them how much to cut and where to cut it. The machines were as complex as many of today's robots. They are still used, often in conjunction with other robots.

Another application of computer technology in industry is CAD-CAM. *CAD* is computer-aided design (and drawing). *CAM* means computer-aided manufacturing (and machining).

An engineer can use a computer to design a needed part and draw the plans for production. See Fig. 61-6. At the same time that the part is designed and plans are drawn, a tape or disk is produced that contains all of the needed information to instruct a computer controlled machine (or robot) in how to make the part.

Automated Factories

Even though robots may fall far short of your science-fiction dreams, when they are coupled with machines such as computerized lathes and a comprehensive CAD-CAM system, the result surpasses our imaginations. In an automated factory, the

Fig. 61-6 A CAD system makes design work easier — it is much like a word processor for pictures.

product is designed on a CAD system. The information for ordering raw stock is handle by the computer. The stock is handled by robots. Automated machines and robots perform all production operations. Products are automatically packaged and labeled. Everything is done by machines.

Humans only have to enter the plant when failures occur. The machines can even correct many of their own mistakes! When things are running smoothly, it is even possible to turn off the lights to save a little money. All of the machines and computers share the same data bank. Even quality control inspections are carried out automatically by robots.

Summary

Robots live up to our expectations if they are realistic. They are not mechanical humans. They are very capable machines which are revolutionizing industry. They may be powered by servo or stepper motors. They may use hydraulic systems. Feedback may be used to provide more accurate positioning in closed loop systems. When linked with CAD-CAM equipment in a fully automated factory, robots are very versatile machines.

Important Terms

robot
hydraulics
open loop
closed loop
repeatability
teach pendant

Review Questions

1. Name a robot you have encountered in a popular science-fiction movie or book. List the differences between it and the industrial robot pictured in Fig. 61-1.

2. Why is it better to use several simple robots in a factory instead of a few complex ones?

3. What are the advantages and disadvantages of a closed loop system?

Chapter 62
Electronics in the Automobile

The Starter Circuit

Few conveniences have had as much impact on the American lifestyle as the automobile. Cars serve transportation needs for business and pleasure. Electricity and electronics serve both crucially important and whimsically frivolous functions in the automobile. This chapter will explore the uses of electricity and electronics in cars.

Before the automobile can go anywhere, you must start its engine. Many people call the engine the "motor." This is really not correct. There are motors in every car, but they are not the main driving power source of the car. They operate accessories and help to start the engine. Figure 62-1 shows the starting circuit of an automobile. Switch S_1 is the key switch (ignition) in the car's interior. Switch S_2 is placed so that it can detect which gear the car's transmission is in. If the automatic transmission is not in neutral or park, this switch will be open and will not permit the car to start. Cars with standard transmissions use a clutch switch, too.

If S_2 is closed and the key switch is also closed, current is permitted to go through the coil of the **solenoid**. In reality, the solenoid is a big relay. It has very heavy duty contacts which can carry the large amount of current needed to start the car (as much as 200 to 300 amperes). When the solenoid closes, this high current powers the starter motor to turn the car's engine. The engine turns — just as a lawn mower engine does when you pull the cord — until it starts.

The reason for using the solenoid is that the wires to the starter motor must be very large (they are actually cables). Likewise, much more resistance would be placed in the starter circuit if you routed the heavy starter motor cable all the way into the passenger compartment just to reach the key switch. The key switch would also have to be very large and cumbersome to operate. The best way to handle the high

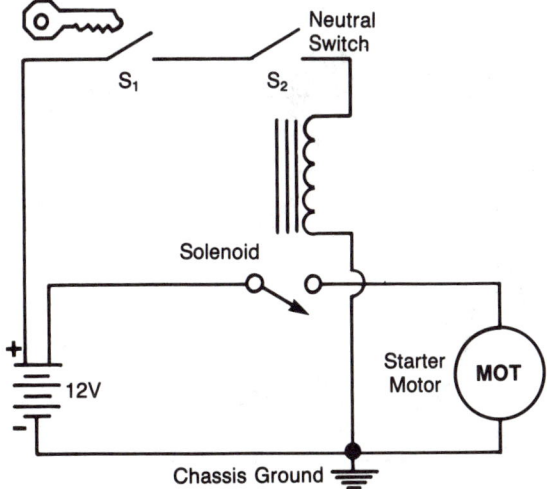

Fig. 62-1 Automobile starting circuit. The solenoid in many models is mounted with or near the starter motor and both the motor and solenoid use their metal cases for ground connections.

starter circuit current is to use a small switch controlling the heavy duty relay (solenoid).

Notice that the battery's negative terminal connects to ground (actually, the frame, metal body, and engine block of the car). Almost all circuits in the car will have their switches on the positive side of the circuit. Many will use the metal case of their motors or other devices as the ground contact by bolting it directly to the engine or frame. This is done in the starter motor.

The Charging Circuit

The heart of the charging circuit is the alternator. The alternator is an AC generator powered by a fan belt driven by the engine. DC charges the battery. Automobiles used to use DC generators, but they had many problems. They did not charge very well at low speeds. The alternator overcomes these problems and can even charge the battery at low idle speeds.

The AC is changed to DC by a diode bridge inside the alternator. Figure 62-2 shows the rest of the charging circuit. The ammeter shows which direction current is flowing and how much. If a light current is flowing from the alternator into the negative battery terminal (colored arrows), then the battery is being charged. The meter will indicate this. When the car is demanding more current than the alternator can produce, current will flow in the direction shown by the gray arrows. This indicates that the battery is discharging. Many cars have a warning light instead of a meter. The warning light comes on any time the battery is discharging.

The regulator controls the output of the alternator. The alternator should produce just a little over 14 volts to charge the 12.6 volt storage battery properly. The regulator adjusts the alternator's output to accommodate extra loads, such as headlights, and still charge the battery. It also ensures that the battery is not damaged by overcharging.

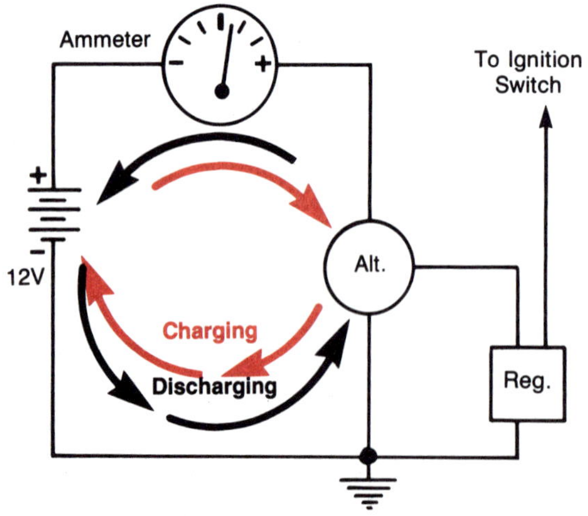

Fig. 62-2 The charging circuit. Older cars had DC generators, but alternators produce more current at low speeds.

Ignition System

The automobile's gasoline engine develops power by burning gasoline in tigtly closed cylinders. The expansion caused by the burning forces a piston (at the bottom of the cylinder) down with great force. This motion is used to power the crankshaft. Most cars have 4, 6, or 8 cylinders so that they are powered smoothly instead of in a pulsing manner like a lawn mower engine. Each cylinder needs a spark plug. The spark plug permits a high voltage arc to be developed in the cylinder at just the right time to burn the gasoline and air mixture that comes from the carburetor.

Most cars, today, use an electronic **ignition system**. A 4-cylinder electronic ignition system is shown in Fig. 62-3. Switch S_1 is the key switch, but it is a different pair of contacts from the ones used in the starter circuit. Once the ignition is turned on, current can flow from the battery through the primary of the **ignition coil** (really a transformer with a large secondary) and through the transistorized control unit.

This current does not flow all of the time. It is only permitted to flow when the control unit allows it to flow. The control unit will allow current to flow through the ignition coil's primary only when the pickup coil senses that one of the teeth on the rotating reluctor is positioned in exactly the right position. The reluctor's teeth are made of iron. It rotates in perfect time with the rotor (on the same shaft) which determines which spark plug will be fired at any given moment in time.

When the pickup coil senses the presence of one of the reluctor's iron teeth, it signals the control unit which allows a quick pulse of current to flow through the ignition coil's primary. The quick pulse in the primary develops a much higher voltage pulse in the secondary. A large magnetic field is built up around the secondary, for a brief instant. When the pulse of current that built this field ends, the field collapses and creates a much higher voltage pulse (caused by counter-counter EMF) in the secondary coil. (Refer back to Chapter 31 if you do not remember how C-CEMF works.)

The pulse enters the distributor. The distributor is a heavy duty rotary switch which can handle the high voltage pulses from the ignition coil secondary. The rotor (rotor button) selects which spark plug wire to send the pulse through. The rotor is turned by the engine's camshaft to select exactly the right **spark plugs** at just the right times. The high voltage current pulse goes through the spark plug wire to the spark plug.

Each spark plug has electrodes separated by a small air gap (about .080 inch). Figure 62-4 shows the construction of a spark plug. When the high voltage pulse reaches the air gap, it jumps right across it and creates a hot, bright blue arc. The arc ignites the fuel-air mixture so that it will burn. The spark plugs are fired one at a time. Each one is fired just as its piston reaches the top of the cylinder with fuel and air ready to be

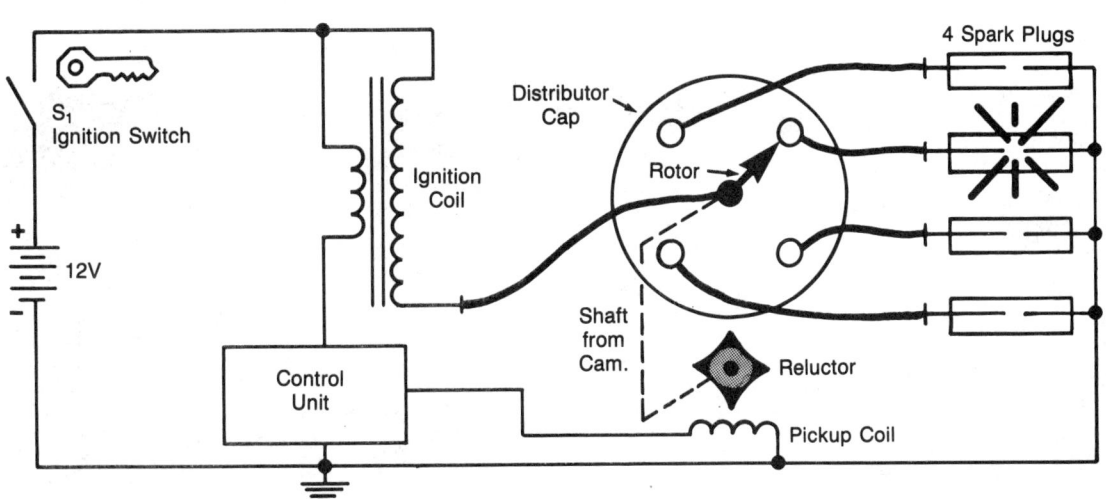

Fig. 62-3 The electronic ignition system in modern cars does not use contact points as older cars had.

Chapter 62 *Electronics in the Automobile*

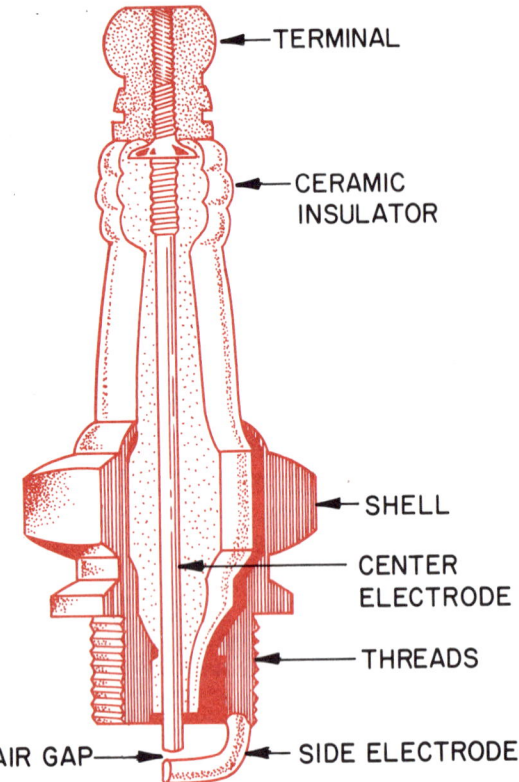

Fig. 62-4 The spark plug has ceramic insulating materials and two electrodes. The ceramic insulator ensures that the spark does not jump across the electrodes before entering the engine — if it becomes cracked, the plug will not function properly.

burned. As the fuel burns, and forces this first piston down, another piston comes to the top of its cylinder. When its gets there, the reluctor will signal the pickup coil to initiate another pulse which will fire that cylinder's spark plug.

Before the electronic ignition system was developed, cars had a pair of contacts (points) which opened and closed. Cars also had a condenser (capacitor) instead of the transistorized control unit, reluctor and pickup coil. The rest of the circuit, however, was basically the same. Many lawn mowers and small engines still use a system very similar to this. Automobile manufacturers are using computer circuits for accurate timing of sparks and to make engines run more efficiently under differing load and weather conditions. Electronic fuel injection is now replacing the carburetor in many models for the same reasons.

Warning System

Modern automobiles have either gauges or indicator lights which warn the driver of potential problems. See Fig. 62-5. In addition to the ammeter (charging) indi-

Fig. 62-5 The dashboard of a modern automobile is busy with gauges, warning lights, and "gadgets" of all sorts.

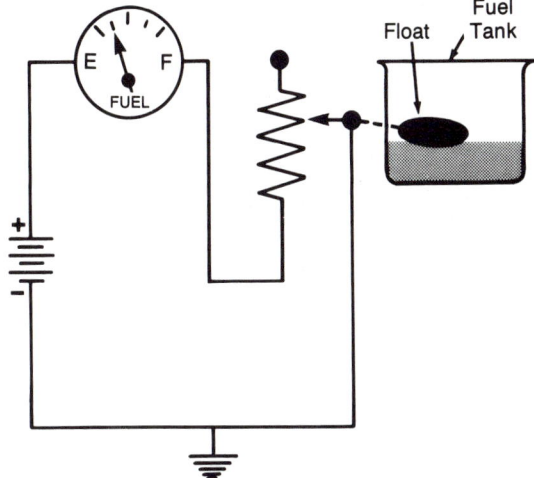

Fig. 62-6 Fuel gauge circuit. Some cars have a warning light in addition to the gauge.

cator, most cars also have a gauge or light to tell when the cooling system fails and the engine begins to overheat. Additionally, cars have a warning indicator for oil pressure. If the engine is operated with low oil pressure or at too high a temperature, it may be severely damaged. A fuel gauge is also used in almost all cars. The circuit for the fuel gauge apears in Fig. 62-6. The wiper (center terminal) of the potentiometer is connected to a float in the fuel tank. When the fuel level changes, the float moves up and down and changes the resistance of the circuit. These resistance changes are indicated by the gauge (meter) which is calibrated to indicate fuel level.

Higher priced automobiles frequently also have warning lights to indicate conditions such as unbuckled seatbelts, doors partially closed, time for maintenance checks, trunk open, and other factors that do not seriously affect the operation of the engine.

Lighting

Second to starting the engine, the system which draws the most current in an automobile is the lighting system. In most cars, there are headlights, tail lights, turn signal lights, brake lights, interior lights, dash lights, and many convenience lights. Most of the lights are wired so that a single wire goes from their switch to the center (button) connector on the lamp. The ground connection may be made directly to the body of the car from the shell portion of the lamp's base.

Accessories

Figure 62-7 shows a simplified wiring diagram for an entire automobile. The diagram looks complicated, but it is actually made up of a lot of simple little circuits like the ones you have already studied. Many of these circuits are for **accessories**. Accessories are extra features which make the car more enjoyable or more comfortable in

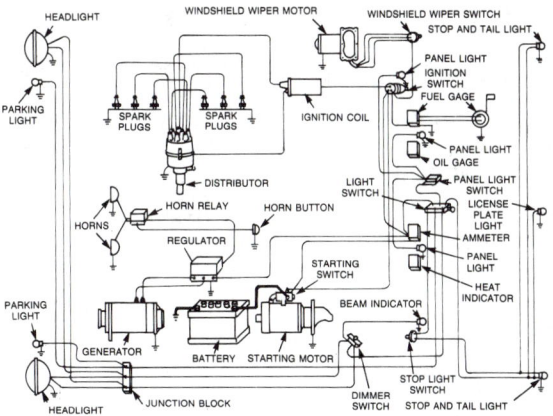

Fig. 62-7 When all of the simple, small circuits in an automobile are shown together it may look confusing.

various ways. They are not needed to make the car run.

The following devices in automobiles have electric motors: windshield wipers, cooling fans, power windows, heater-air conditioning fan, and power seats. Radios and stereo systems have become so popular that few cars are even made without them. Even CB (citizen's band) two-way radio transceivers and car telephones are installed, in many cars. Convenience lighting, power door locks, anti-theft systems, automatic trunk openers, cruise control, cigarette lighters, and warning buzzers are a few of the other accessories which are commonly found in automobiles that use electric current.

Summary

The automobile is an important feature of our modern lifestyle. The starting system is used to start the gasoline engine. The charging circuit keeps the battery freshly charged and powers all of the electrical components and accessories when the engine is running. The ignition system fires the spark plugs to ignite the fuel. Lighting and accessory systems contribute much to the complexity of the automobile's electrical system. The next time you ride in a car, think about the many uses of electricity-electronics in today's automobile.

Important Terms

solenoid
ignition system

ignition coil
spark plug

accessories

Review Questions

1. Refer back to Figs. 62-1 and 62-2 and explain briefly how the starting and charging circuits work. Why do you need the solenoid?

2. Tell the functions of the following parts in the ignition system: ignition key switch, coil, and distributor rotor (button).

Careers: Automotive Mechanic

Mechanics are needed to repair automobiles, trucks, buses, and other equipment. All of these machines now include some electrical-electronic circuits. The jobs available can involve indoor or outdoor work (or both). Most mechanics enjoy a good deal of job satisfaction and good pay, too. It makes you feel good to know that you have skills which other people respect and depend upon. It is also rewarding to see visible evidence of the value of your work as cars "limp" into your shop and then drive out smoothly.

In addition to your electronics class, you should work hard in other subjects to prepare for a career in auto mechanics. Other shop subjects, especially in the power and transportation or auto mechanics areas, will be very helpful. Many trade schools and community colleges offer a full range of courses for mechanics in training. Skills such as good communication are needed for those who wish to advance to management positions.

Project: AC Timing Light

Part List

- 1   100-ohm, 10-watt resistor R_1
- 1   SPST momentary push switch, 120V S_1
- 1   3 Amp, 400 PIV rectifier diode D_1
- 1   50 microfarad, 200 VDC capacitor C_1
- 1   Xenon Flash Tube (RS 272-1145) XT_1
- 1   4 ft. piece, stranded metal spark plug (ignition) wire
- 1   Small spring to connect spark plug and timing light at the same time
- 1   Cardboard core from roll of paper towels (or plastic pipe of similar size)
- 1   AC wall plug with line cord P_1
- Plasticized contact paper to cover tube, electrical tape or shrink wrap to protect connections, heavily insulated wire, and hardware as required.

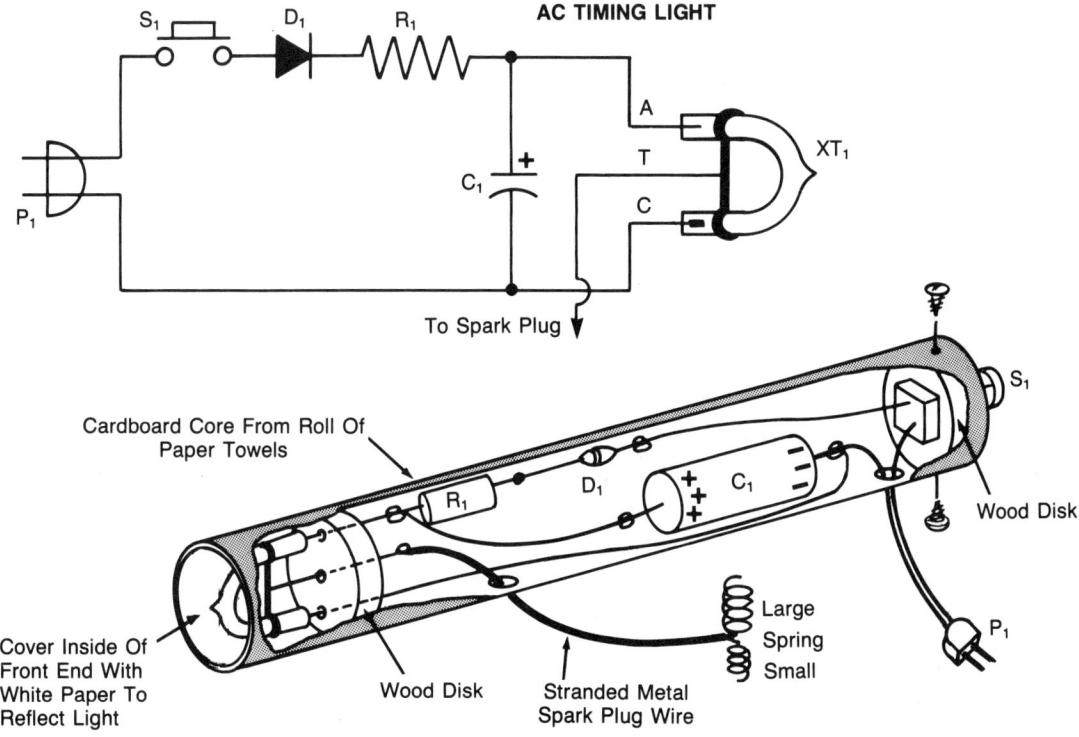

Fig. 62-8   Many of the parts for this AC Timing Light may be salvaged from surplus items.

Chapter 62   *Electronics in the Automobile*

Notes

This timing light is simple to build and very useful. Insulate everything well because there is very high voltage involved. The long tube makes it easy and safe to get the light source nearer the timing mark than with many professional gunstyle timing lights. The soft cardboard tube is stiffened by the two wood disks, tape around the ends and the areas where the two cables exit, and the contact paper covering. The push button switch is easy to operate from the end farthest away from the moving fanbelts of the engine. A PC board may be used, but it is really not needed for this simple project.

To use the light, plug the power cord into an AC wall outlet and connect the spark plug cable to the spark plug used for timing your car. This is usually the number one cylinder. (By firing order, this may not be the first cylinder physically in some cars!) Put a small white spot on the timing mark and on the desired timing indicator for advance or retard. Make sure that no wires are hanging near the fan, belts, or other moving parts.

Start the engine. Aim the light at the timing mark and press the button. If the two white spots align with each other, the engine is properly timed. If the marks do not align, make the needed adjustment to align them while the engine idles. On many cars, this is done by turning the distributor and then clamping it in the new position. Check a shop manual if you are not sure of the proper procedure and setting for your car. The light is quite bright, but strong sunlight may need to be shaded to make it easier to see the timing marks.

Caution: *The way a timing light works is like a strobe light which is perfectly timed with the speed of rotation of the engine. This means that the light will make it appear that the engine — and all of its moving parts — has stopped turning. But the engine will still be turning very fast. Do not touch anything in the engine while the timing light is aimed at it. You could be seriously injured by contacting a moving part which appears to be still.* This caution applies to all timing lights.

Chapter 63

House Wiring

Construction Wiring

Many of the circuits you have studied have been low voltage and low current circuits. The circuits in your home that power lights, appliances, and other devices are either 120- or 240-volt circuits. They also can carry high currents. The circuit to power an electric dryer must be able to carry 30 amperes! For these reasons, strict codes of good practice are followed in electrical wiring of buildings. This chapter explains some of the common practices used in construction wiring.

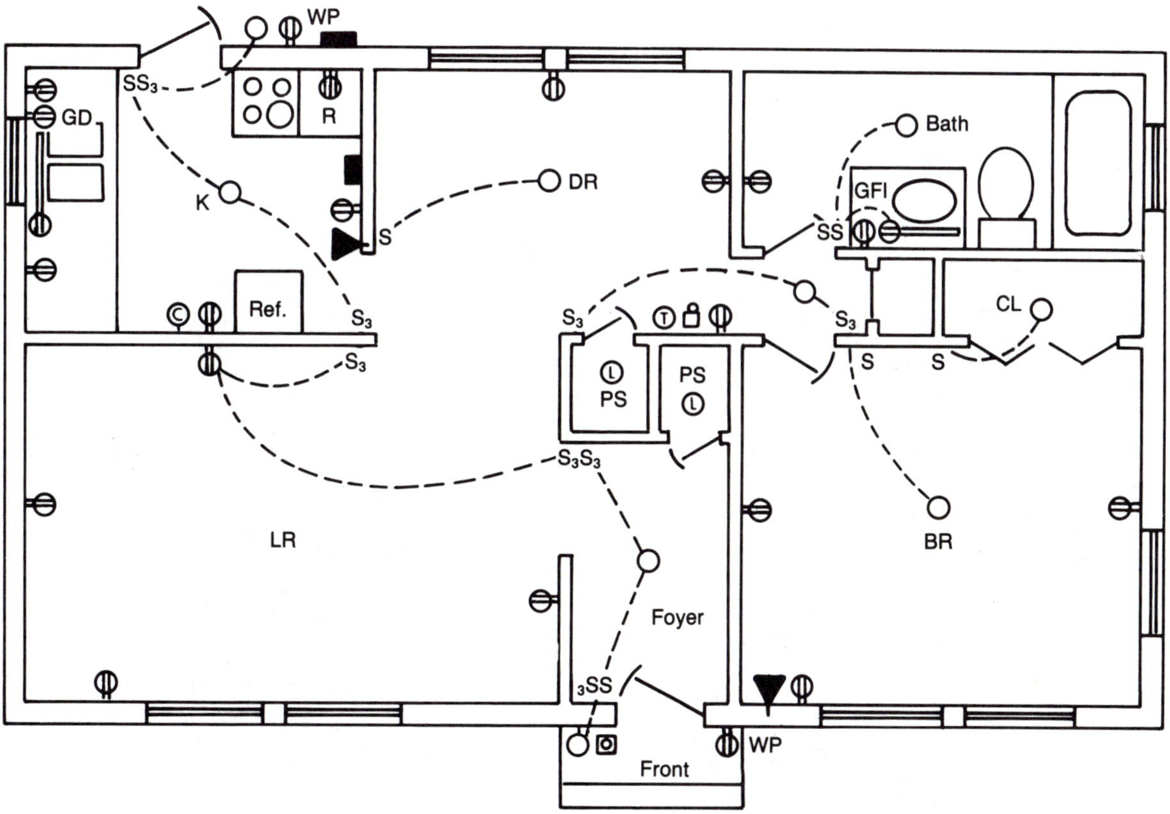

Fig. 63-1 The floor plan shows the locations of electrical fixtures and outlets.

Floor Plans

If all of the walls of a house were sliced about 5 feet from the floor, the top section was removed, and you drew what you would see from a helicopter above the opened house, the drawing would be a floor plan of that house. When architects draw a **floor plan** for a new house, they "imagine" that they are seeing this type of view of the house. Figure 63-1 shows a floor plan for a simple house with the electrical symbols marked to show the locations of electrical devices. A table of the symbols most commonly used and their meanings appears in Fig. 63-2. It is important to place electrical outlets carefully for safety and convenience reasons.

Roughing In

The first job of the electricians who wire a house is called **roughing in**. Unlike the wires, connections, and components in the experiments you have perfomed in class, the outlets and devices, and all connections and splices, must be housed inside protective boxes. The boxes are made of metal or plastic (Fig. 63-3). Additionally, in many urban areas, even the wires (cables) must be placed inside metal shields. When the electrician roughs in a job, he or she mounts the boxes for all outlets and junctions and then runs the cables and wires. The electrician does not install the devices or connect the wires.

Electrical Codes

The **National Electrical Code** (NEC) sets the minimum safety standards for electrical construction wiring in the United States. The code is not a set of laws, but local areas have **local codes** that are almost always the same as the NEC or more rigorous. When no stricter code than the NEC applies, the courts generally use the NEC as a guideline in settling legal cases such as lawsuits. The NEC code manual is revised every three

Sym	Description	Sym	Description
◯	Ceiling Outlet	▬	Fuse Panel
⊖	Duplex Receptacle (Convenience Outlet)	⊖GD	Garbage Disposal
		⊖WP	Weather Proof Outlet
▲	Special Purpose Outlet (Heater, For Example)	T	Thermostat
		S	Switch
⊖D	Clothes Dryer Outlet	S₃	3-Way Switch
⊖R	Range Outlet	----	Wire Or Cable
⊖GFI	Duplex Receptacle With Ground Fault Interrupter	⊸◯	Floodlight
		⊏⊐	Valance Light
▣	Push Button	⊸C	Clock Receptacle
🔔	Bell Or Chimes	L	Keyless Lampholder
◀	Telephone (Not Intercom)	L PS	Pull Chain Lampholder
		⊖	Double Receptacle, Split Wired

Fig. 63-2 Electrical symbols used on floor plans. How many of these can you find in Fig. 63-1?

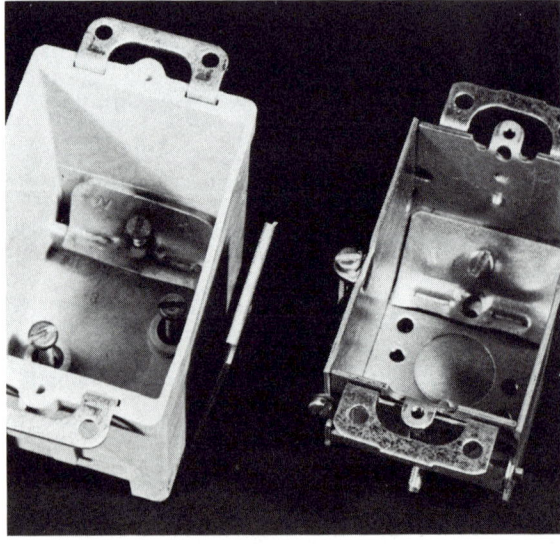

Fig. 63-3 Metal or plastic boxes protect all connections and fixtures.

years to include new findings and recently developed materials.

Another safety precaution is to ensure that all materials and devices used have been approved by the **Underwriters' Laboratories Inc**. You have probably noticed the UL seal on appliances and electrical equipment in your home. Underwriters' Laboratories is a testing agency that performs laboratory experiments on products to see if they are safe for their intended uses.

The codes and UL approval are aimed at ensuring safety. Personal injury and loss of property due to fires and electrical shock can result from the use of inferior materials or substandard workmanship. Local codes in urban centers tend to be stricter than those in rural areas because of the greater danger of fire spreading from one building to another. Likewise, regulations are often stricter for commercial and public buildings than they are for private (single-family) homes. Even the proper applications of grounds, fuses, circuit breakers, and ground fault interrupters are included in code manuals.

Wiring Methods

Here are four basic wiring methods. Some of the methods may not be permitted in certain areas because of local codes and regulations.

Romex Cable

The simplest widely used wiring system installed inside walls uses **Romex cables**. Actually, Romex is a brand name for one company's product. The proper name for this type of cable is **nonmetallic sheathed cable**. It is called Romex by many users because that brand became widely used first. Nonmetallic sheathed cable is often designated by the initials NM or NMC. (They are about the same, but NMC is better for moist and corrosive locations.) Figure 63-4 shows some NM cable.

BX Cable

Another widely used cable is **BX Cable**. Like Romex, BX is an original brand name. The generic name of this cable is "armored cable." Armored cable is also shown in Fig. 63-4. BX is a little more difficult to work with than Romex. It has a flexible metal sheath which protects it from being cut or crushed easily. Special clamps and connectors are used to fasten BX to electrical boxes. This system is more expensive than the Romex system.

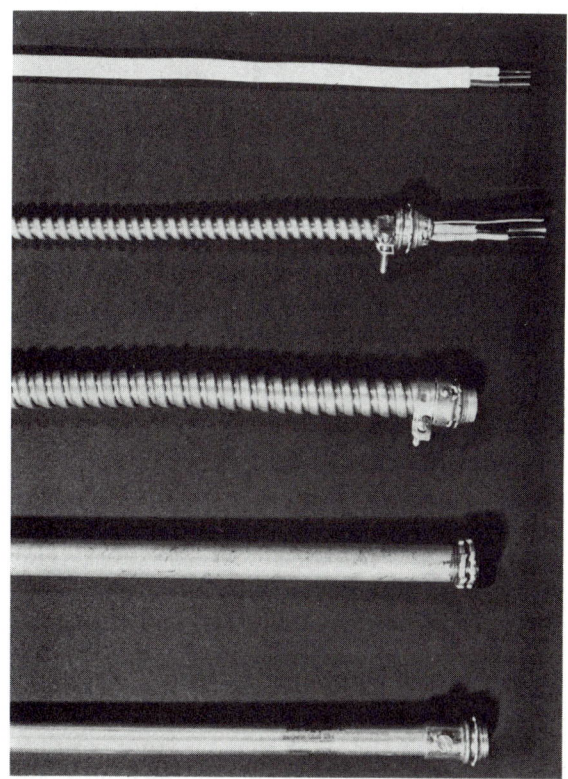

Fig. 63-4 Romex, or NM, cable is used in homes unless local codes require metal sheathed cables. BX cable, also known as armored cable, has a flexible metal covering. Three types of conduit are flexible, rigid, and electrical metallic tubing.

Chapter 63 *House Wiring*

Conduit

There are three common types of conduit. (See again Fig. 63-4.) At first glance **flexible conduit** looks like BX. The flexible tubing is very similar to that used in BX, but the flexible conduit does not already have wires inside it when you buy it (as BX does). All three types of conduit are placed and fastened without wires, then the wires are pulled through them with a fishtape.

The **fishtape** is a stiff, springlike metal tape which can be forced through the conduits, even through bends and curves. When it is all the way through from one box to another, the wires to be installed are carefully attached to the hooked end and then the tape is used to pull them into (and through) the conduit. On large jobs, a fishline may be used. This is a more flexible line which may be sucked into the conduit with a vacuum machine on the opposite end.

The other two types of conduit are rigid conduit and electrical metal tubing. **Rigid conduit** is very much like the iron pipes used to carry water and natural gas. It must be threaded and joined with special fasteners. The strength of rigid conduit and the tightness of the threaded joints make it the best choice for applications which demand maximum protection. Rigid conduit may be bent, but not as easily as other types of conduit and cable.

Electrical metal tubing, or **EMT**, is often called "thinwall conduit." This conduit is actually a thin metal pipe which is joined with special fasteners and connectors. The walls of the conduit are too thin to be threaded. EMT is frequently bent to make installation easier. It is bent with a tool called a **hickey** (Fig. 63-5). A heavier duty hickey, or a power driven bending machine, may be used to bend rigid conduit in a similar manner. The fishtape is used with all three types of conduit. Some plastic conduits are also available for use in corrosive environments (such as chemical plants).

Fig. 63-5 This worker is bending EMT with a hickey.

Surface Wiring

Another wiring system, which is especially helpful for adding circuits to existing structures is called **surface wiring**. Fig. 63-6 shows some surface wiring. Square-looking surface wiring is called "raceway." Special surface mounted boxes are used. Surface wiring has the advantage of great flexibility. It is easy to add new circuits, but it is unattractive in some settings.

Finishing the Job

Regardless of which wiring method is used, when the job is properly roughed in, there will be wires extending about 8 inches from each box. The roughing in is done before the walls are enclosed. Inspections are made, and the walls are enclosed with plaster, paneling, gypsum board (sheet rock), or other materials. After the walls are enclosed, the actual fixtures, outlets, and other devices are installed. The installation must be done carefully to avoid shorts, scrapes on insulation, and loose connections.

Common Residential Circuits

A simple circuit including a duplex (convenience) outlet, a light, and a switch is shown in Fig. 63-7. The colors of the wires in the conduits follow standard practice. Black wires are usually called "hot," because they eventually connect to the "hot" side of the AC generator. The other side of the generator is connected to ground. A set of white wires in our circuit connects to this "neutral" side of the generator. Still another set of wires is a safety "ground" circuit. These safety ground wires are either green or uninsulated.

Fig. 63-6 Raceway is one commonly used surface wiring material.

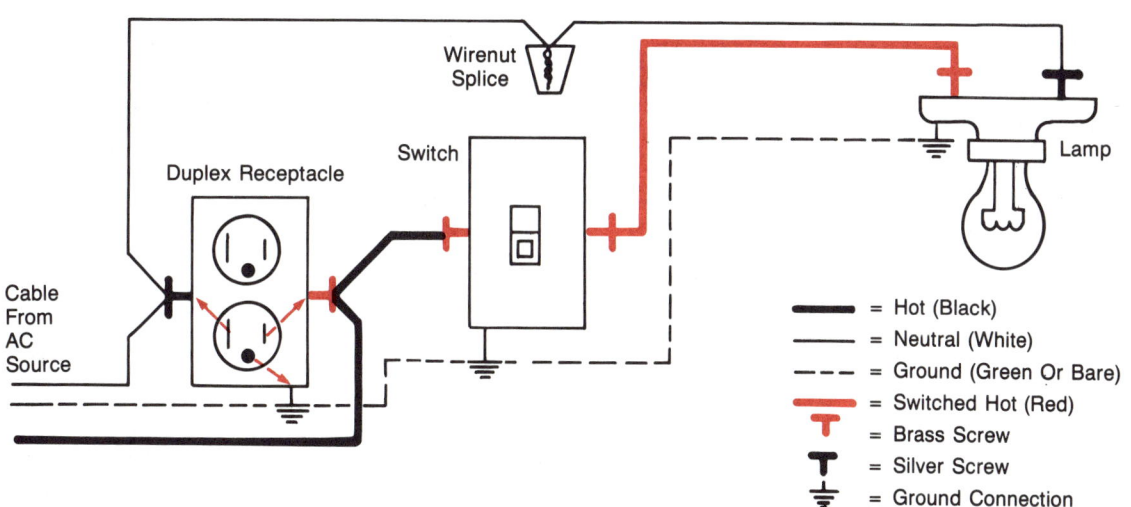

Fig. 63-7 The neutral (white) wire does not connect to the switch at all. In this circuit, the duplex receptacle will be "hot" all of the time, but the lamp will only burn when the switch is on.

Chapter 63 House Wiring

It is important to always connect the wires according to this color code. One good rule of thumb is to connect black wires to brass (colored) screws and white wires to silver (colored) screws. Ground wires connect to green screws or the metal box. Most fixtures and outlets are fitted with the correct colors of screws for this purpose.

Notice the switch in the circuit. It has a black wire coming to it from the power source's "hot" side. A red wire is leaving it to go to the lamp it controls. Do not use white wires on a switch, because white wires should always be neutral (ground side of the line). The red (or other color) signifies that this wire is "hot" (at least parts of the time).

Figure 63-8 shows the installation of two 3-way switches to control the light in the stairwell from either end. You probably have at least one circuit in your home that works this way. The wires that run between the two switches are called **travelers**. They only carry current when both switches are in the correct positions to light the lamp.

Summary

Construction wiring involves roughing in and finishing the installation of circuits, conduits, cables, and fixtures in buildings. The roughing in is done before the walls are enclosed. The major wiring systems used today are Romex, BX, EMT, rigid conduit, and surface raceway wiring. Wiring must be done carefully in buildings because the high levels of voltage and current used present both fire and shock hazards. To promote safety, the National Electrical Code and stricter local codes set standards of good practice in construction wiring.

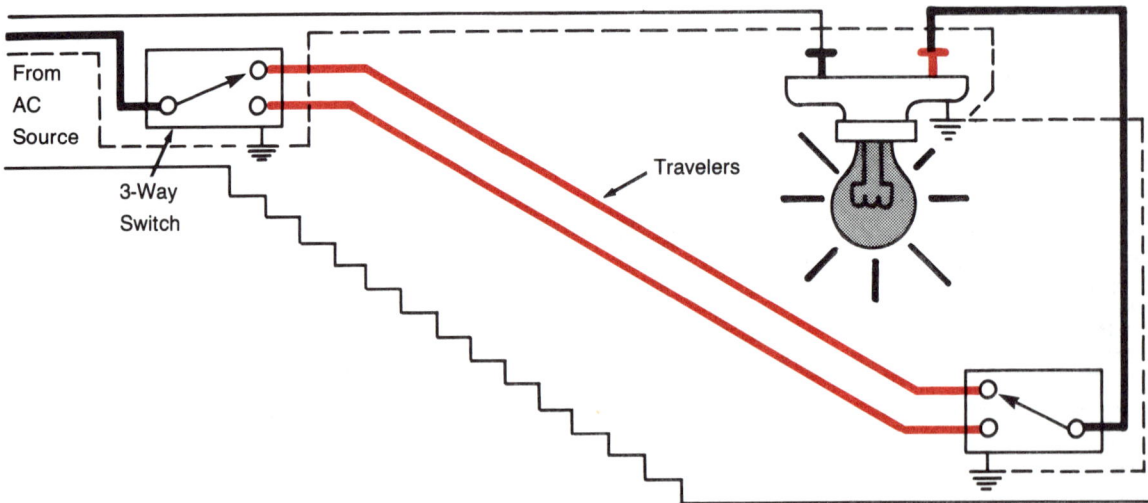

Fig. 63-8 The lamp lights only when both switches are in the same position.

Important Terms

floor plan
roughing in
NEC
local codes
U-L seal
Romex
BX
flexible conduit
fishtape
rigid conduit
EMT
hickey

Review Questions

1. Draw a diagram of a room in your home, perhaps your own bedroom, and fill in the proper symbols to show how it is wired.

2. Explain what is done during the roughing in and finishing stages of the electrical wiring on a construction job.

3. Explain the differences between wiring with each of these systems: Romex, BX, EMT, rigid conduit, and surface wiring.

Chapter 64

Appliance Repair

Working on Appliances

Many of the electrical appliances in your home use parts and components you have already studied. How these components are combined to make appliances and some of the repair techniques used when they fail are the topics of this chapter. The main things to consider when repairing appliances are: safety, keen observation, logical approach, high-level skill and thorough testing.

Safety

Whenever possible, the power should be turned off and disconnected when servicing equipment. True, some devices cannot be tested unless they have power flowing to them. However, there is no need for the power to be on when you are opening the case of an appliance or while parts are being exchanged. Before opening an appliance cabinet, check the schematic and any other available diagrams for safety warnings and locations of hazardous parts.

Always use tools with insulated handles. Also, avoid working on electrical sections of appliances when water is present. For instance, if the washing machine leaks all over the floor (and you are fairly sure it's just a leaking hose — the most probable cause), clean up the water and disconnect the power before opening the machine. Fig. 64-1 shows what could happen if these simple steps are not taken. Also, make sure that all leaks are stopped before turning the power back on. Water which gets into motors and other electrical devices presents a serious shock hazard and can burn up the devices, too.

Make sure that faulty parts are replaced with ones of high quality and the proper ratings. Consult technical parts lists and data sheets carefully when selecting replacement parts. Make sure that electrical parts and the main power cord are properly grounded and that insulation is in good condition. Carefully follow wiring diagrams and color codes.

Fig. 64-1 Water is a very direct path to ground (where the current of all AC "hot" lines is trying to get)! It also reduces the skin's natural resistance so that much more current flows through the body than under normal (dry) conditions. Clean it up before you work!

Keen Observation

The only way to tell if an appliance is in need of repair is to know what it is supposed to do. Many repair persons regret charging costly minimum repair bills (due to company policies) just to tell appliance users that their dishwasher is supposed to take that "2-minute rest break after the power spray cycle."

If the users had been observant, they would have known this to be a normal function of their particular model. Then, they would not have turned the machine off and called the repair shop when they first noticed the pause.

So, carefully observe all appliances while they are running correctly so that you will know and recognize faulty operation. At this time, listen for normal sounds. Some strange sounding noises are perfectly normal in many appliances. They will trick you if the first time you take note of them is when you are trying to isolate a problem, however. Watch what the machine does in each of its cycles. Time the duration of cycles.

Read the entire owner's manual for an appliance. Some people call the service department to complain that the washing machine water is not hot when the switch is set to the hot position. When the service person gets to the site, the water is plenty hot. The user replies that it seems to be hot sometimes and not others. Almost all washers use cold water (or warm) for rinse cycles regardless of the switch setting. If the user had read the manual, a costly repair call would be avoided.

No test instrument or special tool you can buy is as helpful as a good knowledge of what the machine is supposed to do (when operating properly) and when it is to do it. This information can only be gained by keen **observation** of the machine in operation.

Another type of observation is used when there is a problem. Again, watch, listen, feel, smell, and watch the time as the machine tries to do its job. The clues you gather in this "smoke test" will help a great deal. Of course, if there is a strong burning smell, and possibly real smoke, turn the machine off immediately. But as long as the machine seems to be able to cope with its problems without damaging parts, just observing it is the most valuable first step in **troubleshooting**.

Logic

Next, combine all of the information you have gathered from user's manuals, previous observations, what the user has told you about the machine's problems, and the things you found in your testing observations. Take a **logical approach** in isolating the problem area. For instance, if the lights on the front panel don't even light, the problem might be in the power supply. If a loud grunting noise is the only problem with a machine that works normally in every other way, it is senseless to even suspect the power cord or supply.

Some service manuals have troubleshooting flow charts which can lead you through an orderly, logical sequence of tests to help isolate the problem.

Craftsmanship

Replacing defective parts is usually not too difficult. Sometimes they are in difficult-to-reach places or special tools are required. Generally this part of the job, **craftsmanship,** is within the ability range

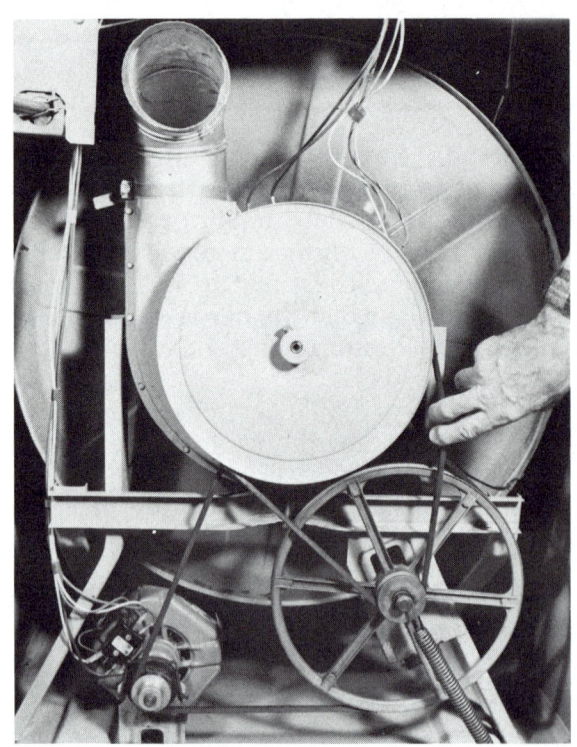

Fig. 64-2 Check the tension of belts and make sure parts are free to move correctly.

of most people who have general tools and are fairly skilled with their hands. It is important that the parts be replaced properly. All connections must be made carefully to prevent future problems.

New gaskets and seals should always be used when water-carrying parts are loosened. It's a shame to successfully replace the $75 motor yourself, and then watch it burn up immediately because you tried to save 75 cents by using the old seals on the water pump! Be careful to line up any belts and pulleys properly and check their tension. See Fig. 64-2. Make certain that linkages to solenoids and other moving parts are free to move properly. Lubricate parts only as specified in the repair manual. Some parts actually need friction to operate properly.

Be careful not to lose or forget important parts. Make sure that you get the right replacement part. Clean used parts and inspect them carefully before re-installing them. Be careful to tighten screws and nuts properly so that they hold. Do not wring them off.

Testing

Once the needed parts are repaired or replaced, and the machine is reassembled, the machine must be run through its entire cycle as a followup "smoke" test. During this second testing, the machine should be observed very carefully. Listen for unusual noises and watch carefully for other clues that could reveal mistakes made during the repair job, or previously undetected problems. It is unfortunate to do a repair job and then later have to redo it because only a **symptom** was repaired instead of the real cause.

If the machine passes this post test, it is ready to return to service. It is still a good idea to check up on the machine after a few uses to ensure that the repair is holding up well.

The other most common complaint with dryers is "it runs, but it doesn't get hot." List the possible causes of this problem. Do any of the things on our earlier list of "not running" causes apply to this symptom? After you finish your checklist, compare it to the list at the end of this chapter to see if you thought of them all.

A Case History

Suppose you are an appliance repair person called to a home to repair an electric dryer that is not functioning. First of all, from your knowledge of electric dryers and what they do, list in order exactly what should happen when a dryer operates. Include the functions of all of the following parts: timer, power cord, motor, control switches, and heating elements. The wiring diagram with many appliances may be helpful to you. Now list the operating sequence of the machine when it works normally.

Back on the job site, the owner tells you that the dryer makes a sound like it is running, but it does not dry the clothes. When you test the machine, you see that it does not turn. You do clearly hear the motor running. What is the most probable cause? The fan belt or other linkage that connects between the motor and the tub is probably the cause of the problem. This is the most common mechanical problem in dryers.

What if there was no noise, but the other symptoms were the same? Now you might suspect the following: power disconnected or circuit breaker tripped, timer broken and not letting current get to motor, motor burned up or jammed (it should hum or smell bad if it is jammed), safety switch or thermostat shutting machine down to prevent damage from another cause.

Should You Ever Need Service

STEP 1 . . .
If your dishwasher should fail to operate, CHECK the following list before calling your dealer. *Save the cost of a service call* because the following situations are not covered by the Warranty.

a. Is fuse blown or circuit breaker open? On portable models, is cord plugged into a grounded outlet?
b. Is the door closed and latched?
c. Is the water supply to the dishwasher turned on?
d. Have you read your Use and Care Guide for complete instructions?

If your dishwasher is not washing properly, please see Problem, Cause, Solution Chart, pages 16 thru 18.

STEP 2 . . .
WHO TO CALL FOR SERVICE . . .

a. First, call your dealer or the repair service he recommends.
b. If for some reason he is not available, or if you now reside in an area other than where the unit was purchased, look in the yellow pages of your phone book under DISHWASHING MACHINES for authorized KitchenAid servicing outlets.
c. If you are unable to obtain satisfactory service in this manner, contact the KitchenAid wholesale distributor. He is listed in the yellow pages of all metro area phone books under DISHWASHING MACHINES. A complete distributor listing is also provided in this book on the inside back page.

STEP 3 . . .
All service should be handled locally by the dealer from whom you purchased the unit, or an authorized KitchenAid servicing outlet or wholesale distributor. If your local service is not satisfactory, contact the Customer Relations Department, KitchenAid Division, Hobart Corporation, Troy, Ohio 45374. Area Code 513 335-7171.

STEP 4 . . .
If you are not satisfied with the action taken, contact MACAP—the Major Appliance Consumer Action Panel. This is a group of independent consumer experts, voicing consumer views at the highest levels of the major appliance industry. Please contact them only when the dealer, authorized servicing outlet, wholesale distributor and Hobart Corporation have failed to resolve your problem.
ADDRESS: Major Appliance Consumer Action Panel, 20 North Wacker Drive, Chicago, Illinois 60606. Phone: A.C. 312 236-3165. MACAP will in turn inform us of your action.

IN ALL CORRESPONDENCE, PLEASE GIVE THE FOLLOWING INFORMATION: DATE, YOUR NAME, ADDRESS AND TELEPHONE NUMBER, KIND OF APPLIANCE, BRAND, MODEL NUMBER, SERIAL NUMBER, DATE PURCHASED, DEALER FROM WHICH PURCHASED (INCLUDE ADDRESS), SERVICING AGENT'S NAME AND ADDRESS IF DIFFERENT FROM DEALER'S, LOCATION OF APPLIANCE, AND CLEARLY DESCRIBE PROBLEM AND SERVICE REQUIRED.

Fig. 64-3 Most owner's manuals have a page like this one which should be followed carefully before the service shop is called.

Summary

Appliance repair requires five main things: safe work habits, keen observation, application of logic, good craftsmanship, and careful followup testing. Since many electric appliances in the home also handle water, the potential of shocks and further damage is increased greatly. Care must be taken at all times. Work should never be done around water. The power should be disconnected at all times except when making specific tests that require the machine to be under power. Keen observation and a logical approach are the key ingredients of successful troubleshooting.

Troubleshooting Sequence for the Dryer

The most likely causes of the no-heating problem are:
- Burned out heating coils (element).
- Safety thermostat defective and not letting power get to coil.
- Bad timer.
- Bad connection or burned wire, especially near the heating coil connection because heat and high current both cause deterioration.
- Switch or other control for adjusting temperature defective.
- Overloaded or clogged lint filter causing thermostat to cut off too quickly.

Important Terms

observation   logical approach   symptom
troubleshooting   craftsmanship

Review Questions

1. What are the main hazards caused by water for the appliance service person? How can these problems be overcome?
2. List the four main steps for repairing appliances. Tell what each step means.
3. Can you remember a time when someone you know has been embarrassed because of an unnecessary service call? If so, describe it. If not, describe an imaginary situation in which this could happen. What is the most important thing for the owner of a new appliance to do to avoid such events in the future?

Careers: Appliance Repair Work

The appliance repair person enjoys many of the same rewards as the automobile mechanic, such as seeing your work appreciated by those who depend on the equipment you service and good pay. Work is done in well equipped shops and in customer's homes. Most good service people are able to repair several types and brands of machines.

Courses in shop, science, mathematics, and communications are all important in preparing for careers in appliance repair. Trade school and community college courses are also available. Your electronics course will help you a great deal in this career.

Index

A

AC (alternating current), 43, 171-75, 191, 194, 206
   circuits, 191, 203-4, 209-10, 232-33
   induction motors, 349-50, 351
   meters, 129-30, 135
Accessories, 397
Alkaline cells, 158
Alternating current. See AC
Alternations, 172
Alternators, 164-65, 168, 394
Ammeters, 129, 131, 132, 394, 396
Amperes (A), 41
Amplification
   defined, 269, 272, 276
   operation of, 269, 270, 271 275, 278-79, 288
Amplifiers. See also Circuits
   operation of, 321, 323, 329
   parts of, 316, 317-18, 319, 320, 321, 323-24, 325
   types of, 306, 315-16, 317-19, 321
Amplitude, 173
AM (amplitude modulation) radios, 358-59, 361, 362
Analog
   control devices, 22, 179
   meters, 122, 125, 150, 151
AND gates, 294-95
Anodes, 254, 267, 268
Appliance repair, 407-12
Arcs, 337, 341-42
Armatures, 166, 347. See also Rotors
Armored cables, 403
Armstrong oscillators, 329-30
Atoms, 14-18
Automobiles, 158, 160-61, 393-98
Auxiliary circuits
   defined, 124
   uses of, 127-30, 134-35, 144, 150, 153
Average voltage, 174, 175
Axial leads, 47

B

Ballast, 342
Bardeen, John, 274, 280
Bases, 274
Batteries, 28, 32, 157-62, 394
   See also Cells
Bell, Alexander, 352, 356
Binary code, 294, 299, 301-2
Bio-electricity, 31, 32
Bipolar devices, 287. See also NPN transistors
Bits, 299, 310
Blanking pulses, 368
Branch circuits, 89-92, 96-97, 103. See also Parallel circuits
Brattain, Walter, 274, 280
Breakdown
   potential, 256
   voltage, 212.214, 215

Bridge rectifiers, 260-61
Brushes, 348
Bunsen, Robert, 101
BX cables, 403
Bytes, 310

C

Cables, 403
CAD-CAM systems, 391-92
Calculators, 69-72, 91, 294, 309
Calibrated scopes, 179, 181
Calibration, 123
Capacitration, 123
Cpacitance, 208-15
Capacitive reactance, 223-27
Capacitors
   coils compared to, 216-20, 224
   operation of, 208-9, 212, 215
     217, 219-20
   parts of, 208, 209, 211, 214
   safety with, 209, 215
   types of, 212-13
   uses of, 209-10, 214, 219, 220,
     243, 245
Carbon composition resistors, 47
Careers, 2-5
   communication skills in, 235, 292, 326,
     374, 398, 412
   with computers, 383, 386
   on electronics assembly lines, 56
   electronics skills in, 149, 235, 292, 298,
     412
   as engineers, 32, 73
   in instrumentation, 149
   mathematical skills in, 73, 132, 149,
     235, 292, 326, 386, 412
   in the military, 235
   personal qualities in, 408-9
   as physicists, 132
   as repair persons, 44, 398, 412
   in research, 18
   scientific skills in, 25, 132, 149, 235,
     292, 326, 412
   in space exploration, 292
   technical skills in, 149, 326, 412
   as technicians, 88, 235, 326

Compounds, 13-14
Computers
   parts of, 283, 284, 294, 305, 307 355
   types of, 309-12
   uses of, 383-86
Conduction, 336-37
Conductors
   defined, 17, 18
   insulators compared to, 251
   resistance of, 45
   types of, 24
   uses of, 20-21, 251
Conduits, 404
Constants, 174, 205
Control
   devices, 21-22, 179
   grids, 269
Convection, 337
Convection current, 255, 258
Copiers, 372-73
Cores, 114, 192, 194
Coulombs, 40
Counter-counter-EMF and counter-EMC,
   190-91, 194, 195
Covalent bonds, 251
Craftsmanship, 408-9
Crystal-controlled oscillators, 330-31
Current, 13-18, 40-43. See also Phase
   shifts; specific current
   characteristics of, 20-21, 172-73, 197-
     98
   in circuits, 82, 87, 89-91, 99, 247
   defined, 40, 42, 43, 59
   in diodes, 254-56, 257, 258
   lag, 199
   measuring, 40, 41-43, 59, 97-98, 126-
     32, 147
   sources of, 20
Cycles, 172

D

Damped oscillations, 245
d'Arsenval meters, 122-24, 125
Data processing, 383-86
DC (direct current)
   circuits, 191, 203-4, 206, 207, 209, 210

defined, 43
motors, 348-49, 351
types of, 260
uses of, 189-91, 194, 195
Decimals, 69, 70, 91
De Forest, Lee, 267, 273
Delays, 197, 198-99
Denominators, 225
Depletion zones, 253, 257-58
Detectors, 359
Dials, 352, 353, 355
Dielectrics, 208, 211-12
Digital
circuits, 294, 304-5, 307
controls, 21
electronics, 294, 303
multimeters (DMM), 150, 151-53
readouts, 122, 125, 150
Diodes, 252-58. See also Power:
supplies; Semiconductors; specific
diodes
bridges, 336
defined, 252
operation of, 254-56, 257, 258
parts of, 255-56, 274
uses of, 129, 253, 259-60, 336
Direct couplings, 321, 323, 325
Direct current. See DC
Direct proportions, 60, 225
Discharging, 209
Discriminators, 361
Dissipation, 78
Distortion, 316
Domains, 109, 110
Doping, 252, 257, 304
DPDT (double-pole, double-throw)
switches, 377
Drop. See Electricity
sources; Voltage
Dry cells, 157-58
Dual trace scopes, 180

E

Eddy currents, 185, 186, 335, 338
Edison, Thomas, 273, 346, 356
Edison effect, 267, 273, 346

Einstein, Albert, 220
Electricity sources, 26-32
Electric motors
operation of, 347-51
parts of, 114, 347, 349
types of, 348-50, 351, 390
Electricians, 3
Electrodes, 28
Electrolytes, 28, 157
Electrolytic capacitors, 213
Electromagnets and electromagnetism
defined, 113
parts of, 113-13
safety with, 117-18
strength of, 114, 118, 122
uses of, 114-17, 118, 122, 123,
163, 168, 378
Electromotive force (EMF). See
Electricity sources; Voltage
Electronic meters, 137-38, 150-53
Electrons, 14, 15-17, 251, 368
Electrostatic copiers, 372-73
Elements, 13
EMF. See Electricity sources;
Voltage
Emitters, 274
EMT (electrical metal tubing), 404
Engineering careers, 3
EPROM (erasable programable ROM
chips), 307, 310
Equations, 60, 61-62
E Rn, 84
Exciters, 164-65
Expressions, 225

F

Failure rate, 54
Faraday, Michael, 220
Farads, 211
Feedback, 328-29, 333
FET (field-effect transistors), 151, 287
Fiber optic cables, 284
Fields, 347. See also Stators
Fishtape conduits, 404
Fixed resistors, 46, 47, 49
Fleming, John, 267, 273

Flexible conduits, 404
Flip-flops, 297-300
Floor plans, 401-2
Floppy disks, 310
Fluorescent lighting, 342-43
Flux fields, 108, 110
Flyback transformers, 369
FM (frequency modulation) radios 360-61, 362
Form of the equation, 60, 61-62
Forward bias, 254, 258
Free electrons, 15-16
Frequency, 172-73
   of capacitive reactance, 226-27
   of circuits, 232-33, 238
   defined, 172, 175
   types of, 234, 246, 359
Fuel cells, 162
Full-scale deflection, 127
Full voltage, 199
Full-wave rectifiers, 260, 261-62, 264
Fuses, 102-3, 105, 106

G

Gates, 294-97, 303
Generators
   alternators compared to, 164, 165
   parts of, 166-67
   types of, 167
   uses of, 26, 32, 164, 394
Given values, 65
Graticules, 178, 179
Ground fault interrupters (G.F.I), 105, 106
Grounding, 104-5, 405, 406. See also Safety
Gutenberg, Johann, 372

H

Half-wave rectifiers, 260, 264
Hard disks, 310
Hartley oscillators, 330
Heat, 29, 32, 335-39
Henry, Joseph, 201, 363
Henrys (H), 192, 194, 205

Hertz (Hz), 172
Hickeys, 404
Holes
   defined, 253
   uses of, 274, 275-76, 277-78, 281, 282
Horizontal deflection coils, 369
Horsepower, 76, 81
Hot wires, 104, 405, 406
Hydraulics, 390
Hydrometers, 160
Hysteresis, 185, 186, 335, 338

I

IGFET (insulated gate field-effect transistors), 287, 288-89
Ignition systems, 394-96, 398
Impedance, 230-33
   calculating, 236-38
   defined, 231, 235
   uses of, 243, 247, 318, 319, 321, 324
Implosion, 370
Incandescent lighting, 341
Incomplete circuits, 23, 24
Independently excited field generators, 167
Inductance, 189-94
Induction heating, 338-39
Inductive couplings, 323-24
Inductive reactance, 191, 203-7, 225. See also Resistance
Inductors. See Coils
Industrial control, 376-80
Insulators, 17, 18, 105, 251
Integrated circuits (IC), 304-7, 336. See also specific integrated circuits
Interface devices, 307
Interlaced scanning, 368
Intermediate frequency, (IF), 359
Intersections, 231
Inverse proportions, 60, 225
Inverters, 296-97
Isolation, 283

J

JFET (junction field-effect transistors), 287-89, 291
Jump starting, 160-61
Junctions, 99

K

Kilovolts (KV), 36
Kilowatthours (kWh), 78, 81
Kirchoff, Gustav, 101
Kircholff's laws, 99-100
Knowns, 66

L

Labeling, 66
Languages (computers), 312
Lasers, 284
LCD (liquid crystal displays), 151-52
Lead-acid batteries, 160
LED (light-emitting diodes), 151-52, 257 281, 283
Lenz, Heinrich, 119
Lighting, 28, 32, 341-44, 397
Limiters, 361, 362
Linear circuits, 305, 307
Lissajous patterns, 180
Lithium cells, 158
Loading, 21, 45, 134, 138. See also Resistance
Local codes, 402
Logic, 65, 66, 70-71, 408
LSI (large-scale integration) chips, 305, 307

M

Magnetostriction, 114
Magnets and magnetism, See also specific magnets
  care of, 110
  characteristics of, 108-10
  defined, 107
  fields, 163
  induction, 107-8
  parts of, 108-9, 110, 376-77
  uses of, 26-27, 32, 163, 164, 168
Mainframes, 309-12
Marconi, Guglielmo, 363
Matter, 13
Maxwell, James, 220
Medical careers, 4
Megavolts (MV), 36
Memory, 91
Mercury cells, 158
Mercury switch-type thermostats, 379
Mercury-vapor lighting, 343
Meters. See also specific meters
  auxiliary circuits in, 127-30, 134-35, 144, 150, 153
  characteristics of, 123, 127
  current in, 123-24
  parts of, 127, 132, 134-35
  reading, 124-25
  uses of, 121, 125, 126
Microamps (A), 42
Microcomputers, 309-12
Microfarads (µf), 211
Microvolts (µV), 36
Microwaves, 339
Milliamps (mA), 42
Millivolts (mV), 36
Milliwatts (mW), 79
Minicomputers, 312
Mirror scales, 148
Modems, 355
Modulators, 358, 362
Molecules, 13
Morse, Samuel, 363
Morse code, 363, 371
MOSFET (metal-oxide semiconductors), 288, 289, 290, 291
Motors. See Electric motors
Multimeters
  advantages and disadvantages of, 145, 150-51
  defined, 144
  operation of, 145-47
  parts of, 144, 148, 153
  types of, 150, 151-53
  uses of, 144-45, 148

Multiple-stage amplifiers, 321
Multipliers, 135, 138
Multishunt meters, 128-29
Multivibrators, 331-333

N

NAND gates, 296
N.C. switches, 377
NEC (National Electrical Code), 402
Negligence, 102, 103. See also
  Safety
Negligible, 45
Neon lighting, 343
Neutral wires, 104, 405, 406
Neutrons, 14
Nickel-cadmium (Ni-Cad) cells, 162
Nonbasic units, 66
Nonmetallic sheathed cables, 403
NOR gates, 297
N.O. switches, 377
NPN transistors, 274, 275-76, 277-79.
  See also Bipolar devices
N-type materials, 252
Nucleus, 14

O

Observation, 408
Oersted, Christian, 113
Ohm, Georg, 60, 64
Ohmmeters, 139-43
Ohms (Ω), 46
Ohm's law
  defined, 60
  equation for, 60, 61-62, 67
  uses of, 60-63, 64
  Watt's law compared to, 78, 79-80
Open circuits, 23, 24
Open loops, 390
Operating systems, 312
Operational amplifiers (OP amps), 306
Operators (telephone), 353, 355
Opto-couplers (OCI), 283
Opto-electronic devices, 281-84
OR gates, 296
Oscillation, 245

Oscillators, 328-33, 357
Oscilloscopes, 173, 177-81
Overloading, 93, 103. See also
  Safety

P

Parallax errors, 124-25, 148
Parallel circuits. See also
    Branch circuits
  defined, 82
  disadvantages of, 85
  operation of, 82, 89-92, 94, 207,
    214, 227
  parts of, 89-92, 130, 214
  series circuits compared to, 82, 89, 94
  uses of, 92-93, 94
Parallel-resonant circuits
  operation of, 243-46, 247
  parts of, 243, 245
  series-resonant circuits compared to,
    243, 247
  uses of, 247-48, 324, 358, 361
Peak voltage measurement, 173, 174, 175
Pegging, 127
Pentode tubes, 271
Permanent magnets, 107, 163, 164, 168
Permeability, 109-10
Phases and phase shifts, 175, 229. See
    also Current
Photoconductive diodes, 281-82
Photodiodes, 281, 284
Photons, 281
Photoresistors, 282
Phototransistors, 282
Phototubes, 282
Photovoltaic cells, 28, 29
Pi (π), 205
Picofarads, (pf), 211
Piezoelectric effect, 29-30, 330-31
Pinning, 127
Plates
  operation of, 209
  parts of, 208
  uses of, 211, 212-13, 214
PN junctions. See Diodes
PNP transistors, 274, 277-78, 319.

See also Bipolar devices
Polarity, 171-72
Poles, 108-9, 376-77
Potential. See Electricity sources; Voltage
Potentiometers (pots)
   defined, 48
   uses of, 48, 49, 139, 179, 217, 264
Power. See also Diodes; Resistors; Transformers
   defined, 81
   measuring, 76, 78-79
   rating, 55
   supplies, 259-64, 268
Primary cells, 157-58
Primary coils, 183
Printing, 371-72
Programming, 312
Protons, 14
P-type materials, 252, 253
Push-pull circuits, 317, 320, 325
Pythagorean theorem, 237

R

Raceways, 404
Radar, 235, 373
Radiation, 337
Radio frequency (RF) probes, 153
Radios, 357-62
RAM (random-access memory), 310
Range, 128
Rate. See Current
RC (resistive-capacitive)
   circuits, 216-20
   couplings, 324
Reactance, 205
Receivers, 352, 358-59, 361, 366
Reciprocals, 91
Recordkeeping, 383
Rectification, 135, 260-61
Relays, 115-17, 377-78
Reluctance, 114
Repairperson, 2-3
Repeatability, 391
Resistance (R), 45-49. See also Inductive reactance; Loading; specific resistance
   calculating, 84
   in circuits, 83-84, 87, 91-92, 94, 96-97
   of conductors, 45
   defined, 59
   disadvantages of, 49, 185, 186, 335, 338
   heating, 336
   measuring, 46, 49, 59, 139-43, 147, 150
Resistors. See also Power: supplies
   characteristics of, 53-55, 78
   color coding of, 51-56
   types of, 46-48, 49, 282
   uses of, 49, 259
Resonance, 229-35
Resonant circuits
   defined, 234, 235
   types of, 236-41, 243-48
   uses of, 234-35
Resonant frequency, 234, 246
Reverse bias, 255, 258
Reverse polarity, 109
Reverse Polish notation (RPN), 70
Rheostats, 48, 49, 139
Rigid conduits, 404
Ringers (telephones), 352
RL circuits, 196-200
Robots, 387-92
ROM (read-ony memory), 310
Romex cables, 403
Root-mean-square (RMS) voltage, 174-75
Rotors, 164, 347, 349
Roughing in, 402, 404-5, 406

S

Safety, 6-8, 12, 102-6, 337
   with appliance repair, 369-70, 407, 411
   with automobile repair, 160-61
   with capacitors, 209, 215
   with circuits, 24
   with electromagnets, 117-18, 338, 39
   with lighting, 344
Saturation, 109
Schematics, 22
Science careers, 4

Scientific notation, 40-41
Screen grids, 271
Secondary cells, 157, 158, 160, 161-62
Secondary coils, 183
Self-induced currents, 189-94
Semiconductors. See also Diodes
   defined, 17, 18
   doping, 252, 257
   parts of, 251-52
   types of, 257, 287-91
   uses of, 269, 336, 378
Sensitivity, 127, 132
Series circuits
   advantages and disadvantages of, 85
   defined, 82
   operation of, 82, 82-84, 87, 89, 91, 206-7, 215, 227
   parallel circuits compared to, 82, 89, 94
   parts of, 214
Series generators, 167
Series loops, 99
Series ohmmeters, 140, 143
Series-parallel circuits, 96-101
Series-resonant circuits, 236-41, 243, 247
Series-wound motors, 348-49
Service careers, 2-3
Servo motors, 351, 390
Shells (atoms), 14, 18
Shockley, William, 274, 280
Short circuits, 23-24, 103. See also Safety
Shunt generators, 167
Shunt ohmmeters, 140-41, 143
Shunts, 127, 132, 134-35
Shunt-wound motors, 348, 349
Significant digits, 52
Silicon-controlled rectifiers (SCR), 290, 291
Silver-oxide cells, 158
Sine waves, 171-72
Slide rules, 69
Smoke detectors, 106. See also Safety
Sodium-vapor lighting, 344
Software, 312, 313
Solar cells, 28, 29

Solenoids, 114-17, 393-94
Space charges, 268
Spark plugs, 395
Spectroscopes, 101
Spike protectors, 105-6. See also Safety
SPST (single-pole, single-throw) switches, 376
Squares and square roots, 70
Static electricity, 30
Statistics, 384-85
Stators, 164, 347, 349
Step-down transformers, 183, 261
Stepper motors, 351, 390
Step-up transformers, 183
Substitution, 66
Superheterodyne receivers, 359, 361, 362
Suppressor grids, 271
Surface wiring, 404
Surge protectors, 105-6. See also Safety
Switches, 350, 351, 376-77
Symptoms, 409
Synchronization pulses, 368
Synchronous motors, 351

T

Tank circuits. See Parallel-resonant circuits
Teach pendants, 391
Technology, 69
Telegraphs, 371
Telemetry, 378
Telephones, 352-56
Televisions, 278-79, 335-36, 366-70
Temporary magnets, 107
Tetrode tubes, 271
Thermionic emission, 267, 273, 346
Thermocouples, 29, 378, 380
Thermostats, 378-79
Theta (θ), 237
Thin film resistors, 47
Thinwall conduits, 404
Throwing (switches), 377
Time constants, 195-200, 204
Toggles, 298

Tolerance, 53-54
Total resistance, 83-84
   in circuits, 91-92, 94, 96-87
      defined, 82, 87
      equation for, 83, 91, 92
Touch-tone telephones, 355
Transducers, 378
Transformers. See also Power:
    supplies
   defined, 182
   operation of, 131, 183-83, 185, 186, 335-36
   parts of, 114, 182-83, 323-24
   types of, 183-84, 185, 261, 369
   uses of, 182, 186, 395
Transistors. See also
    specific transistors
   advantages of, 279
   bias, 316, 317-18, 320
   operation of, 274, 275-76, 277-79
   parts of, 274
   uses of, 328, 336
   voltmeters (TVM), 150, 151
Transmitters, 352, 366
Travelers, 406
Triacs, 291
Trickle charging, 161
Triggered scopes, 179-90, 181
Triodes, 269-71
Troubleshooting, 408
Truth tables, 294
Tunnel diodes, 256
Turns, 114, 183
Tutorials, 386

U

UL (Underwriters' Laboratories) seal, 403
Unipolar transistors, 151, 287
Universal motors, 348-49
Unknowns, 61

V

Vacuums, 211
Vacuum tubes
   advantages and disadvantages of, 279
   diodes, 267-68
   operation of, 268, 269-71
   types of, 269-72, 282
   uses of, 267, 269, 272, 319, 325, 335
   voltmeters, (VTVM), 251
Valance electrons, 251
Varactors, 256-57
Variable resistors, 46, 47-48
Vector diagrams, 230, 235
Vertical deflection coils, 369
Volta, Alessandro, 34, 38
Voltage. See also Electricity
    sources
   in circuits, 83, 84, 89-90, 98, 99 239-41
   defined, 34, 38
   doublers, 262, 336
   equation for, 84
   measuring, 36-37, 134-38, 146-47 150 173-75
   polarity of, 171-72
   regulators, 263-64
   types of, 34-35, 89-90, 174-75, 199
   uses of, 215, 219, 369-70
Voltmeters, 134-38, 150, 151
Volt-ohm-milliammeters (VOM). See Multimeters
Volts, 36, 38

W

Watt, James, 76, 81
Watts (W), 76, 77, 81
Watt's law
   defined, 76
   equation for, 76, 81
   Ohm's law compared to, 78, 79-80
   uses of, 77-78, 85
Wirewound resistors, 47
Wiring, 401-6
Word size, 310
Working voltage (W.V.), 212, 214, 215

X

Xenon lighting, 344
XNOR and XOR gates, 297

Y

Yokes, 368-69. See also Coils

Z

Zener diodes, 256, 336
Zero adjust controls, 139

Credits

EMC Publishing would like to thank all those who contributed to *Electricity and Electronics Today*. We would especially like to thank Gary Burton (technical illustrations), Slater Studio (design and cover photography), and Sam Alesevich and Chris Bakes of Norwalk High School, (consultants). The following is a list of contributors:

AutoCAD® by Autodesk, Inc., 391
Bureau of Reclamation, U.S. Department of the Interior, 156, 168
Beckman Industrial Corporation, a subsidiary of Emerson Electric
   Company, 121 (left), 152 (top), 153
Chartpak, 69
Chicago Convention and Visitors Bureau, Inc., 5, 334
Cray Research, Inc., 293
Dale Electronics, xiii (bottom right), 33, 46
Frank Schroder/Tannenblick Studio, 37, 78, 104, 105, 108, 145-
   147, 161, 337, 351, 402-403, 405, 410
Graco Robotics, Inc., Graco OM5000™, 387
Harvard Apparatus, 385 (top)
Heath Company, xiii (bottom left), 120, 388
Kraft, Wolfgang, 4, 309, 310, 372, 383, 396
Lee Data Corporation, 384
Littelfuse Tracor, 102
Medtronic, Inc., 3
Minnesota Educational Computing Corporation, 75, 385 (bottom)
Minnesota Historical Society, p. 354
National Aeronautics and Space Administration, 375
Pentair, Inc., 11 (bottom left)
Permission granted by Whirlpool Corporation, 411
PREP, Inc., 389
Radio Shack, 9, 10, 11 (middle right), 121 (right), 148, 150,
   152 (bottom), 250, 267, 283, 390

Ramsey Clinic, xiii (top), St. Paul, MN, 2
Roper Whitney Company, 11 (bottom right)
St. Paul Technical Vocational Institute, St. Paul, MN, 1,6
Sharp Electronics Corporation, 70
Simpson Electric Company, 144
Snap-on Tools, 11 (middle left)
Texas Instruments, 29
The Bureau of Engraving, Minneapolis, MN, 202, 306